Folded Unipole Antennas: Theory and Applications

Jeremy Keith Raines,
Ph.D., P.E.

McGraw Hill

New York Chicago San Francisco Lisbon London Madrid
Mexico City Milan New Delhi San Juan Seoul
Singapore Sydney Toronto

The McGraw-Hill Companies

Cataloging-in-Publication Data is on file with the Library of Congress

Folded Unipole Antennas: Theory and Applications

1 2 3 4 5 6 7 8 9 0 DOC DOC 0 1 9 8 7

ISBN-13: 978-0-07-147485-6
ISBN-10: 0-07-147485-4

Sponsoring Editor
Wendy Rinaldi

Editorial Supervisor
Janet Walden

Project Manager
Arushi Chawla

Acquisitions Coordinator
Mandy Canales

Copy Editor
Raina Trivedi

Proofreader
Benodini Banerjei

Indexer
Steve Ingle

Production Supervisor
Jean Bodeaux

Composition
International Typesetting and Composition

Illustration
International Typesetting and Composition

Art Director, Cover
Anthony Landi

In memory of
John H. Mullaney, P.E.
(1920–1994)
Friend & Mentor

ABOUT THE AUTHOR

JEREMY KEITH RAINES was born in Washington, D.C. He received the B.S. in Electrical Science & Engineering from M.I.T., the M.S. in Applied Physics from Harvard University, and the Ph.D. in Electromagnetics from M.I.T. He is a registered Professional Engineer in the State of Maryland. He is a senior member of IEEE and has served as President of the Association of Federal Communications Consulting Engineers.

Since 1972, he has been a consulting engineer in electromagnetics. Projects for government and private industry have concerned antennas, radio wave propagation, radar, architectural electromagnetics, and bioelectromagnetics. Antennas designed by him span the spectrum from ELF through SHF, and they may be found on satellites deep in space, on ships, on submarines, on aircraft, and at a variety of terrestrial sites. Modeling antennas on complicated platforms and in complex environments are chief interests.

Dr. Raines enjoys sharing his knowledge. He was an instructor in the faculty of electrical engineering at M.I.T. He has also taught courses about antennas at the Naval Research Laboratory and at George Washington University. He was a guest lecturer on the subject of bioelectromagnetics at Rockefeller University.

Contents

Preface

Purpose of This Book

The chief motivation for writing this book was to preserve a body of knowledge that the author believes is unique and valuable, with respect to both antenna design and mathematical techniques. Some parts of it already exist, widely dispersed in the technical literature; however, many others, if not most, parts are original research. In any case, to the best of the author's knowledge, all of the parts have never before been brought together in a single volume such as this one.

The subject, folded unipole antennas, is still timely. Folded antennas and arrays of folded antennas can be extremely compact. They can also be broadband compared with other antennas. Conducting obstacles that cause electromagnetic interference can be reconfigured as folded unipoles and thus detuned, or rendered electronically invisible. Many additional applications remain to be explored and developed.

The mathematical techniques, which are mostly analytic and not numerical, are also still timely. Today's computers are both powerful and affordable; however, they are not always the first choice for design purposes. Sometimes exploring an elegant formula from electromagnetic theory provides more insight and is otherwise more productive. In the author's experience, the folded unipole antenna is an outstanding example in this respect.

The book is mostly theoretical. That the folded unipole antenna actually does work has been well established since the 1940s. Thousands of them have been constructed and are presently operating in the AM broadcast and military bands. Rigorous theoretical analyses have been performed for a very few specific configurations. On the other hand, a comprehensive theory explaining why the general class of folded unipole antennas works, and how their designs can be systematically modified, improved, and applied elsewhere, seemed elusive.

Intended Audience

This book has been written with at least three different readers in mind. They are: (1) the systems engineer who is seeking new options for antennas and arrays, especially those that offer extreme compactness and broadband performance; (2) the antenna engineer who is not satisfied by sifting through the mountains of data generated by general purpose, wire antenna, numerical analysis programs; and (3) the electromagnetic generalist who is intrigued by the synergy that results when antennas and transmission lines are integrated into a single radiating device. At one time or another, the author has identified with each of these prospective readers.

For the systems engineer, this book should fill a void. When seeking candidate antennas, his first impulse is to consult a handbook or compendium about antennas. At best, he will find a few pages about folded antennas, and they will only touch on the simplest designs. They may not use the term "folded unipole" at all. The discussion will likely be a section or subsection within a chapter about VLF antennas or AM broadcast antennas, giving the wrong impression that they are limited to use in those bands. In fact, the folded unipole embodies design principles that apply throughout the electromagnetic spectrum from ELF through EHF.

For the antenna engineer, this book should provide some refreshing insights concerning the physics of antenna operation, harkening back to the Golden Age of Antenna Analysis before general purpose, numerical programs were widely available. In the opinion of the author, these programs have largely displaced antenna *design* by numerical antenna *analysis*. In the case of design, mathematical formulas, often elegant, can be manipulated to obtain either an optimal configuration, or at least a well-reasoned starting point. In contrast, numerical analysis involves subjecting countless combinations of wires to number crunching until an acceptable configuration is eventually obtained. Why the configuration happens to work, or if it even approaches an optimum, remains a mystery buried under a mountain of numerical data. To be fair, sometimes this is the only way to go, but it should not be the only choice, or even the first choice, in the tool kit of the antenna designer. The elegance of electromagnetic theory is still an excellent alternative.

For the electromagnetic generalist, this book explores in depth what happens when linear antennas are merged with transmission lines into a single device. New formulas and mathematical approaches are developed as needed. Some of them will have application beyond the folded unipole antenna. For example, in Chap. 9 we will derive the exact formulas for the electromagnetic fields of traveling waves of current, valid both in the near and radiation zones. We will also take a critical look at the definition and calculation of characteristic impedance in Chap. 7. In the literature, the characteristic impedance of antennas has been a topic of controversy for decades. We hope to resolve that controversy here.

Notes Concerning the Examples

The reader will observe that the contour plots in Chaps. 2 through 5, and in Chap. 9, are in shades of gray; however, the captions and text refer to colors. The contour plots were in color for the original manuscript, but they were converted to gray scale to control publication costs. There should be little or no confusion, however, because the "priority color" (i.e., the color of the contour corresponding to optimal design values) is always the darkest gray, nearly black.

The original color contour plots are available at two different websites. They are the author's website, **www.rainesengineering.com**, and the publisher's website, **www.mhprofessional.com**.

Readers may also observe that the frequency for most examples is 984 kHz. This is entirely for numerical convenience. It does *not* imply that the folded unipole is primarily intended for the MF band. This particular frequency was selected simply because a wavelength is 1,000 feet in free space. Thus, it is especially easy to mentally translate linear dimensions into fractions of a wavelength when reading abcissas and ordinates from the graphs and plots. For antennas less complicated than the folded unipole, dimensions are often normalized so that the results are independent of frequency. Folded unipoles, however, integrate linear dimensions, lumped impedances, characteristic impedances, and wavelength into a single electromagnetic device. Therefore, formulating a dimensionless, universal parameter is impractical, if not impossible.

Readers will likely be curious concerning the software used to create the examples. They were all generated using a program called GENIPOL (GENeral unIPOL). GENIPOL incorporates the very same mathematical equations that appear in the text.

The Brief History of This Book

More often than not, an engineering text is introduced as, "This book began as notes for a course offered at the University of ______." This book is nothing of the kind. It concerns a subject that was at the cutting edge of military communications and commercial broadcasting for many decades. The folded unipole antenna still has much unrealized potential and a bright future, in the opinion of the author. He was just lucky enough to enter the world of antenna design at a time when interest in the folded unipole was cycling through a peak. He was also lucky enough to be mentored by some of the eminent contributors to antenna and electromagnetic theory, and by some of the most colorful innovators in industry. One goal of this book is to preserve and share that experience.

The author first encountered folded unipole antenna technology in the summer of 1966, when he was an engineering intern at Multronics, Inc.,

Rockville, Maryland, between undergraduate years at M.I.T. The president of Multronics was **John H. Mullaney, P.E.**, who was awarded some of the early patents for the folded unipole. Its surprising capabilities and the challenge to understand its detailed workings recurred throughout the years. Somehow the prolonged and concentrated study that the folded unipole antenna deserved and required was always sidelined by larger and more urgent projects in communications and radar.

About 35 years later, **James F. Pinkham**, a colleague from those first years at Multronics, mentioned in a chance telephone conversation that someone ought to write a book about the folded unipole antenna. For several years thereafter, the author dismissed the suggestion out of hand because he realized it would take twice as long as anyone expected. Also, engineers are fond of saying, "The Devil is in the details." Filling in those theoretical details concerning the folded unipole antenna would require twice as much formula deriving and computing of examples as anyone expected. Despite these reservations, the idea eventually took hold, and the author wrote a sample chapter and shopped it around. To his pleasant surprise, the publisher immediately expressed "intense interest."

So, the author began writing in earnest. He was correct in the assumption that it would take twice as long as anyone imagined. But it was a grand learning experience and a rare opportunity to consolidate, organize, and preserve a worthwhile technology that connects to many diverse topics in electromagnetic theory. He has no regrets.

That the author embarked on this project at all was perhaps unavoidable destiny. His initials are JKR, and *jkR* is one of the most omnipresent terms in antenna design and electromagnetic theory in general. This term occurs most frequently in the exressions e^{jkR} and e^{-jkR}.

Acknowledgments

This book is the logical result of 40 years of education and experience, from that first summer at Multronics in 1966, to the completion of the first draft in the summer of 2006. During that period the author was mentored, assisted, and otherwise influenced by individuals in both industry and academia.

The book itself would never have happened were it not for the encouragement of two individuals. First, **James F. Pinkham**, a colleague from the first years at Multronics, with firsthand experience of designing and installing folded unipoles, planted the idea of the book in a chance telephone conversation circa 2001. Otherwise, the author would never have considered it. Second, once a sample chapter was written, the idea was enthusiastically endorsed by **Ron Nott**, VP Engineering of Nott Ltd., Farmington, New Mexico. Ron has made a career from installing folded unipole antennas and has hands-on knowledge concerning how well they perform. His constant encouragement and comments sustained the author's creative energy long enough to complete the book, and, as mentioned in the previous section, that was twice as long as anyone anticipated.

The book would never have materialized, of course, without a willing and enthusiastic publisher. For that, the author thanks **Wendy Rinaldi**, Editorial Director at the Professional Books division of McGraw-Hill. When Wendy received a sample chapter, she quickly recognized the potential contribution a book could make to the engineering community. For the ensuing 18 months of intense writing, Wendy provided the rare combination of constant encouragement plus freedom to write and organize with minimal constraints.

There would have been nothing to write were it not for numerous conversations and joint projects with colleagues concerning the folded unipole antenna. **James J. Karaganis** was the engineer who initially hired the author as a summer intern at Multronics. Under the leadership of the company president, the late **John H. Mullaney, P.E.**, a talented team of engineers pursued cutting-edge applications of the folded unipole antenna for military communications. That team included the late **George P. Howard**, who was especially intrigued with the use of folded unipoles for multiplexing, and **Lawrence Behr**, who is now president of Lawrence Behr Associates in Greenville, North Carolina.

Multronics evolved into a broadcast consulting firm, Mullaney Engineering, in Gaithersburg, Maryland. As a result, application of the folded unipole antenna continued with emphasis in the AM radio band. The author pursued a variety of projects with John H. Mullaney and the engineers at Mullaney Engineering. A pivotal one occurred in 1973, with the creation of a computer program to model some of the simplest configurations of the folded unipole. It was remarkably accurate, sometimes to four significant figures, and it is still in use today.

The 1990s was a sad decade in the history of the folded unipole with the passing of two of its pioneers. George P. Howard passed away in 1993, and John H. Mullaney followed in 1994. Mullaney Engineering, Inc., Gaithersburg, Maryland, continues to provide innovative folded unipole designs to the broadcast industry. The author has been privileged to participate in projects and discussions with **Jack Mullaney**, President of MEI, and with **Alan E. Gearing, P.E.**, Senior Engineer, MEI, concerning the folded unipole antenna. The author is especially grateful to Jack Mullaney for researching the fascinating histories of both the configuration and name of the NORD antenna. These are related in Sect. 1.2.

Before his death, John H. Mullaney shared many historical anecdotes and design techniques concerning the folded unipole with **R. Morgan Burrow, Jr., P.E.**, formerly with Mullaney Engineering, and now principal of his own consulting firm. During several conversations, Morgan shared those reminiscences with the author. For example, in 1953, Mullaney was responsible for the first U.S. broadcast antenna to be licensed by the FCC as a folded unipole, Station WLOU.

When she learned of the author's book project, **Nellie Fleet Mullaney**, widow of John H. Mullaney, generously provided original notes and circa 1953, patent applications concerning the folded unipole antenna.

In addition to the knowledge acquired from industrial experience, the author was privileged to stand on the shoulders of some academic giants. They provided a solid foundation in electromagnetic theory for this book.

The author pursued both his B.S. and Ph.D. degrees in the M.I.T. Dept. of Electrical Engineering. During that time, he was especially influenced by two of the faculty. The late **Prof. Robert L. Kyhl** was his dissertation supervisor. He was also a brilliant microwave engineer and veteran of the M.I.T. Radiation Laboratory. While the author was consumed with the deriviation and numerical solution the most formidable vector, integro-differential, eigenvalue equation he has ever encountered before or since, Prof. Kyhl became a good-humored thorn in his side. He often remarked to the author that a little physical intuition and insight was worth a thousand rigorous equations. It took decades for that wisdom to sink in; however, that philosophy should be apparent throughout this book. We have eschewed the general purpose, numerical, wire antenna analysis programs, which can generate matrices with millions of elements, for a much simpler boundary value approach.

The other great influence at M.I.T. was **Prof. Steven K. Burns**. For many years, the author taught Prof. Burns' circuit design Course 6.702, *Bioelectronics Project Laboratory*, and this experience influenced him in at least two ways. First, analyzing, building, and testing circuits for medical instrumentation provided both hands-on experience with and a critical understanding of the distinction between common mode and differential mode currents, voltages, and impedances. That distinction is the cornerstone of the theory of folded unipole antennas. Second, it generated a lasting interest concerning electromagnetic fields and their interaction with living systems. That interest is apparent especially in Chap. 10 of this book.

The author pursued his M.S. degree at Harvard University's Division of Applied Physics. There he was greatly influenced by the late **Prof. Ronold W. P. King**, an eminent authority well known and highly regarded in the antenna community. Circa 1958, Prof. King and his student **Charles W. Harrison, Jr.**, were among the first electromagnetic theorists to subject a version of the folded unipole antenna to a rigorous analysis using integral equations. Though an alternative mathematical formulation is pursued in this book, Prof. King articulated the essential physics for our approach. That physical foundation is that there are two modes of current on the folded unipole. They are the common mode (also known as the antenna mode or unbalanced mode) and the differential mode (also known as the transmission line mode or balanced mode).

With respect to the production details of this book, the author certainly would have thrown in the towel had it not been for the technical know-how of some very skilled individuals who generously shared their time and experience. Engineering colleague **Michael Millard**, Managing Partner, TowerSwitch, LLC, Deerfield Beach, Florida, guided the author along the many meandering learning curves for desktop publication, and especially for Microsoft Equations 3.0, which is a formidable tool, indeed, once it is mastered. **Leslie A. Gelman**, Falls Church, Virginia, similarly guided the author along a learning curve for converting hard copy illustrations and graphics into a digital format suitable for publication. **Jean Bubley**, Manager of the Esther Bubley Photography estate, Brooklyn, New York, and **Johanna Kovitz**, Allston, Massachusetts, are experts concerning Adobe Photoshop, and they labored tirelessly to convert color contour plots into publishable gray scale presentations.

The author also wishes to thank, in advance, readers who discover and report errors in this book. Beyond the simplest configuration of the folded unipole, most of the material here is original research. Numerous sanity checks have been performed along the way, and they have been satisfied and documented here. Still, as engineers are fond of saying, "The Devil is in the details." As a result, errors may yet linger, and the author would appreciate being informed of them.

jkR
Rockville, Maryland
2006

Chapter

1

Introduction

1.1 Introducing the Folded Unipole Antenna

The different ways to introduce this book are practically unlimited, which is appropriate considering the versatility of the folded unipole antenna. They would all be accurate and brimming with optimism for the future of this antenna.

Perhaps the simplest way is: "This is an antenna that for over 50 years has been well enough understood to be practically deployed, but not studied sufficiently to be optimized or more widely embraced."

A grander way is: "All antennas fall into two broad categories. They are linear antennas and aperture antennas. The folded unipole antenna is in many ways the superset of all linear antennas."

Yet another broad introduction might be: "The folded unipole antenna is the superset of linear antennas and transmission lines. While countless books and articles have been written separately about antennas, and separately about transmission lines, no book has been written about what happens when the two are combined into a single electromagnetic device. When this happens, some remarkable performance is possible."

All of the above introductions are truthful. The difference between them is simply perspective or, as the author's late thesis advisor Prof. Robert L. Kyhl was fond of saying, "It all depends upon your point of view." In this book, we will take advantage of many points of view. Our goal is to better understand the folded unipole antenna. We wish to mathematically model it in a way that we can efficiently explore its desirable properties and practical applications. We wish to understand how it fits into the nearly endless selection of antennas available to the design engineer. More generally, we wish to understand how it fits into the universe of electromagnetic devices.

The reader who is encountering the folded unipole antenna for the very first time on this page is by now scratching his or her head and wondering, "What *is* the folded unipole? What does it look like?" The answers, of course, should become increasingly clear with each chapter; however, for now, we emphasize that the essence of the folded unipole is the integration of linear radiators and transmission lines. Visually, the result can take on practically unlimited forms, from a vertical cage of parallel cylinders and wires, to tiny U-shaped patterns on a printed circuit board. The original attraction of the folded unipole was its extremely compact size, especially at long wavelengths; however, it has both historical and potential applications throughout the spectrum from ELF through EHF.

1.2 A Brief History

Figure 1-1 shows the likely evolution of the folded unipole antenna. In fact, however, its detailed early history, prior to 1947, is obscure. The pioneers who participated in that early history, and who might remember it clearly, are long departed. Most likely, the folded unipole evolved from experiments to improve the series-fed monopole in the AM broadcast industry. As early as the 1920s, it was known that a monopole near one-quarter wavelength in height exhibited a practical input impedance of

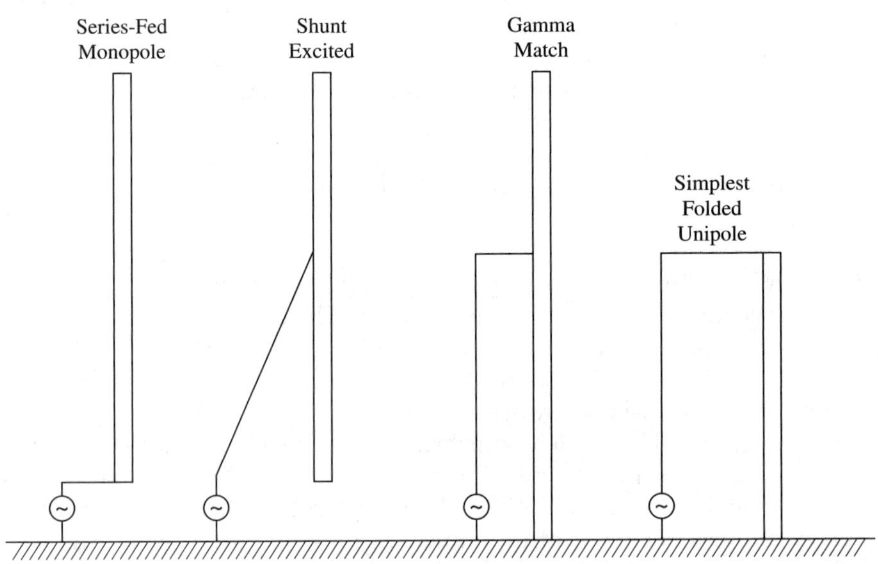

Figure 1-1 The detailed early history of the folded unipole antenna, prior to 1947, is obscure; however, it probably evolved as shown here.

about 36 ohms resistance and 0 ohm reactance. It was easy to couple high power radio transmitters into such an impedance. Monopoles with other heights were less convenient, however. Those shorter than a quarter wavelength exhibited much lower resistance and a significant capacitive reactance that could be difficult to tune out. Monopoles higher than a quarter wavelength exhibited increasingly large resistances and nonzero reactances that also had to be tuned out.

In 1932, a landmark theoretical analysis of linear antennas by P. S. Carter [1] appeared in the *Proceedings of the IRE*. Using well-known mathematical functions, he obtained closed form expressions for the input impedance of the series-fed monopole. These results confirmed what was already known, namely, that a quarter-wave monopole exhibited the most desirable input impedance, and that others required a lot more engineering effort to tame.

By circa 1937, articles were appearing in the *Proceedings of the IRE* concerning shunt fed antennas. It was hoped that by exciting a monopole at some point above its base, using an auxiliary wire, more desirable input impedances would be found. The shunt feed described by Morrison and Smith [2] was not parallel to the main tower. It was slanted.

By circa 1949, articles appeared concerning the Gamma Match Antenna (GMA) [3]. The GMA was similar to the shunt feed except that the auxiliary wire was parallel to the main tower. In fact, the GMA was one special case of the folded unipole antenna.

When an actual folded unipole was first deployed does not seem to be documented. One likely possibility is that radio amateurs experimented with them prior to World War II. Another possibility is that the Army Signal Corps discovered them out of necessity on the beaches in the South Pacific, where signalmen had to quickly erect low profile LF and HF antennas for communications under hostile conditions.

In any case, a formal though limited analysis of the folded unipole was published by van Roberts in the *RCA Review* in 1947 [4]. The discussion implied that the basic impedance properties of the folded unipole were already well known, although references were not provided. This suggests that the information came out of wartime use and had been recently declassified as one of many innovations in the RF and microwave bands.

The AM broadcast industry was eager to adopt the folded unipole. After seeing service in the Army Signal Corps in the South Pacific, radio consultant John H. Mullaney (1920–1994) designed and supervised the installation of these antennas for between 1,000 and 2,000 domestic radio stations, beginning in the late 1940s. He was instrumental in convincing the FCC to formally recognize and grant licenses for the folded unipole antenna. There were many anecdotal reports of significant advantages in addition to a good impedance match. The folded unipole was reportedly more efficient than a series-fed monopole. That is, for a radio transmitter of a given power, the folded unipole broadcast a stronger signal.

Meantime, the advantages of the folded unipole were not lost on the military, especially the United States Navy. In the 1960s, there was a concerted effort to improve communication with its vast fleet of submarines. Increasing numbers of submarines were nuclear powered, and they could stay at sea for many months. It was more important than ever to maintain communication with them. Unfortunately, the best frequencies for this appeared to be in the VLF (3–30 kHz) and LF (30–300 kHz) bands. At 3 kHz, a wavelength is 62 miles, and so a quarter-wave monopole was clearly not a practical option.

John H. Mullaney proposed a configuration called the NORD Antenna to solve the Navy's communication problem. In fact, the NORD was a folded unipole antenna with extreme top loading. It was only about 0.05 wavelengths tall. In the mid 1960s, his company, Multronics, constructed them at various shore stations in the Pacific, including Hawaii, Guam, Japan, and the Philippines. The origins of the name NORD have almost been lost over the years. In fact, it is the call letters of the pirate broadcast station Radio NORD. The station operated from a German freighter called the *MV Bon-Jour* moored in international waters off the coast of Sweden circa 1961. It broadcast 20 kW in the MF band to listeners in Scandinavia and behind the Iron Curtain. The broadcast antenna was improvised from the mast of the ship. It was reconfigured as a folded unipole. The stays of the mast functioned as top loading. Although this prototype operated at much shorter wavelengths, Mullaney suggested that it could be successfully scaled for use in the bands of interest to the United States Navy.

The NORD antenna project generated much controversy. In fact, the antennas exhibited useful input impedances, and submariners reported successful communication with the shore stations. Antenna theorists at official Navy laboratories, however, asserted that the NORD could not possibly work as well as reported. According to the labs, it was just too short to radiate that efficiently and with sufficient bandwidth to provide useful communication. Without a definitive mathematical model of the NORD, it was difficult to defend either side of the debate.

The United States Air Force was also interested in the folded unipole. Its low profile was well suited for telemetry applications on missiles and aircraft where streamlining and low air resistance were paramount concerns.

The folded unipole antenna inevitably diversified away from its original applications in the AM broadcast and military radio bands. For example, folded unipoles are used extensively to detune structures such as water towers and communication towers that would otherwise scatter or reradiate significant amounts of electromagnetic energy. There is an interesting analogy in connection with this application. Microwave research during World War II produced the quarter-wave choke. This was embedded in the flanges at the ends of sections of waveguide. The chokes prevented currents from migrating from the inside to the outside of the waveguide at the joints between sections. Typically, in the RF bands, a section of quarter-wave folded unipole is used for detuning purposes. In analogy to the microwave choke, it prevents currents from flowing on the detuned structure.

There are other diverse applications of the folded unipoles. For example, tiny, printed circuit versions are described in the literature [5].

For such a useful and diverse electromagnetic device, it is surprising that it is barely mentioned in compendiums about antennas. Seldom does it receive more than a few pages. Perhaps the reason is that it has never been extensively modeled and investigated beyond a few simple cases.

1.3 A Golden Age of Antenna Analysis

The folded unipole antenna has evaded extensive analysis because its modeling requires a careful balance between closed form mathematical formulas and numerical evaluation of those formulas. Before the availability of high speed, low cost computers, the examples in this book could be elegantly formulated but not easily evaluated. After the availability of those computers, however, the folded unipole stubbornly refused to yield consistently accurate results to brute force, general purpose, wire antenna analysis programs. To understand how the successful analysis became such an interesting blend of traditional mathematical tools and modern numerical algorithms, it is useful to trace the history of antenna analysis.

A Golden Age of Antenna Analysis arguably spans over a century, from 1854 to 1968. In 1854, William Thomson (1824–1907) published the first analysis of transmission lines [6]. The article appeared 10 years before the publication of the Maxwell Equations. Thomson would eventually be knighted as Lord Kelvin and would amass a vast fortune for his efforts. In 1968, a thin but revolutionary monograph was published by Prof. Roger F. Harrington of Syracuse University [7]. With the title *Field Computation by Moment Methods*, it ushered in an enthusiasm for numerical techniques that ultimately edged out the elegant mathematical formulas of previous years.

Between those years, much brilliant thought was devoted to electromagnetism in general, and, in particular, to transmission lines and antennas, the two fundamental components of the folded unipole. In a series of papers published between 1885 and 1887, Oliver Heaviside (1850–1925) showed that electromagnetic waves propagated along telegraph cables at the speed of light [8]. In 1897, Henry Cabourn Pocklington (1870–1952) derived an integral equation for wire antennas, and used it to show that electromagnetic waves also propagated along the antenna at the speed of light [9]. Thus, the analogy or similarity was established between transmission lines and antennas.

Heaviside and Pocklington established a theoretical unity between antennas and transmission lines, but they also established a schism, although it was not deliberate on their part. On one side of the schism, many brilliant minds continued to develop mathematical models of antennas that bore remarkable resemblance to those for transmission lines. On the other side, as many brilliant minds refined the integral equation approach, with solutions for antenna currents that appeared much more complicated than transmission line currents.

The most prolific proponent of the transmission line analogy was almost certainly Sergei A. Schelkunoff (1896–1992), a brilliant applied mathematician at Bell Laboratories. In a landmark 1941 paper, published in the *Proceedings of the IRE* [10], he derived formulas for the input impedance of antennas with arbitrary size and shape. Those formulas bore remarkable resemblance to those for the input impedance of transmission lines.

An enthusiastic following for the integral equation approach was ignited by Erik Hallén in a 1938 article that presented what is today known as *Hallén's Integral Equation* [11]. Clearly the most prolific of that following was Prof. Ronold W. P. King (1905–2006) of Harvard University. Many of King's analyses are collected in his 1959 classic reference *Theory of Linear Antennas* [12]. The two schools of thought competed vigorously for decades. The transmission line school offered elegant mathematical formulas that could be evaluated without electronic computers, although mathematical tables and slide rules were routinely required. The integral equation school claimed greater rigor with respect to determining the exact current distribution along the antenna, even though the solutions to those integral equations were relatively difficult to obtain. The debate between the two schools often centered around the precise value for the input impedance of a half wave, series-fed monopole. Both agreed that it was a large value because the antenna current approached a minimal value at the antenna input terminals.

The debate was never resolved academically. Instead, it simply fizzled with the publication of Harrington's book concerning the method of moments (MoM). MoM definitely weighed in on the side of integral equations. Further, MoM promised great generality. It could be applied to ensembles of wires that followed arbitrary curves. By that time, high speed computers were readily available to students and engineers. Within a short time, the latest generation of students lost interest in the special mathematical functions of antenna theory. MoM could make quicker work of complicated geometries. The analogies to transmission lines were no longer essential analytic tools.

At the conclusion of the above debate, the folded unipole antenna was an inevitable casualty. Actual transmission lines are part of its configuration, and to ignore that is to invite computational error. Accordingly, when purely numerical techniques are applied to the folded unipole, the results are sometimes nonsensical.

With the above overview, we can revisit the Golden Age of Antenna Analysis with more attention to detail, and with an introduction to some of the mathematical tools we will encounter in later chapters.

1.4 The Beginning of the Golden Age: Transmission Line Theory

As we shall discover repeatedly in later chapters, the folded unipole antenna combines the best features of transmission lines and linear antennas. In this section, we will introduce some simple mathematical formulas for transmission lines that are important in our mathematical model of the folded unipole.

Theoretical interest in transmission lines began in the 1830s with the invention of the electric telegraph. Bare telegraph wires worked well enough above ground; however, it soon became obvious that they would not perform in water, especially seawater. The solution was an insulated cable, and this solution led to more questions. What were the best dimensions of the cable? What were the performance limitations of the cable? These questions begged for a mathematical model. The first useful model was published in 1854 [6] by William Thomson, who was later knighted as Lord Kelvin.

Figure 1-2 shows the model derived by Kelvin. It is different from modern models of transmission lines, chiefly because the inductance per unit length is missing. Kelvin's model was a good beginning, however, especially for the low-frequency waveforms of Morse Code and similar signals. At low frequencies, the inductive reactance per unit length is small compared with the other circuit components of the transmission line, and so his model explained many practical concerns about telegraph cable. The loss of resistances per unit length explained signal attenuation. The *RC* time constant of the cable predicted a limit to the speed of characters before distortion made them unrecognizable.

Note that Kelvin's model preceded publication of the Maxwell Equations by a decade.

Kelvin went on to invent a very popular telegraph receiver, and this brought him vast wealth. He was able to afford a gentleman's estate in Scotland plus a 126-ton private research vessel. He consistently rejected

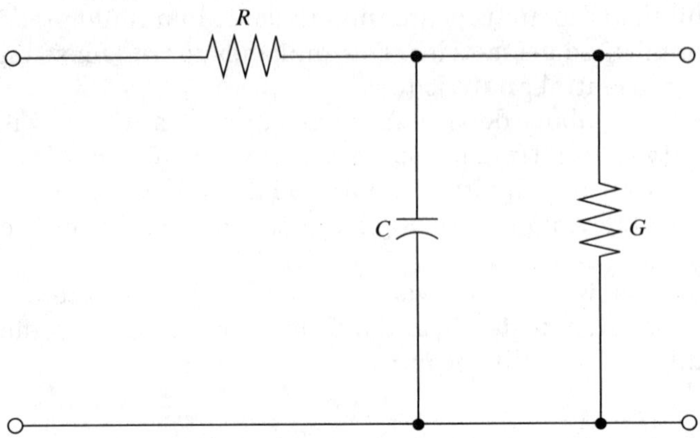

Figure 1-2 Circuit model of the transmission line as proposed by William Thomson, Lord Kelvin, circa 1854. The inductance per unit length was not part of the model.

arguments that inductance per unit length mattered in any model of transmission lines. This proved exasperating to another great mind who would contribute immeasurably to both transmission line and general electromagnetic theory, Oliver Heaviside.

In contrast to the academic background of Lord Kelvin, Heaviside first became interested in electromagnetism while employed as a telegraph operator. By the 1860s, undersea telegraph cables spanned the Atlantic Ocean. Over those great distances, signal distortion became a serious problem. Heaviside insisted that a complete model of transmission lines must include inductance per unit length, as shown in Fig. 1-3. This is the modern model of transmission lines. Based upon this model, Heaviside predicted that distortion would be greatly reduced by periodically loading

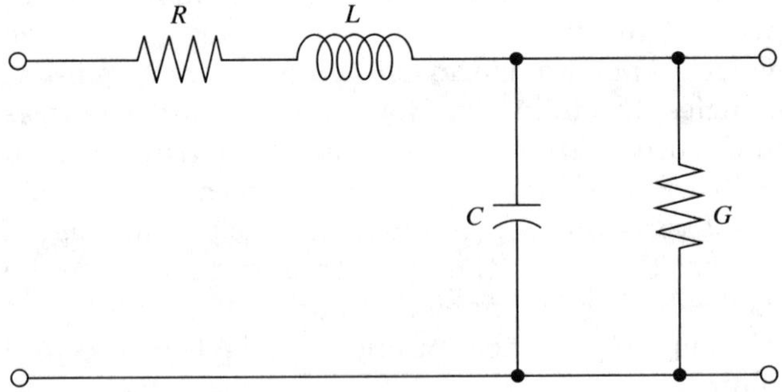

Figure 1-3 Circuit model of the transmission line as proposed by Oliver Heaviside, circa 1885. This is the same model used today.

the Transatlantic Cable with induction coils. He also correctly predicted that electromagnetic waves propagate along the cable at approximately the velocity of light. The inductance per unit length was essential to this prediction.

Heaviside was right on the money as an electromagnetic theorist; however, history records that his recommendation to improve the Transatlantic Cable was ignored "for political reasons." At that time, Kelvin was the highly paid chief consultant to the Transatlantic Cable consortium. His company held the patents for the most popular telegraph receiver. It is plausible that he used his prestige and political clout to quash any challenges to his profitable model of the transmission line. If so, then it was a classic example of vested interests suppressing new technology.

Politics aside, Heaviside's insights concerning transmission lines gave us the essential equations describing their operation. The current is a linear combination of forward and reflected traveling waves:

$$I_d(z) = I_{\mathrm{Fd}}\psi_F(z) + I_{\mathrm{Rd}}\psi_R(z) \tag{1-1}$$

In Eq. 1-1, we introduce notation that we shall use throughout this book. The subscript d denotes the *differential mode*, which is also a balanced mode, or a transmission line mode. The subscript F denotes a forward traveling wave. The subscript R denotes a reflected or reverse traveling wave.

Also in Eq. 1-1, we introduce the forward and reverse traveling wave functions:

$$\psi_F(z) = e^{-\mathrm{jkz}} \tag{1-2}$$

$$\psi_R(z) = e^{+\mathrm{jkz}} \tag{1-3}$$

In Eqs. 1-2 and 1-3, the wave number is:

$$k = \frac{2\pi}{\lambda} \tag{1-4}$$

where λ is the wavelength.

One of the elegant features of transmission line theory is that the voltage between conductors is simply related to the current:

$$V_d(z) = Z_{0d}\left[I_{\mathrm{Fd}}\psi_F(z) - I_{\mathrm{Rd}}\psi_R(z)\right] \tag{1-5}$$

In Eq. 1-5, Z_{0d} is the characteristic impedance of the transmission line.

In terms of Heaviside's circuit model of the transmission line, the characteristic impedance and wave number are:

$$Z_{0d} = \sqrt{\frac{R + j\omega L}{G + j\omega C}} \tag{1-6}$$

$$jk = \sqrt{(R + j\omega L)(G + j\omega C)} \tag{1-7}$$

In our exploration of the folded unipole antenna, the lossy components R and G will be negligible. As a result, Eq. 1-6 reduces to

$$Z_{0d} = \sqrt{\frac{L}{C}} \tag{1-8}$$

and Eq. 1-7 reduces to Eq. 1-4.

Equations 1-1 through 1-8 are so elegant yet simple that they inspired efforts to explain antenna performance in similar terms. Those efforts were largely successful.

1.5 The Zenith of the Golden Age: Antenna Theory

By the 1890s, there was great interest in wireless telegraphy, and experiments strongly suggested its feasibility. It was only natural that the node of the wireless telegraph, the antenna, be modeled as a type of transmission line. Accordingly, in 1897, Henry Cabourn Pocklington set out to prove that the currents flowing along an antenna were similar to those described by Eq. 1-1. To that end, he derived an integral equation describing the current at the surface of the linear wire shown in Fig. 1-4:

$$\int_{-L}^{L} I(u)\left(\frac{\partial^2}{\partial z^2} + k^2\right) G(u,z)\,du = -j\omega\varepsilon_0 E(z) \tag{1-9}$$

In Eq. 1-9, $E(z)$ is the component of the electric field parallel to the wire, and the Green's function is:

$$G(u,z) = \frac{e^{-jk\sqrt{a^2+(z-u)^2}}}{4\pi\sqrt{a^2+(z-u)^2}} \tag{1-10}$$

Arguing that the right side of Eq. 1-9 was zero except between the input terminals of the antenna, Pocklington deduced that the integrand of the left side must also be zero. Then, exploiting the symmetry of $G(u, z)$ with

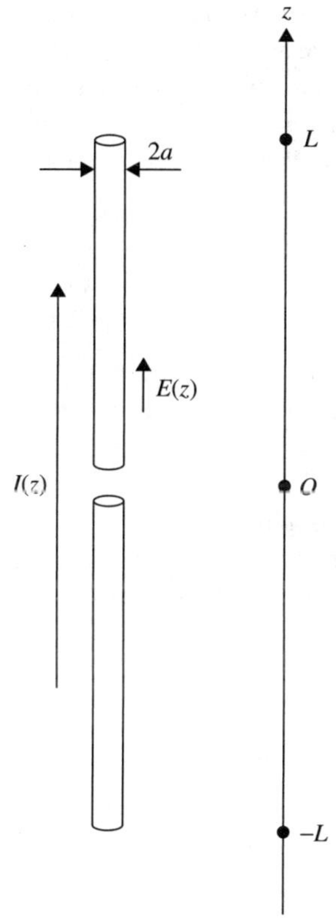

Figure 1-4 The thin linear antenna analyzed by Henry Cabourn Pocklington in 1897. Pocklington's integral equation showed that the current propagated sinusoidally along the antenna at approximately the velocity of light.

respect to the variables u and z, and integration by parts, he found that the current distribution must satisfy the differential equation,

$$\left(\frac{\partial^2}{\partial z^2}+k^2\right)I(z)=0 \tag{1-11}$$

Equation 1-11 is an elementary differential equation with solutions just like Eq. 1-1:

$$I_c(z)=I_{\mathrm{Fc}}\psi_F(z)+I_{\mathrm{Rc}}\psi_R(z) \tag{1-12}$$

The only difference between Eqs. 1-1 and 1-12 is the subscript c. That subscript denotes the *common mode*, which is also an unbalanced mode or antenna mode.

After deriving Eq. 1-12, it is irresistably tempting to define a voltage analogous to Eq. 1-5:

$$V_c(z)=Z_{0c}\left[I_{\mathrm{Fc}}\psi_F(z)-I_{\mathrm{Rc}}\psi_R(z)\right] \tag{1-13}$$

In Eq. 1-13, Z_{0c} is the common mode characteristic impedance of the antenna. In fact, Eq. 1-13 can be made quite rigorous if the voltage can be properly related to the electric field in the vicinity of the antenna. When the length of the antenna is not small compared with a wavelength, then the voltage is not uniquely defined. It depends upon the path of integration along the electric field:

$$V_c(z) = -\oint_C \mathbf{E} \cdot \mathbf{ds} \tag{1-14}$$

Figure 1-5 shows a biconical antenna, and for this geometry Eq. 1-14 can be evaluated in convenient mathematical terms. Using the biconical antenna as a prototype linear radiator was the idea of Schelkunoff in his landmark paper [10]. He went on to show how the results for a biconical antenna could be extended to any linear radiator.

Equations 1-12 and 1-13 show how a linear antenna may be described analogously to a transmission line. This will greatly simplify our

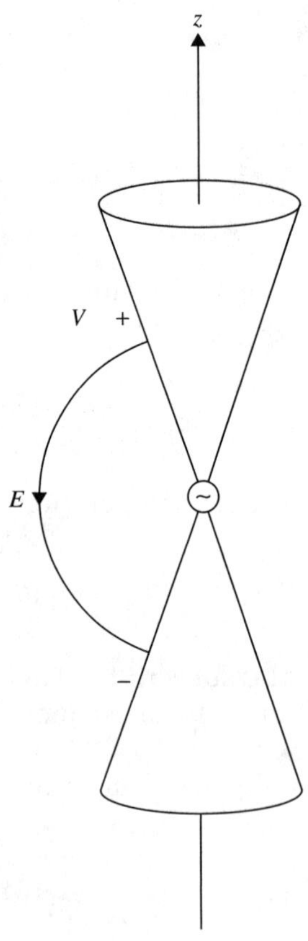

Figure 1-5 The biconical antenna analyzed by Sergei A. Schelkunoff in 1941. This geometry was mathematically convenient for demonstrating the strong similarities between antennas and transmission lines.

mathematical model of the folded unipole antenna, beginning in the next chapter. For now, however, we can demonstrate the simple elegance of these equations by applying them to some monopole antennas.

1.6 The Simple Elegance of Antenna Theory

Figure 1-6 shows a vertical monopole antenna, and that will serve as a good example of how its performance can be quickly and elegantly modeled by the transmission line analogy described in the previous section. Once we obtain the current distribution along the monopole, then the radiation pattern and input impedance readily follow. So, we will focus on deriving the current distribution.

To that end, the current and voltage on the monopole are described by Eqs. 1-12 and 1-13. To determine the amplitudes I_{Fc} and I_{Rc} of the forward and reverse traveling waves, we must derive two independent equations. These equations follow from boundary conditions at the top and at the base of the monopole.

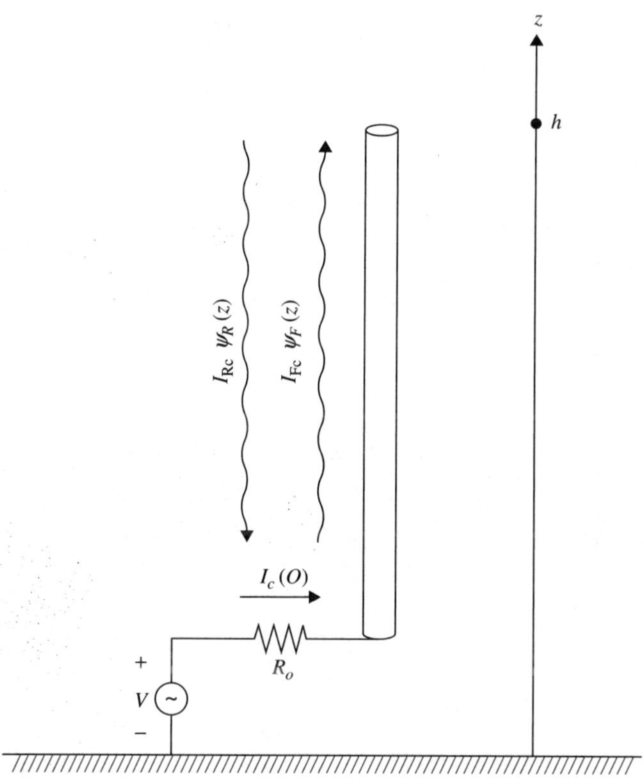

Figure 1-6 Essential features of the series-fed monopole antenna.

At the top of the monopole, the current must vanish because it has nowhere to go:

$$I_c(h)=0 \tag{1-15}$$

In fact, Eq. 1-15 is not as accurate as it could be. Due to *antenna end effect*, part of the current actually does have somewhere to go. We will revisit this phenomenon in Sec. 5.7. As a practical matter, however, for many cases it is not important.

At the base of the monopole, Kirchhoff's voltage law (KVL) requires that:

$$V - I_c(0)R_0 = V_c(0) \tag{1-16}$$

In Eq. 1-16, R_0 is the output impedance of the voltage source V.

If we substitute Eqs. 1-12 and 1-13 into Eqs. 1-15 and 1-16, then after a bit of algebra we obtain a general formula for the current $I_{c(z)}$ everywhere along the monopole:

$$I_c(z) = \frac{V}{(R_0 + Z_{0c}) - \dfrac{\psi_F(h)}{\psi_R(h)}(R_0 - Z_{0c})}\left[\psi_F(z) - \frac{\psi_F(h)}{\psi_R(h)}\psi_R(z)\right] \tag{1-17}$$

Equation 1-17 simplifies greatly for some interesting special cases. For a quarter-wave monopole, $h = \lambda/4$. From Eqs. 1-2 and 1-3, we see that the traveling wave functions become:

$$\psi_F(\lambda/4) = -\psi_R(\lambda/4) = -j \tag{1-18}$$

Applying Eq. 1-18 to Eq. 1-17, the latter simplifies to:

$$I_c(z) = \frac{V\cos(kz)}{R_0} \qquad h = \lambda/4 \tag{1-19}$$

Another special case with a simple result is the half-wave monopole, for which $h = \lambda/2$. In that case the traveling wave functions become:

$$\psi_F(\lambda/2) = \psi_R(\lambda/2) = -1 \tag{1-20}$$

Applying Eq. 1-20 to Eq. 1-17, we obtain:

$$I_c(z) = \frac{-jV\sin(kz)}{Z_{0c}} \qquad h = \lambda/2 \tag{1-21}$$

The vertical monopole was an especially simple example, thanks to the elegance of the transmission line analogy in antenna theory. With that introduction, we will tackle something a bit more challenging in the next section.

1.7 The Loaded Monopole Antenna

The loaded monopole antenna is an excellent intermediate step between the monopole of the previous section and the simplest configuration of the folded unipole, both from geometric and mathematical points of view. Figure 1-7 shows its essential features. The load Z_L is located at an arbitrary distance h_1 along the antenna. The characteristic impedances above and below the load, Z_{0c1} and Z_{0c2}, are not necessarily equal.

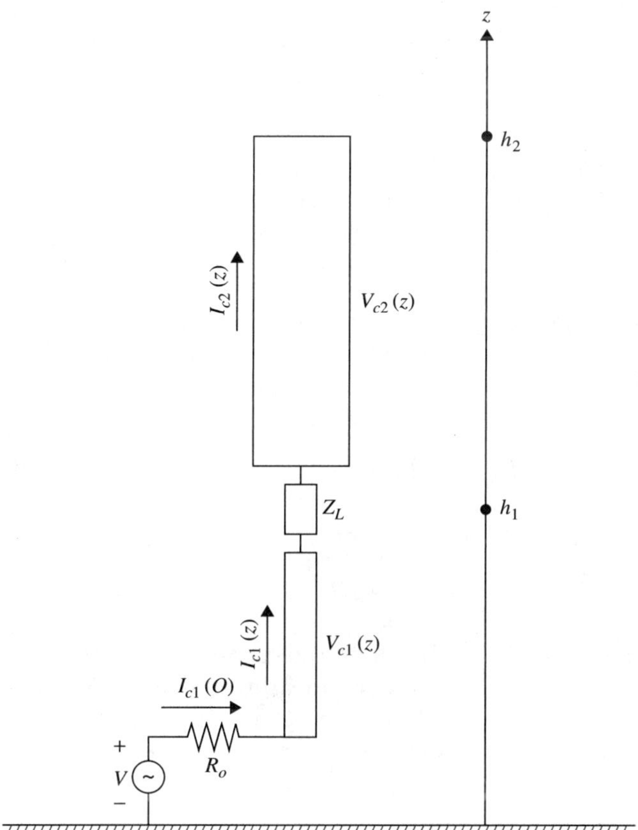

Figure 1-7 Essential features of the series-fed, loaded monopole antenna. For purposes of mathematical modeling, this antenna is an intermediate step between the monopole and the simplest configuration of the folded unipole antenna.

The loaded monopole has been the subject of complicated numerical analyses; however, we will make quick work of it using the same approach as that of the previous section. On each section of the loaded monopole, the currents and voltages are of the same form as Eqs. 1-12 and 1-13:

$$I_{cn}(z) = I_{Fcn}\psi_F(z) + I_{Rcn}\psi_R(z) \qquad n=1,2 \tag{1-22}$$

$$V_{cn}(z) = Z_{0cn}\left[I_{Fcn}\psi_F(z) - I_{Rcn}\psi_R(z)\right] \qquad n=1,2 \tag{1-23}$$

In Eqs. 1-22 and 1-23, the subscripts 1 and 2 denote the sections of the antenna below and above the load, respectively.

From Eqs. 1-22 and 1-23, we see that there are four unknown current amplitudes, I_{Fc1}, I_{Rc1}, I_{Fc2}, and I_{Rc2}. To determine them, we must derive four independent equations. These will take the form of boundary conditions at the top of the antenna, at the base, and at the load. At the top of the antenna, if we neglect the antenna end effect mentioned in the previous section, then the current must vanish:

$$I_{c2}(h_2) = 0 \tag{1-24}$$

At the base of the antenna, KVL requires that:

$$V - I_{c1}(0)R_0 = V_{c1}(0) \tag{1-25}$$

With Eqs. 1-24 and 1-25, we have two of the four required independent equations. The remaining two are boundary conditions at $z = h_1$, where the load Z_L is located. First, the current must be continuous through the load:

$$I_{c1}(h_1) = I_{c2}(h_1) \tag{1-26}$$

Second, the voltage drop across the load is:

$$V_{c1}(h_1) - V_{c2}(h_1) = I_{c1}(h_1)Z_L \tag{1-27}$$

Equations 1-24 through 1-27 are the four independent equations we need to determine the four unknown current amplitudes. If Eqs. 1-22 and 1-23 are substituted into those equations, then after some algebra, we can consolidate the results as a single matrix equation:

$$\mathbf{M} \cdot \mathbf{I} = \begin{pmatrix} V \\ 0 \\ 0 \\ 0 \end{pmatrix} \tag{1-28}$$

In Eq. 1-28, the unknown is the current vector,

$$\mathbf{I} = \begin{pmatrix} I_{Fc1} \\ I_{Rc1} \\ I_{Fc2} \\ I_{Rc2} \end{pmatrix} \tag{1-29}$$

M is a matrix with dimensions 4 × 4:

$$\mathbf{M} = \begin{pmatrix} R_0 + Z_{0c1} & R_0 - Z_{0c1} & 0 & 0 \\ 0 & 0 & \psi_F(h_2) & \psi_R(h_2) \\ (Z_{0c1} - Z_L)\psi_F(h_1) & -(Z_{0c1} + Z_L)\psi_R(h_1) & -Z_{0c2}\psi_F(h_1) & Z_{0c2}\psi_R(h_1) \\ \psi_F(h_1) & \psi_R(h_1) & -\psi_F(h_1) & -\psi_R(h_1) \end{pmatrix} \tag{1-30}$$

We could resolve Eq. 1-30 into four formulas for the unknown current amplitudes; however, we choose to leave it in its present form. It is a form that is rapidly evaluated by desktop computer. Further, it is a form that we shall encounter again and again in later chapters of this book. So, it is a good demonstration of the procedures we will follow for the folded unipole antenna.

The geometry of the loaded monopole antenna is also an excellent introduction to the folded unipole. As shown in Fig. 1-8, if we bend the

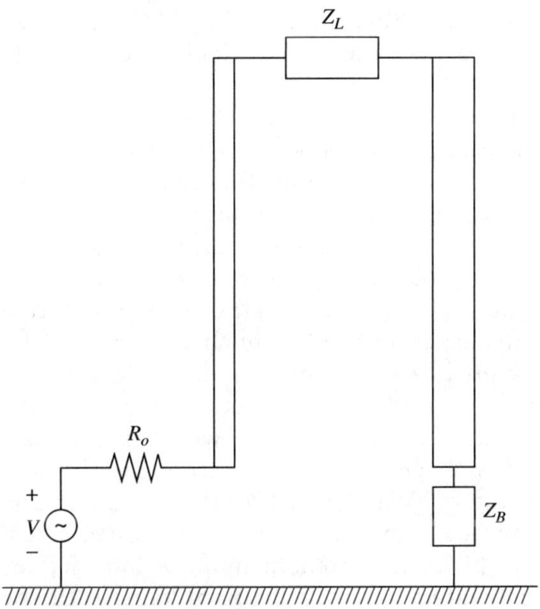

Figure 1-8 The series-fed, loaded monopole is easily deformed into the simplest configuration of the folded unipole antenna.

monopole at the load, and then add a terminating base impedance, the result is the simplest configuration of the folded unipole antenna. We will model that configuration in the next chapter. In the matrix equation for the current on it, **M** will also have dimensions 4×4; however, the elements will be different from those in Eq. 1-30.

Equation 1-30 is notable because of its small dimensions, only 4×4. In the next section, we will take a quick look at the foundations for general purpose, numerical, wire antenna analysis computer programs. We will find that those programs are also based upon matrix equations, but the dimensions are often orders of magnitude greater. This leads to increased computation time and greater numerical noise.

1.8 The End of the Golden Age: Numerical Methods

Prof. Roger F. Harrington's landmark book, *Field Computation by Moment Methods*, caused a revolution in electromagnetic problem solving when it was published in 1968 [7]. Suddenly, a huge backlog of problems, which had been accumulating for over a century, became mathematically tractable, if not actually easy to solve. Further, the timing could not have been better. High speed computers were becoming widely available, first as dial-up time-sharing services and, within a decade, as low cost desktop models. The current generation of engineering and science students were routinely educated in the use of FORTRAN and other scientific programming languages. Electromagnetic field analysis would never be the same.

MoM proved to be a double-edged sword, however. A long legacy of applied mathematics was quickly tossed out the window. It did not make much sense to study clever ways to exploit Bessel functions in cylindrical coordinates, Mathieu functions in elliptical coordinates, or Lamé wave functions in ellipsoidal coordinates if an electronic computer could rapidly evaluate electromagnetic fields in regions of space that did not conform to *any* coordinate system. Unfortunately, those special mathematical functions of applied physics incorporated much physical insight. In fact, they were created explicitly to explore physical phenomena, many of them in electromagnetism. When they were relegated to the academic attic, a certain amount of physical insight went with them.

So, today when confronted with problems in electromagnetics and antenna design, we face a trade-off. We can rush to the computer and exercise general-purpose numerical analysis programs, or we can invest some time beforehand to simplify the problem using a long legacy of elegant mathematics formulated by some brilliant minds. In the case of the folded unipole antenna, the latter approach offers definite advantages.

With the above as prelude, we will briefly explore the foundations of MoM. In so doing, we will find how errors can creep into numerical methods, especially in the case of the folded unipole antenna.

Many, if not most, problems in electromagnetic field analysis, are reduced by MoM to a matrix equation of the form,

$$\mathbf{Z} \cdot \mathbf{I} = \mathbf{V} \tag{1-31}$$

Equation 1-31 is identical in form to Eq. 1-28 which we derived for the loaded monopole antenna in the previous section. The chief difference is the size of the matrices. In Eq. 1-28 the matrix **M** has dimensions 4×4. In Eq. 1-31, the dimensions of **Z** can run into the hundreds or thousands. To better understand how this happens, we will develop the MoM approach for the analysis of general wire antennas. This is a relatively easy procedure which will demonstrate the great attraction of MoM in addition to its subtle limitations. Note there are many possible MoM formulations. The one chosen here follows a path made easy by drawing upon analogies with electronic circuit analysis.

Figure 1-9 shows one segment of a general wire antenna. It is short compared with a wavelength so that KVL and KCL may be applied locally.

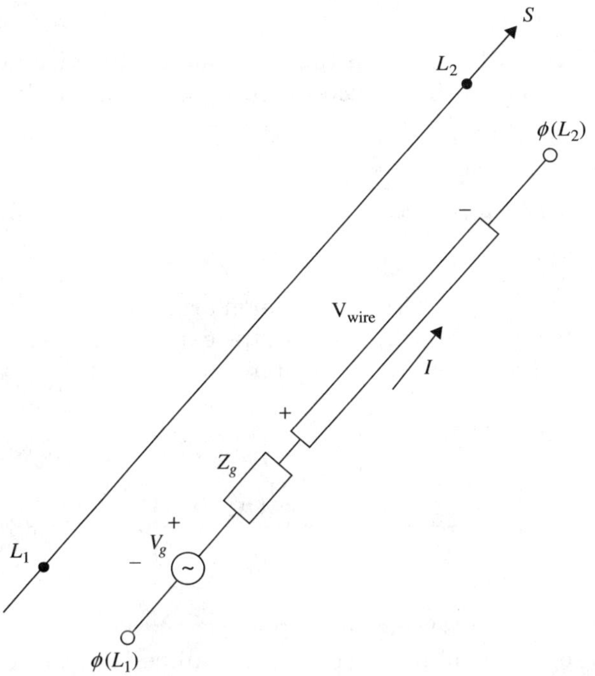

Figure 1-9 The general wire antenna analyzed using the Method of Moments is composed of small segments like this one. Each segment may have a voltage source and lumped circuit impedance as shown, although they are not required. The ends of the segment are at potentials $\Phi(\boldsymbol{L}_1)$ and $\Phi(\boldsymbol{L}_2)$.

It may include a voltage source and lumped circuit impedance, as shown, although they are not required on all segments. The ends of the segment are at unknown potentials $\Phi(L_1)$ and $\Phi(L_2)$.

KVL applied along the segment requires that:

$$V_g - IZ_g = V_{\text{wire}} + \Phi(L_2) - \Phi(L_1) \tag{1-32}$$

If the wire is a good conductor, then on the right side of Eq. 1-32, $V_{\text{wire}} = 0$, and we need only compute the potentials $\Phi(L_1)$ and $\Phi(L_2)$. To that end, rigorous electromagnetic theory tells us that they are related by the Lorentz Gage to the magnetic potential function:

$$\Phi(s) = \frac{j}{\omega\varepsilon_0\mu_0}\frac{\partial A(s)}{\partial s} \tag{1-33}$$

The magnetic vector potential tangential to the wire at position s is:

$$A(s) = \mu_0 \int_{\text{all wires}} \mathbf{u}_s \cdot \mathbf{u}_{s'} I(s')G(s,s')ds' \tag{1-34}$$

In Eq. 1-34, $\mathbf{u}_s$ and $\mathbf{u}_s'$ are unit vectors in the direction of the wire at positions s and s', respectively. The free space Green's function is similar to that in Eq. 1-10:

$$G(s,s') = \frac{e^{-jk|\mathbf{s}-\mathbf{s}'|}}{4\pi|\mathbf{s}-\mathbf{s}'|} \tag{1-35}$$

where **s** and **s′** are vectors extending from a common origin to the positions **s** and **s′**, respectively along the wire. That is, the expression $|\mathbf{s}-\mathbf{s}'|$ on the right side of Eq. 1-35 is simply the scalar distance between **s** and **s′**.

Combining Eqs. 1-33 and 1-34,

$$\Phi(s) = \frac{j}{\omega\varepsilon_0} \int_{\text{all wires}} I(s')\left[\mathbf{u}_s \cdot \mathbf{u}_{s'}\frac{\partial G(s,s')}{\partial s} + G(s,s')\frac{\partial(\mathbf{u}_s \cdot \mathbf{u}_{s'})}{\partial s}\right]ds' \tag{1-36}$$

On the right side of Eq. 1-36, the second term inside the brackets vanishes because the wire segments of interest here will all be straight.

Using Eq. 1-36, Eq. 1-32 becomes:

$$V = V_g - IZ_g \tag{1-37}$$

In Eq. 1-37,

$$V = \frac{j}{\omega\varepsilon_0} \int_{\text{all wires}} \mathbf{u}_s \cdot \mathbf{u}_{s'} I(s') \left[\frac{\partial G(L_2, s')}{\partial s} - \frac{\partial G(L_1, s')}{\partial s} \right] ds' \qquad (1\text{-}38)$$

where,

$$\frac{\partial G(L_2, s')}{\partial s} = \frac{\partial G(s, s')}{\partial s} \Big|_{s = L_2} \qquad (1\text{-}39)$$

and similarly for L_1.

For arbitrarily curved wires, evaluating the integral analytically on the right side of Eq. 1-38 is impossible; however, we can approximate the integral by a discrete sum over N straight segments:

$$V = \frac{j}{\omega\varepsilon_0} \sum_{n=1}^{N} \mathbf{u}_s \cdot \mathbf{u}_{s'} \int_{L_{1n}}^{L_{2n}} I_n(s') \left[\frac{\partial G(L_2, s')}{\partial s} - \frac{\partial G(L_1, s')}{\partial s} \right] ds' \qquad (1\text{-}40)$$

Our ability to evaluate the integral on the right side of Eq. 1-40 depends upon the choice of representation for the current $I_n(s')$ on the nth segment:

$$I_n(s') = I_n f_n(s') \qquad (1\text{-}41)$$

In Eq. 1-40, I_n is an unknown constant, and $f_n(s')$ is a function which is mathematically convenient for evaluating the integral in Eq. 1-40. Hopefully, it is also a reasonably realistic representation of the actual current.

Using Eqs. 1-40 and 1-41, Eq. 1-37 becomes:

$$V_g - IZ_g = \sum_{n=1}^{N} I_n Z_n \qquad (1\text{-}42)$$

Since the wire segment of Fig. 1-9 is one of many such segments used to construct the general wire antenna, let's add the subscript m to Eq. 1-37 to denote the mth segment:

$$V_{\text{gm}} - I_m Z_{\text{gm}} = \sum_{n=1}^{N} Z_{\text{nm}} I_n \qquad m = 1, 2, \ldots N \qquad (1\text{-}43)$$

where the mutual impedance Z_{nm} follows directly from Eqs. 1-40 and 1-41:

$$Z_{\text{nm}} = \frac{j}{\omega\varepsilon_0} \mathbf{u}_s \cdot \mathbf{u}_{s'} \int_{L_{1n}}^{L_{2n}} f_n(s') \left[\frac{\partial G(L_{2m}, s')}{\partial s} - \frac{\partial G(L_{1m}, s')}{\partial s} \right] ds' \qquad (1\text{-}44)$$

Our MoM formulation is almost complete. Rearranging terms in Eq. 1-43, that equation becomes:

$$\sum_{n=1}^{N}(Z_{nm} + \delta_{nm} Z_{gm}) I_n = V_{gm} \qquad m = 1, 2, \ldots N \tag{1-45}$$

In Eq. 1-45, δ_{nm} is the Kronecker delta function, which is simply zero except when $n = m$.

Equation 1-45 is just another way of writing our original matrix equation, Eq. 1-31. By comparing the two, it is clear what the elements of the matrix **Z** must be, and also the elements of the column vector **V**. The elements of the unknown column vector **I** are the amplitudes of the currents I_n on each antenna segment.

1.9 Method of Moments for Simple Antennas and Antenna Arrays

In this section, we will discuss how Eq. 1-31 can be applied to some simple antenna configurations. By simple, we mean that there are no junctions of multiple wires. Monopole antennas and arrays of monopole antennas are included. So are log periodic and Yagi antennas. The folded unipole is excluded because multiple folds may be joined to the main conductor or tower; however, we will include them in the next section. Without carrying out any specific examples, we will discover limitations of the MoM approach.

The difficulty of solving Eq. 1-31 for the unknown currents on each antenna segment is determined by the number and complexity of the matrix elements in **Z**. According to Eq. 1-45, the element in the nth row and mth column is:

$$\mathbf{Z} = [Z_{nm} + \delta_{nm} Z_{gm}] \tag{1-46}$$

On the right side of Eq. 1-46, Z_{nm} is the mutual impedance between filament or segment n and m, and Z_{gm} is an optional lumped circuit impedance in series with filament m. So, the chief challenge in filling the matrix **Z** is computation of the mutual impedances Z_{nm} according to Eq. 1-44. The difficulty of that computation depends upon at least two interdependent choices.

The first choice is the length of each segment. The straight segments must be a reasonable approximation to the arbitrarily curved wire of interest.

The second choice is the representation for the current distribution $f_n(s')$ on each segment, according to Eq. 1-41. In many or most simple MoM computer programs, it is simply a constant, that is:

$$I_n(s') = I_n \tag{1-47}$$

Besides constants, the literature contains discussions of triangle functions, sinusoids, and others. If we choose $f_n(s')$ wisely, then we can compute at least some of the Z_{nm} analytically instead of numerically. This will reduce the accumulation of numerical noise which will limit the accuracy of our results.

For example, we might choose

$$I(s') = I_m \cos(ks') \tag{1-48}$$

as an approximation to the current on segment m for all $m = 1,2, \ldots N$. This choice has at least two advantages. First, we will be able to calculate the Z_{nm} analytically in terms of the classic sine integral and cosine integral functions from antenna theory. Second, our choice is a smooth, continuously differentiable function that probably bears close resemblance to the actual current on that segment.

Whatever our choice for representing the current $I_n(s) = I_n f_n(s)$, filling the **Z** matrix amounts to calculating the mutual impedances according to Eq. 1-44. After doing so, the solution **I** to the matrix equation Eq. 1-31 is a column vector of the coefficients I_n.

From our discussion so far, we expect that the **Z** matrix of the MoM approach will be much larger than the **M** matrix derived in Sec. 1.7. The dimensions of **M** in Eq. 1-28 were 4 × 4. That is, the matrix contained 16 elements, for any length of antenna. For the MoM approach, the dimensions of the matrix **Z** in Eq. 1-31 are N × N, where N is the number of segments. N in turn depends on how long the antenna is compared with a wavelength. Typically, we would choose about 30 segments per wavelength. So, for a half-wave loaded monopole, the **Z** matrix might have dimensions 15 × 15, for a total of 225 elements. This is an order of magnitude greater than the size of the **M** matrix. For modern desktop computers, neither matrix is a challenge; however, the comparison illustrates a trend that will become significant in the case of the folded unipole and ultimately become a limitation.

There is a more fundamental limitation to consider. In our formulation of the MoM approach, we have required that KVL be satisfied, according to Eq. 1-32; however, we have done nothing to satisfy KCL. That is, we have not explicitly required that the current be continuous between segments, or conserved at junctions. Fortunately, for simple antennas, KCL is usually satisfied implicitly by our present version of MoM. For antennas with multiple wire junctions, including the folded unipole, however, the MoM in the form of Eq. 1-31 falls apart and gives nonsensical results.

Aside from the above technical limitations, there is a much more serious philosophical difference to consider. From the discussion in this section, it is clear that the MoM is a *bottoms up* approach to antenna design. That is, we start with a collection of wires and apply the MoM to it. If the results are not satisfactory, then we perturb the collection of wires and repeat the process. After hundreds or thousands of perturbations, this can be a frustrating design approach, indeed.

In contrast, the mathematical model developed in this book may be regarded as a *top down* design approach. That is, we will start with design parameters such as characteristic impedance and antenna length, and these are much less complicated to deal with mathematically than a detailed ensemble of wires. At the final stage of design, we will arrange the wires to obtain the desired values for characteristic impedance and other parameters.

1.10 MoM for Antennas with Multiple Wire Junctions

Figure 1-10 shows a multiple wire junction that could be part of a folded unipole, a top loaded antenna, or any number of complicated wire geometries. To avoid nonsensical answers for these geometries, we must explicitly satisfy KCL at such junctions:

$$\sum_{\text{all wires connected to junction}} I_m = 0 \tag{1-49}$$

We must satisfy Eq. 1-49 simultaneously with Eq. 1-31. A consequence is that our representation for the current on each antenna segment becomes more complicated. The one-term representation in Eq. 1-41, 1-47, or 1-48 only supplied enough unknown constants I_m to satisfy Eq. 1-31. If we wish to satisfy Eq. 1-49 in addition, then we must use a two-term representation for current. There are unlimited choices for such representations; however, one that is mathematically convenient and appeals to our physical intuition is:

$$I(s') = I_{\text{Fm}}\psi_F(s') + I_{\text{Rm}}\psi_R(s') \tag{1-50}$$

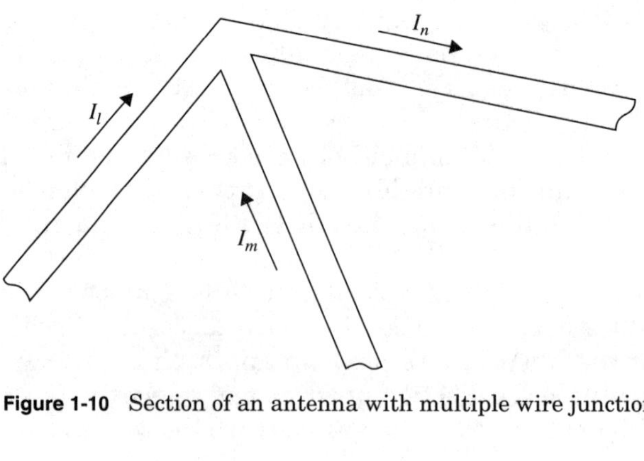

Figure 1-10 Section of an antenna with multiple wire junctions.

In Eq. 1-50, the traveling wave functions are the same as we defined in Eqs. 1-2 and 1-3 when we discussed transmission line theory. Our current representation for MoM now bears a remarkable resemblance to the transmission line currents in Eq. 1-1, and to the antenna currents in Eq. 1-12. In that sense, we have now come full circle, from the simple elegance of transmission lines to the geometrically complicated MoM, and back again.

Equation 1-50 provides the $2N$ unknown constants required to simultaneously satisfy Eqs. 1-31 and 1-49. Those constants are I_{Fn} and I_{Rn}, for $n = 1,2,\ldots N$.

Equations 1-31 and 1-49 can be combined into a single matrix equation:

$$\left(\frac{\mathbf{Z}}{\mathbf{J}}\right) \cdot \mathbf{I} = \left(\frac{\mathbf{V}}{\mathbf{0}}\right) \tag{1-51}$$

On the left side of Eq. 1-51 the **ZJ** matrix has dimensions $2N \times 2N$. That is, it has four times as many elements as the comparable matrix for simple antennas and arrays. In the example for the loaded monopole discussed in the previous section, that amounts to 900 matrix elements. That amount is becoming significant in terms of computation time and numerical noise. It is almost two orders of magnitude greater than the 16 elements needed in Sec. 1.7 for the same antenna.

1.11 The Perils of Large Matrices

In the previous section, we saw how MoM matrices can easily expand to contain hundreds of elements, even for antennas that are not too complicated. It is easy to envision more complicated antennas that require matrices with thousands of elements. When this happens, the odds favor serious numerical problems. Most of these are due to *ill-conditioned* matrices.

An ill-conditioned matrix has a determinant that is almost zero. Formally, then, it is at least very difficult, and as a practical matter perhaps impossible, to obtain solutions to the matrix equation.

Ill-conditioned matrices result from equations that are almost redundant, that is, almost not independent. A simple example of two such equations is:

$$x + y = 1 \tag{1-52}$$

$$x + (1 - \varepsilon)y = 1 - \delta \tag{1-53}$$

In Eq. 1-53, ε and δ are very small numbers, for example, between 10^{-7} and 10^{-6}. Let's assume they are numerical noise, and that we don't truly know where in that range they might be.

The solutions to Eqs. 1-52 and 1-53 are:

$$y = \frac{\delta}{\varepsilon} \tag{1-54}$$

$$x = 1 - \frac{\delta}{\varepsilon} \tag{1-55}$$

For the uncertain range of δ and ε, *y* can take on a value between 0.1 and 10, a hundred-to-one range. *X* can take on a value between –9 and 0.9, a ten-to-one range. So, the opportunity for error is rampant.

It is not difficult to extrapolate the above example to large MoM matrices. Each equation (i.e., row) in the matrix corresponds to a small antenna segment. The more segments there are, the greater the odds are that two or more of them will be described by very similar equations. The equations will be so similar that the terms will only differ by small quantities like ε and δ in our example. When that happens, and if ε and δ are significantly corrupted by numerical noise, then the solutions to the MoM matrix equation will be in doubt.

Sometimes, when the MoM matrix is corrupted by numerical noise the answers are obviously nonsense. At other times, they may appear plausible, but they are wrong just the same!

The MoM matrix is especially prone to be ill-conditioned when the method is applied to the folded unipole antenna. The reason is that there are many wire segments within a space small, compared with a wavelength. The KVL and KCL equations that are consolidated within Eq. 1-51 may change very little from segment to segment. For example, within a transverse plane containing the antenna cross section, there may be three or more fold wires plus the main conductor or tower, all separated by a small fraction of a wavelength. In contrast, a similar transverse plane for a monopole would contain only the tower.

Consistent with the above discussion, reports of actual experience with MoM relating to folded unipole design are mixed. At best, general purpose, numerical, wire antenna analysis computer programs "get the design into the ball park." At worst, users had "no luck at all designing folded unipoles." Between these extremes, there are many reports maintaining that, "the results are full of numerical noise." These observations should come as no surprise!

1.12 Conclusions

The folded unipole antenna integrates transmission lines with linear radiators, combining the best features of both. It is extremely versatile and capable of exhibiting very desirable performance. As early as the 1940s, engineers observed that, compared with vertical monopole antennas, folded unipoles could provide superior input impedance.

In the 1960s, folded unipoles with lengths only a tiny fraction of a wavelength provided VLF and LF communications for the United States Navy. About the same time, the United States Air Force recognized that the low profile of the folded unipole was well suited to the streamlining requirements of missiles and aircraft.

Despite the above advantages, the folded unipole has neither been fully explored nor optimized. In compendiums about antennas, it seldom rates more than a few paragraphs. The likely reason is that a mathematical model was a bit too complicated for precomputer mathematics. After computers became widely available, however, the folded unipole stubbornly resisted modeling with general, numerical techniques because its geometry tends to generate *ill-conditioned* matrices. So, successful modeling and exploration of the folded unipole requires a careful balance between the classical mathematical functions of antenna theory and computerized equation solving.

To better understand the details of that balance, in this chapter we reviewed both classic mathematical and numerical techniques. During the Golden Age of Antenna Analysis, spanning approximately the years 1854–1968, elegant formulas for both transmission lines and antennas were derived. After 1968, a revolution in electromagnetic problem solving was ushered in by MoM combined with the widespread availability of electronic computers.

The Golden Age of Antenna Analysis featured contributions from some brilliant and colorful individuals. They include physicists and mathematicians such as Lord Kelvin (1854–1907), Henry Cabourn Pocklington (1870–1952), Sergei A. Schelkunoff (1896–1992), and Ronold W. P. King (1905–2006). They also include innovative engineers such as Oliver Heaviside (1850–1925) and John H. Mullaney (1920–1994). From the very beginning in the mid-nineteenth century, the history reads much like a soap opera, with heated debates concerning the best mathematical models, and vested interests suppressing new technology.

We briefly examined how the MoM might be used to create a general purpose, numerical, wire antenna analysis program. We compared that approach with the one we will pursue in this book, to model the folded unipole. Both approaches result in matrix equations; however, the MoM matrix will be larger by orders of magnitude. At least as important are the different design philosophies implicit in the two approaches. The general purpose, numerical, wire antenna analysis program is a *bottoms up* approach. That is, we begin with a detailed ensemble of wires and analyze its performance. If the result is not satisfactory, then we perturb the wires and repeat the process. This can require many arduous iterations. In contrast, in this book we follow a *top down* approach. We begin by specifying parameters such as characteristic impedances and antenna length, and these are much less tedious to deal with mathematically than a detailed ensemble of wires. At the final stage of design, we arrange the wires to obtain the desired values for characteristic impedance and the other parameters.

This book is intended to fill the gap between the Golden Age and modern numerical techniques. The folded unipole antenna fell into that gap, and for many decades it lay there neglected, waiting to be completely modeled and developed.

References

1. Carter, P. S., Circuit Relationships in Radiating Systems and Applications to Antenna Problems, *Proc. of the IRE*, **20**, 1932, pp. 1004–1041.
2. Morrison, J. F., and P. H. Smith., The Shunt Excited Antenna, *Proc. of the IRE*, **25,** 1937, pp. 673–696.
3. Washburn, H. H., The Gamma Match, *QST*, **25**, September, 1949, pp. 20–21, 102.
4. Roberts, W. van B., Input Impedance of a Folded Dipole, *RCA Review*, **8**, June, 1947, pp. 289–300.
5. Pozar, D. M. (ed.), D. H. Schaubert (ed.), Microstrip Antennas, *IEEE Press*, 1995, pp. 37–38.
6. Thomson, W., On the Theory of the Electric Telegraph, *Philosophical Magazine*, **9**, 1854.
7. Harrington, R. F., Field Computation by Moment Methods, Prentice Hall, 1968.
8. Heaviside, O., Electromagnetic Induction and Its Propagation, *Three-Part Article in the Electrician*, **23**, 1885, **24**, 1886, **25**, 1887.
9. Pocklington, H. C., Electrical Oscillations in Wires, *Proc. of the Camb. Phil. Soc.,* **9**, 1897, pp. 324–332.
10. Schelkunoff, S. A., Theory of Antennas of Arbitrary Size and Shape, *Proc. of the IRE,* **29,** 1941, pp. 493–521.
11. Hallén, E., Theoretical Investigations into Transmitting and Receiving Antennas, *Nova Acta Regiae Soc. Sci. Upaliensis*, **11**, 1938, pp.1–44.
12. King, R. W. P., Theory of Linear Antennas, Harvard University Press, 1959.

Symbols Used in This Chapter

$A(s)$	Component of magnetic vector potential parallel to a segment of wire that lies along the s-axis
C	Capacitance per unit length of a transmission line
$E(z)$	Component of electric field parallel to the z-axis
E_n	Component of electric field radiated by the current on the nth wire segment, and parallel to the mth wire segment
$f_n(s')$	Function used to represent current on a wire segment in the MoM analysis.
G	Conductance per unit length of a transmission line
$G(s, s')$	Green's function, used to weight current when computing magnetic vector potential and electric field
$G(u, z)$	Green's function, same as $G(s,s')$
h	z-coordinate of top of monopole antenna
h_1	z-coordinate of top of lower section of loaded monopole antenna
h_2	z-coordinate of top of upper section of loaded monopole antenna
$\mathbf{I}$	Vector of unknown current coefficients in matrix equations for loaded monopole antenna, and in general MoM formulation
$I(s_m)$	Current flowing in mth antenna segment
$I_c(z)$	Common mode current flowing on monopole antenna
$I_{c1}(z)$	Common mode current flowing on lower section of loaded monopole antenna
$I_{c2}(z)$	Common mode current flowing on upper section of loaded monopole antenna

$I_{cn}(z)$	Common mode current flowing on *n*th section of loaded monopole antenna
$I_d(z)$	Differential mode current flowing on transmission line
I_{Fc}	Amplitude of common mode, forward traveling wave currrent
I_{Fcn}	Amplitude of common mode, forward traveling wave current on *n*th section of loaded monopole antenna
I_{Fd}	Amplitude of differential mode, forward traveling wave current
I_{Fm}	Amplitude of forward traveling wave current on *m*th segment of wire, in the MoM formulation
I_m	Amplitude of current on the *m*th segment of wire, in the MoM formulation
I_n	Amplitude of current on the *n*th segment of wire, in the MoM formulation
I_{Rc}	Amplitude of common mode, reverse traveling wave current on monopole antenna
I_{Rcn}	Amplitude of common mode, reverse traveling wave current on *n*th section of loaded monopole antenna
I_{Rd}	Amplitude of differential mode, reverse traveling wave current on monopole antenna
I_{Rm}	Amplitude of reverse traveling wave current on *m*th wire segment, in the MoM formulation
$\mathbf{J}$	Matrix of equations satisfying Kirchhoff's Current Law in the MoM formulation
j	Square root of −1
k	Wave number
L	Inductance per unit length of a transmission line
$\mathbf{M}$	Matrix in the equation for currents on the loaded monopole antenna
R	Resistance per unit length of a transmission line
R_0	Output impedance of a voltage source
s	Linear coordinate along the axis of a wire antenna
s'	Variable of integration that follows the axis of a wire antenna
$\mathbf{u}_s$	Unit vector in direction of wire at position *s*
$\mathbf{u}_s'$	Unit vector in direction of wire at position *s'*
$\mathbf{V}$	Vector in the matrix equation for antenna currents, both on the loaded monopole and in the MoM formulation
V	Amplitude of voltage source
$V_c(z)$	Common mode voltage on the monopole antenna
$V_{c1}(z)$	Common mode voltage on the lower section of the loaded monopole antenna
$V_{c2}(z)$	Common mode voltage on the upper section of the loaded monopole antenna
$V_{cn}(z)$	Common mode voltage on the *n*th section of the loaded monopole antenna
$V_d(z)$	Differential mode voltage on a transmission line
V_g	Generator or source voltage on a wire segment in the MoM formulation
V_{gm}	Generator or source voltage on the *m*th wire segment in the MoM formulation
V_m	Voltage between the ends of the *m*th wire segment in the MoM formulation
x	Unknown variable in an example of ill conditioned equations
y	Unknown variable in an example of ill conditioned equations
$\mathbf{Z}$	Matrix in the equation for antenna currents in the MoM formulation
Z_{0cn}	Common mode characteristic impedance for the *n*th section of the loaded monopole antenna
Z_{0d}	Differential mode characteristic impedance for a transmission line
Z_g	Lumped circuit impedance on a wire segment in the MoM formulation
Z_{nm}	Mutual impedance between the *n*th and *m*th wire segment in the MoM formulation
δ	Small, noisy number in the example of ill-conditioned equations
δ_{nm}	Kronecker Delta Function, one when *n* equals *m*, otherwise zero
ε	Small noisy number in the example of ill-conditioned equations
ε_0	Permittivity of free space

$\Phi(L_1)$	Potential at lower end of wire segment, in the MoM formulation
$\Phi(L_2)$	Potential at upper end of wire segment, in the MoM formulation
$\Phi(s)$	Potential along wire segment, in the MoM formulation
λ	Wavelength
μ_0	Permeability of free space
ω	Frequency in radians per second
$\psi_F(z)$	Forward traveling wave function
$\psi_R(z)$	Reverse traveling wave function

Chapter

2

The Simplest Configuration

2.1 Introduction

The folded unipole combines the best features of the linear antenna with those of transmission lines. The electromagnetic radiation from a linear antenna is predictable, and such antennas can be readily combined into arrays to form a practically unlimited variety of radiation patterns. The stored energy in transmission lines can be harnessed to match impedance, introduce phase shift, and create versatile electronic filters. Clearly, the folded unipole can be designed as a sophisticated antenna, or antenna array element, featuring broadbanding, multifrequency operation, select frequency rejection, and a host of other desirable properties.

In this chapter, we will derive a mathematical model for the simplest folded unipole. In later chapters, we will use the same approach to methodically expand the model to include the most sophisticated versions. So, we are establishing some fundamental analytic tools and principles here.

Our mathematical approach will be remarkably compact and simple compared with general purpose, numerical, wire antenna analysis techniques. While the latter require the solution of matrix equations with matrix dimensions on the order of a hundred or more, the dimension of our matrix will be a mere four. Thus, the difference in the number of matrix elements is about *three orders of magnitude*. The advantages of our approach are many. Notably, they include far less numerical noise, which tends to compromise the solutions of large matrix equations. Also, even in the present age of extremely fast desktop computers, our approach offers an important advantage in speed. Some of our design examples include the simultaneous variation of two design parameters over 10,000 iterations. Yet, these examples are executed in less than a minute. Using the general, numerical, wire antenna analysis computer programs, they would require hours or days!

After deriving the mathematical model, we will use it extensively in this chapter to demonstrate some remarkable properties of the folded unipole. Examples will include antennas shorter than the traditional quarter wavelength. These will notably feature an extremely short unipole that is only 1/20 of a wavelength (i.e., 18 electrical degrees) in height. We will also examine some tall unipoles about one-half wavelength in height, a dimension that traditionally has been challenging for antenna engineers with respect to impedance matching.

We will clearly see that even this most simple configuration of the folded unipole is a very versatile device indeed. It has no fewer than six independent design variables compared with only two (height and radius) for the traditional series-fed antenna. No wonder, then, that the folded unipole can exhibit remarkable performance that sometimes invites disbelief.

2.2 Essential Features of the Configuration

Figure 2-1 shows the simplest configuration of the folded unipole. At first glance, it appears to be little more than a transmission line with terminating impedances Z_T at the top and Z_B at the base. This first impression is misleading, however.

Although the transmission line is depicted as a two-wire line, this is only for simplicity in the drawing. In fact, our mathematical model will

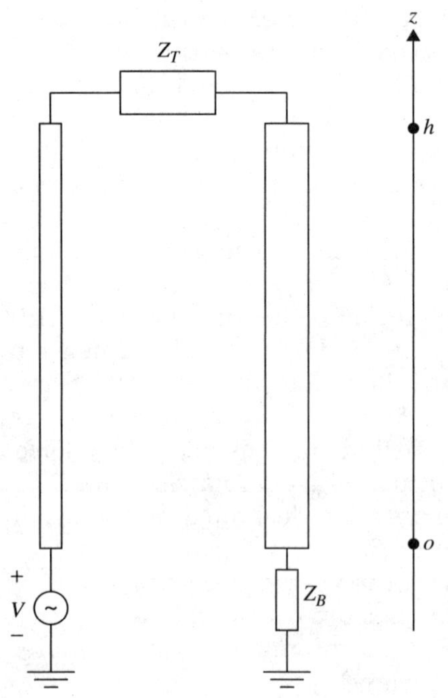

Figure 2-1 Essential features and geometry for the simplest folded unipole configuration. The antenna is terminated with arbitrary impedances at both top and base. Two parallel conductors are shown only for simplicity in the drawing. In fact, the analysis applies to any arbitrary cross section.

apply to arbitrary cross sections with any number of conductors. This is because any cross section can be represented mathematically by two characteristic impedances, a differential (or balanced, or transmission line) mode Z_{0d}, and a common (or unbalanced, or antenna) mode Z_{0c}. In Chap. 7, we will show in detail how to compute these characteristic impedances for arbitrary cross sections.

The two modes mentioned earlier, differential and common, are fundamental to the physics of operation for the folded unipole. Common mode currents are responsible for the electromagnetic radiation and the antenna-related properties of the unipole. Differential mode currents are responsible for the transmission line related properties, and they provide the means to design filters, impedance matching networks, and phase shifters as part of the unipole.

It is easy to write down simple mathematical expressions for the currents in each of the two modes. For the common mode, the current is:

$$I_c(z) = I_{\mathrm{Fc}} e^{-\mathrm{jkz}} + I_{\mathrm{Rc}} e^{+\mathrm{jkz}} \tag{2-1}$$

In Eq. 2-1, I_{Fc} is the amplitude of a forward traveling wave, and I_{Rc} is the amplitude of a reflected traveling wave. In general, both amplitudes can be complex, that is, they have nonzero amplitude and phase. Similarly for the differential mode, the current is:

$$I_d(z) = I_{\mathrm{Fd}} e^{-\mathrm{jkz}} + I_{\mathrm{Rd}} e^{+\mathrm{jkz}} \tag{2-2}$$

In Eqs. 2-1 and 2-2, we have written the traveling wave functions $\psi_F(z)$ and $\psi_R(z)$ explicitly as complex exponentials, according to Eqs. 1-2 and 1-3. For the configuration of interest in this chapter, the complex exponentials are slightly more convenient. The wave number k is:

$$k = 2\pi/\lambda \tag{2-3}$$

where λ is the wavelength.

Along with the common mode and differential mode currents, there are also common and differential mode voltages. The voltages and currents are related by the usual transmission line equations.

$$V_c(z) = Z_{0c}\left(I_{\mathrm{Fc}} e^{-\mathrm{jkz}} - I_{\mathrm{Rc}} e^{+\mathrm{jkz}}\right) \tag{2-4}$$

$$V_d(z) = Z_{0d}\left(I_{\mathrm{Fd}} e^{-\mathrm{jkz}} - I_{\mathrm{Rd}} e^{+\mathrm{jkz}}\right) \tag{2-5}$$

From Eqs. 2-1 and 2-2, we see that there are four complex current amplitudes to determine, I_{Fc}, I_{Rc}, I_{Fd}, and I_{Rd}. They are forward and reflected current amplitudes for each of two modes, common and differential.

Before proceeding further, however, we acknowledge that Eqs. 2-1 and 2-4 for the antenna mode currents and potentials are idealizations.

Since circa 1970, much literature and computer code has concerned the exact determination of antenna currents using numerical methods, and the results may show some deviation from pure sinusoidal or traveling waves. These numerical methods are certainly appropriate for the analysis of existing structures or finalized designs for which the detailed geometry is known. For design purposes, however, the traveling wave representation offers significant computational advantages. We have already mentioned them in the Sec. 2.1. They are reduced numerical noise and substantially faster execution.

Before embarking upon our detailed mathematical modeling, it will be interesting and useful to review the history of such models. This history will serve at least two purposes. First, it will confirm the underlying physical principles from which our mathematical model will follow. Second, it will demonstrate the original and continuing need for such a model.

2.3 The Analysis by Roberts

A journal article by W. van B. Roberts in 1947 [1], summarizes the early history of antenna analysis as it relates to the folded antennas in general, and, therefore, the folded unipole in particular. According to Roberts, some useful properties were well known before there was a mathematical model to explain them. He wrote:

> The folded dipole has become a familiar expedient for increasing the input impedance of a dipole antenna... It is well known that this arrangement has four times the impedance of a simple dipole. But if two auxiliary elements are used [instead of one, then] the impedance is multiplied by something more like nine. And finally it is generally known in a qualitative sort of way that if the two dipoles are of different diameters, the impedance multiplier is no longer four, but more or less than four according as the auxiliary element is of greater or less diameter than the driven element...There appear to be no formulas readily available in the literature which give the impedance in the general case of elements of arbitrary size, shape, number, and arrangement.

The above remarks by Roberts suggest that, prior to 1947, a general theory to explain and expand the operation of the folded unipole was yet to be published.

Roberts introduced a novel and clever approach that would be used again in later publications for over a decade. As shown in Fig. 2-2, he conceptually replaced the single source or transmitter with three sources. Thus, he explicitly separated out the common mode and differential mode currents. The common mode currents are excited by the single source connected between ground and the remaining two sources. Those latter sources excite the differential mode currents.

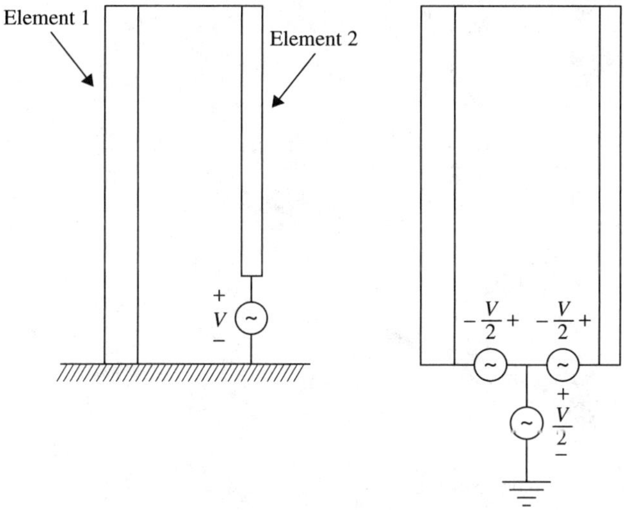

Figure 2-2 The quarter-wave folded unipole analyzed by Roberts in 1947 [1]. Also shown is the innovative way he separated out the common and differential mode currents by substituting three equivalent sources for the actual source.

The ultimate formula at the conclusion of Robert's article is:

$$Z_{\text{in}} = R(1 + Z_1/Z_2)^2 \tag{2-6}$$

In Eq. 2-6, R is the input resistance, and Z_1 and Z_2 are differential mode characteristic impedances. Roberts explains that, "Z_1 is the characteristic impedance of a line formed of conductors each like element 1 and with the same center-to-center spacing as the actual elements, while Z_2 is the same except that the line is formed of conductors each like element 2."

Roberts limited his formula to resonant quarter-wave unipoles and half-wave dipoles for which the input reactance vanishes. Also, no terminating impedances at either the top or base of the antenna were considered. So, formulas for the complex input impedance of antennas of more general lengths would have to wait; however, the quantitative analysis of the folded unipole antenna had definitely begun.

2.4 The Analysis by Leonhard, Mattuck, and Poté

In a 1955, journal article [2], Leonhard, Mattuck, and Poté, extended the analysis of Roberts to folded unipoles shorter than a quarter wavelength, and with terminating impedances at the base. Figure 2-3 shows

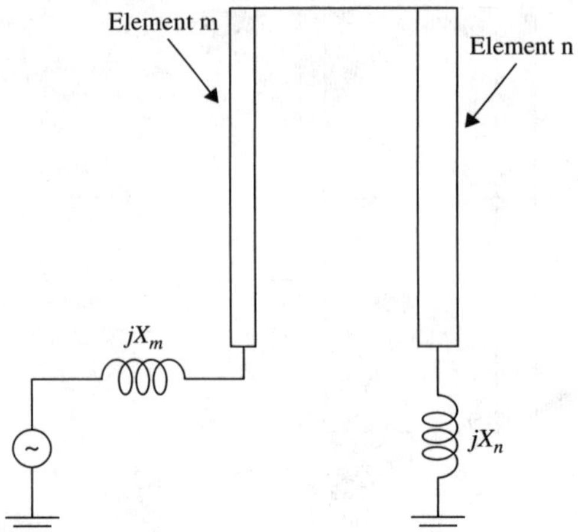

Figure 2-3 The short, folded unipole antenna, with base terminating impedances, analyzed by Leonhard, Mattuck, and Poté in 1955 [2].

the configuration they analyzed. They explicitly recognized the underlying physical principles when they wrote:

> The transformation of input impedance is shown to be due to two factors: (1) the division of radiation mode currents between the driven and undriven portions of the antenna, and (2) the flow of currents in a nonradiating transmission-line mode... The general procedure used will be an extension of the method employed by Roberts to analyze the quarter-wave case.

According to the above remarks, Poté *et al* were well aware of the importance of differential mode currents, in addition to the radiating common mode currents.

In the course of their analysis, they introduced a parameter which would be used in many subsequent articles, and also in this chapter. The parameter is the *current transformation ratio*, that is, the amount of common mode current in the fold connected to the transmitter, relative to the common mode current in the remaining conductor. They credit a previous 1950, article by Guertler [3] for the formula,

$$p = \ln\left(\frac{s}{a_1}\right) \Big/ \ln\left(\frac{s}{a_2}\right) \tag{2-7}$$

In Eq. 2-7, s is the spacing between the two conductors, and a_1 and a_2 are their radii.

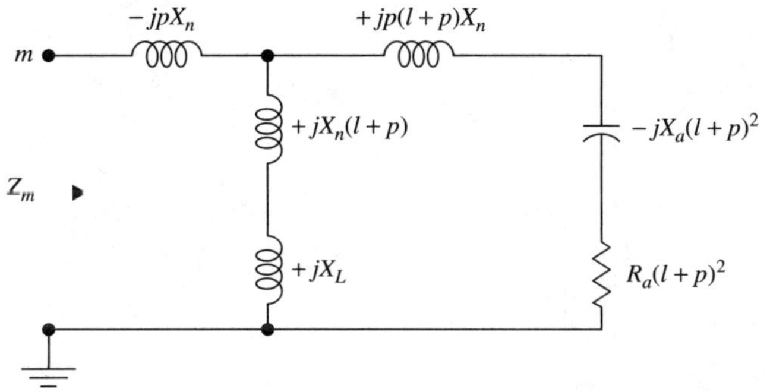

Figure 2-4 The equivalent circuit for the short, folded unipole antenna, derived by Leonhard,Mattuck, and Poté [2].

They derived an equivalent circuit for the folded unipole, shown in Fig. 2-4, and from that a formula for the complex input impedance,

$$Z_m = \left\{ \frac{1}{\left[R_a(1+p)^2 - jX_a(1+p)^2 + jp(1+p)X_n\right]} + \frac{1}{j\left[X_n(1+p) + X_L\right]} \right\}^{-1} - jpX_n \qquad (2\text{-}8)$$

Clearly, Poté *et al* significantly built upon the initial contribution by Roberts. They mathematically modeled folded unipoles shorter than the resonant quarter wavelength. Further, they accounted for the effects of terminating impedances at the base of the unipole. Still needed for a complete theory, however, was a model for unipoles longer than a quarter wavelength, and unipoles with terminating impedances at the top.

2.5 The Analysis by King and Harrison

In a 1960, journal article [4], the eminent team of antenna theorists R. W. P. King and Charles W. Harrison, Jr. analyzed the folded unipole of arbitrary length with a terminating impedance at the top. That configuration is shown in Fig. 2-5. Using an approach based upon integral equations, in which the unknown currents are part of an integrand, they derive the input admittance for the antenna:

$$Y_{in} = \frac{F}{Z_d} + \frac{1}{Z_t} \qquad (2\text{-}9)$$

In Eq. 2-9, "Z_d is the input impedance of a center-driven isolated cylindrical antenna of radius d," where

$$d = \sqrt{b}(a_1 a_2)^{1/4} e^p \qquad (2\text{-}10)$$

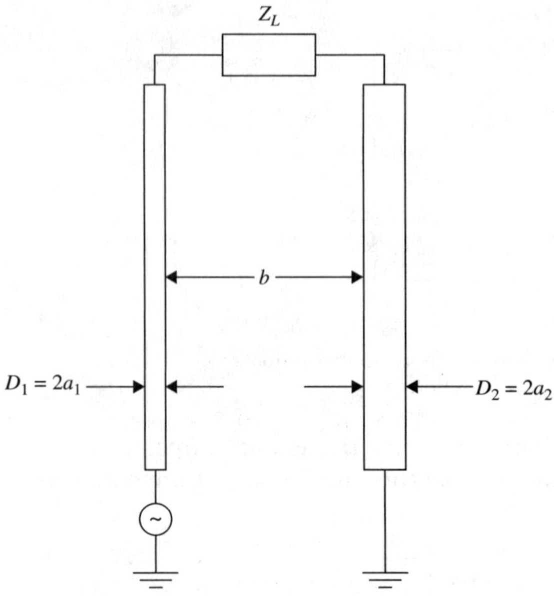

Figure 2-5 The folded unipole antenna, with top terminating impedance, analyzed by C. W. Harrison, Jr., and R. W. P. King in 1960 [4].

and,

$$p = \left[\ln\left(\frac{a_1}{a_2}\right)\right]^2 \Big/ 4\ln\left(\frac{b^2}{a_1 a_2}\right) \tag{2-11}$$

Also, in Eq. 2-9,

$$F = \left[\ln\left(\frac{b}{a_2}\right) \Big/ \ln\left(\frac{b^2}{a_1 a_2}\right)\right]^2 \tag{2-12}$$

and,

$$Z_t = 2Z_c\left(\frac{Z_L + jZ_c \tan\beta h}{Z_c + jZ_L \tan\beta h}\right) \tag{2-13}$$

In Eq. 2-12, the differential mode characteristic impedance is:

$$Z_c = \frac{\zeta_0}{2\pi}\ln\frac{b^2}{a_1 a_2} \tag{2-14}$$

As part of their approach, King and Harrison confirmed the fundamental physics articulated in earlier journal articles and followed in this chapter. They wrote:

> The mathematical formulation of problems that involve unbalanced multiwire lines with series and interconnecting impedances is more easily followed if the underlying principle is first understood. This is simple. It consists of the separation of those currents that can be treated by conventional transmission-line theory... The radiating currents are the... codirectional or unbalanced currents; the currents that contribute negligibly to radiation are the conventional equal and opposite, or balanced, line currents.

Comparing the above remarks with those of Poté *et al* in the previous section, there is definitely a consensus among antenna analysts concerning the important distinction between unbalanced (or common mode, or antenna mode) and balanced (or differential mode, or transmission mode) currents.

The article also mentions the folded unipole with terminating impedances at both the top and the base; however, the results and details are evidently written down in a limited circulation laboratory report [5].

The integral equation approach by King and Harrison is mathematically elegant and theoretically rigorous, and Eqs. 2-9 through 2-13 are easy to evaluate for the particular configuration shown in Fig. 2-5. An approach that applies to more general configurations and also results in equations that can be rapidly evaluated, for designs that may require 10,000 or more iterations, does not appear in the literature.

With the historical background of the preceding three sections, the stage is now set for our mathematical approach. The underlying physics is well established. It remains only to couple those physical principles with tenable mathematical tools.

2.6 Equations at the Top of the Folded Unipole

Before digressing on the history of folded unipole analysis in Sect. 2.3–2.5, we had expressed the voltages and currents on the antenna in terms of four current amplitudes, I_{Fc}, I_{Rc}, I_{Fd}, and I_{Rd}. The subscript F denotes a forward traveling wave, and R denotes a reflected one. The susbcript c denotes a common mode current, and d denotes a differential one. In this and the following two sections, we will develop a method for determining those amplitudes. It will prove worthwhile to develop this method carefully because it will serve as the foundation for analyzing more general unipole configurations in later chapters.

Figure 2-6 shows the configuration of interest with currents and voltages. Note that the common mode currents are not necessarily equal on the different conductors. They are related by the *current transformation ratio*, ν, which we mentioned in Sec. 2.4. It is determined by the cross section of the antenna. In Chap. 7, we will describe how to compute the current transformation ratio for any arbirary cross section.

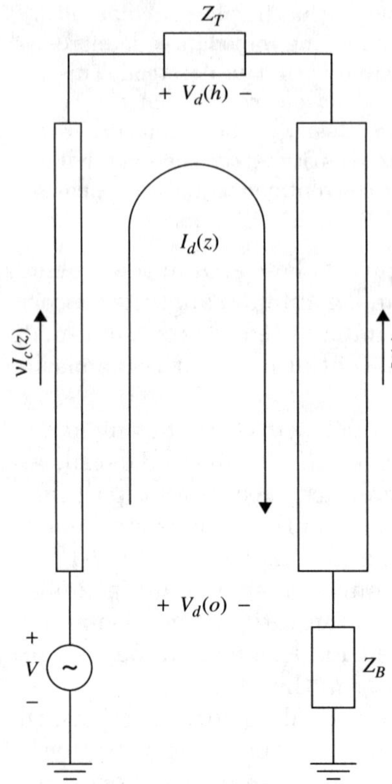

Figure 2-6 Voltages and currents used for analysis of the folded unipole antenna.

At the top of the antenna there are two boundary conditions. First, the common mode current must vanish because it has no place to flow:

$$I_c(h) = 0 \tag{2-15}$$

Second, the differential mode voltage and current are related by Kirchhoff's voltage law (KVL):

$$V_d(h) - Z_T I_d(h) = 0 \tag{2-16}$$

2.7 Equations at the Base of the Folded Unipole

In addition to Eqs. 2-15 and 2-16, we need two more equations to uniquely determine the four unknown current amplitudes. These can be formulated at the base of the antenna. First applying KVL around the loop starting at ground, proceeding across the base terminals, and returning to ground:

$$Z_B[I_d(0) - I_c(0)] + V_d(0) = V \tag{2-17}$$

Second, the common mode voltage at the base is obviously the input voltage:

$$V_c(0) = V \tag{2-18}$$

Equations 2-15 through 2-18 are the four independent equations we need to determine the four current amplitudes.

2.8 Matrix Formulation

Our goal is to solve for the currents in a computationally efficient way that lends itself to rapid, large scale iterations. In that way, we can best explore the effects of the many independent design parameters provided by the folded unipole, especially for the more complicated configurations. To that end, we can rewrite Eqs. 2-15 through 2-18 as a single matrix equation. This is a straightforward procedure, although the algebraic details are typically tedious. We will describe the necessary steps and then write down the results.

The first step is to rewrite Eqs. 2-15 through 2-18 explicitly in terms of the four unknown current amplitudes I_{Fc}, I_{Rc}, I_{Fd}, and I_{Rd}. This is readily accomplished by substituting for the currents $I_c(z)$ and $I_d(z)$ from Eqs. 2-1 and 2-2. Similarly, we can substitute for the voltages

$$\mathbf{M} \cdot \mathbf{I} = \begin{pmatrix} V \\ V \\ 0 \\ 0 \end{pmatrix} \tag{2-19}$$

$V_c(z)$ and $V_d(z)$ from Eqs. 2-4 and 2-5. The result is the matrix equation,

In Eq. 2-19, current vector $\mathbf{I}$ is:

$$\mathbf{I} = \begin{pmatrix} I_{\mathrm{Fc}} \\ I_{\mathrm{Rc}} \\ I_{\mathrm{Fd}} \\ I_{\mathrm{Rd}} \end{pmatrix} \tag{2-20}$$

The matrix $\mathbf{M}$ is:

$$\mathbf{M} = \begin{pmatrix} -Z_B & -Z_B & Z_B + Z_{0d} & Z_B - Z_{0d} \\ Z_{0c} & -Z_{0c} & 0 & 0 \\ \psi_F(h) & \psi_R(h) & 0 & 0 \\ 0 & 0 & (Z_{0d} - Z_T)\psi_F(h) & (-Z_{0d} - Z_T)\psi_R(h) \end{pmatrix} \tag{2-21}$$

In Eq. 2-21, the forward traveling wave function, introducted in Chap. 1, is:

$$\psi_F(h) = e^{-\mathrm{j}kh} \tag{2-22}$$

The reverse traveling wave function is:

$$\psi_R(h) = e^{+\mathrm{j}kh} \tag{2-23}$$

Equation 2-19 is a standard form that can be rapidly evaluated by computer, especially because its dimensions (4×4) are so small. In contrast, a formulation using the general, numerical, wire antenna analysis approach would result in matrices with dimensions on the order of 100×100, or greater. So, our approach results in less numerical noise and increased capability for large scale design iterations.

In any case, once we have solved Eq. 2-19 for the current amplitudes, we can readily calculate the radiated electromagnetic field, radiated power density, and input impedance. These are the subjects of the next two sections.

2.9 Radiated Field and Power

Figure 2-7 shows the geometry for calculating the radiated electromagnetic field from a vertical conductor, such as those that make up the folded unipole antenna and its image in a perfectly conducting ground. The figure is significantly different from the usual textbook example, however, because the currents flowing on the conductor are traveling waves. In most textbooks, they are standing waves, and there is no question concerning the phase or direction of propagation for the image currents in the ground.

For the images of the traveling waves, however, the phase and direction of propagation are not immediately obvious. It turns out that they are correct as shown in the figure; however, it is instructive to verify that this is true. We can do this by showing that, for a monopole antenna, the currents reduce to the well-known sinusoidal standing wave. To that end, the current above ground is:

$$I_c(z) = I_{\mathrm{Fc}} e^{-\mathrm{j}kz} + I_{\mathrm{Rc}} e^{+\mathrm{j}kz} \tag{2-24}$$

The image of the current in the ground is:

$$I_c(\zeta) = I_{\mathrm{Fc}} e^{-\mathrm{j}k\zeta} + I_{\mathrm{Rc}} e^{+\mathrm{j}k\zeta} \tag{2-25}$$

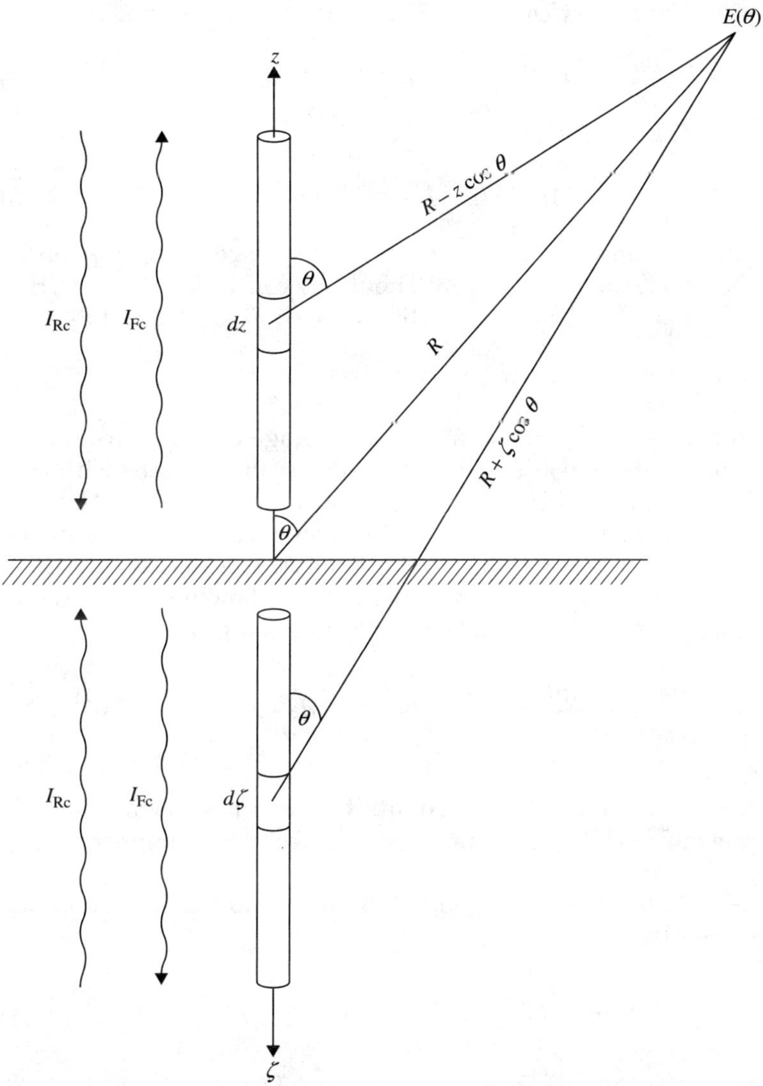

Figure 2-7 Geometry for calculating the radiated electromagnetic field and power from the folded unipole antenna. The mirror images of the traveling wave currents are correct as shown, although this may not be intuitively obvious.

At the ends of both the antenna and its image in the ground,

$$I_c(h) = 0 \tag{2-26}$$

Applying the boundary condition Eq. 2-26, Eq. 2-24 becomes:

$$I_c(z) = I_{Fc}(e^{-jkz} - e^{-j2kh}e^{+jkz}) \tag{2-27}$$

Equation 2-25 becomes:

$$I_c(\zeta) = I_{Fc}(e^{-jk\zeta} - e^{-j2kh}e^{+jk\zeta}) \tag{2-28}$$

If the initial assumption for the form of the image current given by Eq. 2-25 is correct, then the current should be continuous at the drive point of the monopole. That is, according to both Eqs. 2-27 and 2-28,

$$I_c(0) = I_{Fc}(1 - e^{-j2kh}) \tag{2-29}$$

Equation 2-29 shows that the actual and image currents are indeed continuous at the drive point as desired. Now we can proceed with confidence with the calculation of radiated fields and power densities.

As has been mentioned repeatedly in previous sections, only the common mode current contributes to the radiated electromagnetic field. The contributions to that field from the current flowing on the incremental segments dz of the unipole and $d\zeta$ of its image are:

$$dE(\theta) = \frac{(1+\nu)\eta k \sin\theta}{4\pi R} e^{-jkR}\left[I_c(z)e^{-jkz\cos\theta}dz + I_c(\zeta)e^{-jk\zeta\cos\theta}d\zeta\right] \tag{2-30}$$

In Eq. 2-30, the factor (1+ν) accounts for currents flowing in both conductors of our model. η is the impedance of free space, approximately 120π ohms.

The total radiated electromagnetic field is the sum of contributions from all segments:

$$E(\theta) = \frac{(1+\nu)\eta k \sin\theta}{4\pi R} e^{-jkR} F(\theta, 0, h) \tag{2-31}$$

where:

$$F(\theta, a, b) = \int_a^b dz I_c(z) e^{jkz\cos\theta} + \int_a^b d\zeta I_c(\zeta) e^{-jk\zeta\cos\theta} \tag{2-32}$$

In Eq. 2-32, we have deliberately retained the general limits of integration, a and b, instead of 0 and h, because we can obtain a general, exact formula for $F(\theta,a,b)$. That formula will be useful repeatedly in later chapters. Using Eq. 2-1, we can perform the integration in Eq. 2-32 exactly, and thus obtain the desired formula:

$$F(\theta,a,b)=\frac{1}{jk}\left\{\begin{array}{l}\dfrac{-I_{\mathrm{Fc}}\left[e^{-\mathrm{jkb}(1-\cos\theta)}-e^{-\mathrm{jka}\,(1-\cos\theta)}\right]}{(1-\cos\theta)}+\dfrac{I_{\mathrm{Rc}}\left[e^{+\mathrm{jkb}(1+\cos\theta)}-e^{+\mathrm{jka}(1+\cos\theta)}\right]}{(1+\cos\theta)}\\+\dfrac{I_{\mathrm{Rc}}\left[e^{+\mathrm{jkb}(1-\cos\theta)}-e^{+\mathrm{jka}(1-\cos\theta)}\right]}{(1-\cos\theta)}-\dfrac{I_{\mathrm{Fc}}\left[e^{-\mathrm{jkb}(1+\cos\theta)}-e^{-\mathrm{jka}(1+\cos\theta)}\right]}{(1+\cos\theta)}\end{array}\right\} \tag{2-33}$$

Of course, for our specific purposes in this chapter, in Eq. 2-33, $a=0$ and $b=h$.

The radiated power density follows readily from the electromagnetic field:

$$S=\frac{|E(\theta)|^2}{\eta} \tag{2-34}$$

The total radiated power is the summation of the power density over a great hemisphere with origin at the base of the unipole:

$$P=\int_0^{2\pi}d\varphi\int_0^{\pi/2}d\theta R^2\sin\theta\, S(\theta) \tag{2-35}$$

By symmetry, the integration over the bearing φ can be performed independently, leaving only an integration of the coordinate θ:

$$P=2\pi\int_0^{\pi/2}d\theta R^2\sin\theta\, S(\theta) \tag{2-36}$$

We might be able to perform the integration on the right side of Eq. 2-36 analytically using the classic special functions of antenna theory, such as the sine and cosine integrals *Si* and *Ci*, and the exponential integral *Ei*. Before the days of fast and inexpensive computers, extensive tables of these functions were in the libraries of most antenna engineers. Today, however it is just as fast and easy to evaluate the right side of Eq. 2-36 numerically.

2.10 Input Impedance and Current

Now that we have found the currents on the unipole and the radiated power, it is easy to calculate the input impedance. Historically, this was a chief attraction of the folded unipole to antenna engineers. In the LF and broadcast bands, unipoles shorter than a quarter wavelength could be designed with input impedances compatible with powerful radio transmitters.

In terms of the radiated power determined in the previous section, the input resistance is:

$$R_{\text{in}} = \frac{P}{|I_{\text{in}}|^2} \tag{2-37}$$

In terms of the applied voltage shown in Fig. 2-6, the input reactance is:

$$X_{\text{in}} = \frac{V}{I_{\text{in}}} \tag{2-38}$$

In Eqs. 2-37 and 2-38, the input current is:

$$I_{\text{in}} = \nu I_c(0) + I_d(0) \tag{2-39}$$

The currents on the right side of Eq. 2-39 are illustrated in Fig. 2-6. They were determined when we solved the matrix Eq. 2-19.

2.11 Efficiency

We can include ohmic losses in the resistive parts of the two impedances Z_T and Z_B of our model, that is, R_T and R_B. The total power dissipated in those resistances is:

$$P_d = |I_c(0) + I_d(0)|^2 R_B + |I_d(h)|^2 R_T \tag{2-40}$$

In terms of the total radiated power computed in the previous section, the *power efficiency* is:

$$E_p = \frac{P}{P + P_d} \tag{2-41}$$

In some applications, *field efficiency* is the preferred metric:

$$E_f = \sqrt{\frac{P}{P + P_d}} \tag{2-42}$$

2.12 Being Selective About Examples and Displaying Results

With the matrix Eq. 2-19 for the currents on the folded unipole, and related equations for input impedance, we now have the simulation tools necessary to explore the properties of the folded unipole antenna. As Roberts wrote in 1947, many of these were reported anecdotally without the benefit of supporting theory. There are still some phenomena today that are reported from the field, and perhaps they would be better understood with the help of our mathematical model.

The versatility of the model is a double-edged sword. We can easily vary any and all of six independent design parameters. They are: (1) height h, (2) current transformation ratio ν, (3) base terminating impedance Z_B, (4) top terminating impedance Z_T, (5) common mode characteristic impedance Z_{0c}, and (6) differential mode characteristic impedance Z_{0d}. On the other hand, systematically varying all of these would quickly bury us under an overwhelming volume of data, at least for the purposes of this book. We will have to make some arbitrary choices.

A related challenge is displaying our computational results. With up to six independent design parameters to vary simultaneously, what are the display options? They range from varying only one at a time and then plotting the results as a traditional graph, to varying all six simultaneously and creating some sort of novel display. From time to time in the literature, there have been proposals to pursue the latter extreme by creating displays with lines, or "polypoints," as the independent variable, and other polypoints as the dependent variables. This approach is intriguing but requires further refinement at this time.

In all but the first few examples in this chapter, we will focus on the effects of three of the six independent design variables. They are height h, base terminating reactance X_B, and top terminating reactance X_T. These seem to provide the most dramatic changes in antenna performance. We will hold the remaining design variables at constant, "ballpark" values.

For displaying computational results, the compromise pursued here will be to vary up to two independent variables simultaneously. In that case, we can readily display the results as contour plots. We will use contour plots extensively in this and subsequent chapters.

2.13 Examples of Short Folded Unipole Antennas

We'll begin exploring the properties of folded unipoles by focusing on heights in the vicinity of a quarter-wave and shorter. Historically, these first attracted antenna engineers because it was much easier to couple high power radio transmitters to short folded unipoles than to short series-fed monopoles. Monopoles provided only two independent design variables, height and radius. With up to six independent design variables, it was much easier to obtain the desired input impedance of 50 + j0 ohms with a folded unipole.

Figure 2-8 shows the input impedance of a folded unipole as height is varied from 0 to 300 feet, at a frequency of 984 kHz. The black line is the input resistance, and the gray line is the input reactance. There is

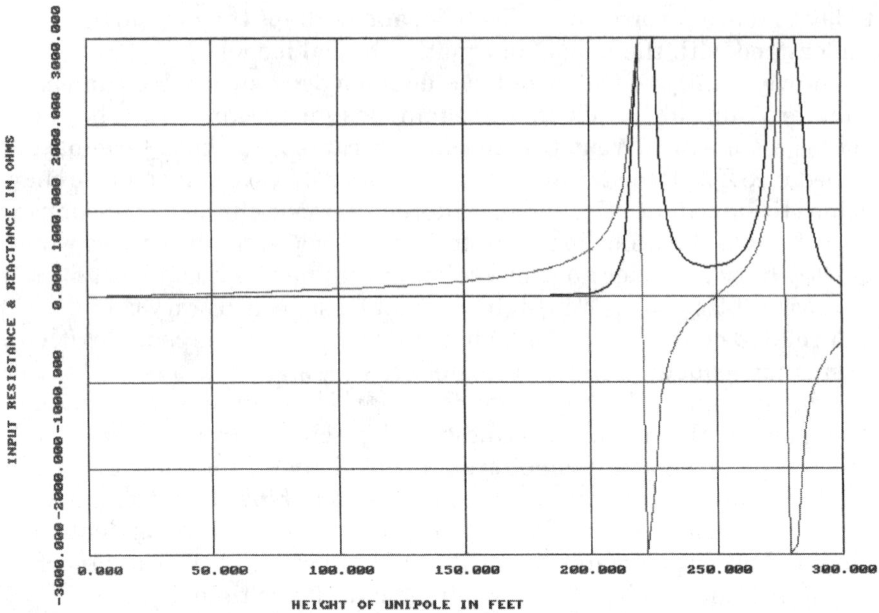

Figure 2-8 The input resistance (black) and reactance (gray) for the folded unipole antenna as height is continuously varied, and the current transformation is held constant at 0.5. There are three separate resonances in the vicinity of a quarter-wave, in contrast to a single resonance for a series-fed monopole.

nothing special about the choice of frequency except that it is mathematically convenient. A wavelength is 1,000 feet, and so 250 feet is one-quarter wavelength. It is similarly easy to mentally convert other dimensions from the displays into wavelengths.

The remaining five independent design variables are held constant. Both terminating impedances are short circuits; however, we will vary those in subsequent figures. The common mode characteristic impedance Z_{0c} is 1,200 ohms. The differential mode characteristic impedance Z_{0d} is 120 ohms. The current transformation ratio ν is 0.5. These are "ballpark" values. Exact values are determined by the detailed cross section of the antenna. In Chap. 7, we will see how to calculate them exactly, and we shall find that the ballpark is rather large. For now, however, these values will serve to demonstrate some dramatic results.

From the figure, there are at least four very distinctive differences compared with the impedance of a series-fed monopole:

- Instead of a single resonance at one-quarter wavelength, there are three resonances. These occur at heights less than, equal to, and greater than one-quarter wavelength.
- For heights shorter than the first resonance, the input reactance is positive or inductive instead of capacitive.

- The input resistance at the quarter-wave resonance, is much greater than the textbook value of 36 ohms for a series-fed monopole at quarter-wave resonance.
- The input resistances at the other two resonances are very great, on the order of thousands of ohms.

Figure 2-9 is similar to Fig. 2-8, except that the current transformation ratio ν has been changed from 0.5 to 1.0. Qualitatively, the results of the two figures are similar. For example, there are three resonances instead of only one for the series-fed monopole.

There are a couple of obvious differences, however. The position of the first resonance has shifted toward a shorter height. The position of the third resonance has shifted toward a taller one. That is, the resonances are more widely separated. Also, the value of the input resistance at the second, or quarter-wave, resonance has decreased. It is about four times the textbook value of 36 ohms for a series-fed monopole. This result is in agreement with previous theories of the folded unipole, including the 1947 article by Roberts.

In Fig. 2-10, the current transformation ratio is 5.0. As before, the results are qualitatively similar to the previous two figures.

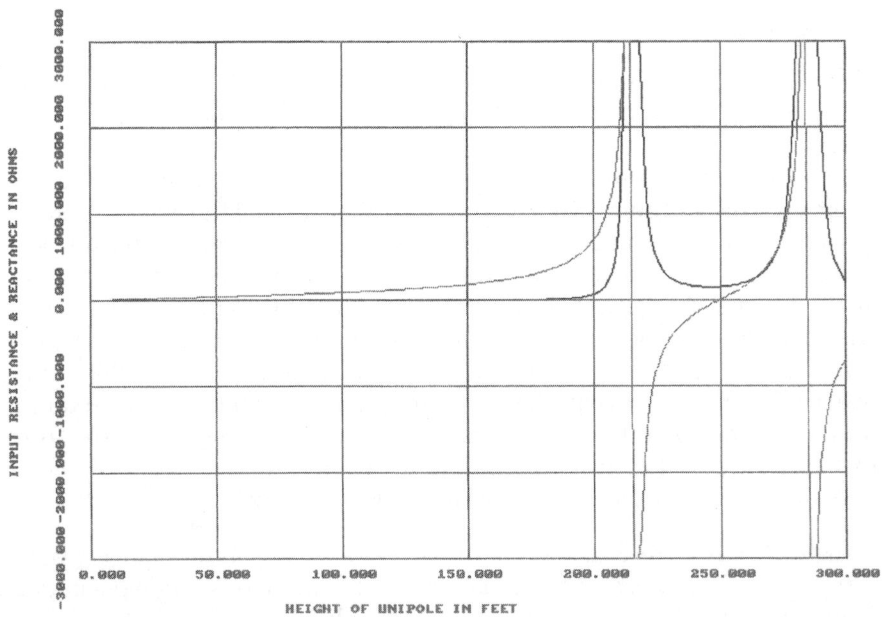

Figure 2-9 The input resistance (black) and reactance (gray) for the folded unipole antenna as height is continuously varied, and the current transformation is held constant at 1.0. The input resistance at quarter-wave resonance is about four times that of a series-fed monopole, in agreement with Roberts [1].

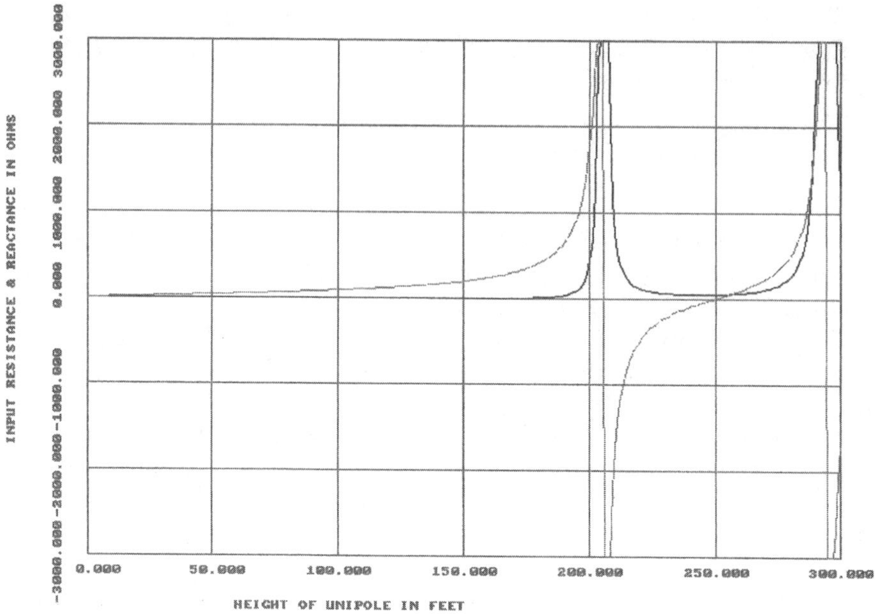

Figure 2-10 The input resistance (black) and reactance (gray) for the folded unipole antenna as height is continuously varied, and the current transformation is held constant at 5.0. The three resonances are more widely separated compared with lesser current transformation ratios.

The resonances are even more widely separated, however. Also, the value of the input resistance at quarter-wave resonance has greatly decreased.

In previous theories of the folded unipole, much emphasis was placed upon the role of the current transformation ratio. That was the main reason for presenting Figs. 2-8 through 2-10; however, we found that more dramatic changes resulted from varying the terminating impedances X_B and X_T. So, in the figures that follow, we will fix the current transformation ratio at 0.5, and focus on the terminating impedances.

Figure 2-11 shows the input resistance and reactance for folded unipoles with heights varying from 0 through 300 feet, all terminated in a base reactance of 1,000 ohms. The results are obviously different from Fig. 2-8, which is for the same antenna with a base reactance of 0 ohm, or a short circuit. Some of the chief differences are:

- Instead of three resonances, there are now four. The classic quarter-wave resonance is still present. Similar to Fig. 2-8, there is a single resonance for heights less than one-quarter wave, although this resonance has shifted to a much shorter height. At heights greater than one-quarter wavelength, there are now two, closely-spaced resonances.

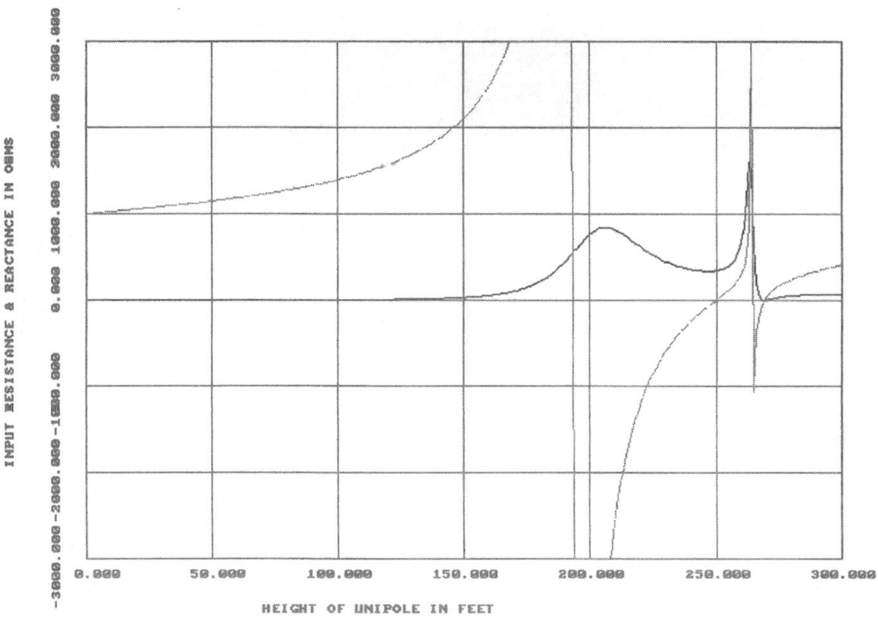

Figure 2-11 The input resistance (black) and reactance (gray) for the folded unipole antenna as height is continuously varied, and the base terminating impedance is $j1{,}000$ ohms. That is, the base terminating reactance is 1,000 ohms. The current transformation is held constant at 0.5. The first resonance has shifted to a much lower height.

- The input resistances at the quarter-wave resonance are about the same; however, at the other resonances they are much lower. At the first resonance, the input impedance is about 535 + j0 ohms, which approaches a practical value for coupling to a high power radio transmitter. This resonance occurs at a height of about 194 feet, or 70 electrical degrees, which is more than 20 percent shorter than one-quarter wavelength.

Figure 2-12 is similar to Fig. 2-11 except that the terminating base reactance has been increased to 2,000 ohms. It is seen that there are still four resonances, including the classic one at a quarter-wavelength. The first resonance occurs at a shorter height, about 163 feet, or 59 electrical degrees. The input impedance at that resonance is about 93 + j0 ohms, increasingly practical for coupling to a radio transmitter.

In Fig. 2-13, the terminating base reactance has been increased to 3,000 ohms. The first resonance occurs at about 135 feet, or 49 electrical degrees. This is only about half the height of a traditional series-fed monopole. The input impedance is about 40 + j0 ohms, not much different from the 50 +j0 ohms required for optimal coupling to a radio transmitter.

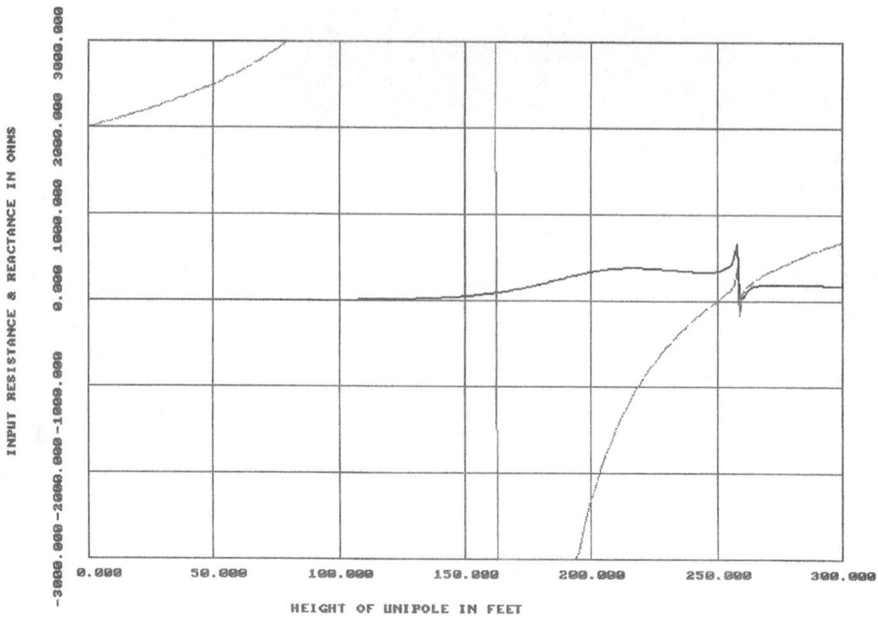

Figure 2-12 The input resistance (black) and reactance (gray) for the folded unipole antenna as height is continuously varied, and the base terminating reactance is 2,000 ohms. The current transformation is held constant at 0.5. The first resonance continues to shift downward.

From Figs. 2-11 through 2-13, it is obvious that a very short folded unipole with the appropriate terminating base reactance can provide an optimal input impedance. There are still other variable parameters to explore, however.

In Figs. 2-14 through 2-16, the top terminating reactance is varied from 100 to 300 ohms. The base terminating reactance is 0 ohm in all three examples. It is seen that there are three different resonances, including the textbook one at a quarter-wavelength. The first resonances occur at decreasing heights as the terminating top reactance increases. For $X_T = 300$ ohms, the first resonance occurs at about 69 feet, or only 25 electrical degrees. The input impedance at that resonance is about $210 + j0$ ohms, well within the practical range.

From the examples thus far, it is clear that extraordinary height reductions can be obtained with the correct choice of terminating reactances, at both the base and top of the folded unipole. At the same time, it is clear that the graphical displays that we have been using are cumbersome at best. We can display only one design variable continuously. In our examples thus far, that variable has been height. Converging on optimal values for terminating reactances is a hit or miss affair. There must be a better approach. We should be able to display the results of

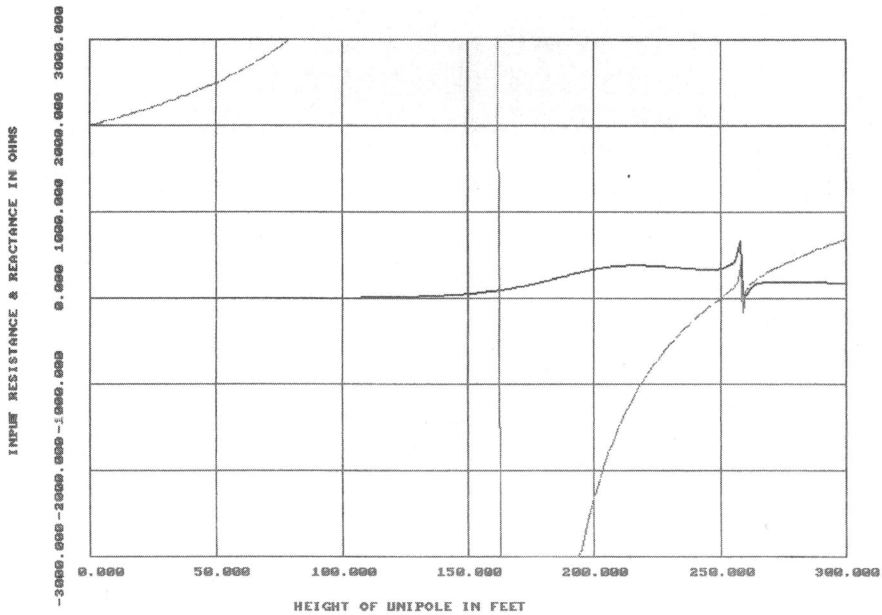

Figure 2-13 The input resistance (black) and reactance (gray) for the folded unipole antenna as height is continuously varied, and the base terminating reactance is 3,000 ohms. The current transformation is held constant at 0.5. The first resonance has shifted lower still.

continuously varying more than design parameter. Fortunately, that is exactly what we can do using contour plots.

Figure 2-17 shows the input impedance as both antenna height h and terminating base reactance X_B are simultaneously and continuously varied. The upper contour plot shows the input resistance. The color green shows resistances in the range 25–75 ohms. Yellow shows 75–150 ohms. Red shows 150–300 ohms. Gray shows 300 ohms and greater. White shows less than 25 ohms. It is seen that there are many combinations of height and terminating base reactance that result in the range of interest.

The lower contour plot shows the input reactance. The color cyan shows reactances in the range 0–1,000 ohms. Yellow shows 1,000–2,000 ohms. Magenta shows 2,000–3,000 ohms. Gray shows 3,000 ohms and greater. White shows negative or capacitive input reactances. The boundary between gray and white is evidently a resonance, where the input reactance changes rapidly from negative to positive. The boundary between white and cyan is also a resonance.

We will use these color conventions in all of our examples.

From Fig. 2-17, it is seen that there is a useful resonance in the vicinity of 150 feet, or about 54 electrical degrees, and with X_B = 2,250 ohms.

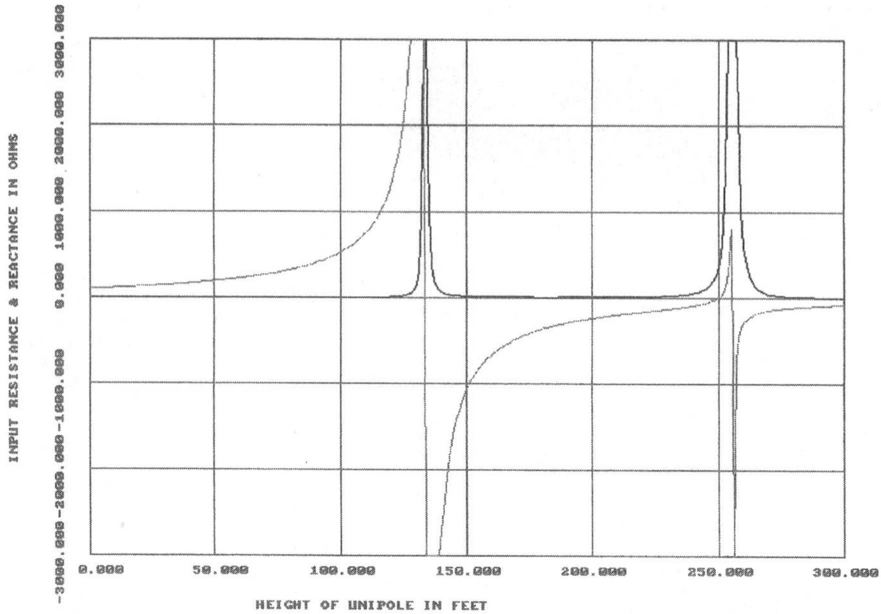

Figure 2-14 The input resistance (black) and reactance (gray) for the folded unipole antenna as height is continuously varied, and the top terminating impedance is $j100$ ohms. That is, the top terminating reactance is 100 ohms. The current transformation is held constant at 0.5. A relatively small top terminating reactance causes the first resonance to shift a similar amount as when the base terminating reactance is an order of magnitude greater.

In this example, the top terminating reactance is always 0 ohm; however, we will soon remove that constraint.

Figure 2-18 shows the input impedance of the folded unipole when both the height h and top terminating reactance X_T are varied simultaneously. The base terminating reactance is a short circuit. There appear to be some useful ranges for heights of 125, or 45 electrical degrees, feet and shorter.

All of the examples so far demonstrate that it is feasible to design folded unipole antennas substantially shorter than one-quarter wavelength. In the examples which follow, we will select a particular height and then adjust the terminating reactances until we have a useable input impedance.

Figure 2-19 shows the input impedance of a 150-foot folded unipole with the two terminating reactances varied simultaneously. It is seen that there are many useable reactance pairs in the vicinity of $X_B = 0$ and $X_T = 65$ ohms. There are also many useable pairs in the vicinity of $X_B = 2{,}300$ and $X_T = 30$ ohms. In that region, the antenna input reactance passes through a resonance, and the input resistance is in the green zone (i.e., between 25 and 50 ohms.)

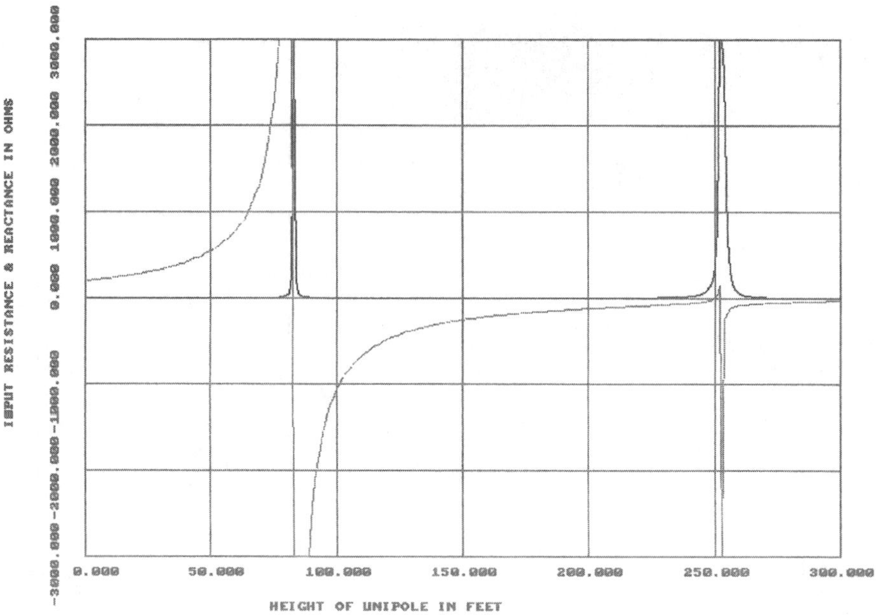

Figure 2-15 The input resistance (black) and reactance (gray) for the folded unipole antenna as height is continuously varied, and the top terminating reactance is 200 ohms. The current transformation is held constant at 0.5. The first resonance has continued to shift downward to less than 1/10 wavelength.

Figure 2-20 shows the input impedance of a 125-foot, or 45 electrical degree, folded unipole. It is seen that there are many useable reactance pairs in the vicinity of X_B = 2,000 and X_T = 103 ohms. In that region, the antenna input reactance passes through a resonance, and the input resistance is in the green zone.

Figure 2-21 shows the input impedance of a 100-foot, or 36 electrical degree, folded unipole. It is seen that there are many useable reactance pairs in the vicinity of X_B = 1,500 and X_T = 153 ohms. As in the previous two examples, the input reactance passes through resonance, and the input resistance is in the green zone.

So far from the examples, it is clear that we can design short folded unipole antennas with useable input impedances for almost any arbitrary height. Let's try for an extremely short one.

Figure 2-22 shows the input impedance of a 50-foot, or 18 electrical degree, folded unipole. This is 1/5 the height of the traditional series-fed quarter-wave monopole. Especially at long wavelengths, this is a significant advantage indeed. It is seen that there are many useable reactance pairs in the vicinity of X_B = 800 and X_T = 355 ohms. As in the previous examples, the input reactance passes through resonance, and the input resistance is in the green zone. There is no indication that this height is the lower limit. We can most likely design even shorter ones!

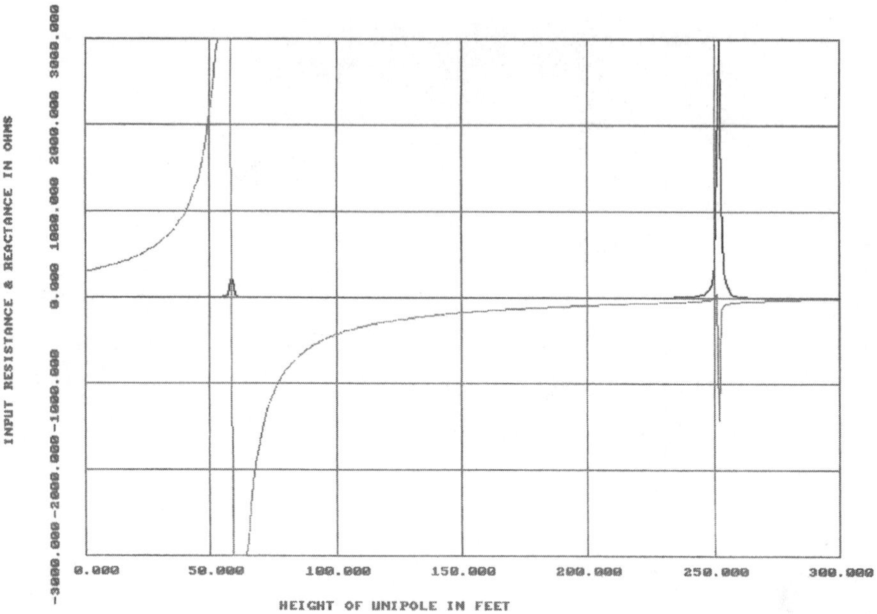

Figure 2-16 The input resistance (black) and reactance (gray) for the folded unipole antenna as height is continuously varied, and the top terminating reactance is 300 ohms. The current transformation is held constant at 0.5. The first resonance has continued to shift downwards, approaching 1/20 wavelength.

From the field there have been anecdotal reports of extremely short folded unipole antennas, and these reports have been met with the raised eyebrow of disbelief by some. The above examples provide firm theoretical support and credibility for these remarkable antennas.

2.14 Examples of Tall Folded Unipole Antennas

In the previous section, we saw that useful input impedances could be obtained from folded unipoles as short as 50 feet at 984 kHz, or 18 electrical degrees. This compares favorably with the series-fed monopole, which traditionally must be about a quarter-wavelength tall, or 250 feet at 984 kHz. The savings in materials and structural complexities are obvious.

We are often stuck with tall antennas, however, for any one of many reasons. For example, the antenna may be used simultaneously at longer wavelengths, and at those it is not tall or might actually be short. Another reason might be that an existing antenna just happens to be available but it is long compared with the wavelength of interest. In any case, configuring the tall antenna as a series-fed monopole can invite problems.

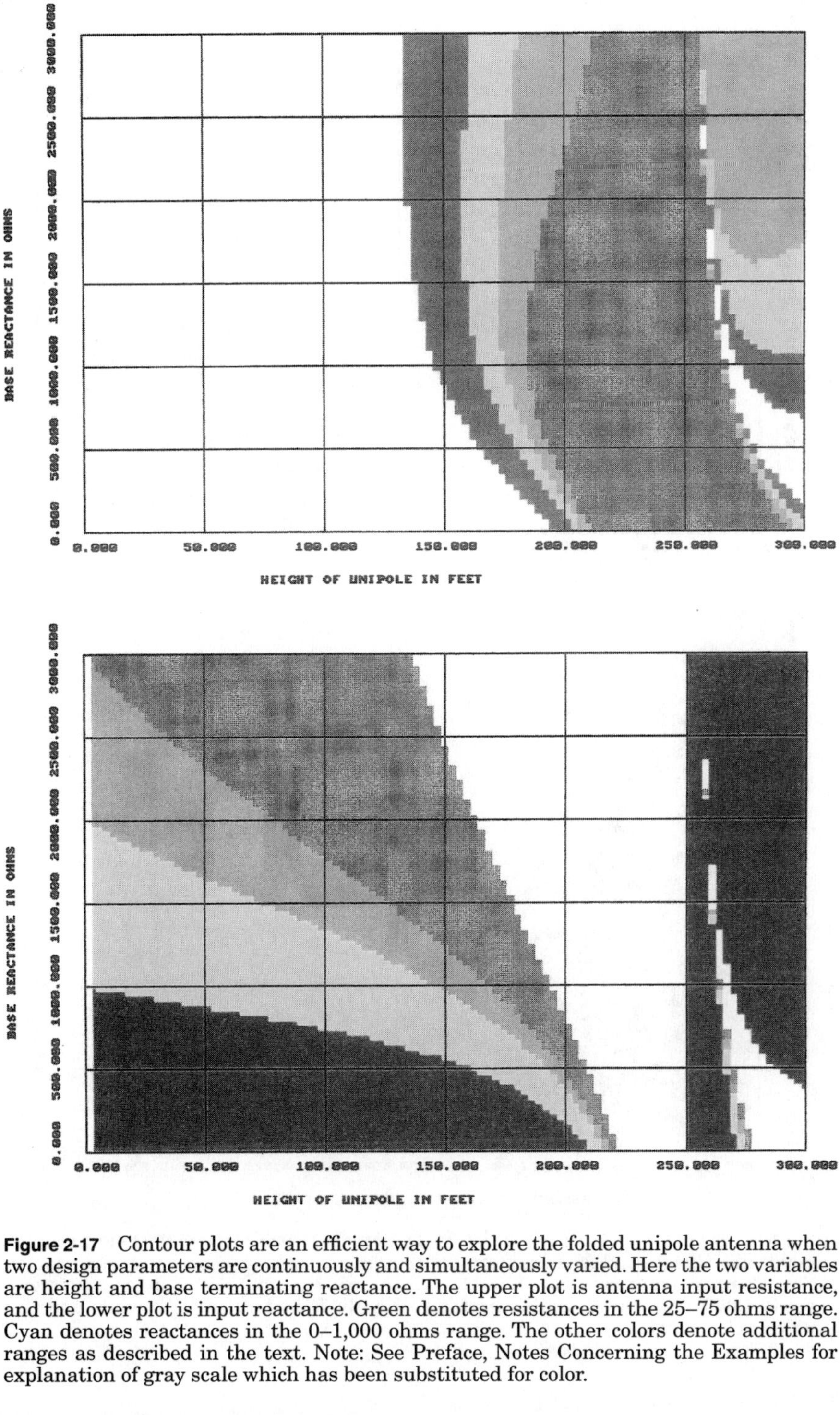

Figure 2-17 Contour plots are an efficient way to explore the folded unipole antenna when two design parameters are continuously and simultaneously varied. Here the two variables are height and base terminating reactance. The upper plot is antenna input resistance, and the lower plot is input reactance. Green denotes resistances in the 25–75 ohms range. Cyan denotes reactances in the 0–1,000 ohms range. The other colors denote additional ranges as described in the text. Note: See Preface, Notes Concerning the Examples for explanation of gray scale which has been substituted for color.

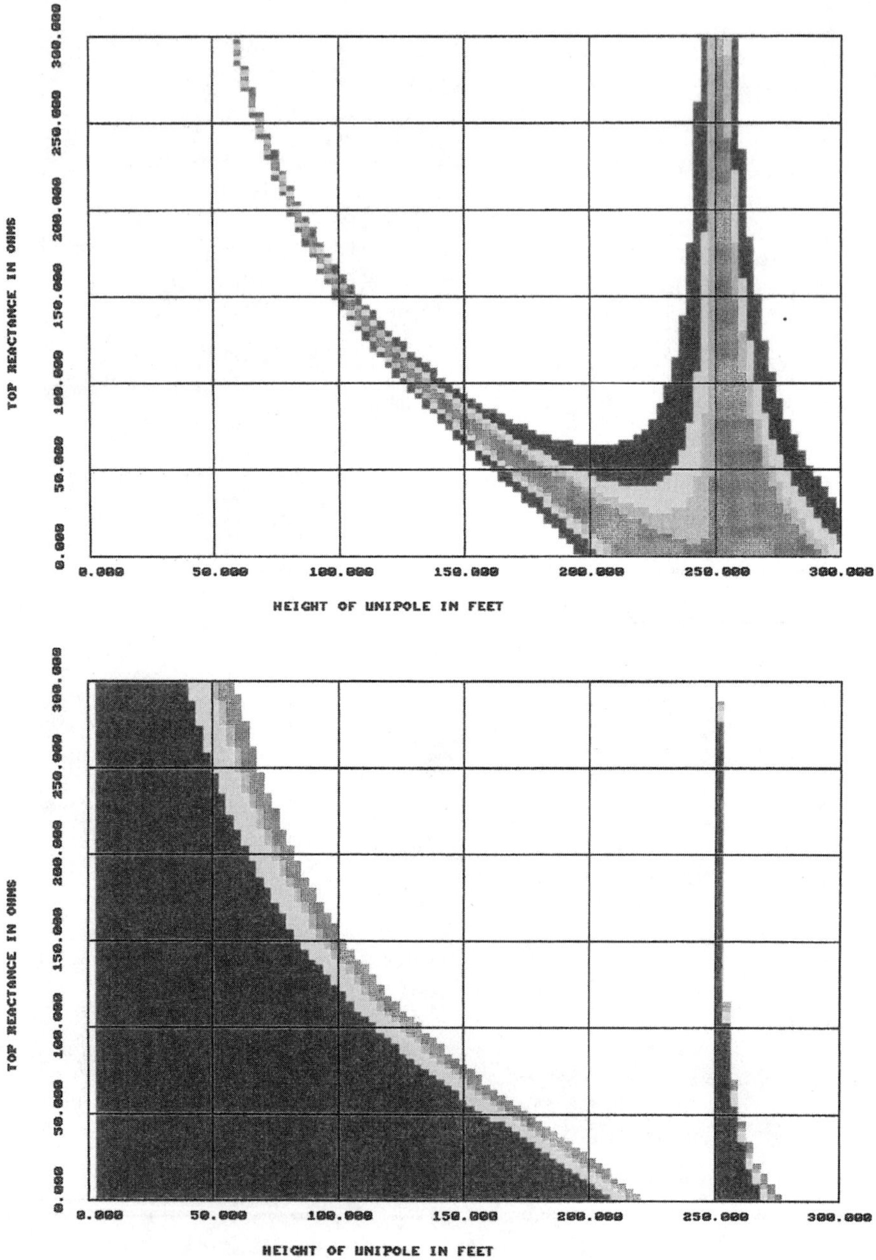

Figure 2-18 Input resistance (upper plot) and reactance (lower plot) for the folded unipole antenna when height and top terminating reactance are simultaneously and continuously varied. Green denotes resistances in the 25–75 ohms range. Cyan denotes reactances in the 0–1,000 ohms range. In the reactance plot, the boundary between white and any other color is a resonance.

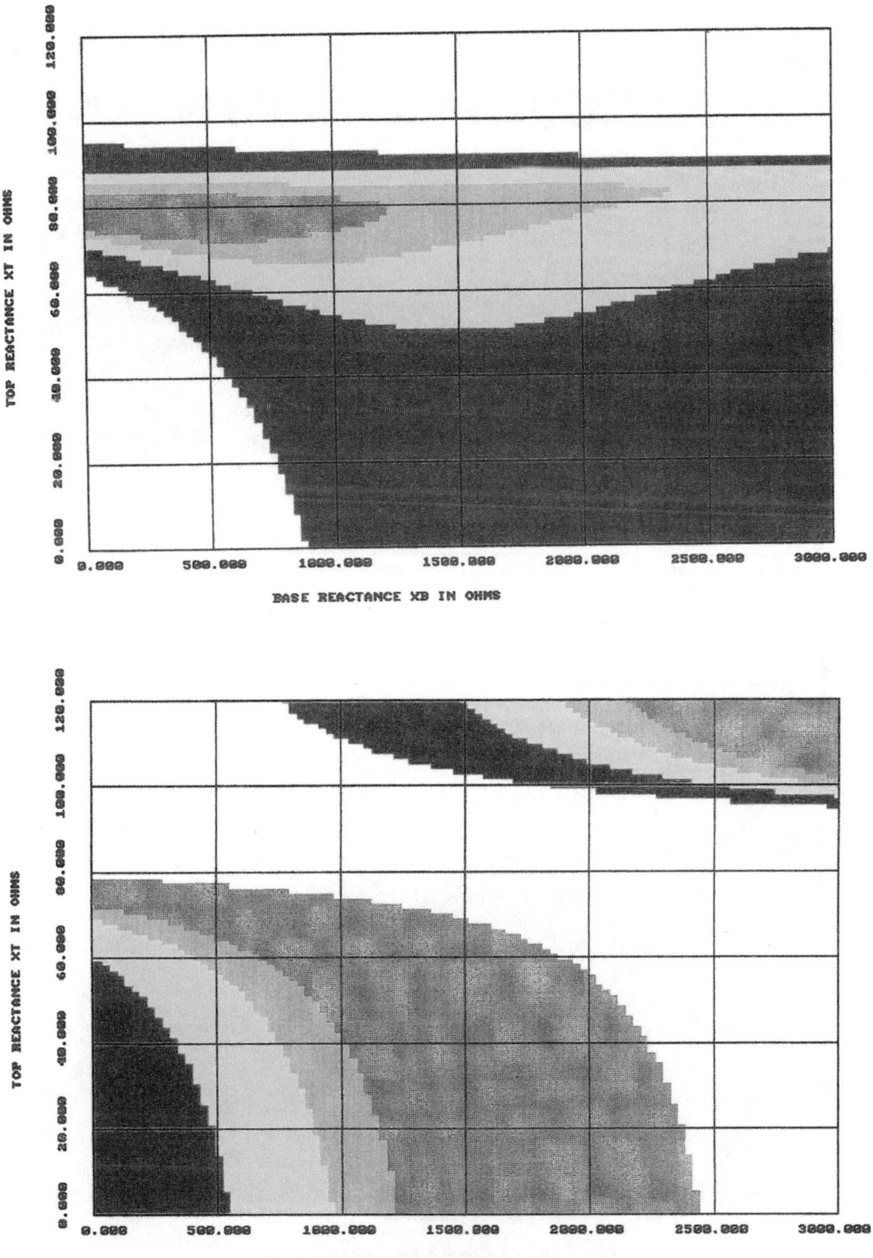

Figure 2-19 Input resistance (upper plot) and reactance (lower plot) for a 150-foot folded unipole antenna. At 984 kHz, it is only 54 electrical degrees tall. Terminating reactances at both base and top are simultaneously and continuously varied. Green denotes resistances in the 25–75 ohms range. Cyan denotes reactances in the 0–1,000 ohms range. It is seen that there are many combinations of terminating reactances that produce useful input impedances.

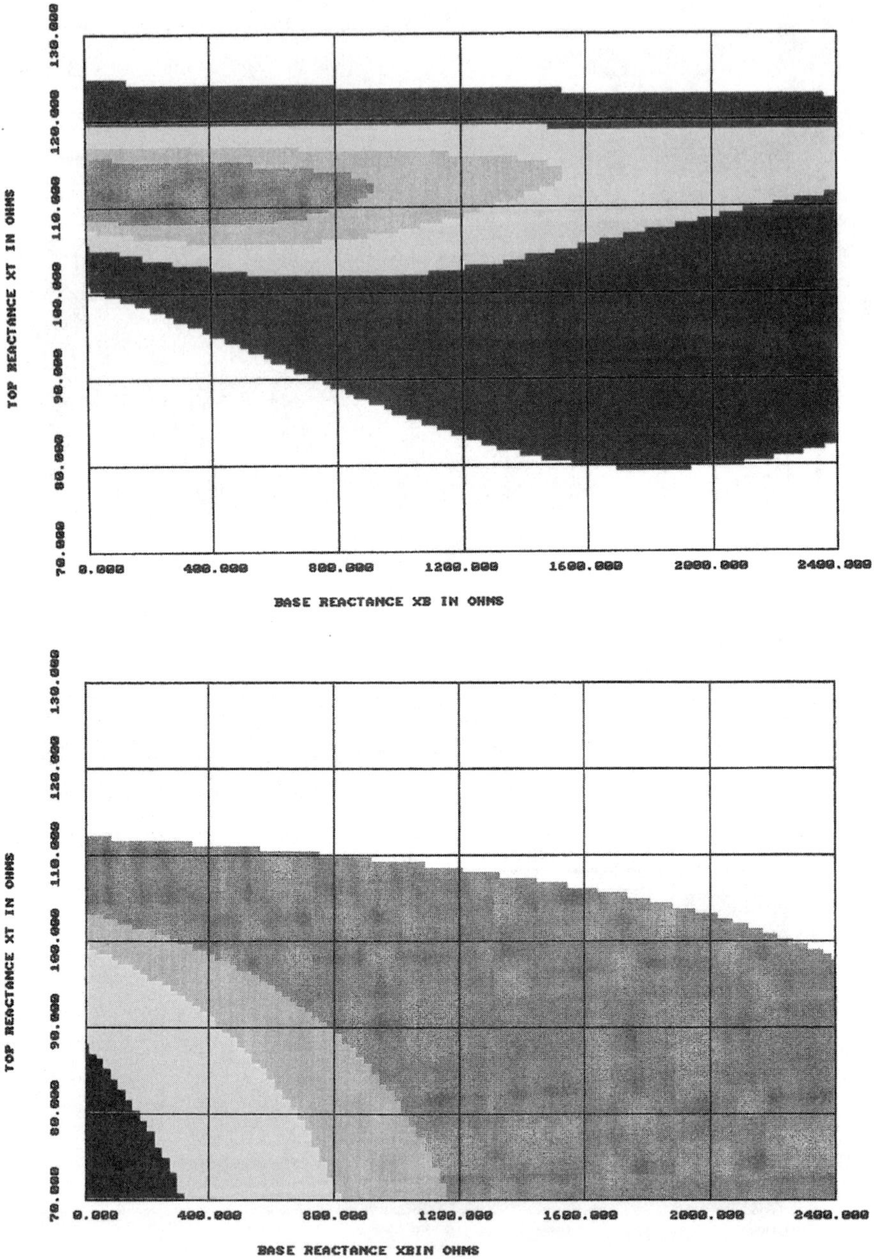

Figure 2-20 Input resistance (upper plot) and reactance (lower plot) for a 125-foot folded unipole antenna. At 984 kHz, it is only 45 electrical degrees tall. Terminating reactances at both base and top are simultaneously and continuously varied. Green denotes resistances in the 25–75 ohms range. Cyan denotes reactances in the 0–1,000 ohms range. It is seen that there are many combinations of terminating reactances that produce useful input impedances.

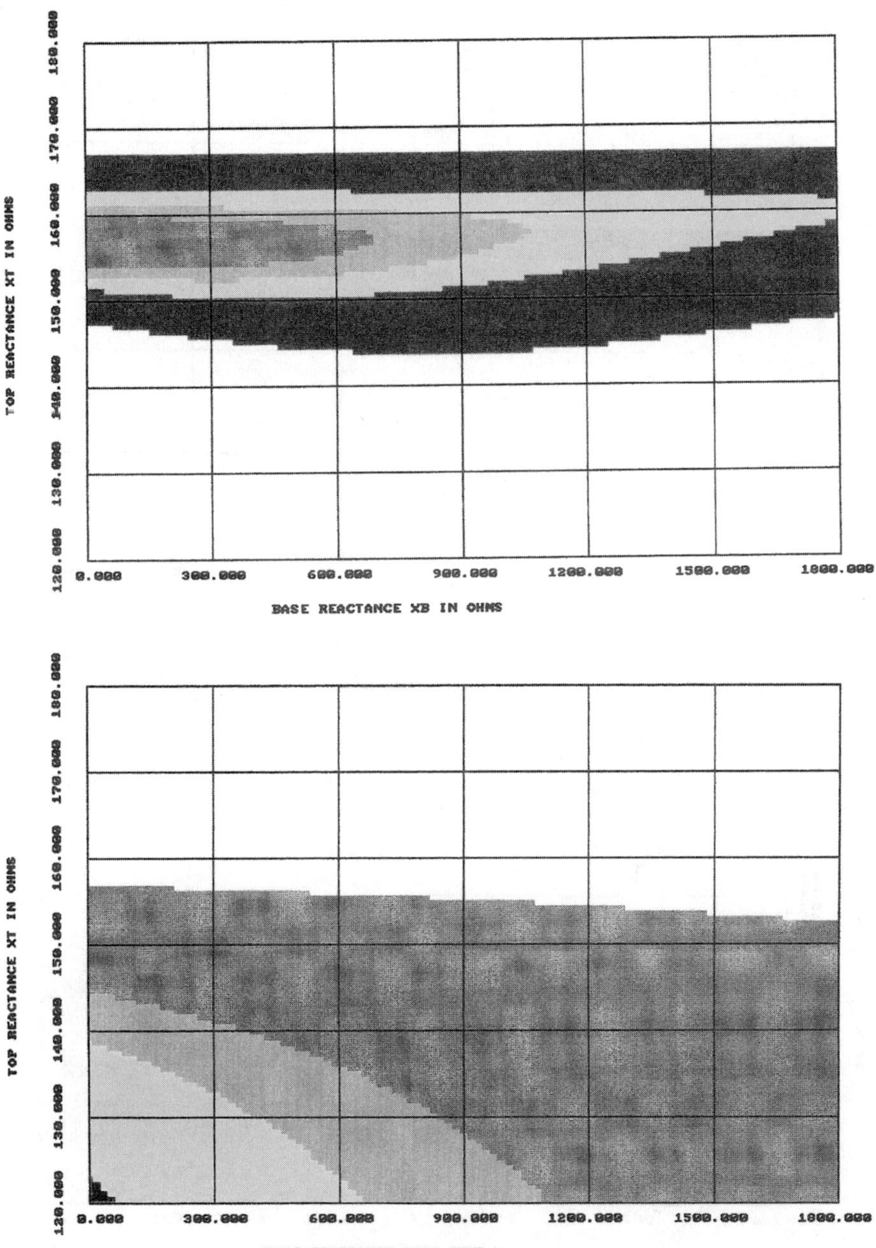

Figure 2-21 Input resistance (upper plot) and reactance (lower plot) for a 100-foot folded unipole antenna. At 984 kHz, it is only 36 electrical degrees tall. Terminating reactances at both base and top are simultaneously and continuously varied. Green denotes resistances in the 25–75 ohms range. Cyan denotes reactances in the 0–1,000 ohms range. It is seen that there are many combinations of terminating reactances that produce useful input impedances.

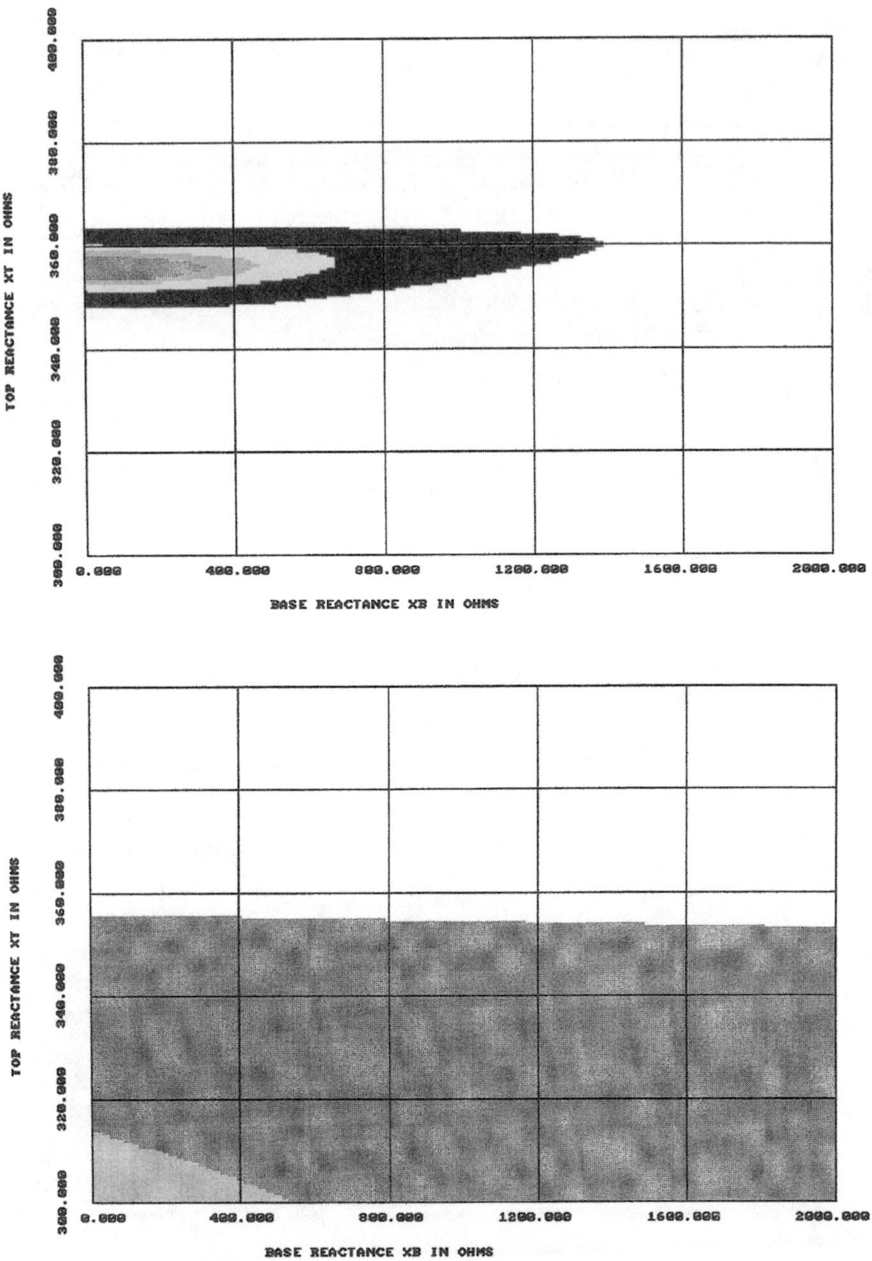

Figure 2-22 Input resistance (upper plot) and reactance (lower plot) for a 50-foot folded unipole antenna. At 984 kHz, it is only 18 electrical degrees tall. Terminating reactances at both base and top are simultaneously and continuously varied. Green denotes resistances in the 25–75 ohms range. Cyan denotes reactances in the 0–1,000 ohms range. Even for this very short antenna, it is seen that there are combinations of terminating reactances that still produce useful input impedances.

Especially in the vicinity of a half-wavelength, the input impedance becomes very large, on the order of thousands of ohms, and it is neither useful nor controllable. In these cases, we will show that the folded unipole can provide useful input impedances.

In this section, we will proceed in a fashion similar to that of the previous one. That is, we will examine a few traditional graphs to obtain a straight forward but narrow view of performance for tall folded unipoles. Then, we will move on to contour plots which are much richer and more comprehensive with respect to displaying information.

Figure 2-23 shows the input impedance for heights varying from 0 through 1,500 feet at 984 kHz. This range is the same as 0 through 1.5 wavelengths, or 0 through 540 electrical degrees. The upper plot shows both input resistance and reactance on a linear scale. The lower plot shows input resistance only, on a logarithmic scale. It is seen that the input impedance is almost a periodic function, repeating itself about every half-wavelength.

In the vicinity of odd multiples of one-quarter wavelength, it is seen that there are three resonances. One of these is at the classic quarter-wavelength observed in the case of series-fed monopoles. Another is at a shorter height. The third is at a taller height. The input resistance varies over a wide range in the vicinity of these resonances, ranging from less than one ohm to over one megohm. As a practical matter, it is doubtful that the latter value would be actually observed or measured. A fundamental phenomenon known as *antenna end effect* ultimately limits the maximum input impedance except for the extreme, conceptual case of the infinitely thin monopole. We will discuss this phenomenon in detail and extend our mathematical model to include it in Sect. 5.7. In any case, the range of resistances include many useful values.

In the vicinity of even multiples of one-quarter wavelength, a single resonance is observed. At these resonances, the input resistance is vanishingly small. Note, however, that for this example, the terminating base and top reactances are both zero. When we change those values, we can change the resonance properties drastically.

Figure 2-24 is the same as the previous figure, except that the terminating base reactance is capacitive, $X_B = -1{,}500$ ohms. In the previous section for short unipoles, we did not consider the possibility of negative terminating reactances. For tall unipoles, however, such reactances may be useful.

The input impedances in this example are obviously very different from those in the previous one. In particular, the input resistances at even multiples of one-quarter wavelength are in the useful range instead of being vanishingly small.

Figure 2-25 is the same as the previous two, except that the terminating base reactance is inductive, $X_B = 1{,}500$ ohms. The input impedances

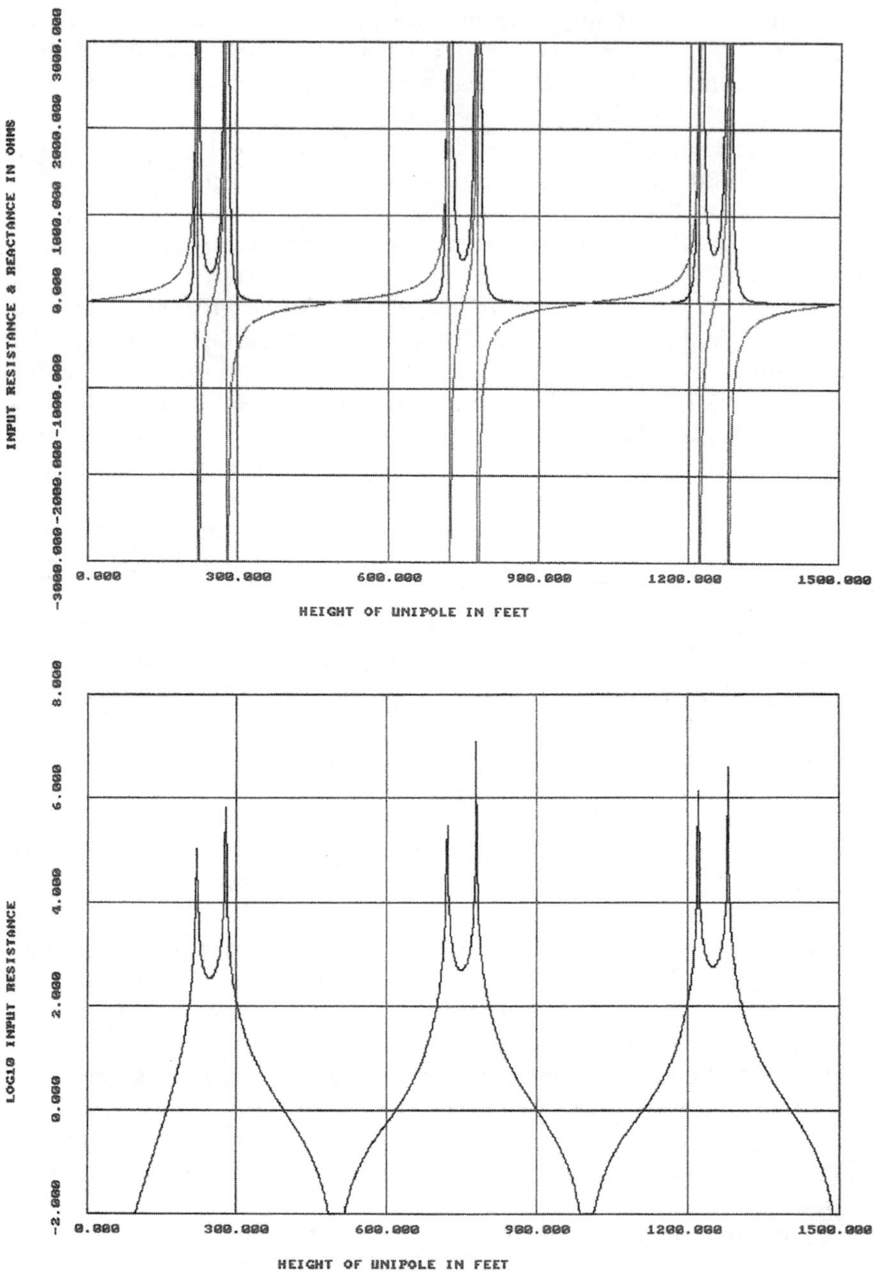

Figure 2-23 Input resistance (black) and reactance (gray) for folded unipoles as tall as 3/2 wavelengths, or 540 electrical degrees. The lower plot shows the input resistance on a logarithmic scale. In practice, we would probably not observe input resistances greater than a few thousand ohms due to the ***antenna end effect,*** discussed in Sect. 5.7.

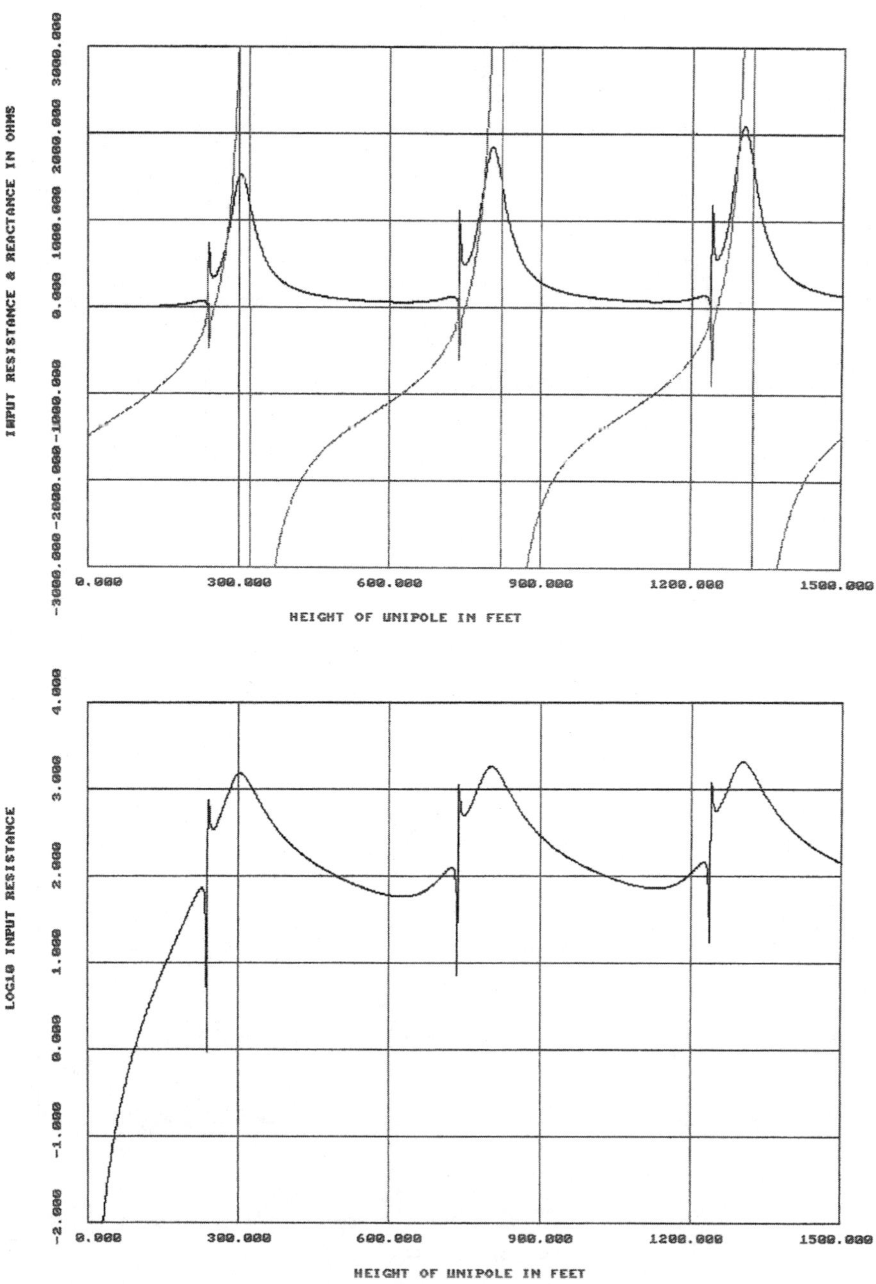

Figure 2-24 Input resistance (black) and reactance (gray) for folded unipoles as tall as 3/2 wavelengths, or 540 electrical degrees, with a terminating base reactance of –1,500 ohms. With tall unipoles, a negative or capacitive terminating reactance can produce useful input impedances. The lower plot shows the input resistance on a logarithmic scale.

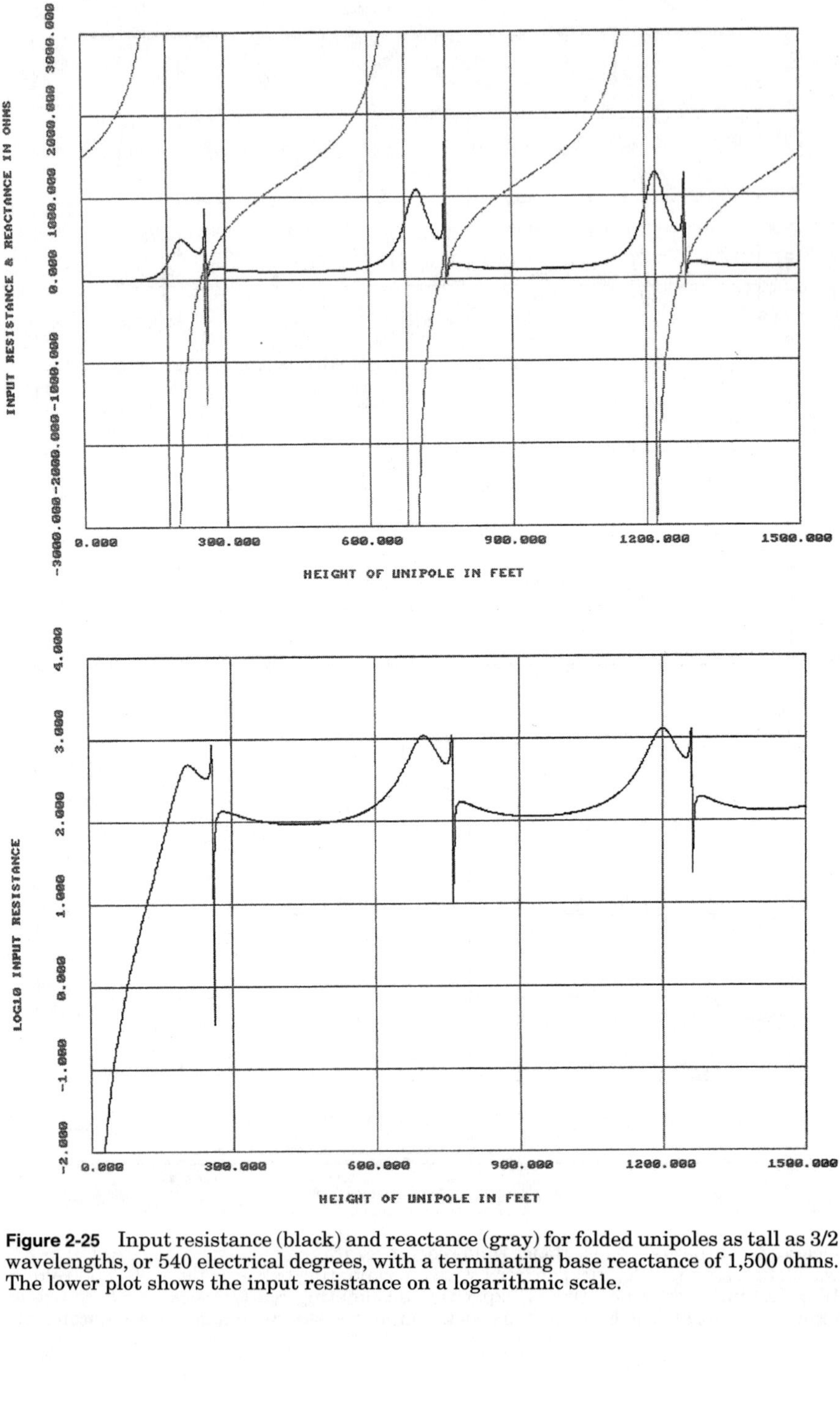

Figure 2-25 Input resistance (black) and reactance (gray) for folded unipoles as tall as 3/2 wavelengths, or 540 electrical degrees, with a terminating base reactance of 1,500 ohms. The lower plot shows the input resistance on a logarithmic scale.

are clearly different compared with the previous examples. Evidently, varying the terminating base reactance has a dramatic effect upon antenna input impedance.

Figure 2-26 shows what happens to antenna input impedance when the terminating top reactance is capacitive, $X_T = -150$ ohms. Figure 2-27 shows what happens when the terminating top reactance is inductive, $X_T = 150$ ohms. Clearly, this terminating reactance also has a dramatic effect.

For the same compelling reasons articulated in the previous section, it is useful to leave behind the traditional graphs, and move on to contour plots for purposes of examining the many useful results obtained when we continuously vary more than one design parameter.

Figure 2-28 shows in the input impedance of the folded unipole as both height h and base terminating reactance X_B are continuously varied. The upper contour shows input resistance. The lower contour shows input reactance. The color convention is the same as used previously. That is, green represents input resistances in the 25–75 ohms, and so forth. Cyan represents reactances in the 0–1,000 ohms range, and so forth. It is seen that there are many useful combinations of tall heights and base terminating reactances.

Figure 2-29 shows the input impedance of the folded unipole as both height $\boldsymbol{h}$ and top terminating reactance X_T are continuously varied. Again, there are many useful combinations of tall heights and top terminating reactances.

It is very likely, based upon the previous two examples, that we can obtain useful input impedances for tall folded unipole antennas of any arbitrary height. The following three examples will show that this is true.

Figure 2-30 shows the input impedance of a 375-foot folded unipole as both terminating reactances are continuously varied. This antenna is 3/8 wavelength or 135 degrees tall. It is seen that there are many useful combinations for X_B in the vincinity of 500 ohms, with X_T having any value greater than 0. In those cases, the input resistance is in the green zone, and the input reactance is in the cyan zone.

What about the pathological case of a half-wave antenna? For series-fed monpoles with that length, the input resistance is very large and unstable. Figure 2-31 shows what happens in the case of a folded unipole. It is seen that there are many useful combinations of terminating reactance that put the input resistance into the green zone while simultaneously putting the input reactance into the cyan zone. Evidently, the folded unipole configuration readily tames the half-wave antenna.

Figure 2-32 shows the input impedance of a 625-foot folded unipole. This antenna is 5/8 wavelength, or 225 electrical degrees tall. By now,

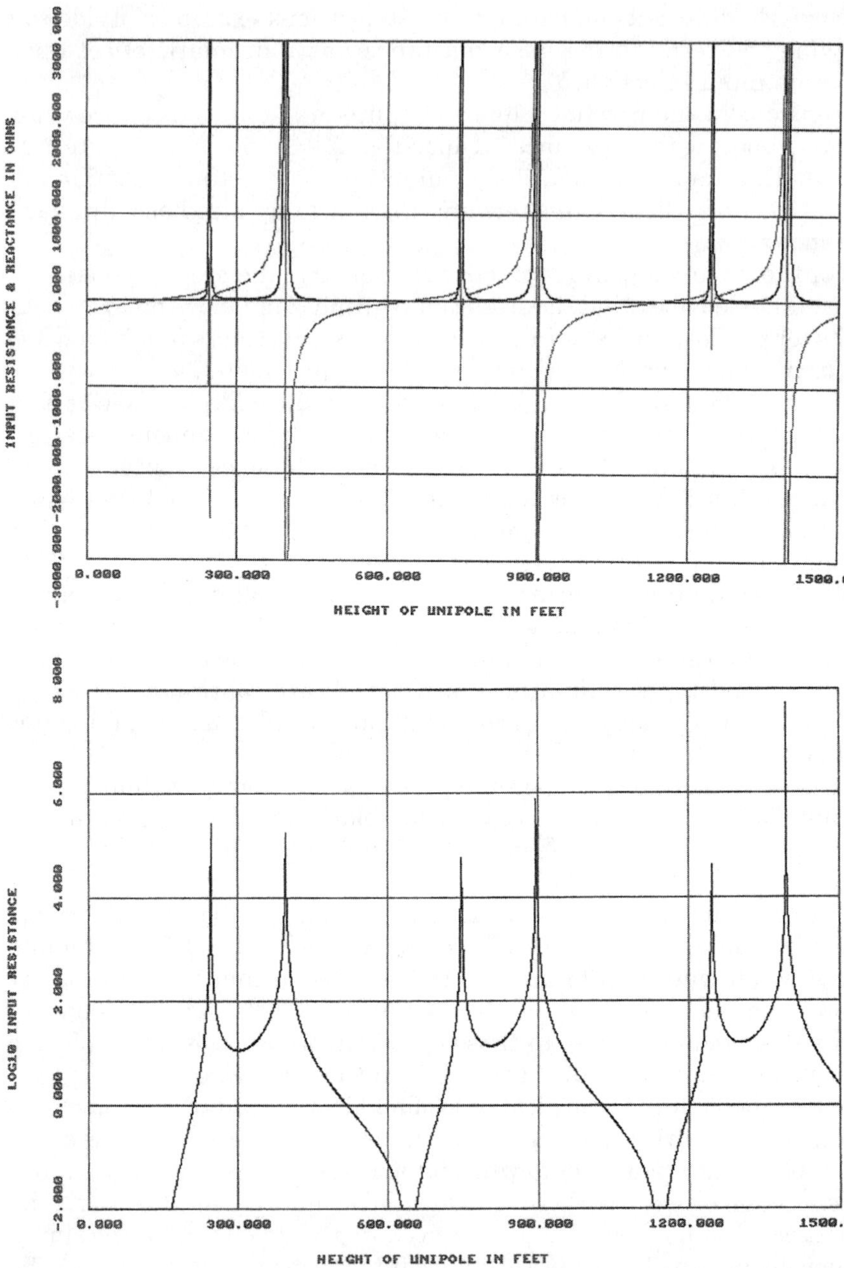

Figure 2-26 Input resistance (black) and reactance (gray) for folded unipoles as tall as 3/2 wavelengths, or 540 electrical degrees, with a terminating top reactance of –150 ohms. With tall unipoles, a negative or capacitive terminating reactance can produce useful input impedances. The lower plot shows the input resistance on a logarithmic scale.

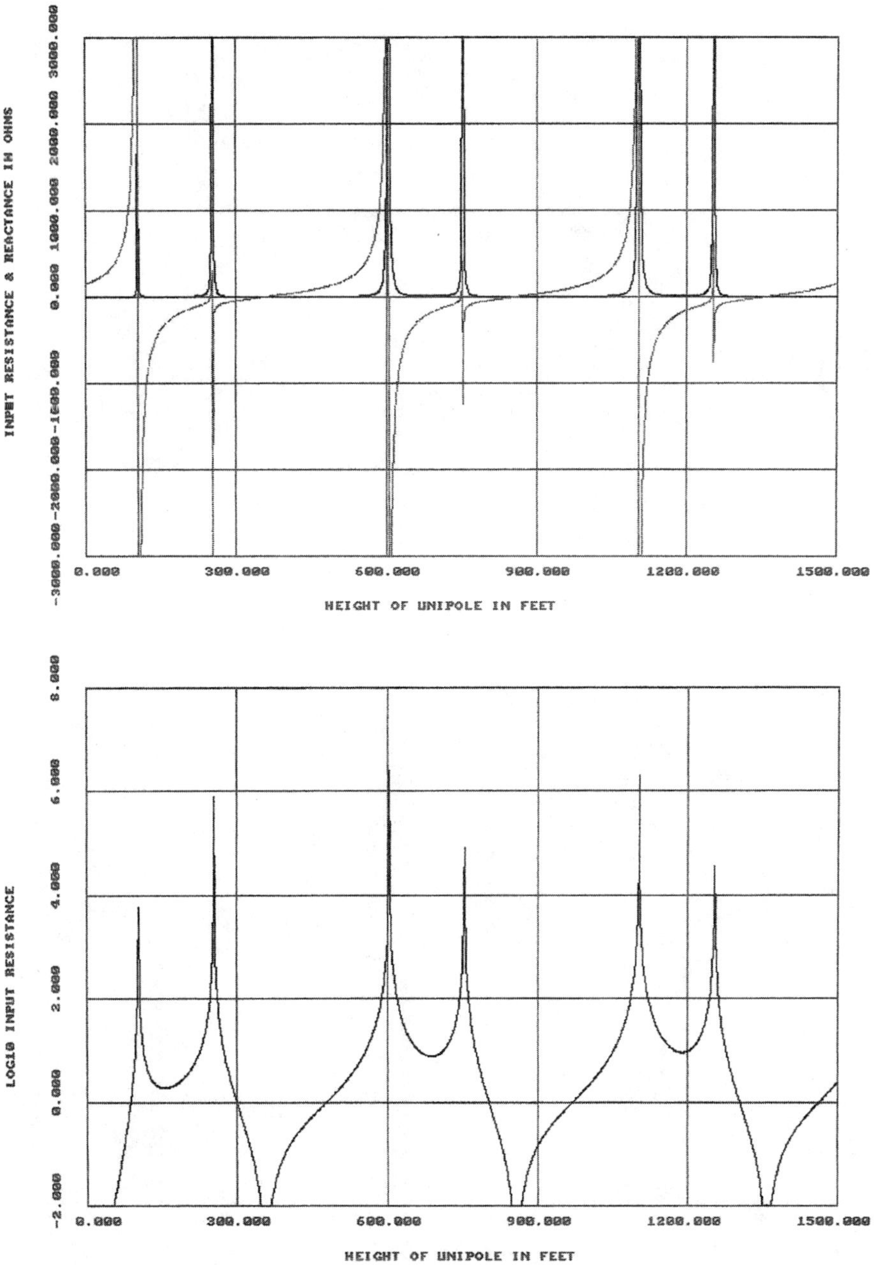

Figure 2-27 Input resistance (black) and reactance (gray) for folded unipoles as tall as 3/2 wavelengths, or 540 electrical degrees, with a terminating top reactance of 150 ohms. The lower plot shows the input resistance on a logarithmic scale.

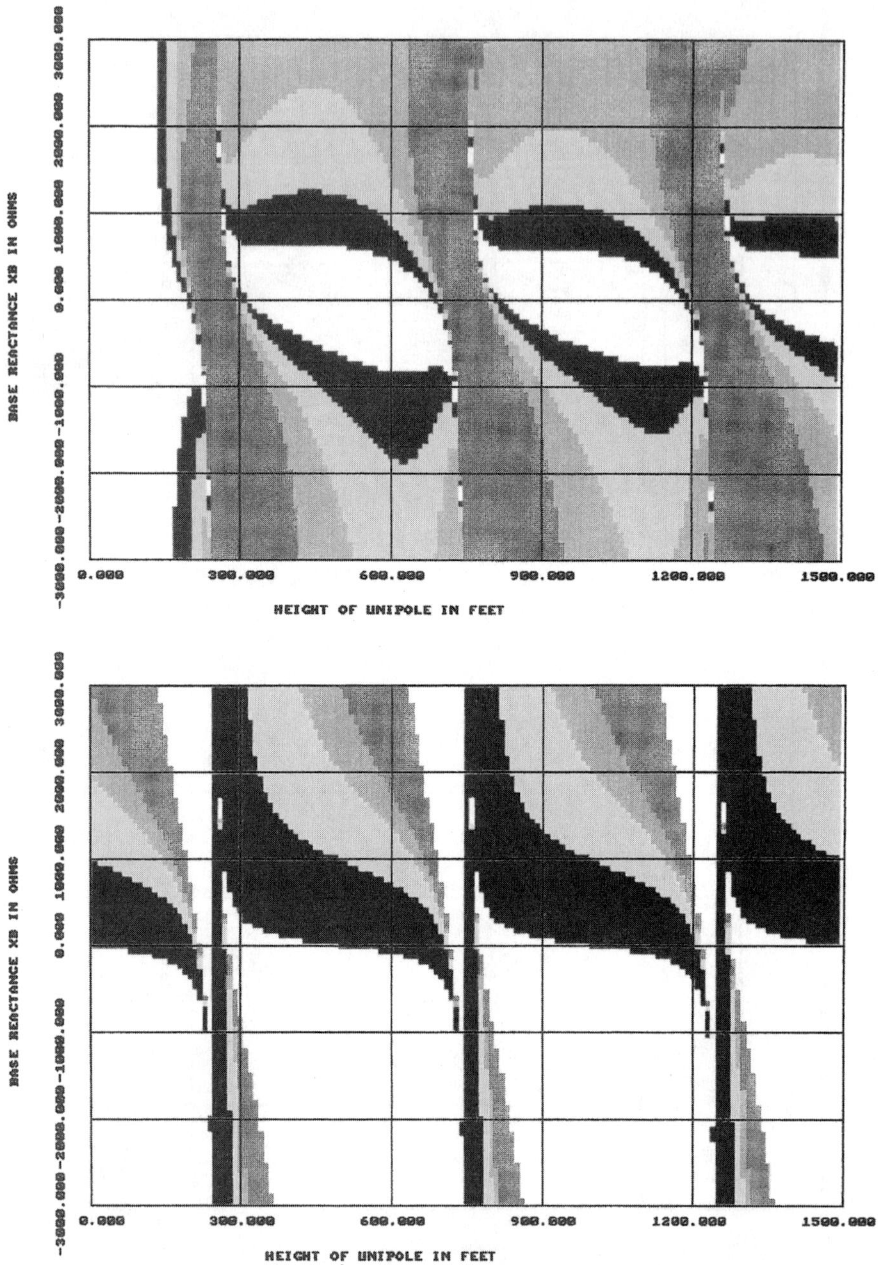

Figure 2-28 Contour plots are an efficient way to explore the folded unipole antenna when two design parameters are continuously and simultaneously varied. Here the two variables are height and base terminating reactance. The upper plot is antenna input resistance, and the lower plot is input reactance. Green denotes resistances in the 25–75 ohms range. Cyan denotes reactances in the 0–1,000 ohms range. The other colors denote additional ranges as described in the text.

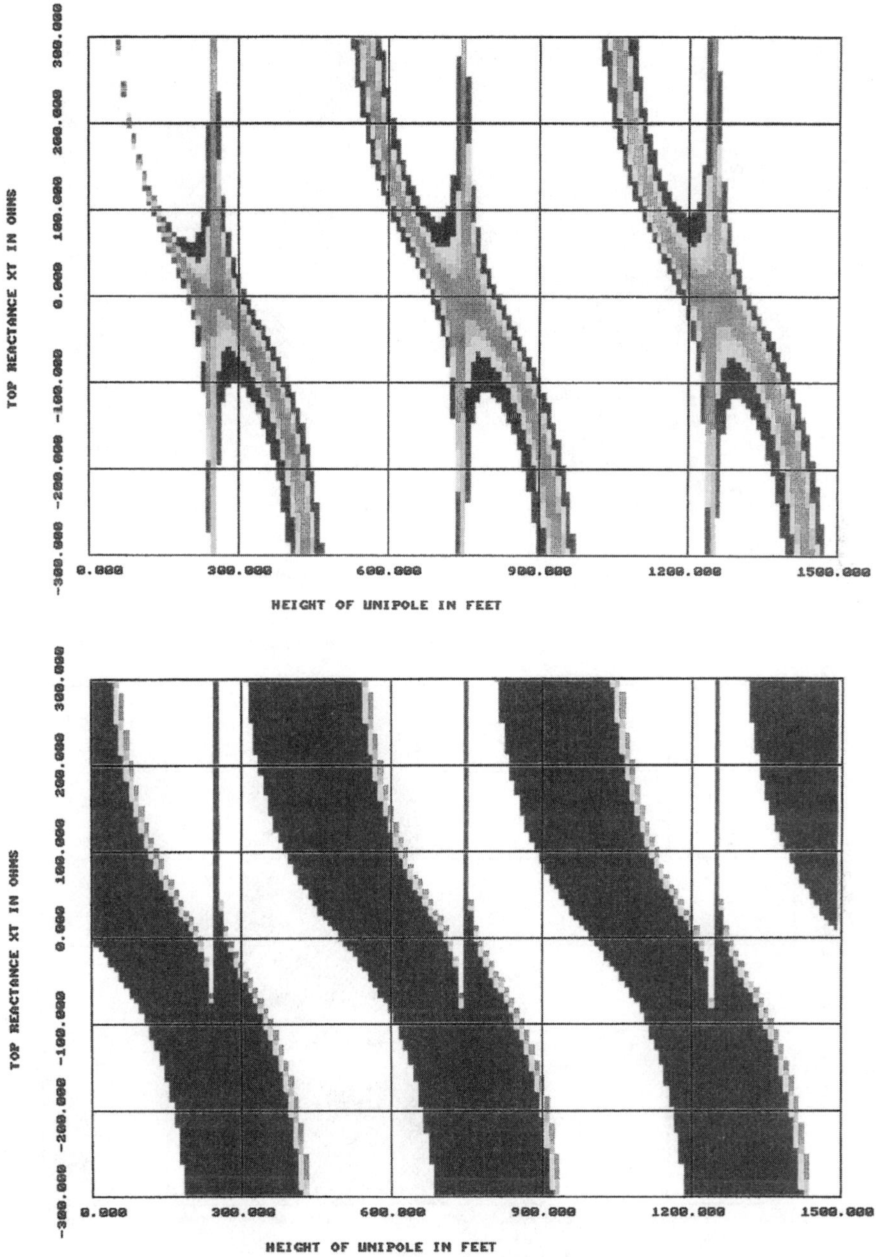

Figure 2-29 Input resistance (upper plot) and reactance (lower plot) for the folded unipole antenna when height and top terminating reactance are simultaneously and continuously varied. Green denotes resistances in the 25–75 ohms range. Cyan denotes reactances in the 0–1,000 ohms range. In the reactance plot, the boundary between white and any other color is a resonance.

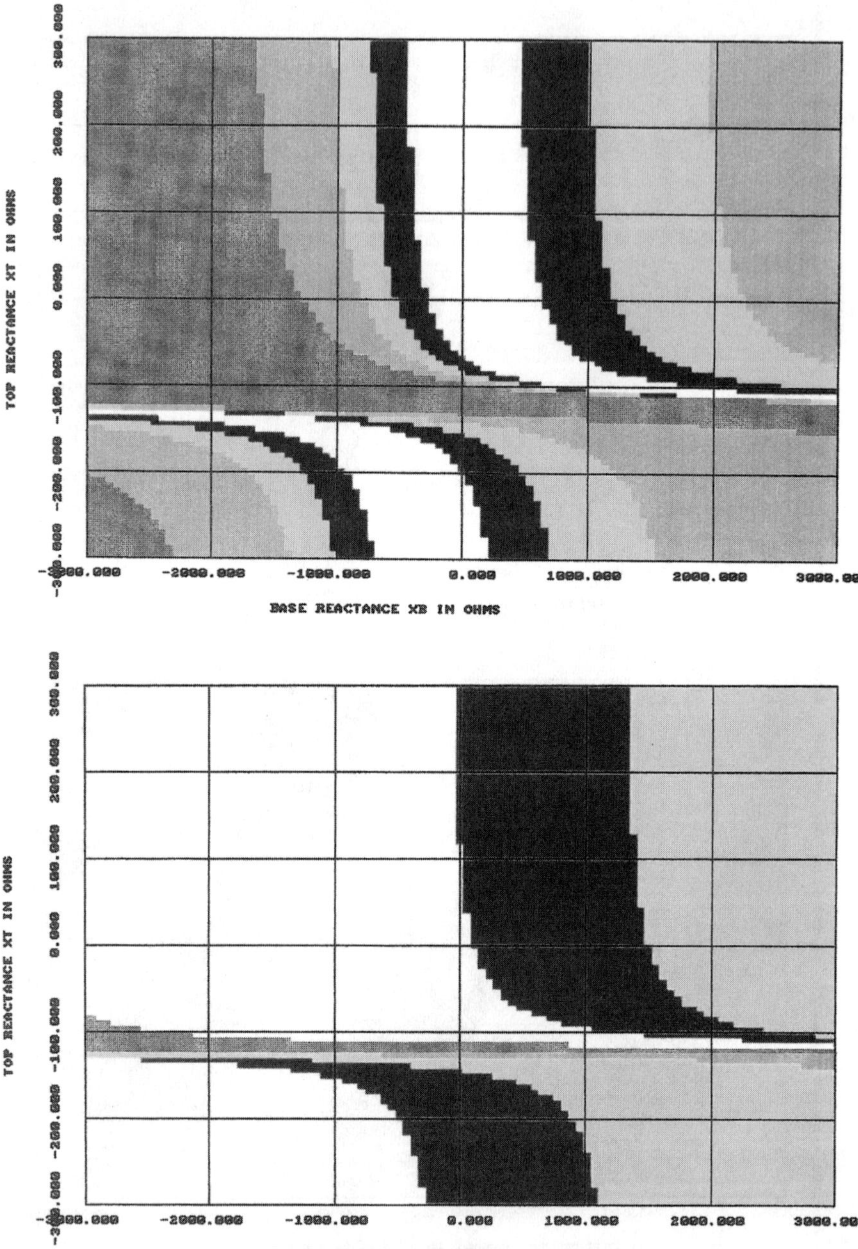

Figure 2-30 Input resistance (upper plot) and reactance (lower plot) for a 375-foot folded unipole antenna. At 984 kHz, it is 135 electrical degrees tall. Terminating reactances at both base and top are simultaneously and continuously varied. Green denotes resistances in the 25–75 ohms range. Cyan denotes reactances in the 0–1,000 ohms range. It is seen that there are many combinations of terminating reactances that produce useful input impedances.

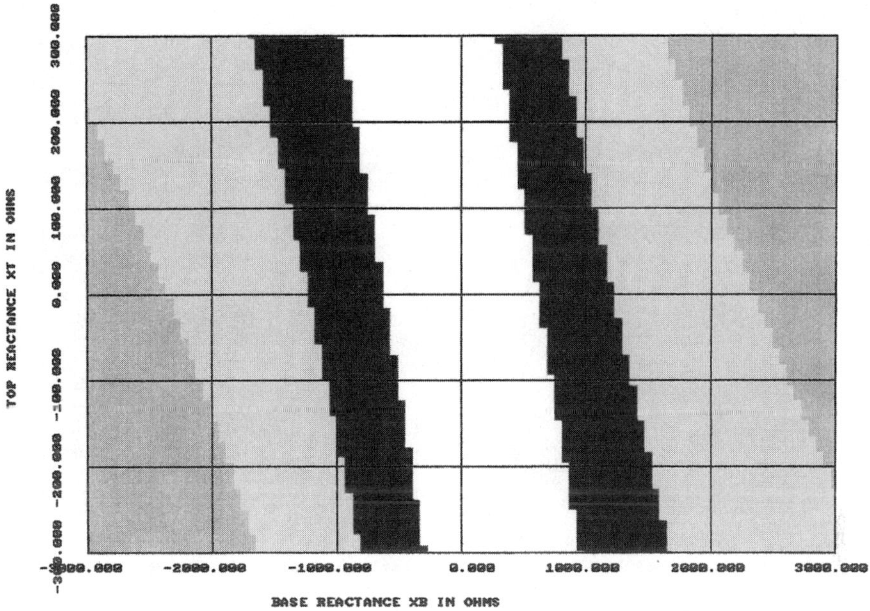

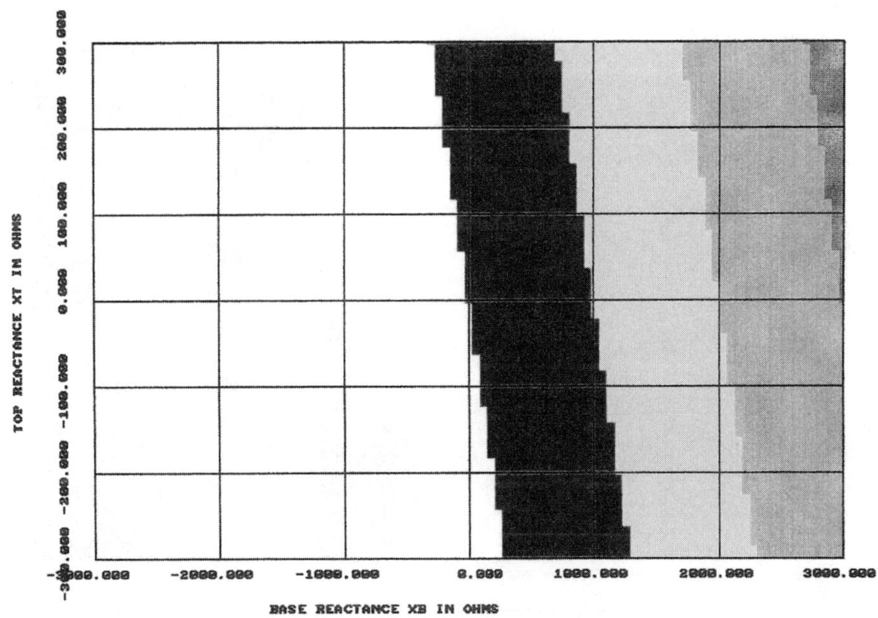

Figure 2-31 Input resistance (upper plot) and reactance (lower plot) for a 500-foot folded unipole antenna. At 984 kHz, it is 180 electrical degrees tall. For series-fed monopoles, this is a very troublesome height because input resistance can be thousands of ohms. Terminating reactances at both base and top are simultaneously and continuously varied. Green denotes resistances in the 25–75 ohms range. Cyan denotes reactances in the 0–1,000 ohms range. It is seen that there are many combinations of terminating reactances that produce useful input impedances.

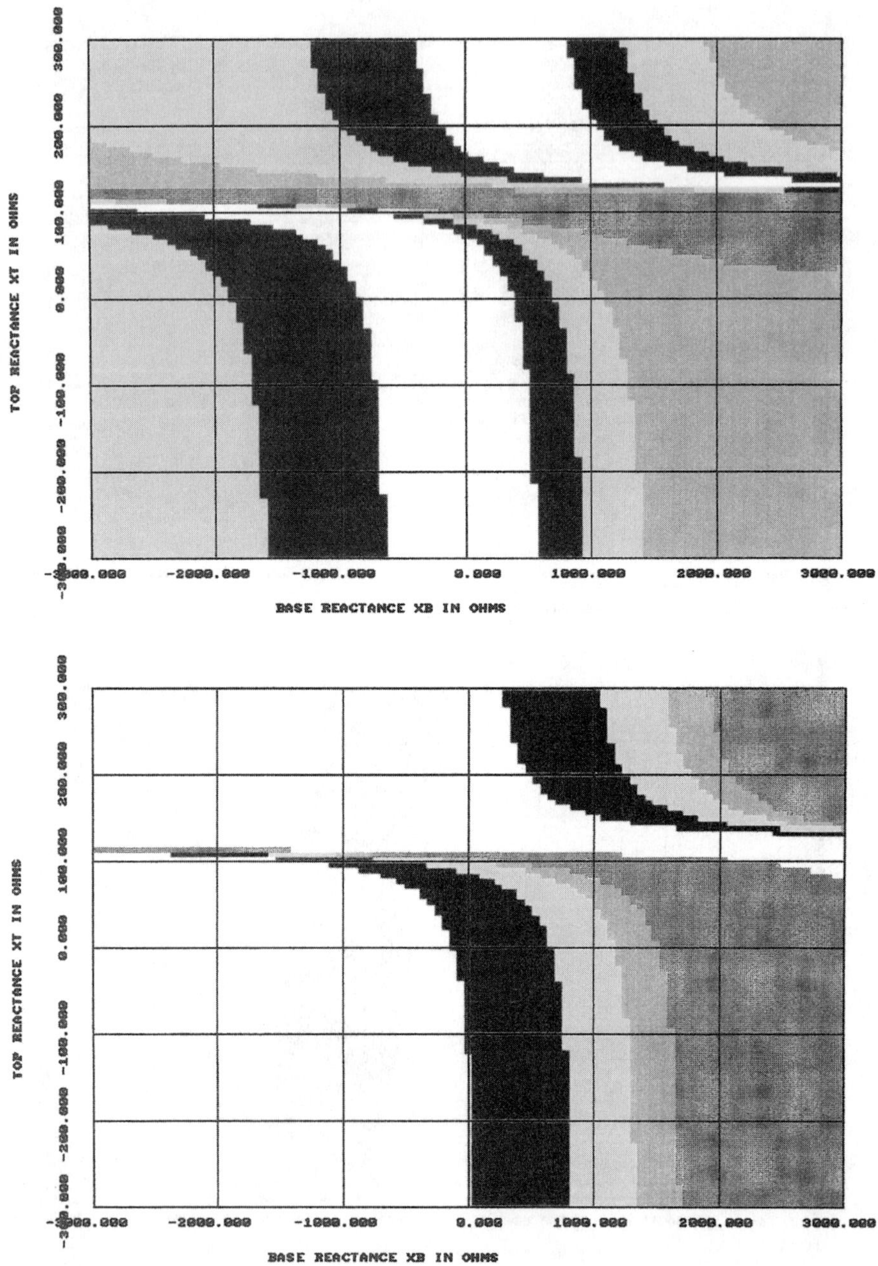

Figure 2-32 Input resistance (upper plot) and reactance (lower plot) for a 625-foot folded unipole antenna. At 984 kHz, it is 225 electrical degrees tall. Terminating reactances at both base and top are simultaneously and continuously varied. Green denotes resistances in the 25–75 ohms range. Cyan denotes reactances in the 0–1,000 ohms range. It is seen that there are many combinations of terminating reactances that produce useful input impedances.

we should not be surprised to find many useful combinations of terminating reactances that put the input resistance into the green zone, and the input reactance into the cyan zone.

2.15 Conclusions

We have developed an original mathematical model for the simplest configuration of folded unipole antenna. We have used the model to explore the remarkable properties of this versatile antenna. Even the simplest configuration provides no fewer than six independent design variables. They are: (1) height h, (2) terminating base impedance Z_B, (3) terminating top impedance Z_T, (4) current transformation ratio ν; (5) common mode characteristic impedance Z_{0c}, and (6) differential mode characteristic impedance Z_{0d}. In contrast, the traditional series-fed monopole provides only two design variables, height and radius.

The advantages of the extra design variables provided by the folded unipole are profound, indeed. Useable input impedances were obtained for very short antennas. In one example, the antenna was only 18 electrical degrees tall, which is 1/5 the height of the traditional quarter-wave, series-fed monopole. Useable input impedances were also obtained for very tall antennas. Examples were presented for antennas 135, 180, and 225 electrical degrees in height.

Our original mathematical model breaks new ground in at least two ways. First, it is the most complete analytic model available. In the published literature, some analytic models are limited to quarter wavelengths, or to short antennas. No other analytic models appear to include terminating impedances at both top and base of the unipole.

Second, our mathematical model is extremely fast compared with general purpose, numerical, wire antenna analysis programs. For example, each contour plot included in this chapter displays the results from 10,000 or more separate design iterations. All of these iterations required about a minute of computation time. In contrast, the same computation using a general purpose, numerical, wire antenna analysis program would require hours or days.

Our mathematical model is computationally fast because its matrices are about *three orders of magnitude* smaller than the matrices in a general purpose, numerical, wire antenna analysis program. For the simplest configuration of folded unipole of interest in this chapter, the matrix in our model has dimensions of 4 × 4, or 16 matrix elements. In contrast, the competing programs require matrices on the order of 100 × 100, or 10,000 matrix elements. Actually, they can be substantially larger, depending upon the height of the antenna, and the complexity of its cross section.

Smaller matrices have another important advantage besides speed, that is, low numerical noise. Large matrices tend to have high computational noise which compromises the solutions.

The novel mathematical approach developed in this chapter can be readily extended to more complicated configurations of the folded unipole antenna. The extensions and resultant examples will be the subjects of chapters to follow.

References

1. Roberts, W. van B., Input Impedance of a Folded Dipole, *RCA Review* **8**, June, 1947, pp. 289–300.
2. Leonhard, J., R. D. Mattuck, and A.J.Poté., Folded Unipole Antennas, *IRE Transactions on Antennas and Propagation*, **3**, July, 1955, pp. 111–116.
3. Guertler, R., Impedance Transformation in Folded Dipoles, *Proc. of the IRE*, **38**, Sept., 1950, pp. 1042–1047.
4. Harrison, C.W., Jr., and R. W. P King., Folded Dipoles and Loops, *IRE Transactions on Antennas and Propagation*, **9**, March, 1961, pp. 171–187.
5. Harrison, C.W., Jr., Hybrid Transmission Lines, *Sandia Corp., Albuquerque, N. M., Tech. Memorandum*, **14**, May, 1958, pp. 157–158.

Symbols Used in This Chapter

a_1	Radius of folded antenna conductor in the analyses of Harrison & King, and Poté *et al.*
a_2	Radius of folded antenna conductor in the analyses of Harrison & King, and Poté *et al.*
d	Equivalent antenna radius in the analysis by Harrison & King
$dE(\theta)$	Electric field radiated by incremental length of the unipole
$E(\theta)$	Electric field radiated by the unipole
E_f	Efficiency of the folded unipole, with respect to radiated electromagnetic field
E_p	Efficiency of the folded unipole, with respect to power
$F(\theta, a, b)$	Weighted integral over the current on the folded unipole
h	Vertical coordinate of the top of the unipole
$I_c(z)$	Common mode current on the unipole
$I_c(\zeta)$	Common mode current on the image of the unipole in the ground
$I_d(z)$	Differential mode current on the unipole
$I_d(\zeta)$	Differential mode current on the image of the unipole in the ground
I_{Fc}	Coefficient of the forward traveling wave component of the common mode current
I_{Fd}	Coefficient of the forward traveling wave component of the differential mode current
$\mathbf{I}$	Vector of the four current coefficients on the unipole
I_{in}	Input current at the terminals of the unipole
I_{Rc}	Coefficient of the reflected traveling wave component of the common mode current
I_{Rd}	Coefficient of the reflected traveling wave component of the differential mode current
j	Square root of -1
k	Wave number
$\mathbf{M}$	Matrix in the equation for currents on the unipole
P	Total radiated power

p	Current transformation ratio in the analysis of Poté *et al.*
R	Input resistance of a center-fed dipole in the analysis by Roberts; spherical coordinate for distance
R_B	Terminating base resistance
R_{in}	Input resistance
R_T	Terminating top resistance
$S(\theta)$	Radiated power density
V	Voltage applied to input terminals of unipole
$V_c(z)$	Common mode voltage along the unipole
$V_d(z)$	Differential mode voltage along the unipole
X_B	Terminating base reactance
X_{in}	Input reactance
X_T	Terminating top reactance
Y_{in}	Input admittance in the analysis of Harrison & King
z	Vertical coordinate along the unipole
Z_B	Terminating base impedance
Z_L	Terminating top impedance in the analysis of Harrison & King
Z_m	Input impedance in the analysis of Poté *et al.*
Z_T	Terminating top impedance
Z_{0c}	Characteristic impedance for common mode on the unipole
Z_{0d}	Characteristic impedance for differential mode on the unipole
Z_1	Characteristic impedance in the analysis by Roberts
Z_2	Characteristic impedance in the analysis by Roberts
β	Wave number in the analysis by Harrison & King
η	Impedance of free space
φ	Spherical coordinate for bearing
λ	Wavelength
ν	Current transformation ratio for common mode currents
θ	Spherical coordinate for azimuth
$\psi_F(h)$	Forward traveling wave function
$\psi_R(h)$	Reverse traveling wave function
ζ	Vertical coordinate along image of the unipole in the ground
ζ_0	Impedance of free space in the analysis by Harrison & King

Chapter

3

The Gamma Match Antenna and the Two-Segment Folded Unipole

3.1 Introduction

We can readily extend the analytic tools and techniques developed in Chap. 2 to explore the more complicated configurations of interest in this chapter. Actually, we will accomplish much more than that. By the end of this chapter, our approach to mathematical modeling will be general enough to design and evaluate arbitrarily complicated versions of the folded unipole antenna.

Chapter 2 firmly established at least three general and powerful principles. We might prefer to call them philosophies, perspectives, or points of view. Whatever our choice of moniker, the first relates to antenna design. The second relates to mathematical modeling of antennas. The third relates to the display of computed results for antenna performance. These three principles will be indispensable guides as we expand our investigation to more complicated configurations in this chapter and in later ones.

Antenna design: The folded unipole is the intentional integration of linear radiators and transmission lines into a single electromagnetic device. The benefit of this integration is a great number of independent design parameters. For the simplest configuration of folded unipole in Chap. 2, there are six such parameters. For more complicated configurations, the number of parameters can be increased until the desired antenna performance is achieved. In contrast, for the series-fed monopole, there are only two independent design parameters, height and radius.

Mathematical modeling: Antenna currents on all folded unipoles can be separated into two modes of traveling waves, a common mode (or antenna mode, or unbalanced mode) and a differential mode (or transmission line mode, or balanced mode). Once this is understood, the mathematical model becomes much simpler than the ones upon which general purpose, numerical, wire antenna analysis computer programs are based. There are at least two advantages of the simpler model. First, the computation time is orders of magnitude faster. This means we can explore tens of thousands of combinations of antenna design parameters in about a minute, while the general-purpose programs would require hours or days to do the same. Second, the simpler model exhibits far less numerical noise that tends to compromise solutions. This is because both models end up solving matrix equations; however, the number of matrix elements in the simpler model is *three to four orders of magnitude* less than that for the general-purpose model.

Displays: Contour plots are a much more efficient way to display calculations of antenna performance than the traditional *xy* graph. Contour plots show us what happens when we continuously and simultaneously vary two design variables. Certain combinations of those variables may result in very desirable antenna performance that otherwise might have been overlooked. In contrast, the *xy* graph allows us to continuously vary only one design variable, raising the risk of skipping over a useful design.

Equipped with the above powerful and general principles, we are prepared to model and evaluate more complicated configurations of the folded unipole antenna.

For the antenna designer seeking options for a variety of configurations, there is easily enough subject matter in this chapter for two separate, lengthy chapters. It is collected here, however, because the Gamma Match Antenna (GMA) is a special case of the 2-segment folded unipole (2SFU). Curiously, the former has been discussed in journal articles since circa 1949, while the latter has not been analyzed at all. This chapter will firmly establish the logical and mathematical similarities between the two.

The number of independent design variables will increase from six for the configuration in Chap. 2 to eight for the GMA, and to eleven for the 2SFU. So, we should expect extraordinary versatility from these configurations. We will not be disappointed.

From an engineering point of view, an important advantage of the 2SFU is the length of the fold. In Chap. 2, the length of the fold had to be the same as the length of the tower or undriven conductor. In this chapter, that constraint is removed. This can be of great convenience and practical importance, especially for towers or undriven conductors that are long compared with a wavelength. In this chapter, we will explore this advantage extensively.

We will discover other significant advantages. Most notably, the extra design variables can provide us with a very broadband antenna.

3.2 Essential Features of the GMA and 2SFU

Figure 3-1 shows the 2SFU. As mentioned above, the GMA is a special case, and we will elaborate on that shortly in this section. Compared with Fig. 2-1, it is readily seen that there are three new impedances (Z_1, Z_{0c2}, and Z_{0d2}) and one new linear dimension (h_2). Though it is not shown in the figure, there is also one new current transformation ratio ν_2. So, there are five new, independent design variables in addition to the six we explored as part of the simpler configuration in Chap. 2. With a total of eleven independent design variables, we should expect versatile and superior performance from the 2SFU.

Although each segment of the 2SFU is depicted as a two-wire line, this is only for simplicity in the drawing. In fact, our mathematical model will apply to arbitrary cross sections with any number of conductors. As we discussed in Chap. 2, any cross section can be represented mathematically by two characteristic impedances, common mode and differential mode. In Chap. 7, we will show in detail how to compute the characteristic impedances for any cross section.

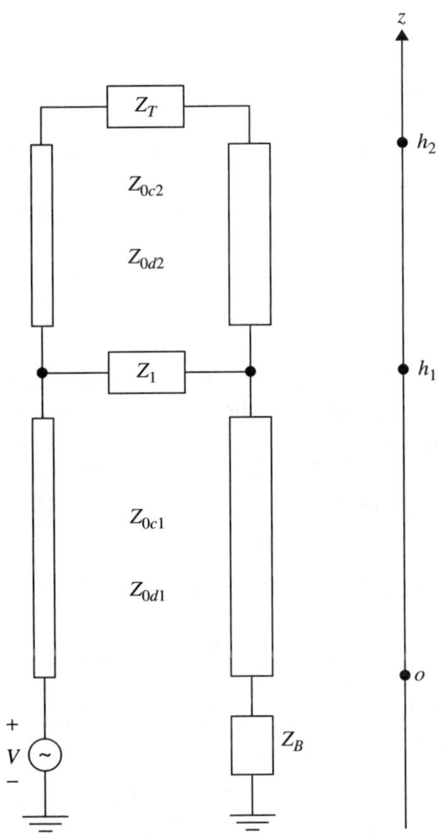

Figure 3-1 Essential features and geometry for the 2-segment folded unipole antenna (2SFU). Two parallel conductors are shown only for simplicity in the drawing. The upper and lower segments, in general, have different characteristic impedances and different current transformation ratios.

The underlying physics for our mathematical model of the 2SFU is a simple extension of that presented in Sec. 2.1. That is, the common mode currents are:

$$I_{cn}(z) = I_{Fcn}\psi_F(z) + I_{Rcn}\psi_R(z) \qquad n = 1,2 \tag{3-1}$$

The differential mode currents are:

$$I_{dn}(z) = I_{Fdn}\psi_F(z) + I_{Rdn}\psi_R(z) \qquad n = 1,2 \tag{3-2}$$

The notation in Eqs. 3-1 and 3-2 is different in two ways from Eqs. 2-1 and 2-2, which described the analogous currents on the simpler folded unipole. First, we have added the subscript n to denote the segment of the 2SFU. The currents, in general, are different on each one. Second, we have returned to using the traveling wave functions introduced in Chap. 1. The forward traveling wave function is:

$$\psi_F(z) = e^{-jkz} \tag{3-3}$$

The reflected or reverse traveling wave function is:

$$\psi_R(z) = e^{+jkz} \tag{3-4}$$

We will use these functions extensively in our mathematical model, and Eqs. 3-3 and 3-4 will prevent the notation from becoming too cluttered. In the further interests of keeping notation simple, it is readily seen that

$$\psi_F(0) = \psi_R(0) = 1 \tag{3-5}$$

Continuing in a similar manner to Chap. 2, the common mode and differential mode potentials are closely related to the currents, according to the usual transmission line equations,

$$V_{cn}(z) = Z_{0cn}\left[I_{Fcn}\psi_F(z) - I_{Rcn}\psi_R(z)\right] \qquad n = 1,2 \tag{3-6}$$

$$V_{dn}(z) = Z_{0dn}\left[I_{Fdn}\psi_F(z) - I_{Rdn}\psi_R(z)\right] \qquad n = 1,2 \tag{3-7}$$

Equations 3-6 and 3-7 are analogous to Eqs. 2-4 and 2-5 for the simpler folded unipole configuration.

As mentioned at the beginning of this section, the GMA is a special case of the 2SFU. In that case,

$$Z_{0d2} \to \infty \tag{3-8}$$

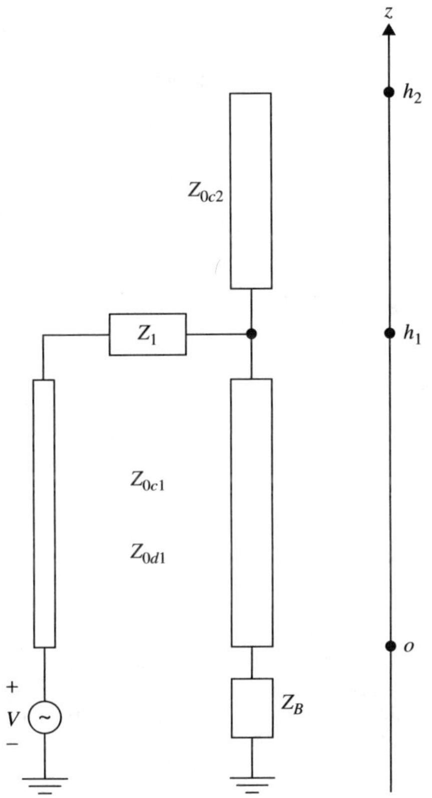

Figure 3-2 Essential features and geometry for the Gamma Match Antenna (GMA). Two parallel conductors are shown only for simplicity in the drawing. In fact, the analysis applies to any arbitrary cross section that can be mathematically described by two characteristic impedances, one for common mode and another for differential mode.

That is, the fold wire vanishes, as shown in Fig. 3-2. Since there is no fold wire, there are no differential mode currents or potentials on the upper segment of the 2SFU. So, both geometrically and mathematically, the 2SFU becomes the GMA.

The GMA has only eight independent design variables instead of the eleven available with the 2SFU. The three design variables lost are Z_{0d2}, Z_T, and ν_2.

To mathematically model the GMA, we could simply apply Eq. 3-8 to the model of the 2SFU; however, this would be more cumbersome than having a specialized model for the GMA. The GMA is sufficiently useful to justify developing a separate model.

3.3 Equation at the Top of the GMA

Figure 3-3 shows the configuration of interest with currents and voltages. According to Eqs. 3-1 and 3-2, there are six unknown, complex current amplitudes to determine. Four of them are the two sets of common mode amplitudes, $I_{\text{Fc}n}$ and $I_{\text{Rc}n}$, for $n = 1$ (the lower segment) and for $n = 2$

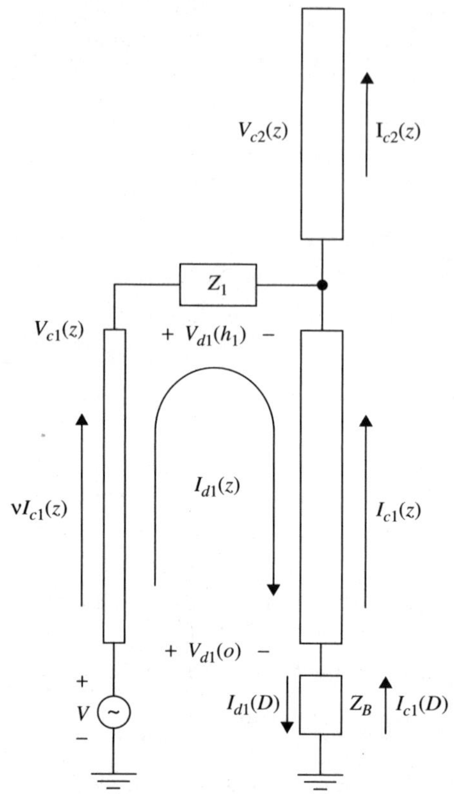

Figure 3-3 Voltages and currents used for analysis of the Gamma Match Antenna (GMA).

(the upper segment). The remaining two are the differential mode amplitudes, I_{Fd1} and I_{Rd1}, on the lower segment. Since the fold has vanished on the upper segment, there are no differential mode currents there. With six unknown current amplitudes, we must formulate six independent equations to determine them.

Constraints at the top of the GMA provide one of the six equations. The common mode current must vanish there because it has nowhere to flow:

$$I_{c2}(h_2) = 0 \tag{3-9}$$

We still need five more independent equations. We'll have to look elsewhere on the GMA to find them.

3.4 Equations at the Base of the GMA

Constraints at the base of the GMA provide two more independent equations. Kirchhoff's voltage law (KVL) requires that

$$Z_B[I_{d1}(0) - I_{c1}(0)] + V_{d1}(0) = V \tag{3-10}$$

Also, by simple inspection of the fold it is clear that

$$V_{c1}(0) = V \tag{3-11}$$

The remaining three equations that we need must come from the boundary at $z = h_1$ between the upper and lower segments of the GMA.

3.5 Equations at the Boundary Between the Segments of the GMA

At the boundary $z = h_1$, there are two constraints on the common mode voltages and currents, and one on the differential mode. The common mode voltages must be continuous across the boundary:

$$V_{c1}(h_1) = V_{c2}(h_1) \tag{3-12}$$

According to Kirchoff's current law (KCL), the common mode currents must also be continuous:

$$(1+\nu)I_{c1}(h_1) = I_{c2}(h_1) \tag{3-13}$$

For the differential mode, the impedance Z_1 defines the relationship between the voltage and current:

$$Z_1 I_{d1}(h_1) = V_{d1}(h_1) \tag{3-14}$$

Equations 3-9 through 3-14 are the six independent equations we need to determine the unknown, complex amplitudes for the common mode and differential mode currents flowing everywhere along the GMA.

3.6 Matrix Equation for the Currents on the GMA

Similar to Chap. 2, a matrix equation for the six unknown, complex current amplitudes would be the most efficient way to explore the many independent design variables provided by the GMA. Such an equation is a standard form that can be solved rapidly using well-established computer algorithms. For matrices with dimensions small enough, we can evaluate the performance of the GMA for thousands of different combinations of the many independent design variables, all in a matter of seconds.

Equations 3-9 through 3-14 can be consolidated into a single matrix equation in a straightforward manner. The details are tedious but the essentials are simply this. In the equations, substitute for the currents using Eqs. 3-1 and 3-2, for the voltages using Eqs. 3-6 and 3-7. Thus, Eqs. 3-9 through 3-14 are rewritten explicitly in terms of the six unknown current amplitudes. The resulting matrix equation may be written down by inspection. It is:

$$\mathbf{M}\cdot\mathbf{I}=\begin{pmatrix}V\\V\\0\\0\\0\\0\end{pmatrix} \tag{3-15}$$

In Eq. 3-15, the current vector is:

$$\mathbf{I}=\begin{pmatrix}I_{\mathrm{Fc1}}\\I_{\mathrm{Rc1}}\\I_{\mathrm{Fd1}}\\I_{\mathrm{Rd1}}\\I_{\mathrm{Fc2}}\\I_{\mathrm{Rc2}}\end{pmatrix} \tag{3-16}$$

The matrix has dimensions 6 × 6. To avoid writing individual matrix elements that are too small to read on the printed page, it is helpful to partition **M** into submatrices:

$$\mathbf{M}=\begin{pmatrix}\mathbf{A}_{11} & \mathbf{A}_{12}\\\mathbf{A}_{21} & \mathbf{A}_{22}\end{pmatrix} \tag{3-17}$$

In Eq. 3-17, the submatrices have dimensions 3 × 3. They are:

$$\mathbf{A}_{11}=\begin{pmatrix}Z_{0c1} & -Z_{0c1} & 0\\-Z_B & -Z_B & Z_B+Z_{0d1}\\0 & 0 & 0\end{pmatrix} \tag{3-18}$$

$$\mathbf{A}_{12}=\begin{pmatrix}0 & 0 & 0\\Z_B-Z_{0d1} & 0 & 0\\0 & \psi_F(h_2) & \psi_R(h_2)\end{pmatrix} \tag{3-19}$$

$$\mathbf{A}_{21}=\begin{pmatrix}(1+\nu)\psi_F(h_1) & (1+\nu)\psi_F(h_1) & 0\\Z_{0c1}\psi_F(h_1) & -Z_{0c1}\psi_R(h_1) & 0\\0 & 0 & (Z_1-Z_{0d1})\,\psi_F(h_1)\end{pmatrix} \tag{3-20}$$

$$\mathbf{A}_{22}=\begin{pmatrix}0 & -\psi_F(h_1) & -\psi_R(h_1)\\0 & -Z_{0c2}\psi_F(h_1) & Z_{0c2}\psi_R(h_1)\\(Z_1+Z_{0d1})\,\psi_R(h_1) & 0 & 0\end{pmatrix} \tag{3-21}$$

In Eqs. 3-18 through 3-21, the forward and reverse traveling wave functions are defined by Eqs. 3-3 and 3-4.

Equation 3-15 is well suited for rapid evaluation by desktop computers. It is practical to solve for the current vector **I** over and over again for thousands of different combinations of the independent design variables.

Compared with general purpose, wire antenna analysis computer programs, Eq. 3-15 offers the same significant advantages discussed in Chap. 2. Those are speed and less numerical noise. The matrix **M** in Eq. 3-15 has dimensions 6 × 6. In contrast, the general purpose, wire antenna program requires matrices with dimensions on the order of 300 × 300 for large and complicated structures. So, while **M** has only 36 elements, the matrix of the general-purpose program has about 100,000. That is a difference of *three to four orders of magnitude.*

Even in the age of fast and inexpensive computers, the speed advantage of Eq. 3-15 is significant. Most of the examples presented in this chapter result from solving Eq. 3-15 10,000 times. Yet all of these iterations required less than a minute on an ordinary desktop computer. In contrast, a general purpose, wire antenna analysis computer program would require hours or days to execute the same examples.

3.7 Radiated Field and Power for the GMA

Equations 2-30 through 2-36 for the radiated field and power of the simpler folded unipole apply to the GMA after some minor modifications. These modifications account for the increase in the number of segments. In particular, Eq. 2-31 must be replaced by:

$$E(\theta)=\frac{\eta k \sin\theta}{4\pi R}e^{-jkR}\sum_{n=1}^{N}(1+v_n)F_n(\theta,a_n,b_n) \tag{3-22}$$

Equation 3-22 can be used for folded unipole antennas with an arbitrary number of segments N. For the GMA of interest here, $N=2$, $v_2=0$, $a_1=0$, $b_1=h_1$, $a_2=h_1$, and $b_2=h_2$.

Similarly, Eq. 2-33 must be replaced by:

$$F_n(\theta,a_n,b_n)=\frac{1}{jk}\left\{\begin{aligned}&-\frac{I_{\mathrm{Fcn}}\left[e^{-jkb_n(1-\cos\theta)}-e^{-jka_n(1-\cos\theta)}\right]}{(1-\cos\theta)}+\frac{I_{\mathrm{Rcn}}\left[e^{+jkb_n(1+\cos\theta)}-e^{+jka_n(1+\cos\theta)}\right]}{(1+\cos\theta)}\\&+\frac{I_{\mathrm{Rcn}}\left[e^{+jkb_n(1-\cos\theta)}-e^{+jka_n(1-\cos\theta)}\right]}{(1-\cos\theta)}-\frac{I_{\mathrm{Fcn}}\left[e^{-jkb_n(1+\cos\theta)}-e^{-jka_n(1+\cos\theta)}\right]}{(1+\cos\theta)}\end{aligned}\right\} \tag{3-23}$$

Like Eq. 3-22, Eq. 3-23 applies to folded unipole antennas with any number of segments.

3.8 Input Impedance and Current for the GMA

From Chap. 2, Eqs. 2-37 through 2-39 also apply to the GMA. Equation 2-39 may be rewritten, if desired, to clarify that the current transformation ratio and currents are all on the base segment:

$$I_{\text{in}} = \nu_1 I_{c1}(0) + I_{d1}(0) \tag{3-24}$$

Equations 2-39 and 3-24 are identical in substance; however, in the latter we have simply added the subscript "1" to emphasize the role of the base segment. Equation 3-24 applies to folded unipole antennas with an arbitrary number of segments.

3.9 Efficiency of the GMA

We can readily extend Eqs. 2-40 through 2-42 for efficiency to include the folded unipole with an arbitrary number of segments N. Equation 2-40 for the power dissipated by ohmic and ground losses must be replaced with:

$$P_d = |I_{c1}(0) - I_{d1}(0)|^2 R_B + \sum_{n=1}^{N-1} |I_{\text{dn}}(h_n) - I_{\text{d,n+1}}(h_n)|^2 R_n + |I_{\text{dN}}(h_N)|^2 R_T \tag{3-25}$$

For the GMA, $N = 2$, and $I_{d2}(h_1) = I_{d2}(h_2) = 0$. So, Eq. 3-25 reduces to:

$$P_d = |I_{c1}(0) - I_{d1}(0)|^2 R_B + |I_{d1}(h_1)|^2 R_T \tag{3-26}$$

As with Eq. 2-40, we observe an intriguing possibility in the first term of Eqs. 3-25 and 3-26. That is, if the common mode and differential mode currents are equal, or nearly equal, then the first term vanishes. So, a large part of the dissipated power is eliminated, creating a surprisingly efficient antenna. Indeed, anecdotal reports from the field concerning actual folded unipoles have described efficiencies exceeding the monopoles that they replaced.

Equations 2-41 and 2-42 are still applicable without modification.

3.10 Examples of the GMA with Fold Height as the Primary Variable

The mathematical model which we began in Chap. 2, and built upon in Sec. 3.2 through 3.9, allows us to explore the many design variables provided by the GMA. Such a powerful tool must be handled with care, however. As discussed in Sec. 2.12, with so many independent design variables, it is easy to be buried under a mountain of examples. Accordingly, we will be selective and attempt to focus on the variables which provide the most dramatic changes in antenna performance. Of those variables, perhaps the most obvious is fold height h_1. The option to make h_1 different from the tower height h_2 is what distinguishes the GMA from simpler configurations.

We will begin our investigation with some traditional *xy* graphs. We will soon find, however, that the contour plots introduced in Chap. 2 are a superior way to continue the investigation.

Figure 3-4 shows the input impedance of a 300-foot tower as the fold height is varied. The frequency in this example, and for all examples in this chapter, is 984 kHz. So, the tower is about 0.3 wavelengths tall. The choice of frequency is purely for mathematical convenience. A wavelength is 1,000 feet, making it especially easy to interpret the linear dimensions of the antenna. We could just as well have picked 98.4 kHz, 984 MHz, 9.84 GHz, or any other frequency. The mathematical model and design lessons would be the same.

We have fixed some other design variables as well so that we can concentrate on the few that provide the most dramatic effect on antenna performance. For example, the current transformation ratio ν_1 is 0.5. The common mode characteristic impedances Z_{0c1} and Z_{0c2} are 1200 ohms and 1600 ohms, respectively. The differential mode impedance Z_{0d1} is 120 ohms. These are all "ballpark" values determined by the detailed cross section of the antenna. In Chap. 7, we will show how to calculate these values exactly for any cross section.

Figure 3-4 shows the usual resonance when the fold height is about 245 feet, just a few feet shorter than a quarter wavelength, even though the tower height is greater. As expected, the input reactance is positive or inductive for heights below resonance. There is a curious maximum input resistance when the fold height is about 75 feet, or 27 electrical degrees. It is also seen that an input resistance of 50 ohms is obtained at a fold height of only 55 feet, or about 20 degrees.

Figure 3-5 shows the input impedance of a 375-foot tower as the fold height is varied. It is seen that the resonance has shifted to a lower fold height, about 234 feet. The input resistance reaches 50 ohms for a fold height of 120 feet, or about 43 electrical degrees.

Figure 3-6 shows the input impedance of a 500-foot, or half-wavelength, tower as the fold height is varied. Resonance has shifted to a fold height slightly greater than a quarter wavelength, about 252 feet. The input resistance reaches 50 ohms for a fold height of 204 feet. In contrast, a half-wavelength monopole would be very difficult to match, with an input resistance on the order of 1000 ohms. So, the GMA configuration is a practical way to impedance match a half-wavelength tower.

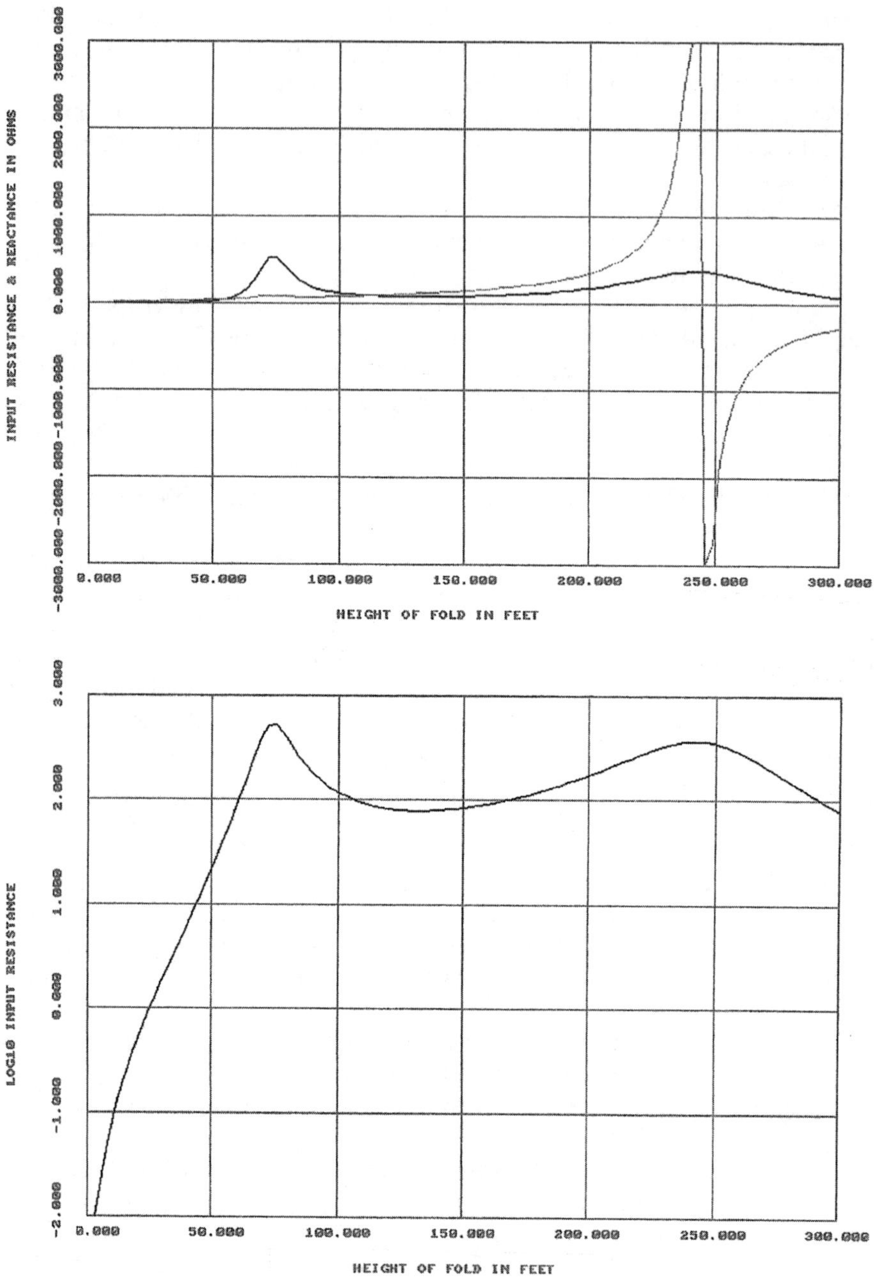

Figure 3-4 Input resistance (black) and reactance (gray) for a 300-foot GMA. At 984 kHz, it is 0.3 wavelength long, or 108 electrical degrees. The lower plot shows the input resistance on a logarithmic scale. The choice of 984 kHz is entirely for mathematical convenience because the wavelength is 1,000 feet. The analysis applies to any frequency ranging from the ELF through the EHF bands.

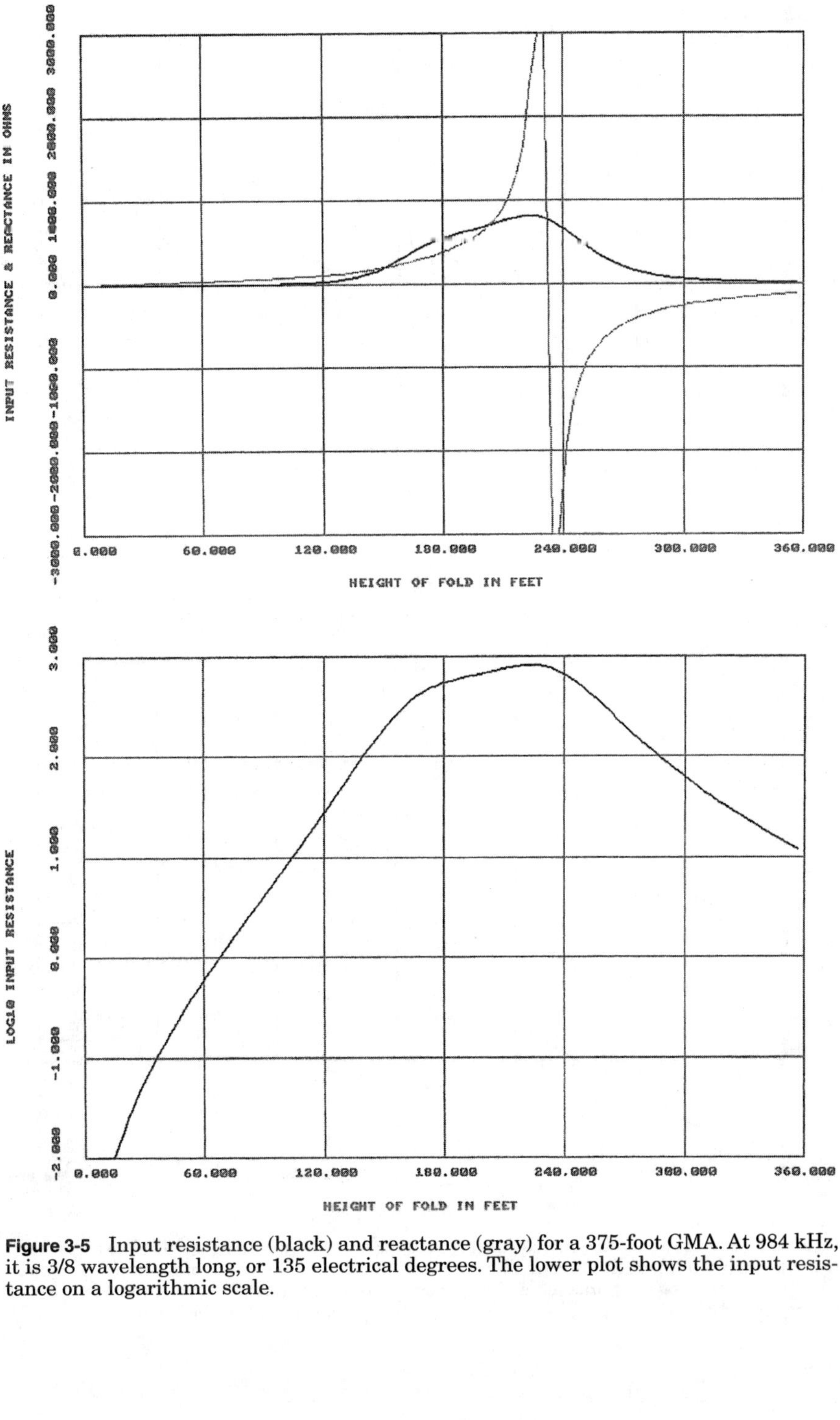

Figure 3-5 Input resistance (black) and reactance (gray) for a 375-foot GMA. At 984 kHz, it is 3/8 wavelength long, or 135 electrical degrees. The lower plot shows the input resistance on a logarithmic scale.

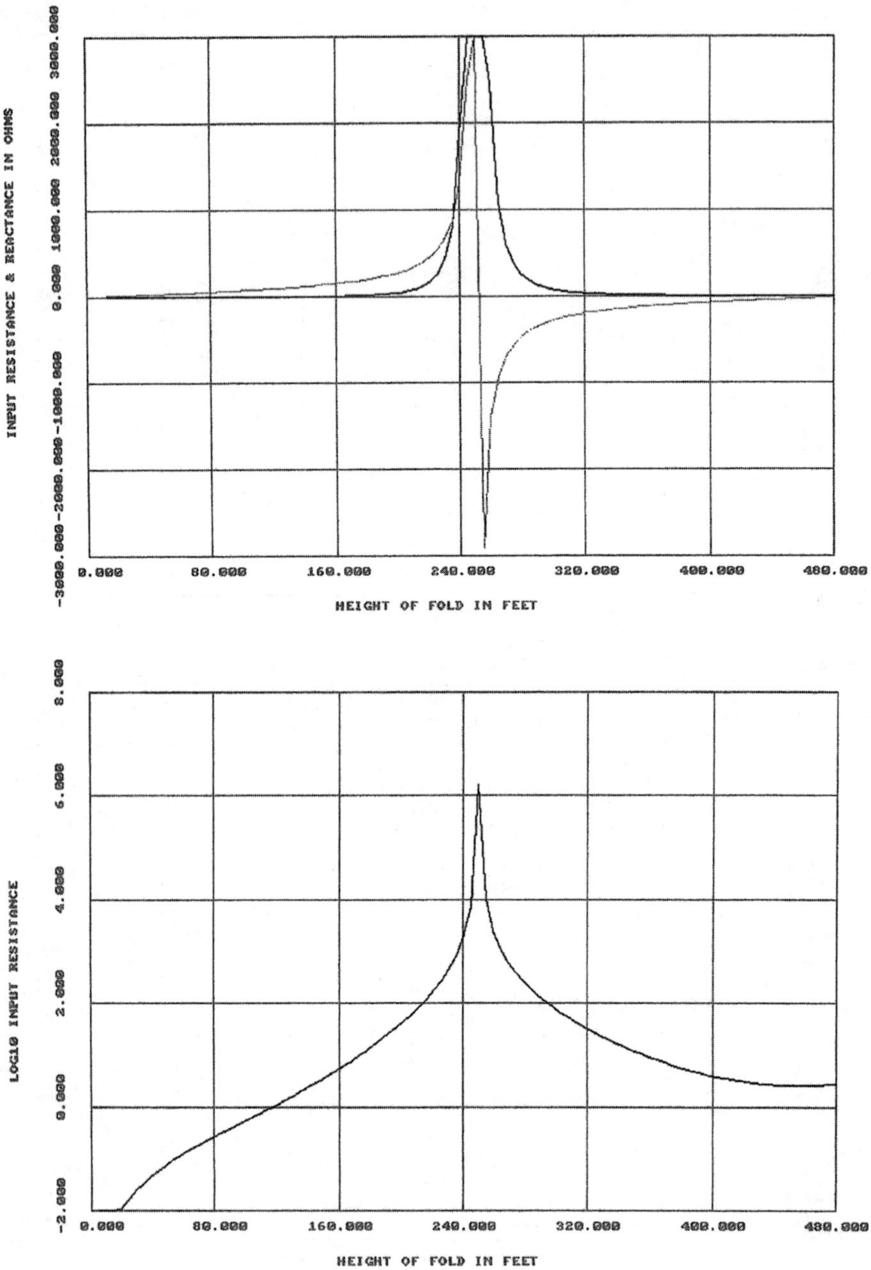

Figure 3-6 Input resistance (black) and reactance (gray) for a 500-foot GMA. At 984 kHz, it is one-half wavelength long, or 180 electrical degrees. The lower plot shows the input resistance on a logarithmic scale. In practice, we would not observe a megohm input resistance because of antenna end effect which is discussed in Sec. 5.7.

Figure 3-7 shows the input impedance of a 625-foot tower as fold height is varied. Resonance occurs at about 268 feet. The input resistance reaches 50 ohms for a fold height of 210 feet.

Figure 3-8 shows the input impedance of a 750-foot tower as fold height is varied. Resonance first occurs at 251 feet, and it reoccurs one-half wavelength later, which is not surprising. Input resistance is 50 ohms at a fold height of 100 feet.

Figure 3-9 shows the input impedance of a 1,500-foot tower as fold height is varied. Resonance first occurs at 255 feet, just a bit beyond one-quarter wavelength, and it reoccurs at one-half wavelength intervals. Input resistance is 50 ohms at a fold height of 105 feet.

There is no obvious pattern from the above examples, although each one demonstrates the practical use of the GMA. The reason, as we discovered in Chap. 2, is that the traditional *xy* graph is not an efficient way to investigate the effect of multiple independent design variables. The contour plot is a much better tool.

In the contour plots for this chapter, we will introduce one new color each for the resistance and reactance displays. This is because the extra design variables provided by the GMA and 2SFU allow us to converge on a narrower range of desirable design values. Accordingly, resistances in the range from 45 to 55 ohms are shown by dark green. Reactances in the range from 0 to 100 ohms are shown by dark cyan.

A summary of contour colors for resistance displays is:

White	0–25 ohms
Dark green	45–55 ohms
Green	25–75 ohms
Yellow	75–150 ohms
Red	150–300 ohms
Gray	300 ohms and greater

A summary of contour colors for reactance displays is:

White	Less than 0 ohms
Dark cyan	0–100 ohms
Cyan	100–1,000 ohms
Yellow	1,000–2,000 ohms
Magenta	2,000–3,000 ohms
Gray	3,000 ohms and greater

With the above convention for colors, we are now ready to explore with contour plots the great versatility of the GMA in a much more meaningful way.

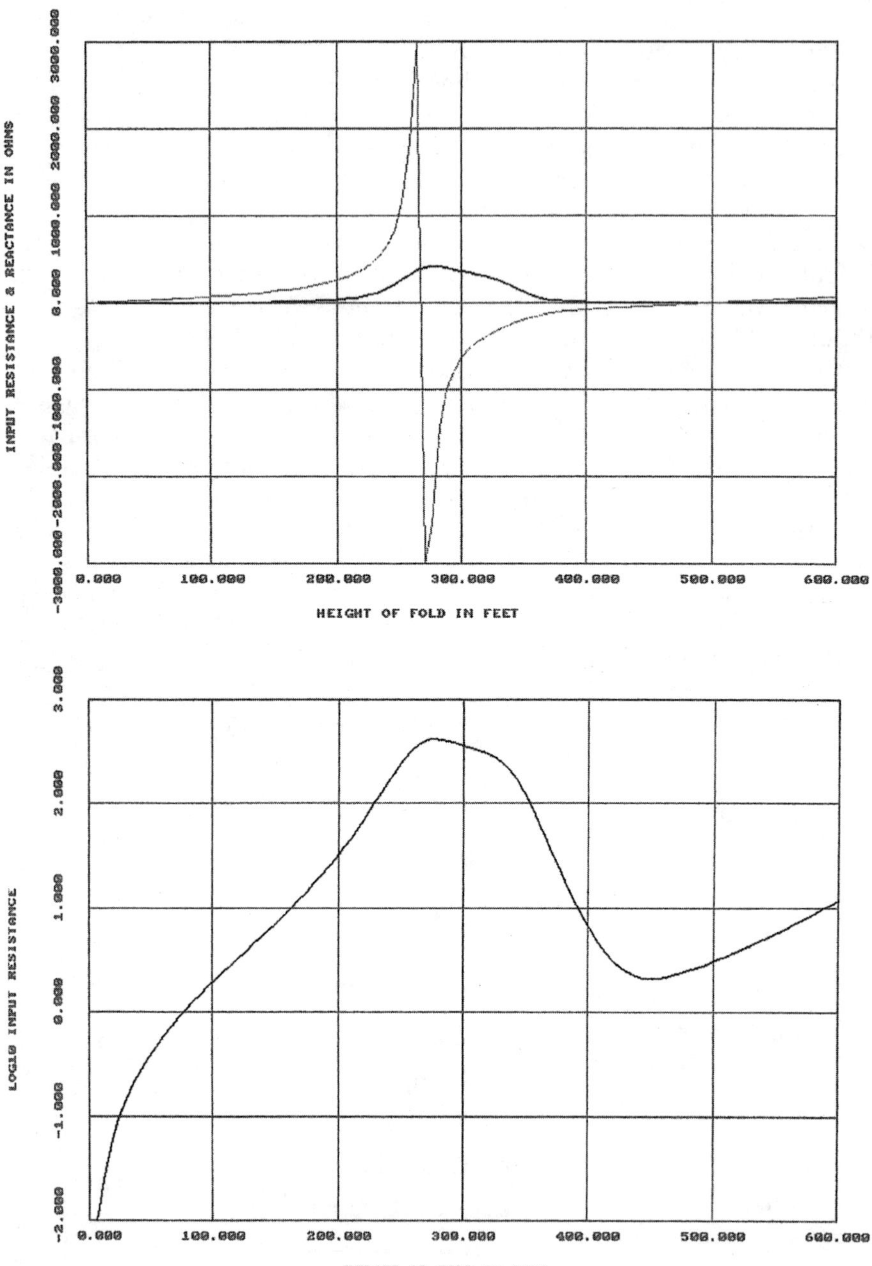

Figure 3-7 Input resistance (black) and reactance (gray) for a 625-foot GMA. At 984 kHz, it is 5/8 wavelength long, or 225 electrical degrees. The lower plot shows the input resistance on a logarithmic scale.

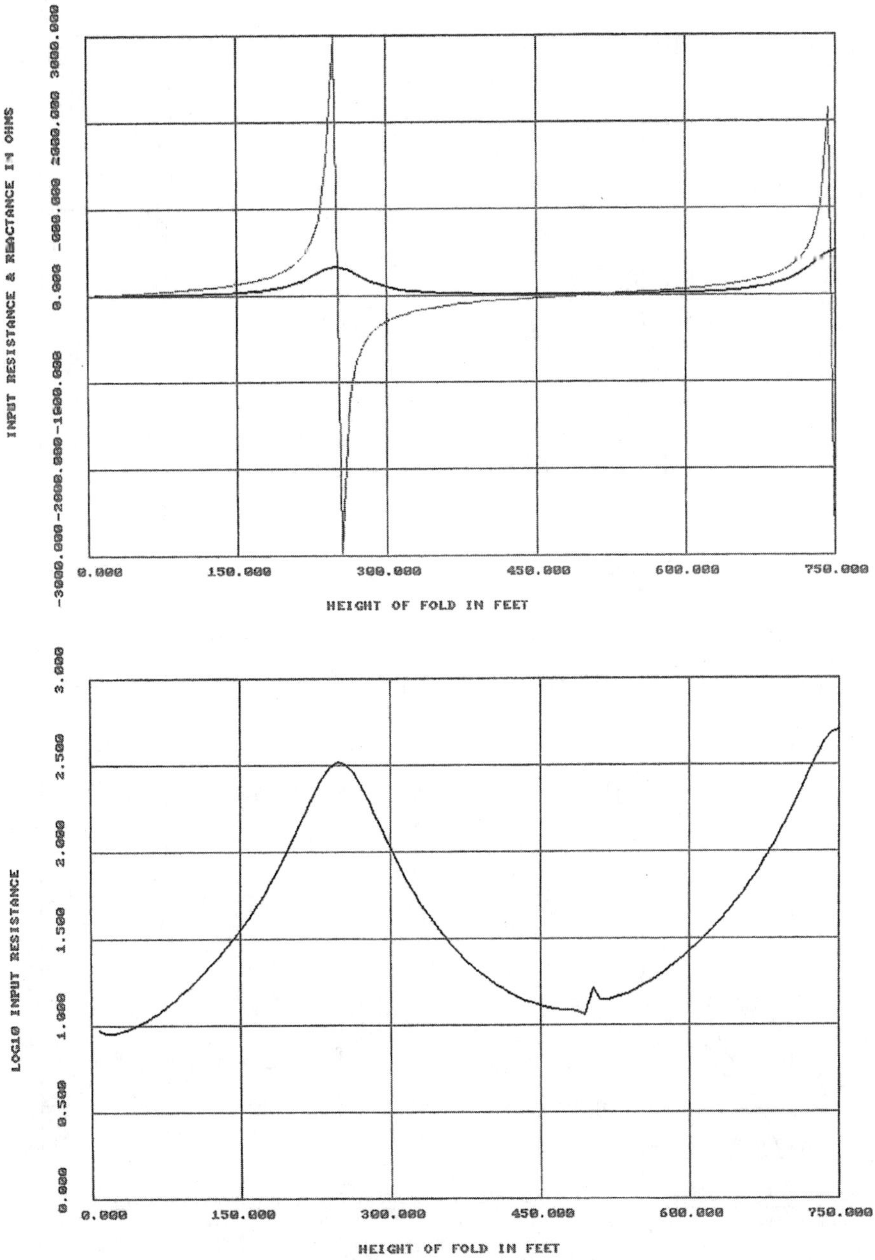

Figure 3-8 Input resistance (black) and reactance (gray) for a 750-foot GMA. At 984 kHz, it is three-quarters wavelength long, or 270 electrical degrees. The lower plot shows the input resistance on a logarithmic scale.

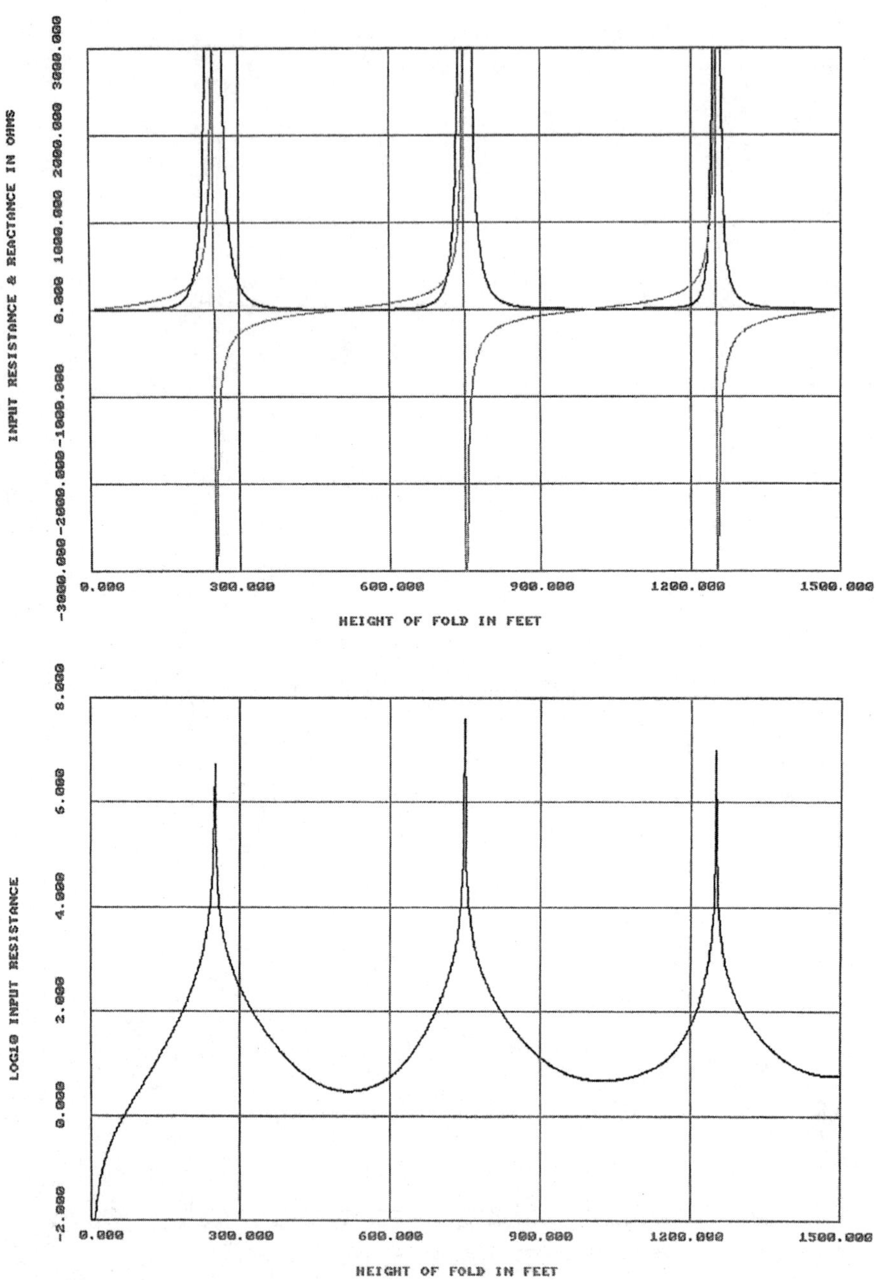

Figure 3-9 Input resistance (black) and reactance (gray) for a 1,500-foot GMA. At 984 kHz, it is 3/2 wavelength long, or 540 electrical degrees. The lower plot shows the input resistance on a logarithmic scale. Periodic behavior is clearly observed. As in Fig. 3-6, input resistances on the order of a megohm will not be observed in practice because of antenna end effect.

Figure 3-10 shows the input impedance for the GMA as both the tower height and the fold height are varied continuously and simultaneously. The upper triangular region of each contour plot is blank because the fold height cannot be greater than the tower height. Repetitive patterns are clearly seen in both resistance and reactance. It is also seen that for many combinations of fold and tower heights, both resistance and reactance in the most desirable ranges (dark green and dark cyan, respectively) can be achieved.

Figure 3-11 shows the input impedance as both tower and fold heights are varied simultaneously. The difference from the Fig. 3-10 is that the base terminating reactance has been changed from 0 ohm to 3,000 ohms. The dramatic change in impedance is clearly seen. Though the most desirable range of resistance is achieved for many combinations, they do not coincide with the most desirable reactance range. They do coincide with the second best choice of reactance range (cyan), however.

Figure 3-12 shows the input impedance when the base terminating reactance is –3,000 ohms. The most desirable range of resistance is achieved for many combinations; however, they do not coincide with a desirable choice of reactance.

Figure 3-13 shows the input impedance when the fold terminating reactance is 600 ohms. The base terminating reactance is 0 ohm. It is seen that there are many combinations that provide both resistance and reactance in the most desirable ranges, even more so than shown in Fig. 3-10.

Figure 3-14 shows the input impedance when the fold terminating reactance is –600 ohms. Observations are similar to those for the previous Fig. 3-13.

From the examples so far, combinations of nonzero terminating reactances with variable fold height provide the best opportunities for achieving the most desirable input impedance, that is, 50 ohms with an inductive reactance less than 100 ohms. We have hardly begun to exploit those combinations, however. We will continue the investigation along those lines in the next section.

3.11 Examples of the GMA with Variable Terminating Reactance *and* Fold Height

Following up the suggestive results of the previous section, we will revisit Figs. 3-4 through 3-9 from the perspective of contour plots. In Sec. 3.10, the *xy* graphs in those figures provided only a one-dimensional slice of antenna performance. In particular, both terminating reactances were fixed at zero, and so we most likely missed many successful designs. The contour plots will add an extra dimension, allowing us to capture those designs. For the same selection of fixed heights h_2 in Figs. 3-4 through 3-9,

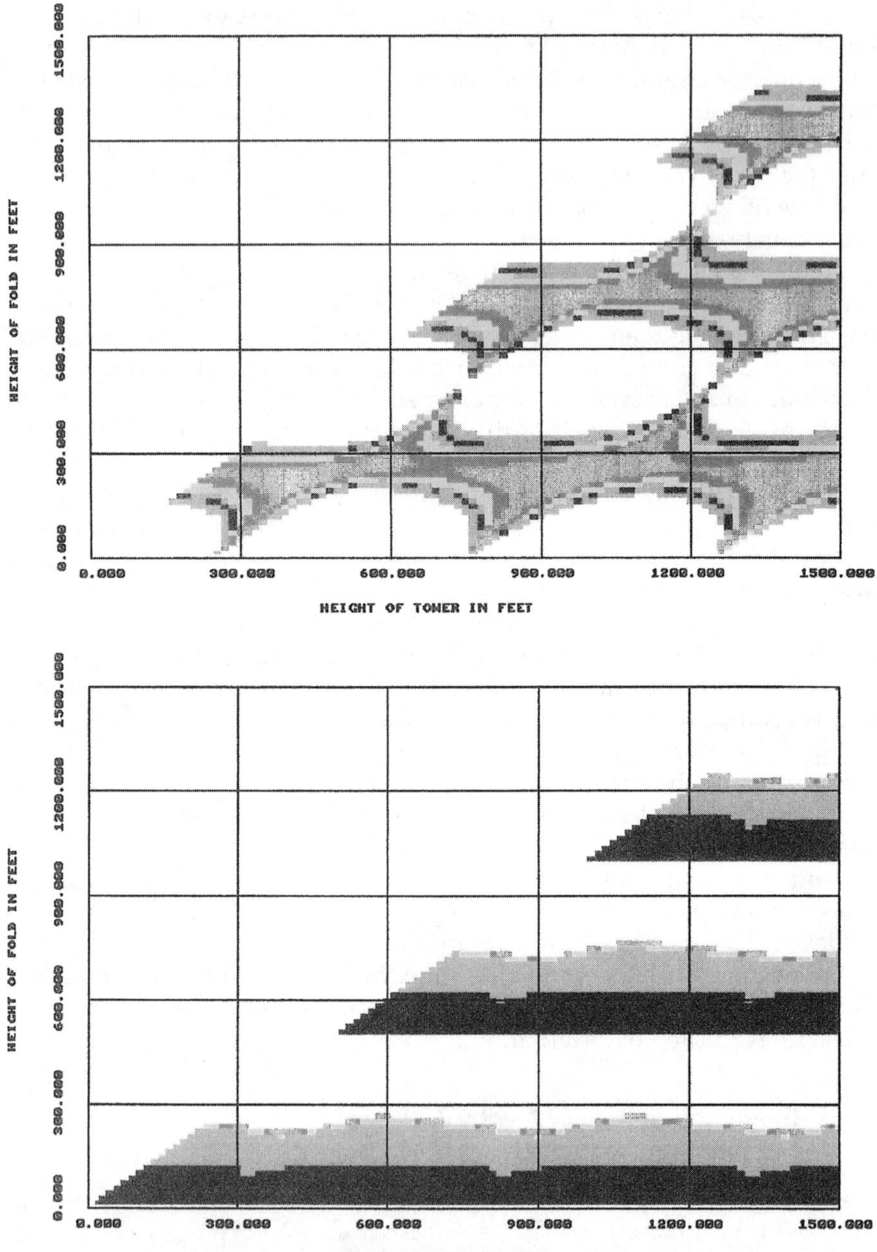

Figure 3-10 Contour plots are an efficient way to explore the Gamma Match Antenna (GMA) when two design parameters are continuously and simultaneously varied. Here the two variables are antenna height h_2 and fold height h_1. The upper plot shows input resistance, and the lower plot shows input reactance. Dark green denotes resistances in the 45–55 ohms range. Dark cyan denotes reactances in the 0–100 ohms range. In Sec. 3.10, the meaning of the other colors is described in detail. Note: See Preface, Notes Concerning the Examples for explanation of the gray scale which has been subsituted for color.

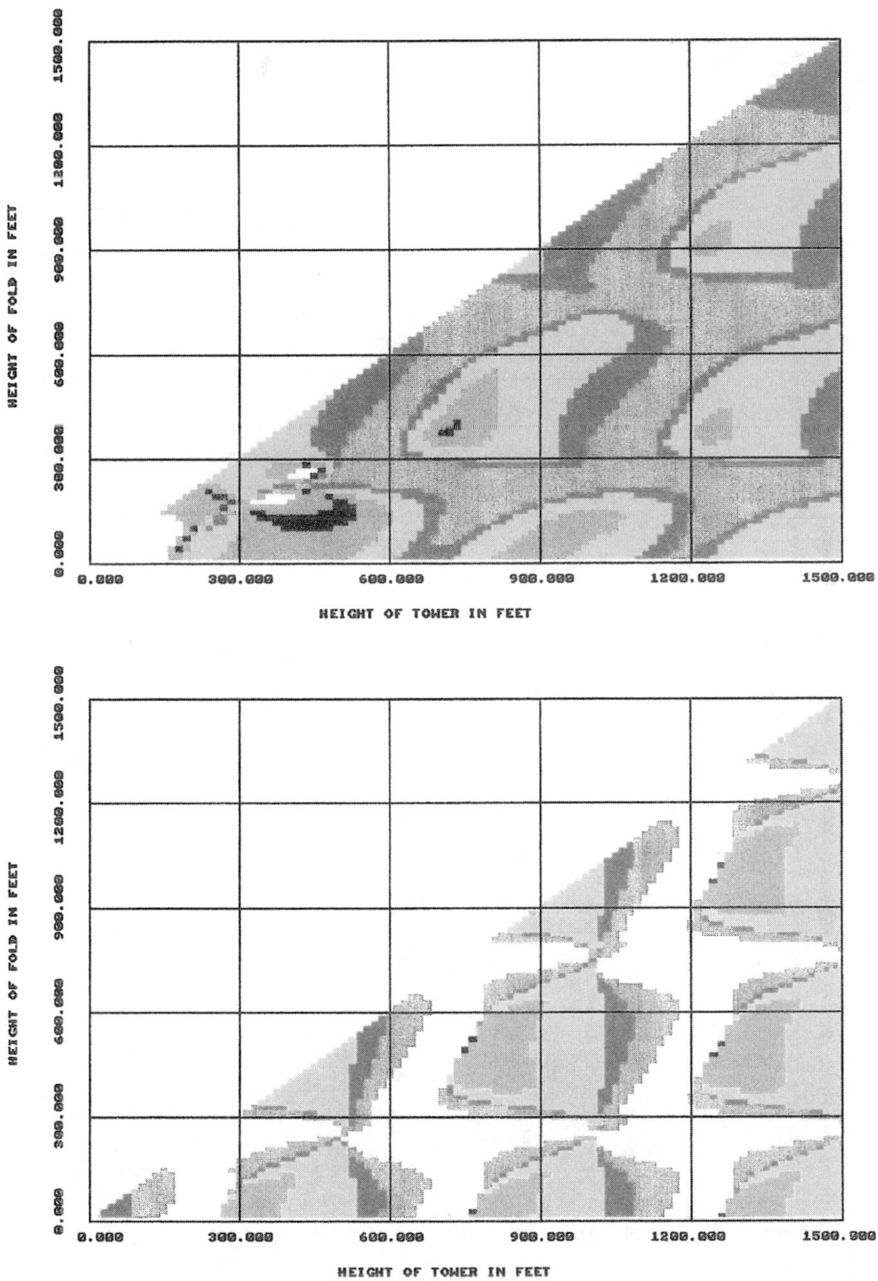

Figure 3-11 Input resistance (upper plot) and reactance (lower plot) for the GMA when antenna height and fold height are simultaneously and continuously varied. The difference from the plots in Fig. 3-10 result from the base terminating reactance X_B being changed from 0 to 3,000 ohms. The upper triangular regions are blank because the fold height cannot be greater than the antenna height.

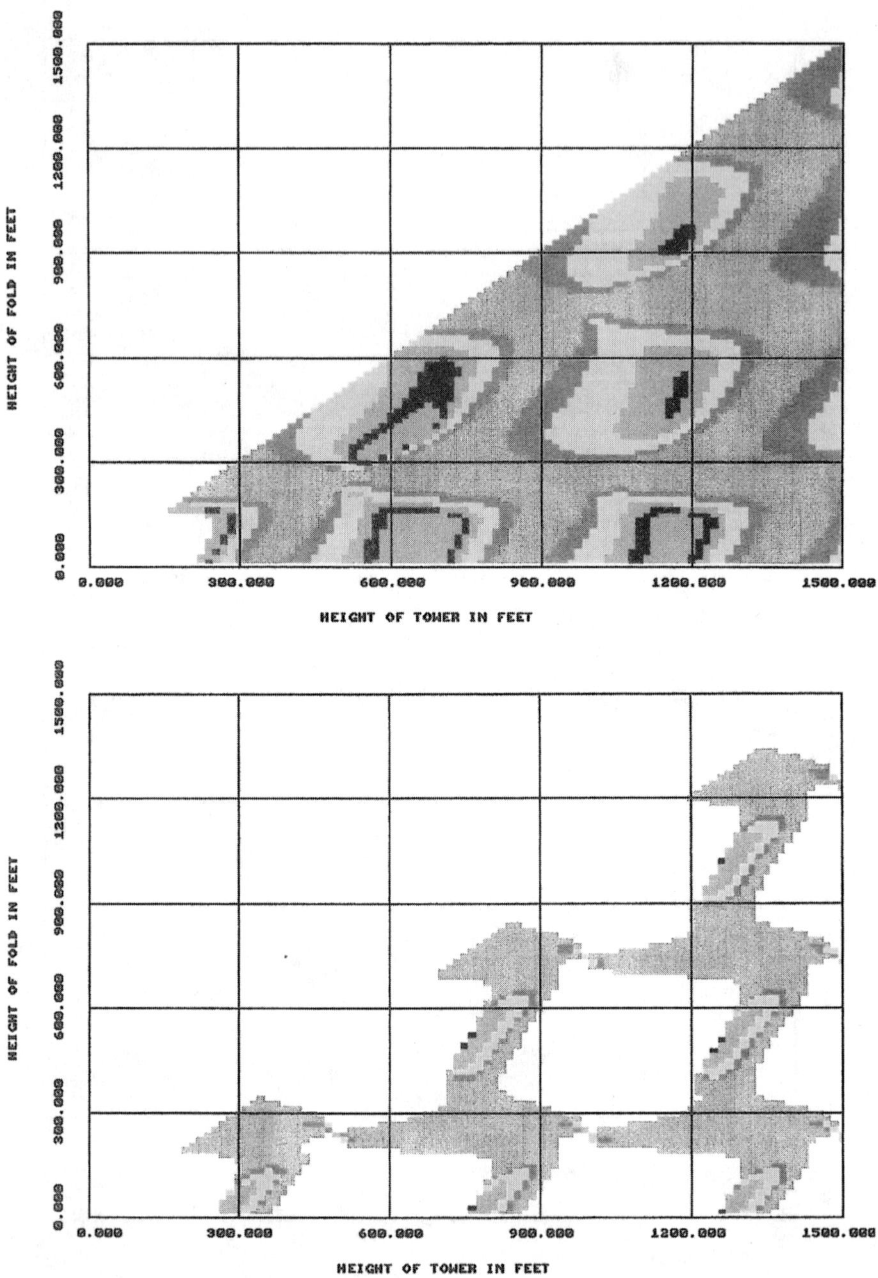

Figure 3-12 Input resistance (upper plot) and reactance (lower plot) for the GMA when antenna height and fold height are simultaneously and continuously varied, with a base terminating reactance of –3,000 ohms.

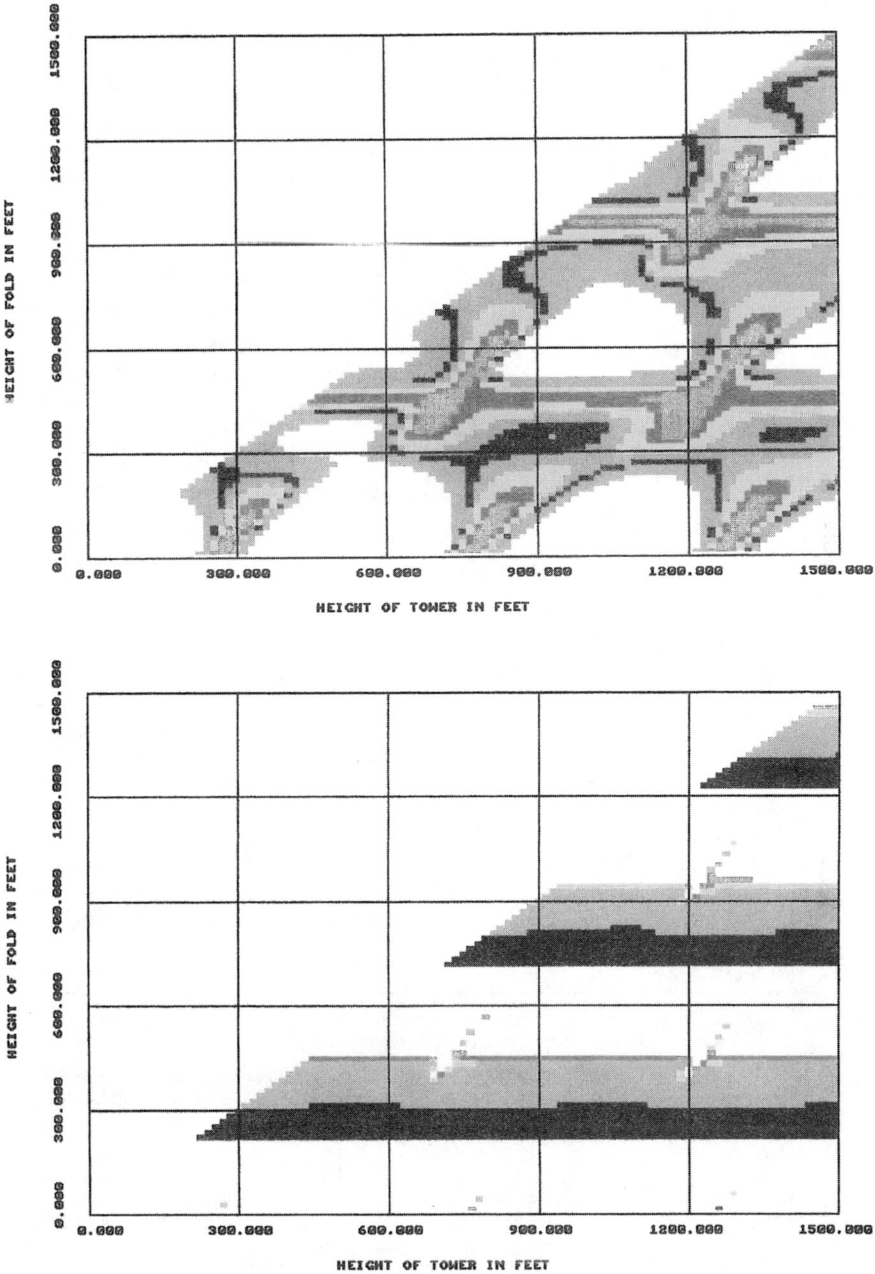

Figure 3-13 Input resistance (upper plot) and reactance (lower plot) for the GMA when antenna height and fold height are simultaneously and continuously varied, with a fold terminating reactance X_1 of –600 ohms. The base terminating reactance has been reset to zero.

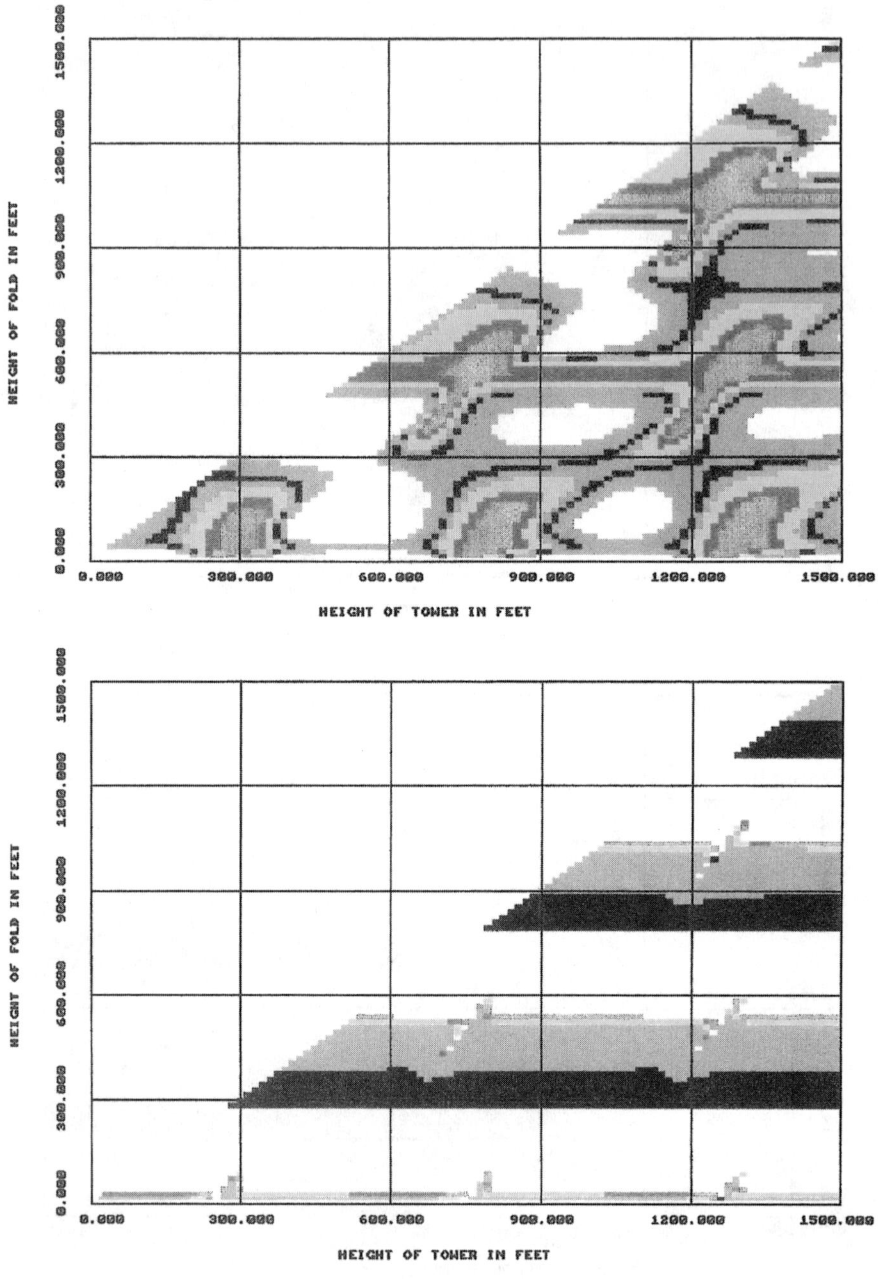

Figure 3-14 Input resistance (upper plot) and reactance (lower plot) for the GMA when antenna height and fold height are simultaneously and continuously varied, with a fold terminating reactance of 600 ohms.

we will simultaneously vary both fold height h_1 and one of the terminating reactances X_B or X_T.

Figures 3-15 and 3-16 are the contour plot versions of Fig. 3-4, for a fixed height $h_2 = 300$ feet. In Fig. 3-15 the base reactance is variable, and in Fig. 3-16 the fold reactance is variable. It is seen that there are many combinations that provide optimal input resistance (shown in dark green), and some of these simultaneously provide the most desirable range of input reactance (shown in dark cyan.) The lower left quadrant of Fig. 3-16 shows an especially broad range of useful combinations.

Figures 3-17 and 3-18 are the contour plot versions of Fig. 3-5, for a fixed height of $h_2 = 375$ feet. As in the above example, the lower left quadrant of Fig. 3-18 shows an especially broad range of desirable combinations of h_2 and X_T.

Figures 3-19 and 3-20 correspond to Fig. 3-6, for a fixed height of $h_2 = 500$ feet, or a one-half wavelength tower. A one-half wavelength, or 180 electrical degree, monopole is notoriously difficult to impedance match because its input resistance is on the order of 1000 ohms. Further, that large resistance is often unstable and fluctuates with nearby environmental changes. In contrast to the monopole, it is seen that the GMA configuration offers many combinations with optimal resistance (shown in dark green) simultaneously with either optimal reactance (dark cyan) or near-optimal reactance (cyan).

Figures 3-21 and 3-22 correspond to Fig. 3-7, for a fixed height of $h_2 = 625$ feet. By now, we should not be surprised to observe many combinations that provide optimal resistance and reactance. The upper center part of Fig. 3-22 is especially rich in such combinations.

Figures 3-23 and 3-24 correspond to Fig. 3-8, for a fixed height of $h_2 = 750$ feet. For this height, we are beginning to see repetitive patterns in the contours. This should not be surprising for heights over one-half wavelength. Recall that transmission line impedances are periodic, and that the differential mode currents on the GMA are, in fact, transmission line mode currents.

Figures 3-25 and 3-26 definitely exhibit periodic patterns. They correspond to Fig. 3-9, for a fixed height of $h_2 = 1{,}500$ feet. The usual observation applies, namely, many combinations of fold height and terminating reactance result in optimal input impedance.

Quite clearly, the investigation pursued in this section has been extremely worthwhile. Our hunch was confirmed. For any given height tower, a wide range of combinations of fold height and terminating reactance readily produce optimal input impedance for the GMA.

There is yet another productive way to look at the problem, and that is the subject of the next section.

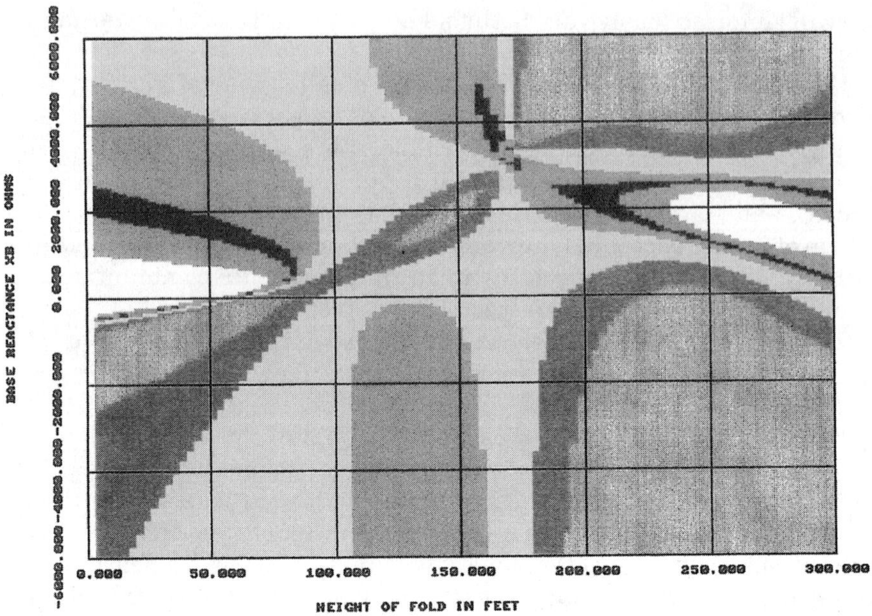

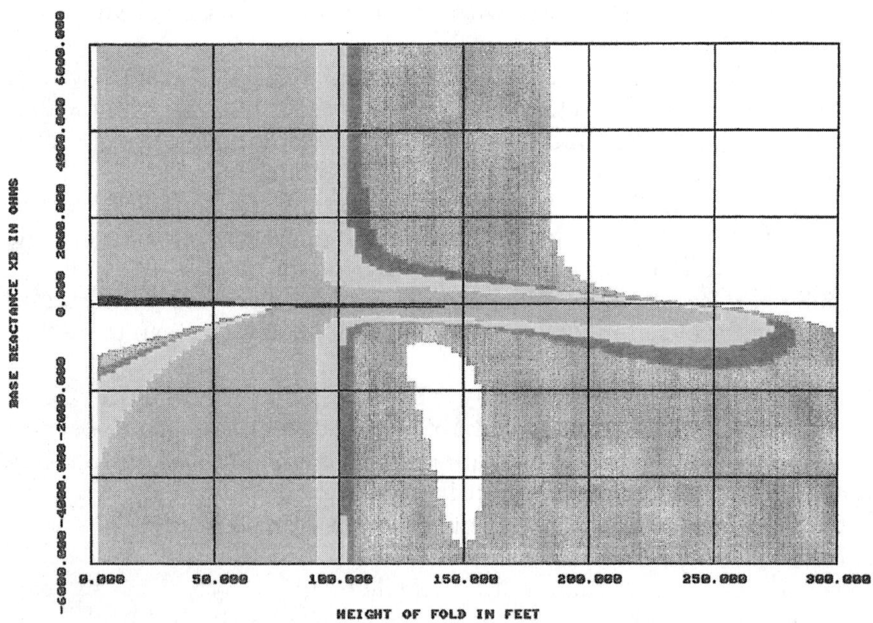

Figure 3-15 Input resistance (upper plot) and reactance (lower plot) for a 300-foot GMA. At 984 kHz, it is 0.3 wavelength long, or 108 electrical degrees. The two continuous variables are fold height h_1 and terminating base reactance X_B.

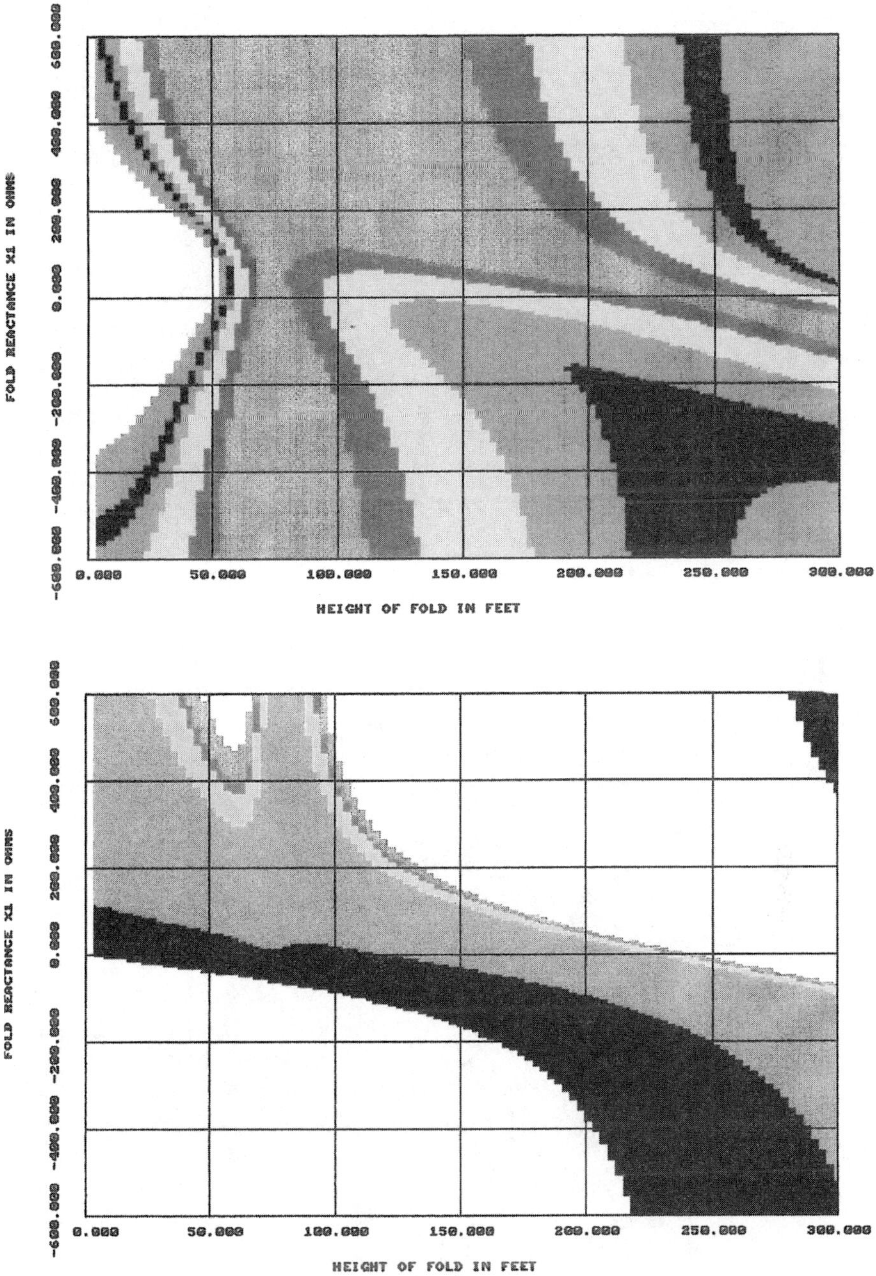

Figure 3-16 Input resistance (upper plot) and reactance (lower plot) for a 300-foot GMA. The two continuous variables are fold height h_1 and terminating fold reactance X_1.

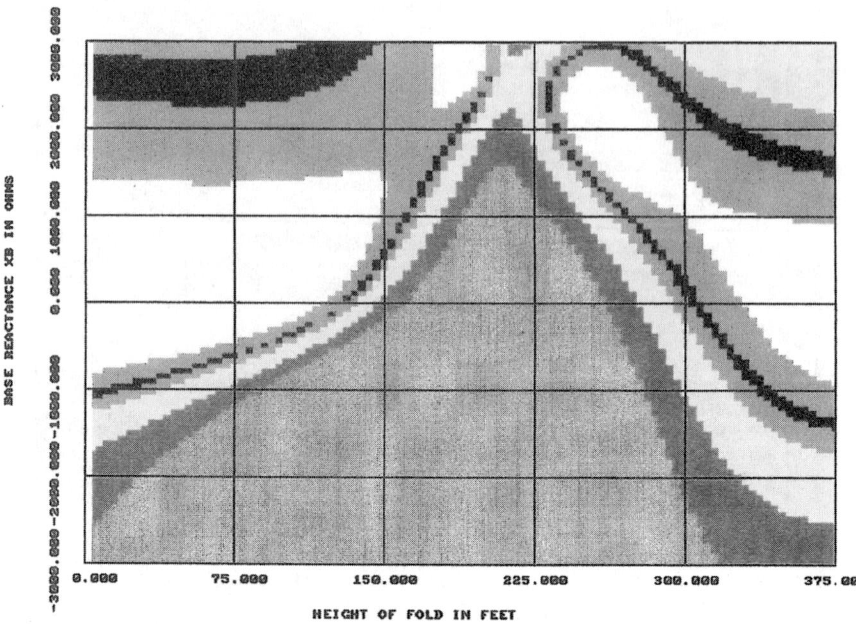

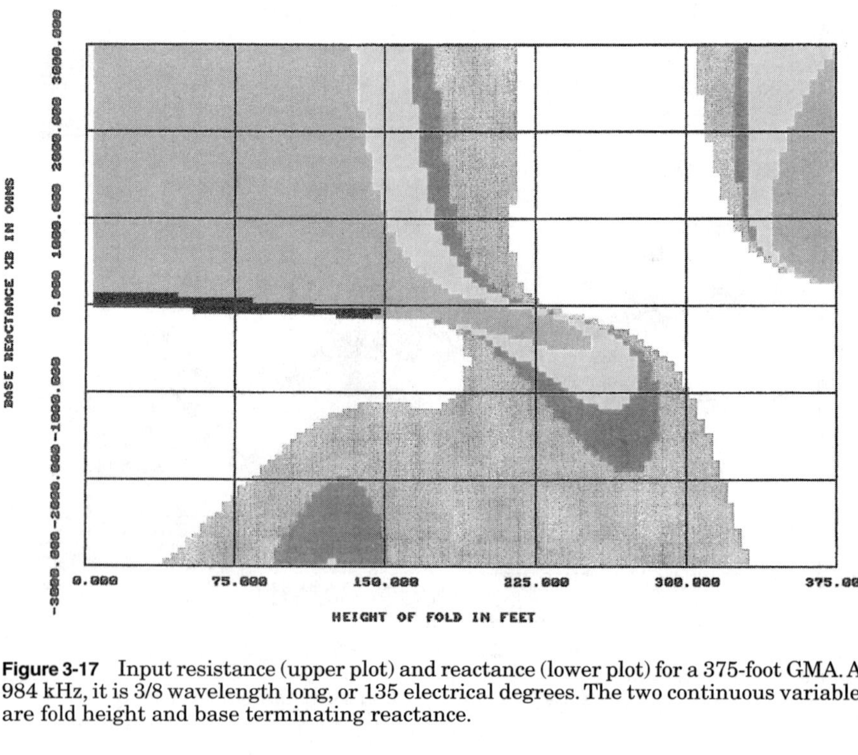

Figure 3-17 Input resistance (upper plot) and reactance (lower plot) for a 375-foot GMA. At 984 kHz, it is 3/8 wavelength long, or 135 electrical degrees. The two continuous variables are fold height and base terminating reactance.

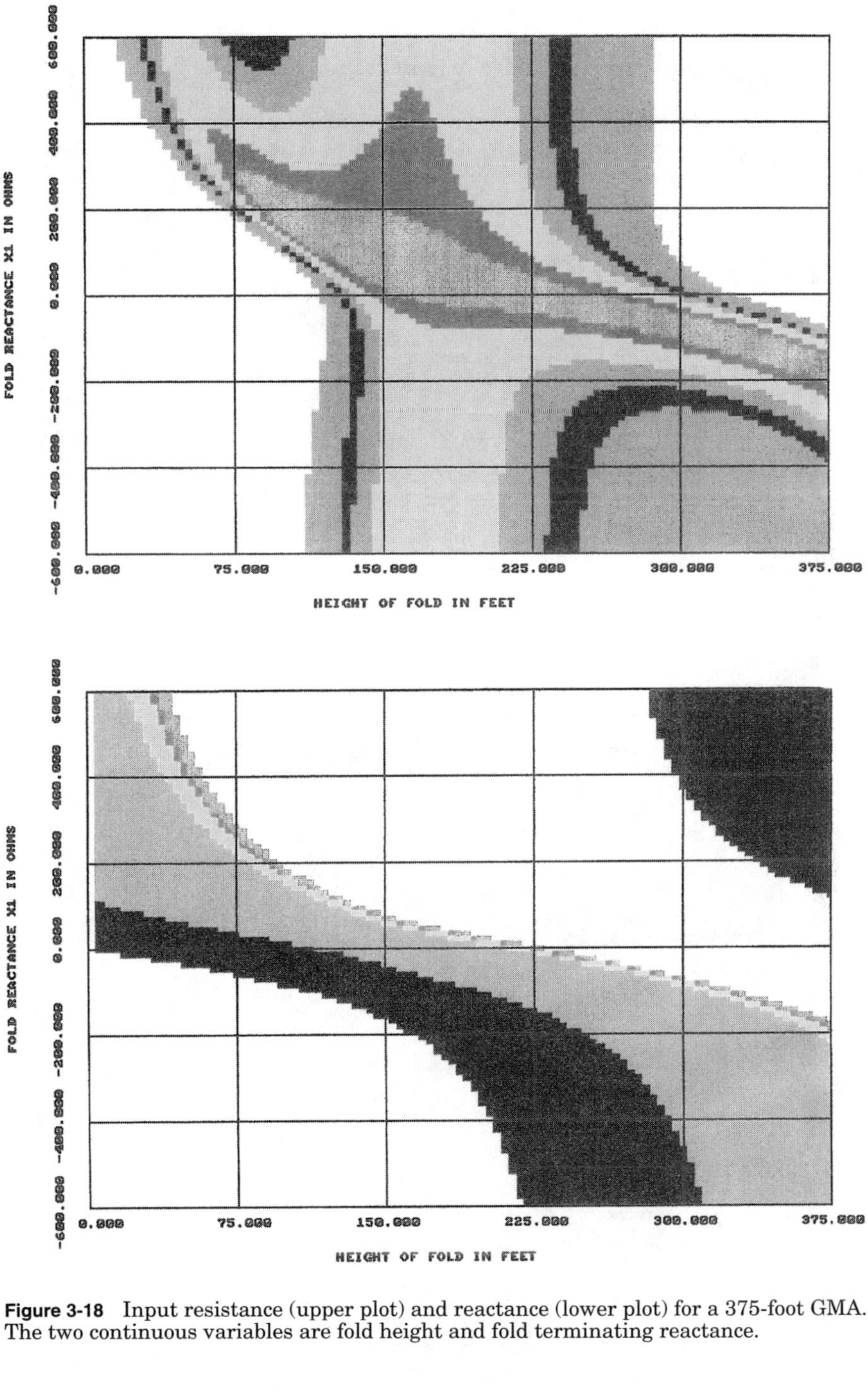

Figure 3-18 Input resistance (upper plot) and reactance (lower plot) for a 375-foot GMA. The two continuous variables are fold height and fold terminating reactance.

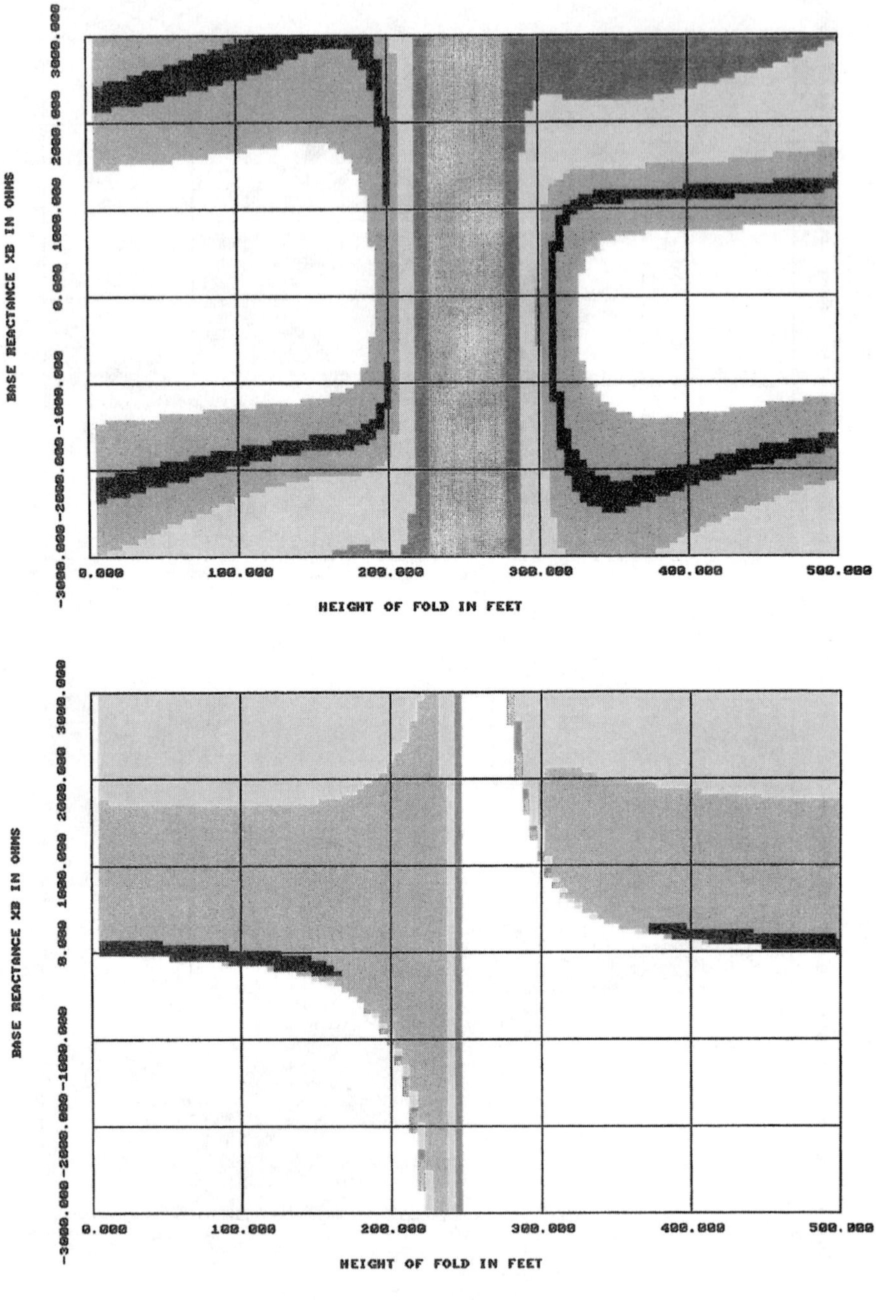

Figure 3-19 Input resistance (upper plot) and reactance (lower plot) for a 500-foot GMA. At 984 kHz, it is one-half wavelength long, or 180 electrical degrees. The two continuous variables are fold height and base terminating reactance.

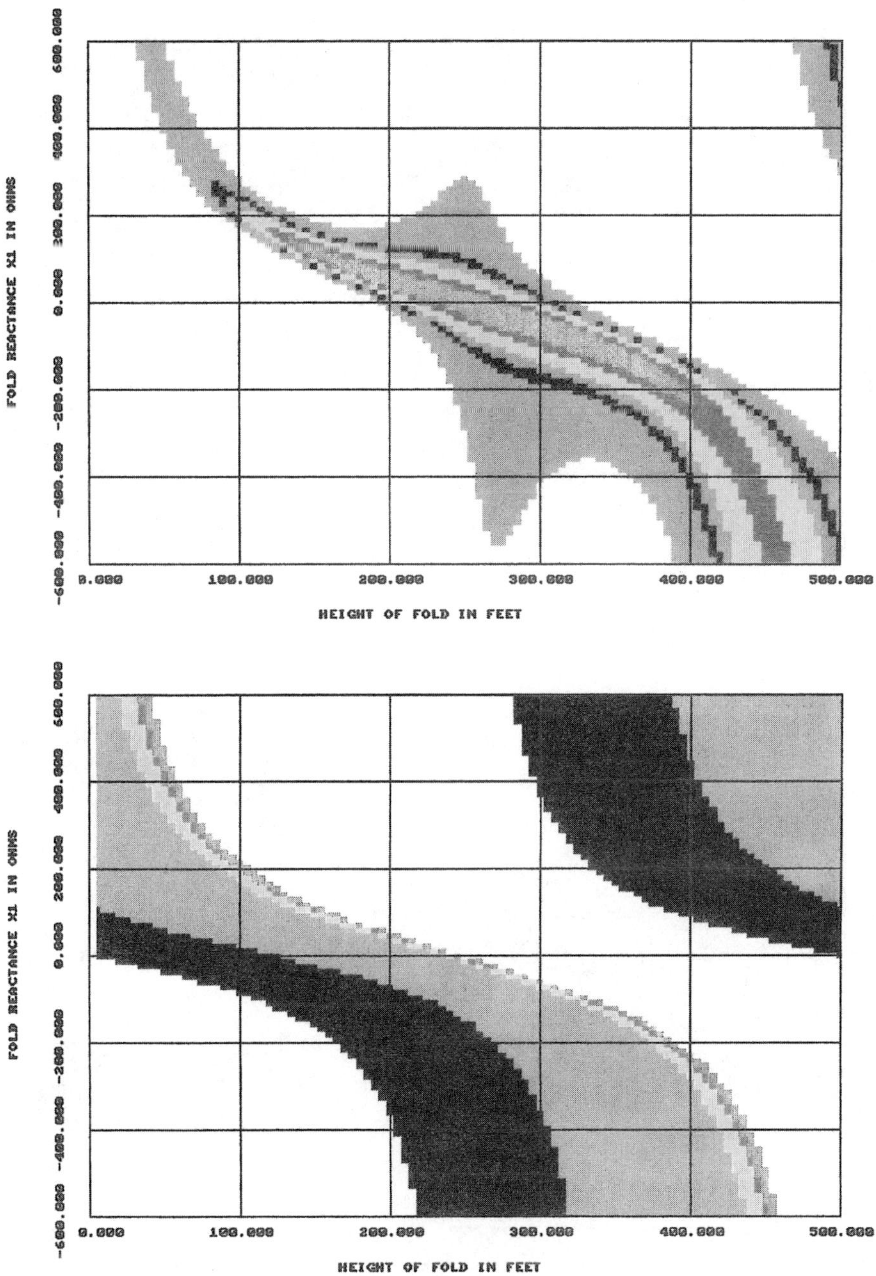

Figure 3-20 Input resistance (upper plot) and reactance (lower plot) for a 500-foot GMA. The two continuous variables are fold height and fold terminating reactance.

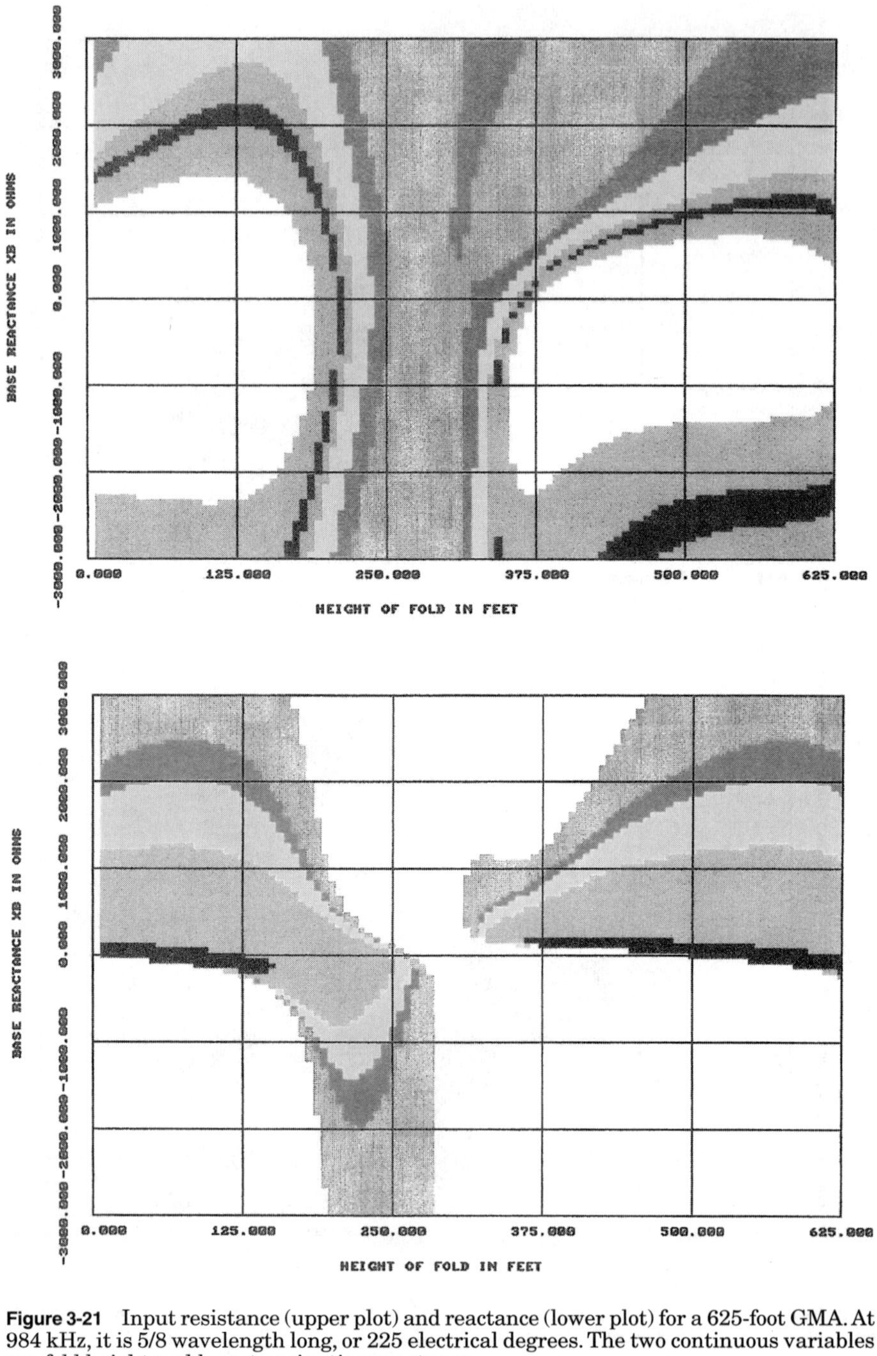

Figure 3-21 Input resistance (upper plot) and reactance (lower plot) for a 625-foot GMA. At 984 kHz, it is 5/8 wavelength long, or 225 electrical degrees. The two continuous variables are fold height and base terminating reactance.

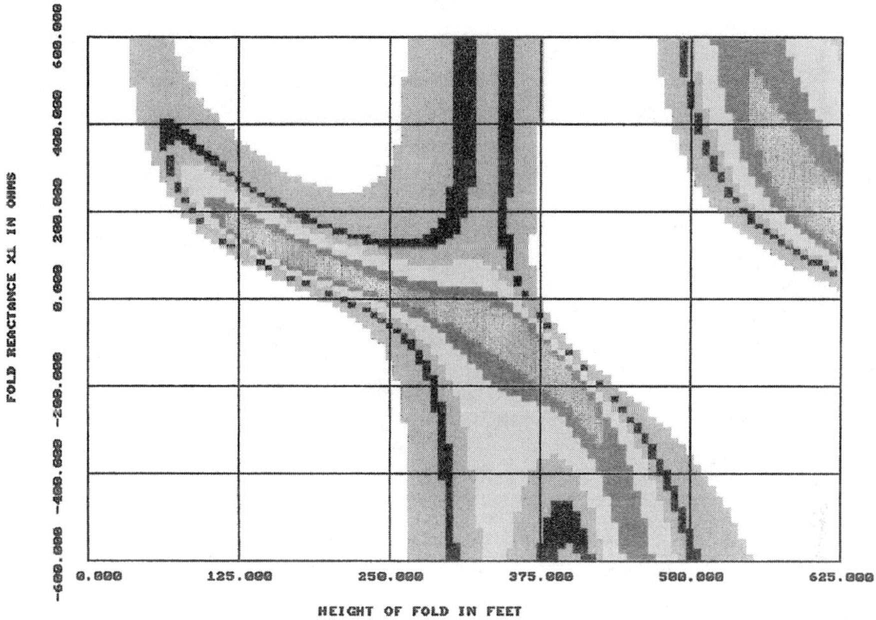

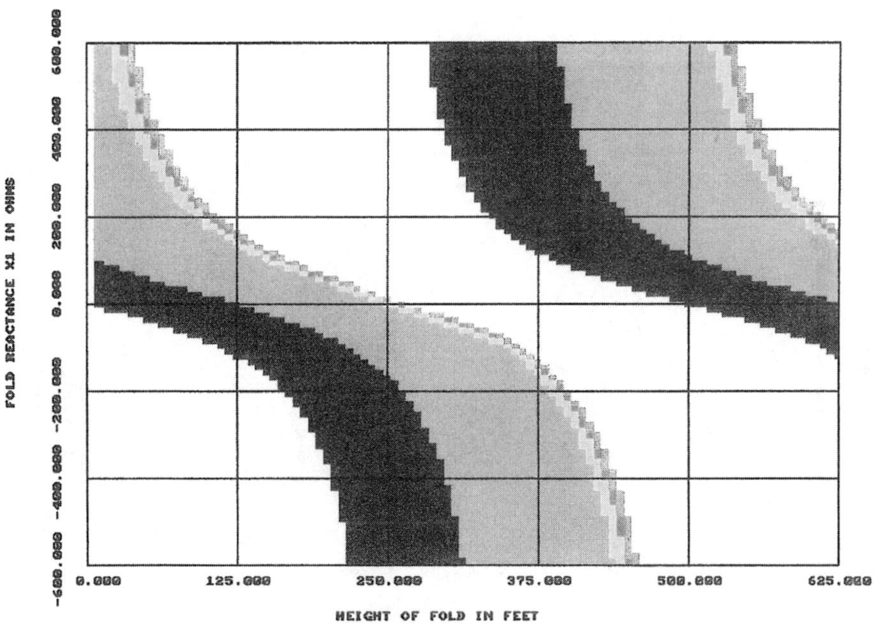

Figure 3-22 Input resistance (upper plot) and reactance (lower plot) for a 625-foot GMA. The two continuous variables are fold height and fold terminating reactance.

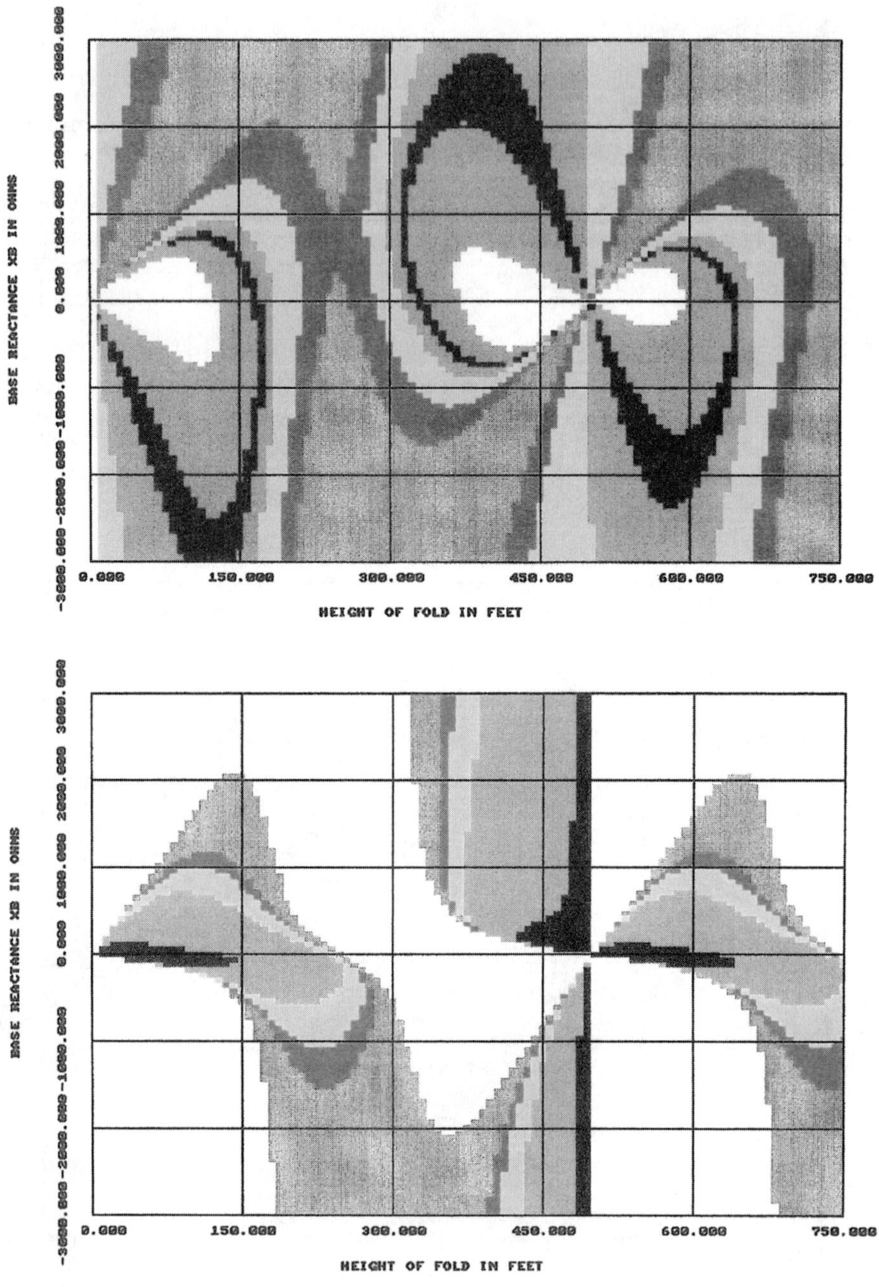

Figure 3-23 Input resistance (upper plot) and reactance (lower plot) for a 750-foot GMA. At 984 kHz, it is three-quarters wavelength long, or 270 electrical degrees. The two continuous variables are fold height and terminating base reactance.

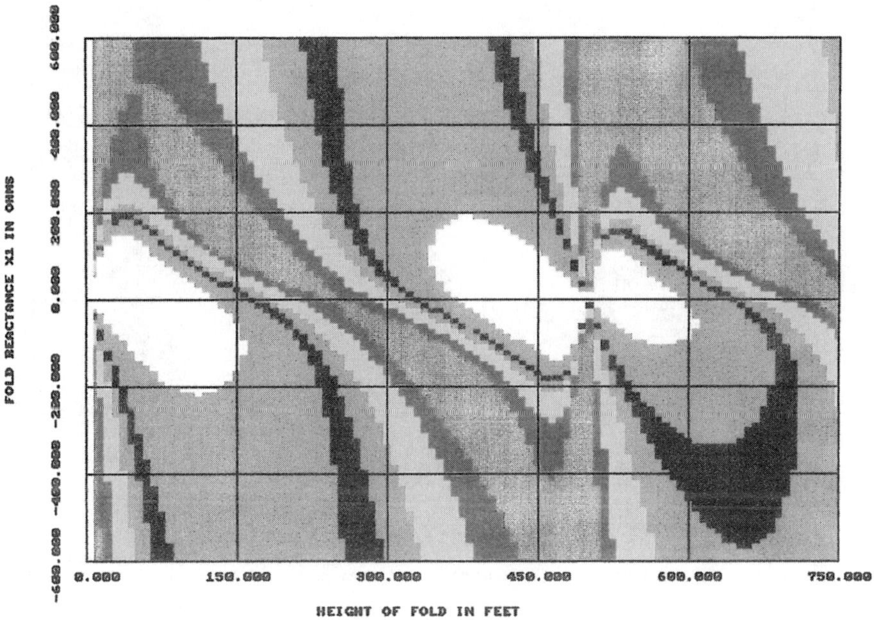

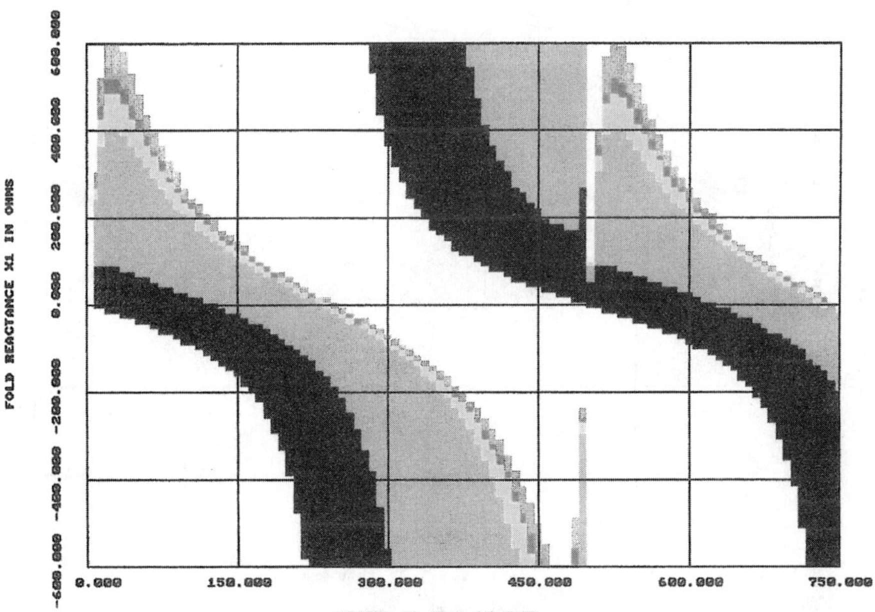

Figure 3-24 Input resistance (upper plot) and reactance (lower plot) for a 750-foot GMA. The two continuous variables are fold height and terminating fold reactance.

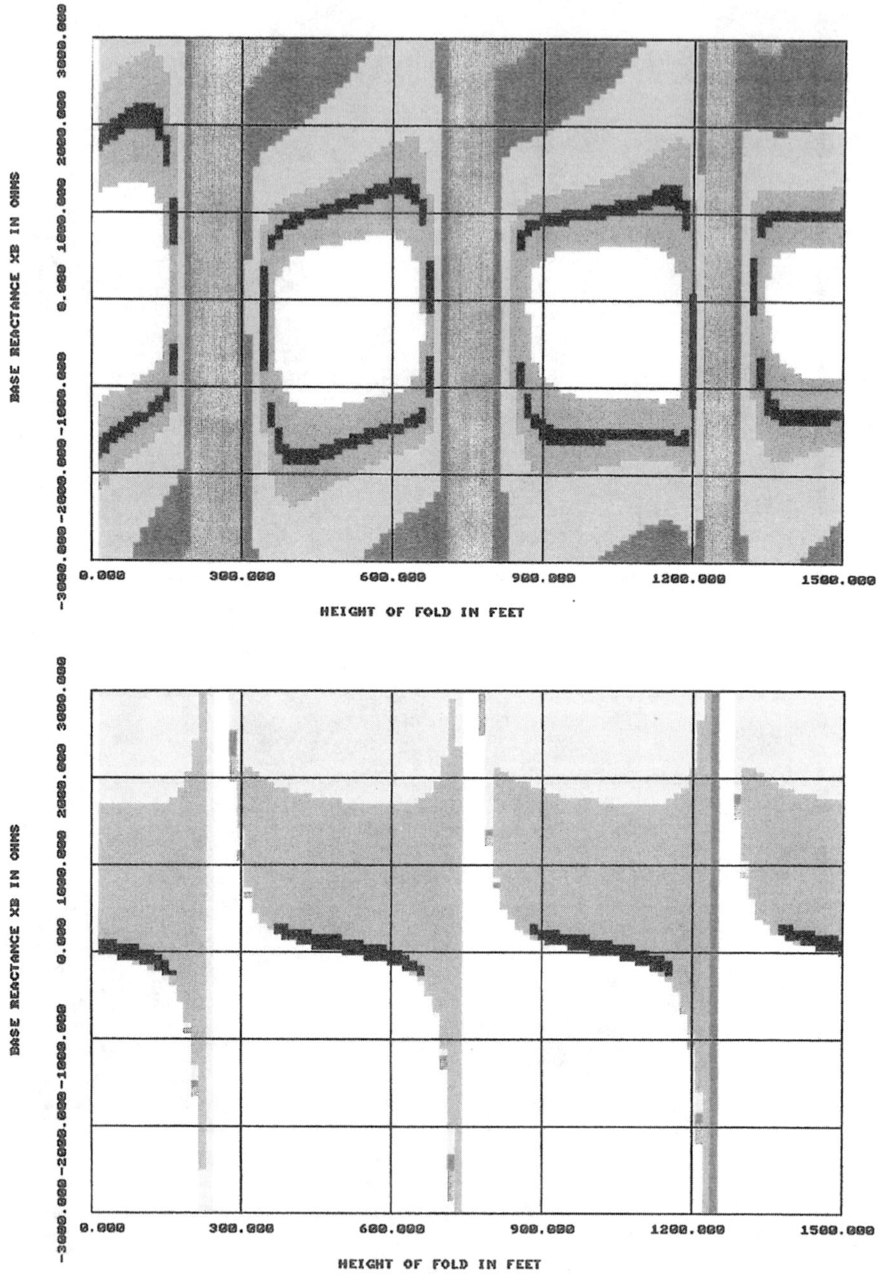

Figure 3-25 Input resistance (upper plot) and reactance (lower plot) for a 1,500-foot GMA. At 984 kHz, it is 3/2 wavelength long, or 540 electrical degrees. The two continuous variables are fold height and terminating base reactance.

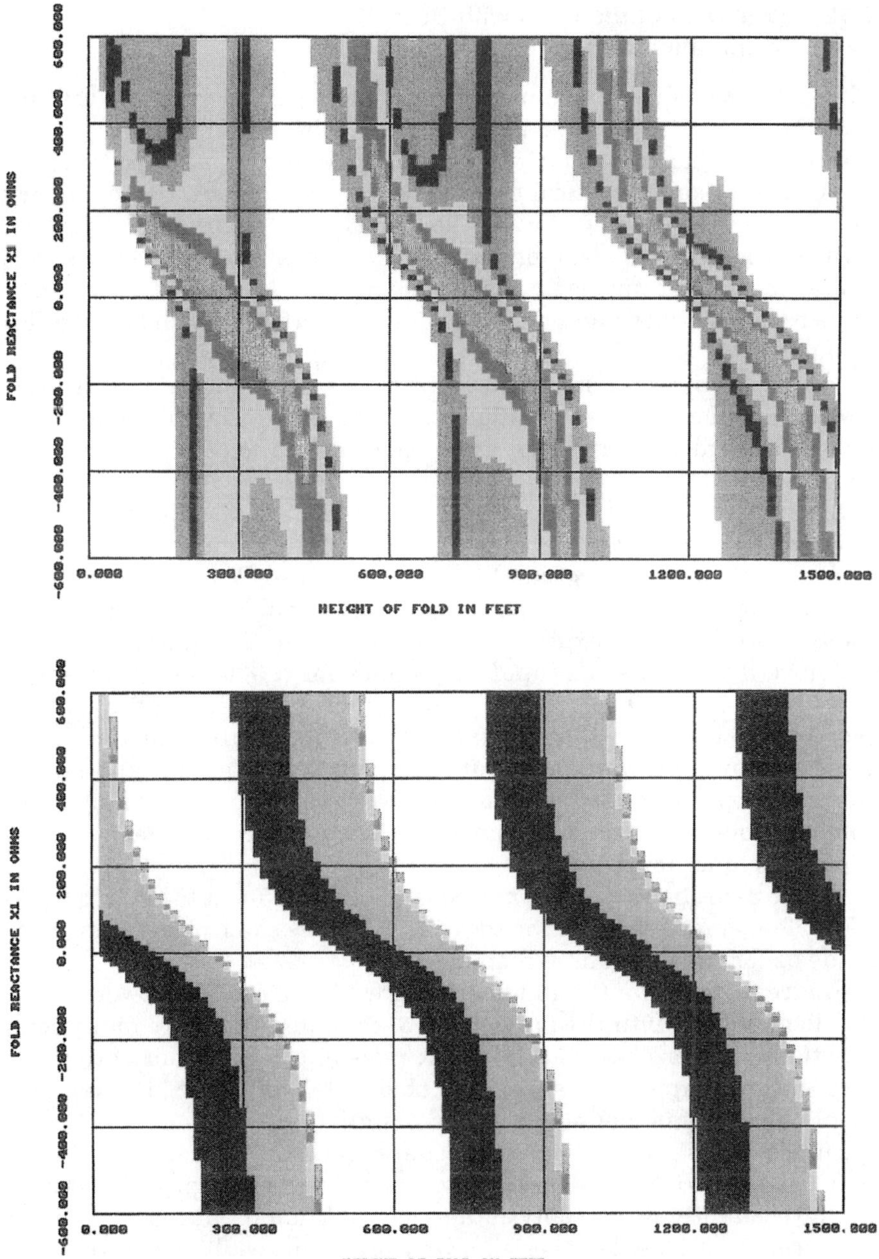

Figure 3-26 Input resistance (upper plot) and reactance (lower plot) for a 1,500-foot GMA. The two continuous variables are fold height and terminating fold reactance.

3.12 Examples of the GMA with Both Terminating Reactances Variable

The examples of the previous section are compelling evidence that for any tower height h_2, we can always find a combination of fold height and terminating reactance that will provide optimum input impedance for the GMA. In some practical cases, however, we do not have the luxury of continuously varying the fold height h_1. For AM broadcast towers, parts of the tower may double as vertical real estate for antennas that operate at shorter wavelengths. The fold may not be permitted to extend anywhere near these antennas for fear of interfering with their radiation patterns.

The restriction on fold height is not necessarily limited to large AM towers. For the tiny GMA configurations on printed circuit boards, the crowded board topology may simply not allow a fold to extend into a certain region.

In any case, for certain applications at both long and short wavelengths, both the tower height h_2 and fold height h_1 must remain fixed. Can we still obtain an optimal input impedance? With the two terminating reactances X_B and X_T as independent variables, the answer most likely is, 'yes'. We will explore that expectation in this section.

Figure 3-27 shows the input impedance for a 300-foot tower and a short fold of only 50 feet. The two continuous variables on the contour plot are the terminating reactances. It is seen that there are combinations that provide optimal input resistance (dark green) and simultaneously optimal input reactance (dark cyan). There are many more combinations that provide optimal input resistance and near-optimal input reactance (cyan).

Figure 3-28 shows the input impedance for a 375-foot tower and a fold of 125 feet, or one-third of the tower height. The same observations are made as for the previous example.

Figure 3-29 shows the input impedance for a 750-foot tower with a 250-foot fold. As with the previous two examples, there are many combinations that provide optimal input resistance and either optimal or near-optimal input reactance. This obviously will reduce the requirements for construction material and installation effort.

Figure 3-30 shows the input impedance for a 1,500-foot tower with a 500-foot fold. While there are many combinations that provide optimal input resistance, we can only obtain near-optimal input reactance with those combinations. This is still a very practical result, however.

The above examples are not exhaustive; however, they do cover a wide range of possibilities. It is a very good bet that with so many independent design variables at our disposal, we can always impedance match a GMA, even though the linear dimensions are fixed to satisfy other constraints.

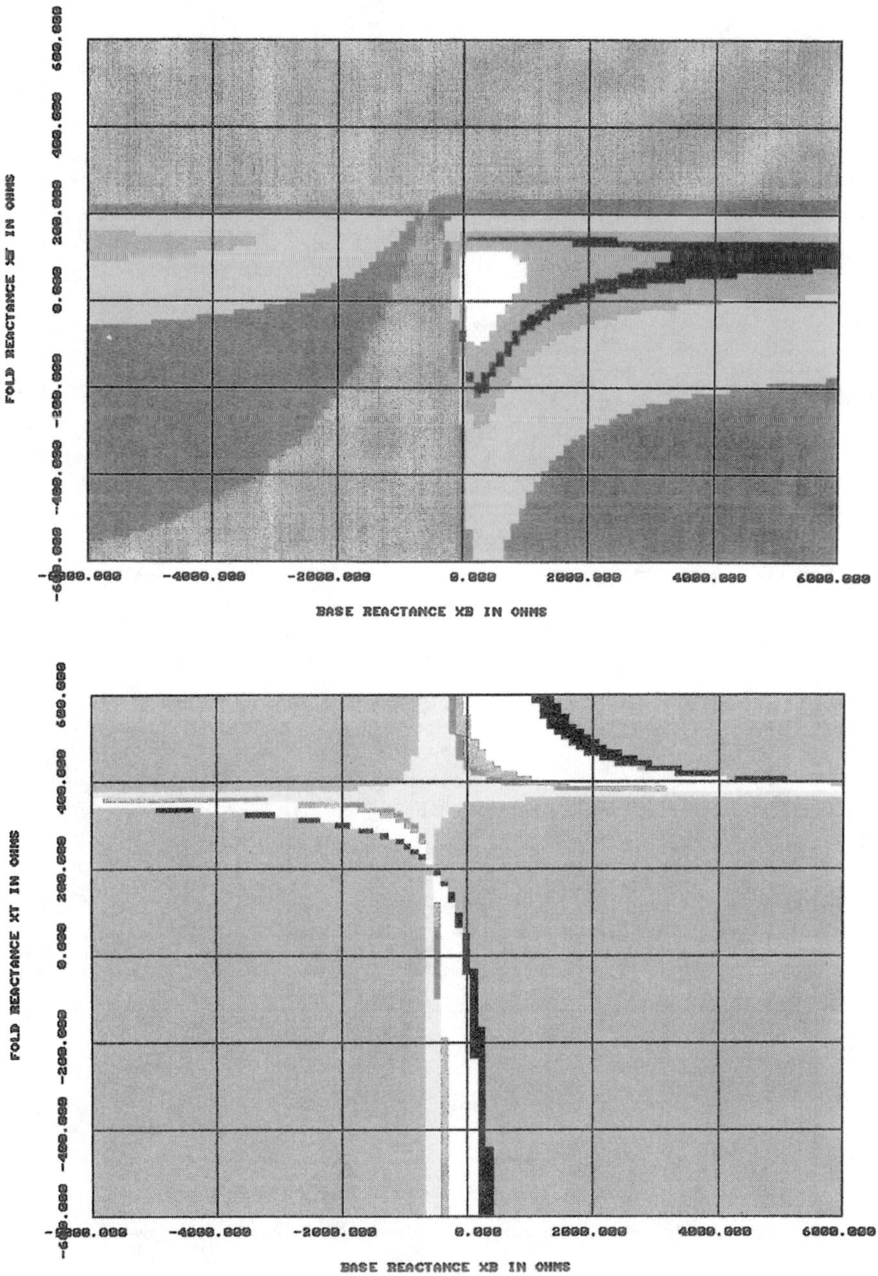

Figure 3-27 Input resistance (upper plot) and reactance (lower plot) for a 300-foot GMA with a 50-foot fold. At 984 kHz, the taller conductor is 108 electrical degrees, and the fold is 18 electrical degrees. The two continuous variables are base terminating reactance X_B and fold terminating reactance X_1.

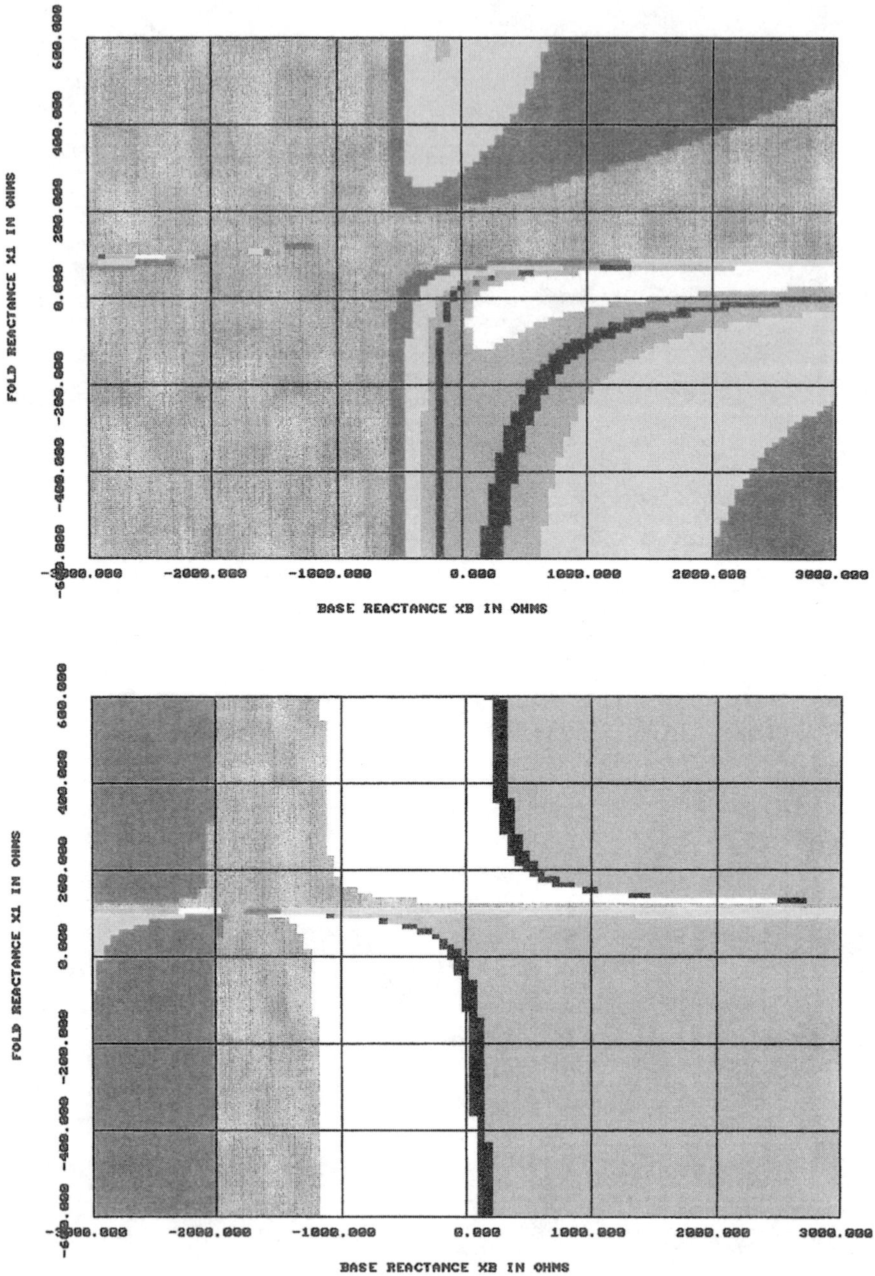

Figure 3-28 Input resistance (upper plot) and reactance (lower plot) for a 375-foot GMA with a 125-foot fold. At 984 kHz, the taller conductor is 135 electrical degrees, and the fold is 45 electrical degrees. The two continuous variables are base terminating reactance and fold terminating reactance.

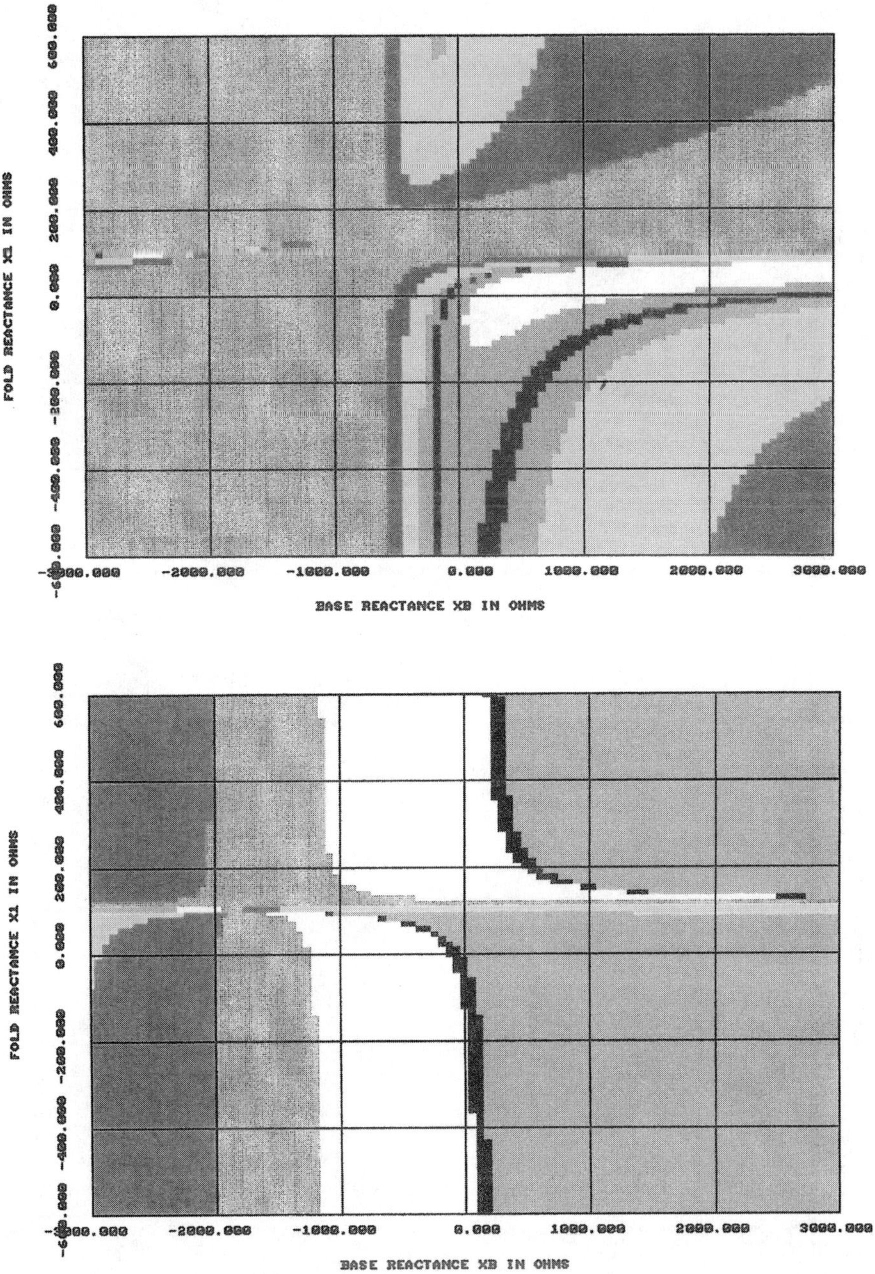

Figure 3-29 Input resistance (upper plot) and reactance (lower plot) for a 750-foot GMA with a 250-foot fold. At 984 kHz, the taller conductor is 270 electrical degrees, and the fold is 90 electrical degrees. The two continuous variables are base terminating reactance and fold terminating reactance.

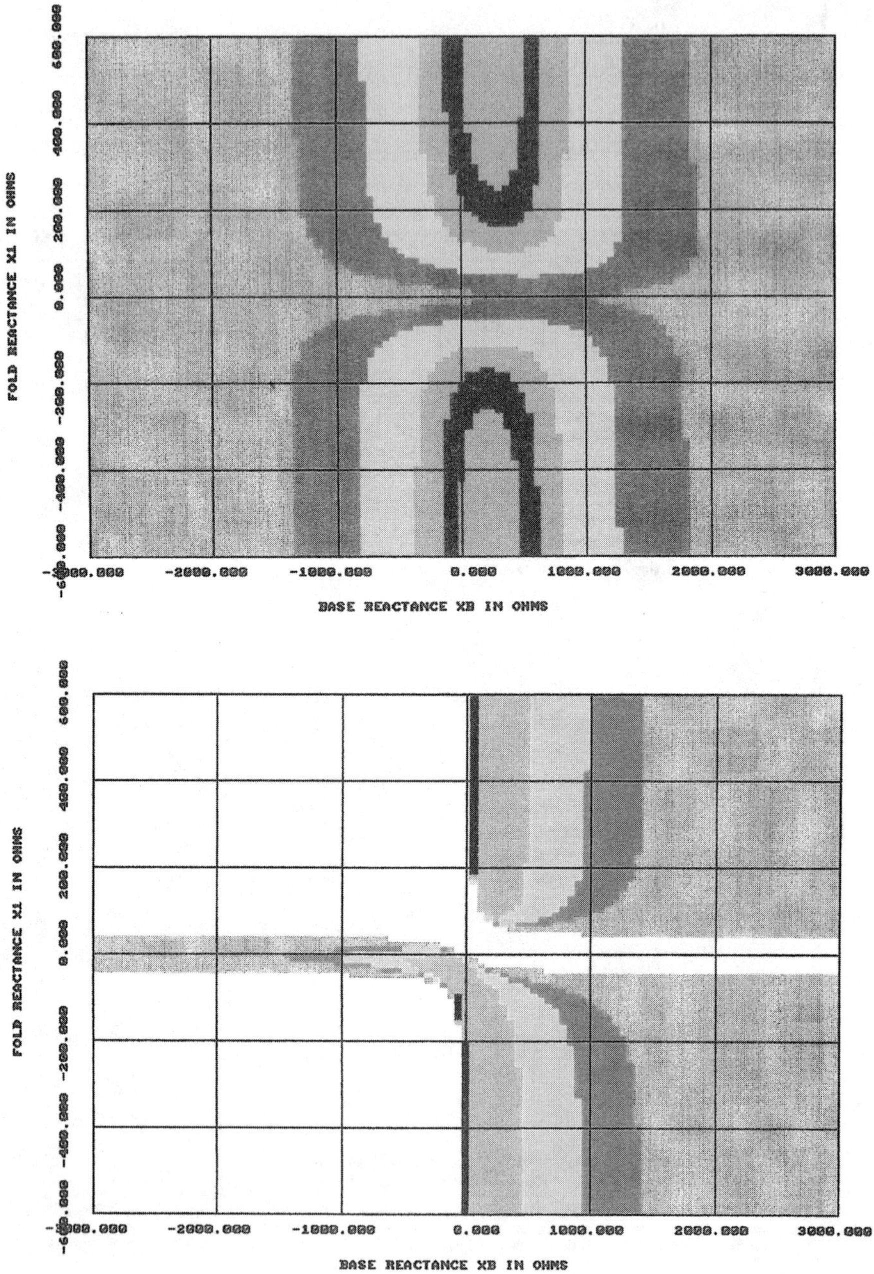

Figure 3-30 Input resistance (upper plot) and reactance (lower plot) for a 1,500-foot GMA with a 500-foot fold. At 984 kHz, the taller conductor is 540 electrical degrees, and the fold is 180 electrical degrees. The two continuous variables are base terminating reactance and fold terminating reactance.

3.13 Equations at the Top of the 2SFU

As has already been discussed, the GMA is just a special case of the 2SFU. The latter provides three more independent design variables. On the other hand, the examples of the previous section show that a good impedance match can always be obtained with a GMA. What more could the 2SFU offer? Perhaps extending our mathematical model of the GMA to include the 2SFU will offer some surprising answers.

Extending the model is a straightforward process. We will rely upon the same underlying physics and boundary conditions that we have been developing methodically beginning with Chap. 2. To that end, Fig. 3-31 shows the configuration of interest with currents and voltages. In contrast to the GMA, it is seen that there are nonzero differential mode currents and voltages on the upper segment of the 2SFU. Also, there is a current transformation ratio ν_2 that, in general, is different from ν_1.

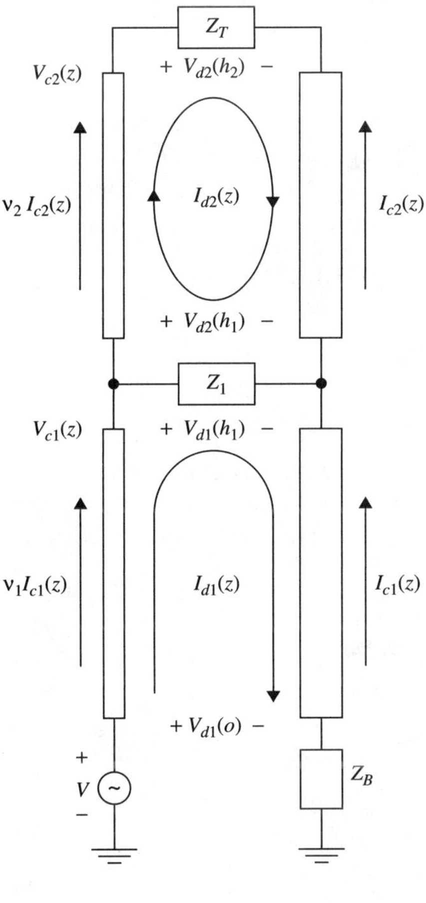

Figure 3-31 Voltages and currents used for analysis of the 2-segment folded unipole antenna (2SFU).

From the figure, we evidently have eight unknown current amplitudes to determine. They are the forward and reflected current amplitudes, for both common and differential modes, and on both upper and lower antenna segments. To determine them, we will need eight independent equations.

Two of the necessary equations follow from boundary conditions at the top of the 2SFU. First, the common mode current must vanish there because it has nowhere to flow:

$$I_{c2}(h_2)=0 \tag{3-27}$$

Second, the top terminating impedance constrains the differential mode voltage and current, according to the formula,

$$-I_{d2}(h_2)Z_T+V_{d2}(h_2)=0 \tag{3-28}$$

We still need six more independent equations.

3.14 Equations at the Base of the 2SFU

After the experience of Sec. 2.7 and 3.4, we should expect two more independent equations from boundary conditions at the base of the antenna. That is indeed the case.

KVL requires that:

$$Z_B[I_{d1}(0)-I_{c1}(0)]+V_{d1}(0)=V \tag{3-29}$$

By simple inspection at the fold,

$$V_{c1}(0)=V \tag{3-30}$$

With Eqs. 3-27 through 3-30, we have accumulated half of the equations necessary to determine the unknown current amplitudes. The remaining four equations will follow from constraints at the boundary between the upper and lower segments.

3.15 Equations at the Boundary Between the Segments of the 2SFU

Both differential and common mode voltages must be continuous across the boundary between the upper and lower segments:

$$V_{d2}(h_1)=V_{d1}(h_1) \tag{3-31}$$

$$V_{c1}(h_1)=V_{c2}(h_1) \tag{3-32}$$

KCL requires that the common mode current also be continuous across the boundary:

$$\nu_1 I_{c1}(h_1) + I_{c1}(h_1) = \nu_2 I_{c2}(h_1) + I_{c2}(h_1) \tag{3-33}$$

Finally, the fold impedance Z_1 constrains the differential mode voltages and currents according to the formula,

$$V_{d1}(h_1) - Z_1[I_{d1}(h_1) - I_{d2}(h_1)] = 0 \tag{3-34}$$

Equations 3-27 through 3-34 are the eight independent equations we need to determine the unknown, complex amplitudes of the common and differential mode currents flowing everywhere on the 2SFU.

3.16 Matrix Equation for the Currents on the 2SFU

In identical fashion to that used in Sec. 2.8 and 3.6, we can consolidate Eqs. 3-27 through 3-34 into a single matrix equation for the unknown current amplitudes. The matrix equation is:

$$\mathbf{M} \cdot \mathbf{I} = \begin{pmatrix} V \\ V \\ 0 \\ 0 \\ 0 \\ 0 \\ 0 \\ 0 \end{pmatrix} \tag{3-35}$$

As discussed in Sec. 3.6, as the matrices become larger, it is necessary to partition them so that the individual elements are not so small that they are unreadable on the printed page. In the case of Eq. 3-35, this is also true of the current vector **I**:

$$\mathbf{I} = \begin{pmatrix} \mathbf{I}_1 \\ \mathbf{I}_2 \end{pmatrix} \tag{3-36}$$

In Eq. 3-36, the subscript denotes the segment of the 2SFU, and the current vectors for the individual segments may be expressed by the general formula,

$$\mathbf{I_n} = \begin{pmatrix} I_{\mathrm{Fcn}} \\ I_{\mathrm{Rcn}} \\ I_{\mathrm{Fdn}} \\ I_{\mathrm{Rdn}} \end{pmatrix} \qquad n = 1,2,\ldots N \tag{3-37}$$

For the 2SFU, $N = 2$ in Eq. 3-37; however, the equation generally applies to folded unipole antennas with any number of segments N.

In Eq. 3-34, the partitioned matrix M has the very same form as Eq. 3-17 in Sec. 3.6. In Eq. 3-34, however, the submatrices have dimensions 4×4. They are:

$$\mathbf{A}_{11} = \begin{pmatrix} -Z_B & -Z_B & Z_B + Z_{0d1} & Z_B - Z_{0d1} \\ Z_{0c1} & -Z_{0c1} & 0 & 0 \\ 0 & 0 & 0 & 0 \\ 0 & 0 & 0 & 0 \end{pmatrix} \tag{3-38}$$

$$\mathbf{A}_{12} = \begin{pmatrix} 0 & 0 & 0 & 0 \\ 0 & 0 & 0 & 0 \\ \psi_F(h_2) & \psi_R(h_2) & 0 & 0 \\ 0 & 0 & (Z_T - Z_{0d2})\psi_F(h_2) & (Z_T + Z_{0d2})\psi_R(h_2) \end{pmatrix} \tag{3-39}$$

$$\mathbf{A}_{21} = \begin{pmatrix} 0 & 0 & -Z_{0d1}\psi_F(h_1) & Z_{0d1}\psi_R(h_1) \\ 0 & 0 & (Z_{0d1} - Z_1)\psi_F(h_1) & (-Z_{0d1} - Z_1)\psi_R(h_1) \\ (1+\upsilon_1)\psi_F(h_1) & (1+\upsilon_1)\psi_R(h_1) & 0 & 0 \\ Z_{0c1}\psi_F(h_1) & -Z_{0c1}\psi_R(h_1) & 0 & 0 \end{pmatrix} \tag{3-40}$$

$$\mathbf{A}_{22} = \begin{pmatrix} 0 & 0 & Z_{0d2}\psi_F(h_1) & -Z_{0d2}\psi_R(h_1) \\ 0 & 0 & Z_1\psi_F(h_1) & Z_1\psi_R(h_1) \\ -(1+\upsilon_1)\psi_F(h_1) & -(1+\upsilon_2)\psi_R(h_1) & 0 & 0 \\ -Z_{0c2}\psi_F(h_1) & Z_{0c2}\psi_R(h_1) & 0 & 0 \end{pmatrix} \tag{3-41}$$

Equations 3-38 through 3-41 show that the matrix **M** is a *sparse matrix*. That is, most of the elements are zero. In general, a sparse matrix, or a matrix that can be reduced to a sparse matrix, is also a robust matrix. Solutions to robust matrix equations tend to be resistant to numerical noise.

Compared with general purpose, numerical, wire antenna analysis computer programs, matrix Eq. 3-35 provides the same advantages discussed in connection with Eqs. 2-19 and 3-15. For the mathematical model of the 2SFU, the matrix **M** has dimensions 8×8, or 64 individual matrix elements. In contrast, the general-purpose programs can require on the order of a million matrix elements. So, Eq. 3-35 continues to provide the significant advantages of speed and low numerical noise.

We have already derived the equations for the radiated electromagnetic field and power, input impedance, and efficiency. There is no need to write down new ones. Equations 3-22 through 3-35 continue to apply to the 2SFU, and for any folded unipole with N segments. Equations 2-41 and 2-42 also apply. So, we are ready to use our model to explore the capabilities of the 2SFU.

3.17 Examples of the 2SFU with Fold Height as the Primary Variable

Figures 3-32 through 3-37 show the input impedance of the 2SFU, for the same range of design variables as Figs. 3-10 through 3-14 for the GMA. The contour plots are different only because the upper segments of the two folded unipoles are different. The 2SFU has nonzero differential mode currents and voltages. This results in significant changes for at least four design variables. They are: (1) In the 2SFU the current transformation ratio ν_2 is nonzero, and we have arbitrarily fixed it at a "ballpark" or typical value of 0.5. (2) In contrast to the GMA, for which the differential mode characteristic impedance is an open circuit, Z_{0d2} is finite for the 2SFU, and we have fixed that at 120 ohms. (3) The common mode impedance Z_{0c2} has been fixed at 1,200 ohms instead of 1,600 ohms for the GMA. (4) There is an additional impedance Z_T terminating the upper segment. With this overview in mind, let's examine the figures one at a time.

Figure 3-32 shows the input impedance with all terminating reactances set to zero. The two continuous variables are the total antenna height h_2, and the height of the first segment h_1. The length of the upper segment clearly must be h_2–h_1. It is seen that there are combinations of h_1 and h_2 that provide optimal input resistance (shown in dark green). The same combinations simultaneously provide near-optimal input reactance (shown in cyan). These results are similar to those in Fig. 3-10 for the GMA. So far, it does not appear that we have benefited from the extra design variables provided by the upper segment of the 2SFU. Perhaps if we persevere, advantages will emerge. Indeed, that will prove to be the case.

Figure 3-33 shows the input impedance of the 2SFU when the base terminating reactance is 3,000 ohms. It is seen that there are many combinations that provide optimal input resistance, and that some of these simultaneously provide optimal input reactance (shown in dark cyan). In general, the contours fill many areas that were blank in the previous figure. Evidently, nonzero terminating reactances can provide significant benefits.

Figures 3-34 shows the input impedance when the base terminating reactance is –3,000 ohms. The observations are similar to those of the previous figure, reinforcing the apparent benefits of nonzero terminating reactances.

In Fig. 3-35, the base terminating reactance is reset to zero, and the fold terminating reactance is set at 600 ohms. It is seen that there are some combinations that simultaneously provide optimal input resistance and near-optimal input reactance. In general, however, the contours do not fill nearly as many blank areas as in the case of Figs. 3-33 and 3-34.

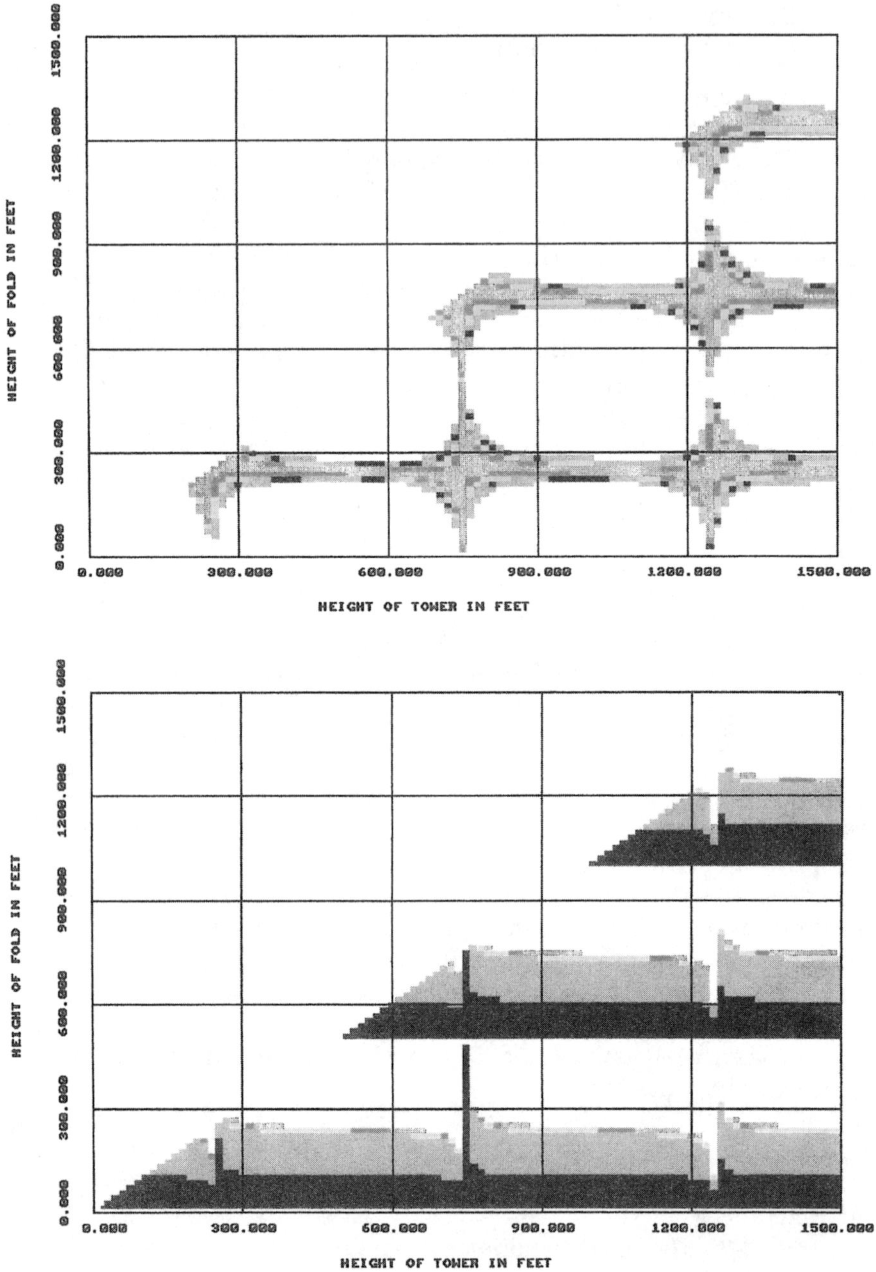

Figure 3-32 Input resistance (upper plot) and reactance (lower plot) for the 2SFU. The two continuous variables are antenna height h_2 and fold height h_1. The upper triangular sections of the plots are blank because the fold height cannot exceed the antenna height. The frequency is 984 kHz purely for mathematical convenience because a wavelength is 1,000 feet. The analysis applies to any frequency in the ELF through EHF bands.

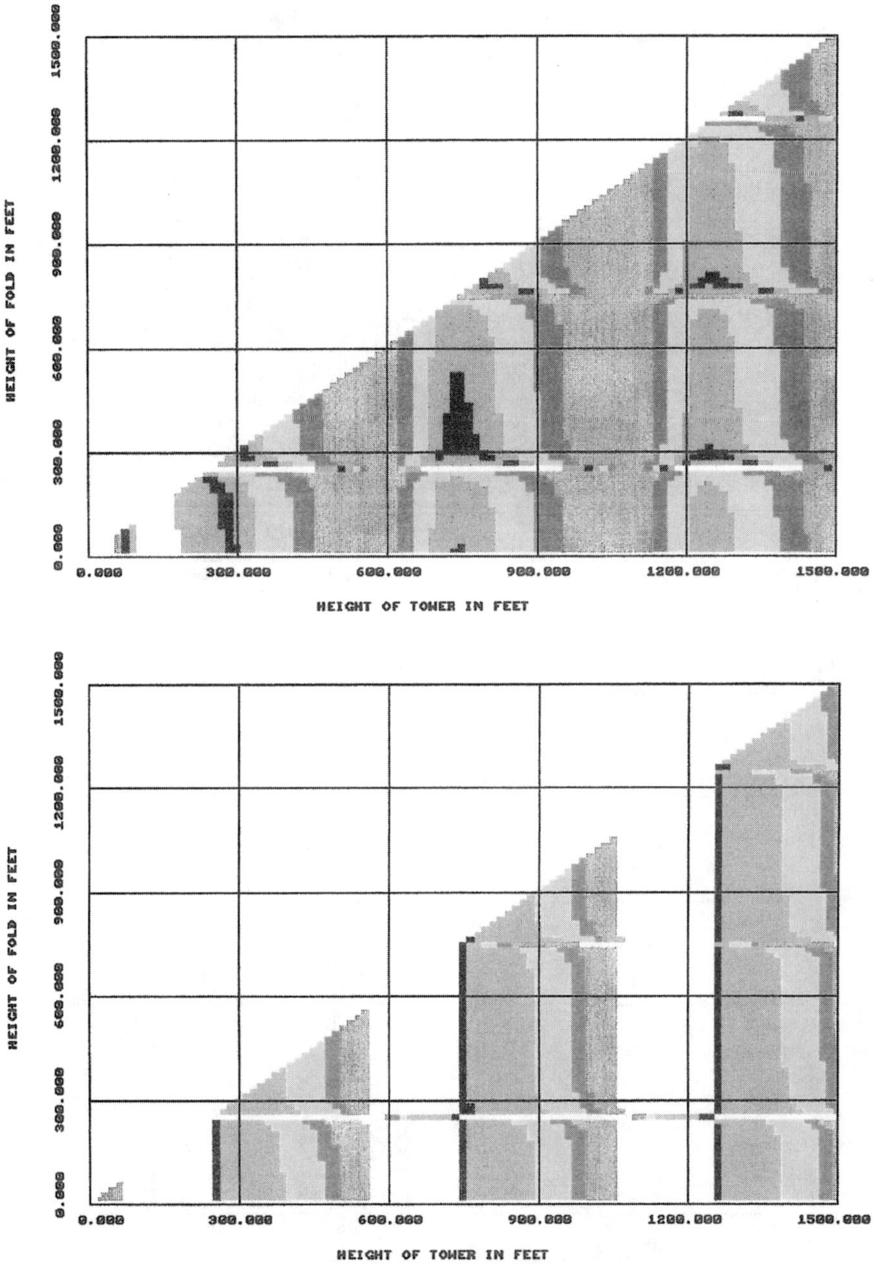

Figure 3-33 Input resistance (upper plot) and reactance (lower plot) for the 2SFU, with a base terminating reactance X_B of 3,000 ohms. The two continuous variables are antenna height h_2 and fold height h_1.

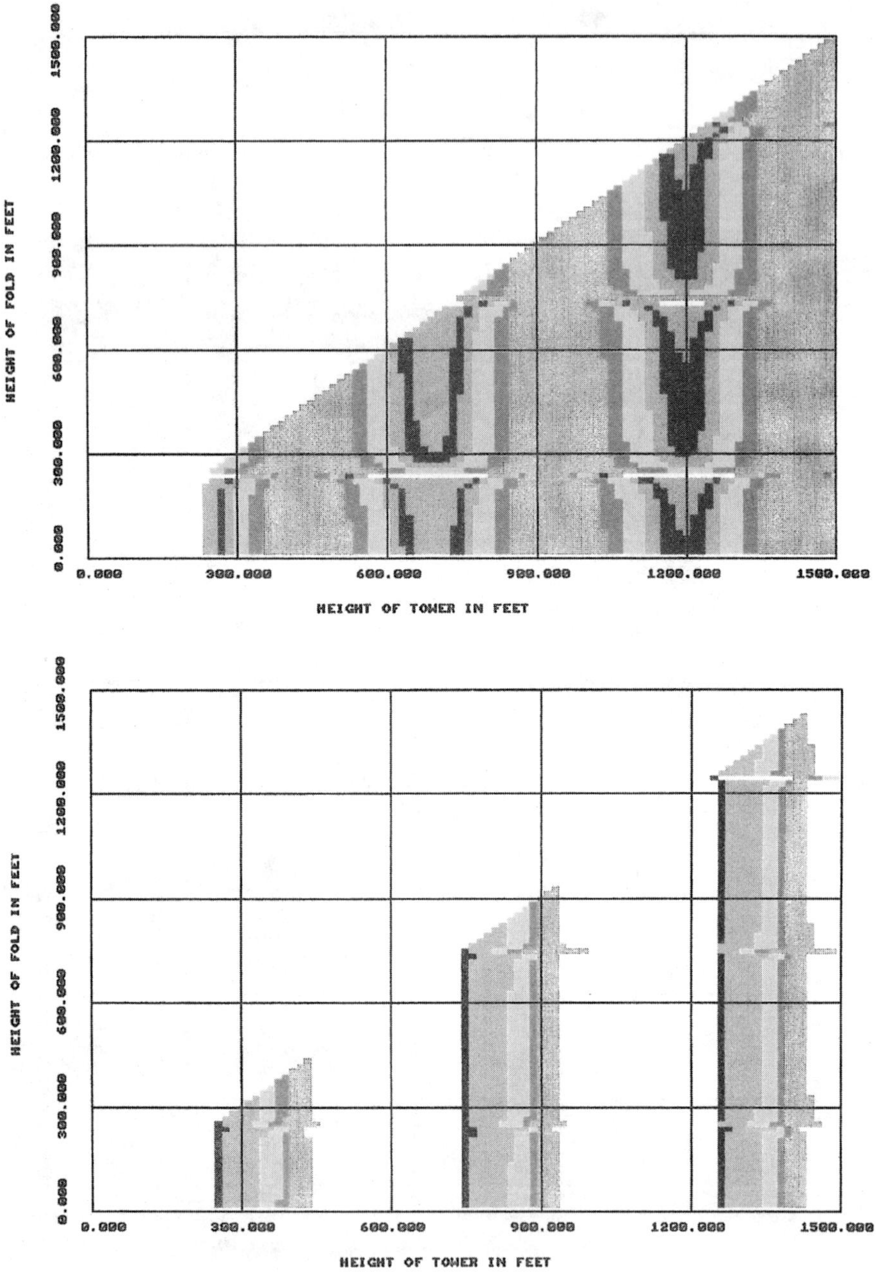

Figure 3-34 Input resistance (upper plot) and reactance (lower plot) for the 2SFU, with a base terminating reactance of –3,000 ohms. The two continuous variables are antenna height and fold height.

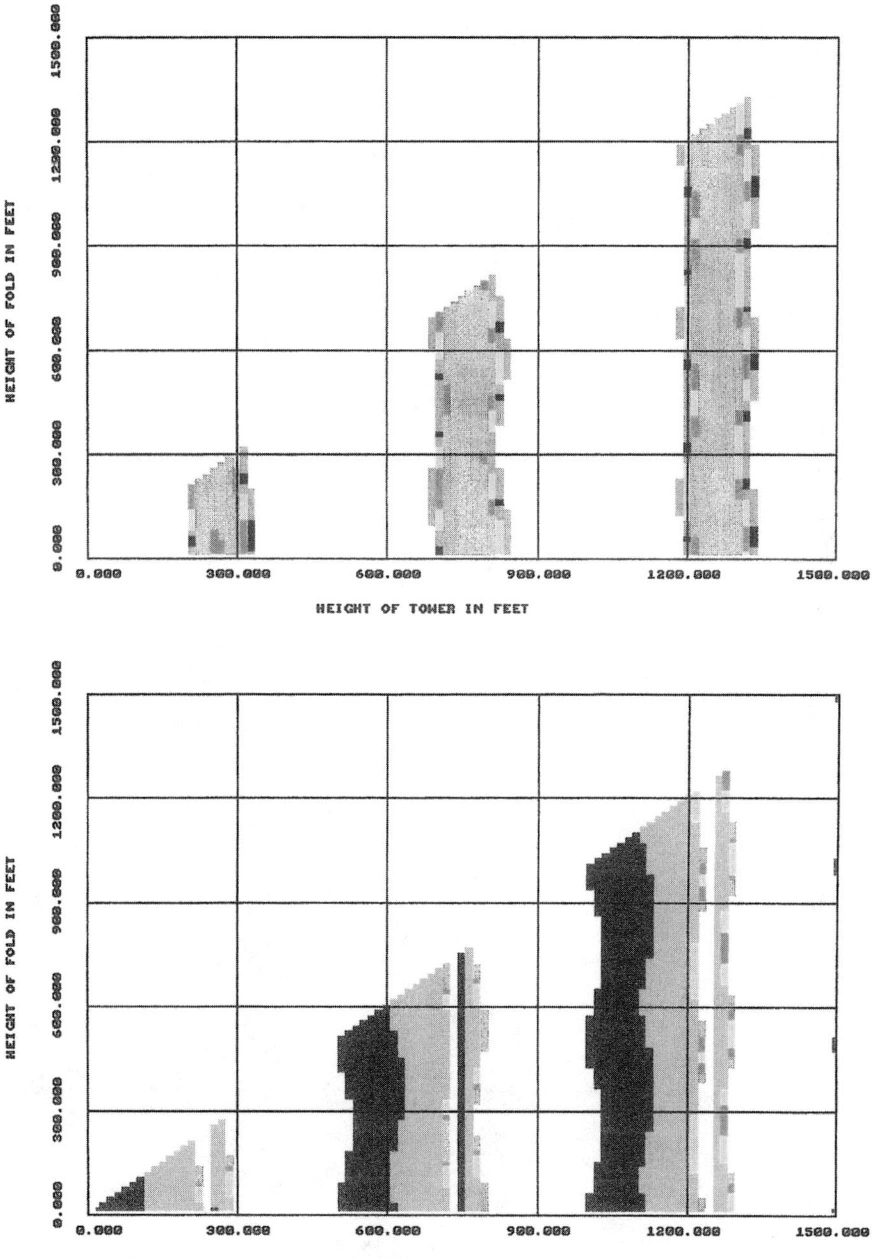

Figure 3-35 Input resistance (upper plot) and reactance (lower plot) for the 2SFU, with a fold terminating reactance X_1 of 600 ohms. The two continuous variables are antenna height h_2 and fold height h_1.

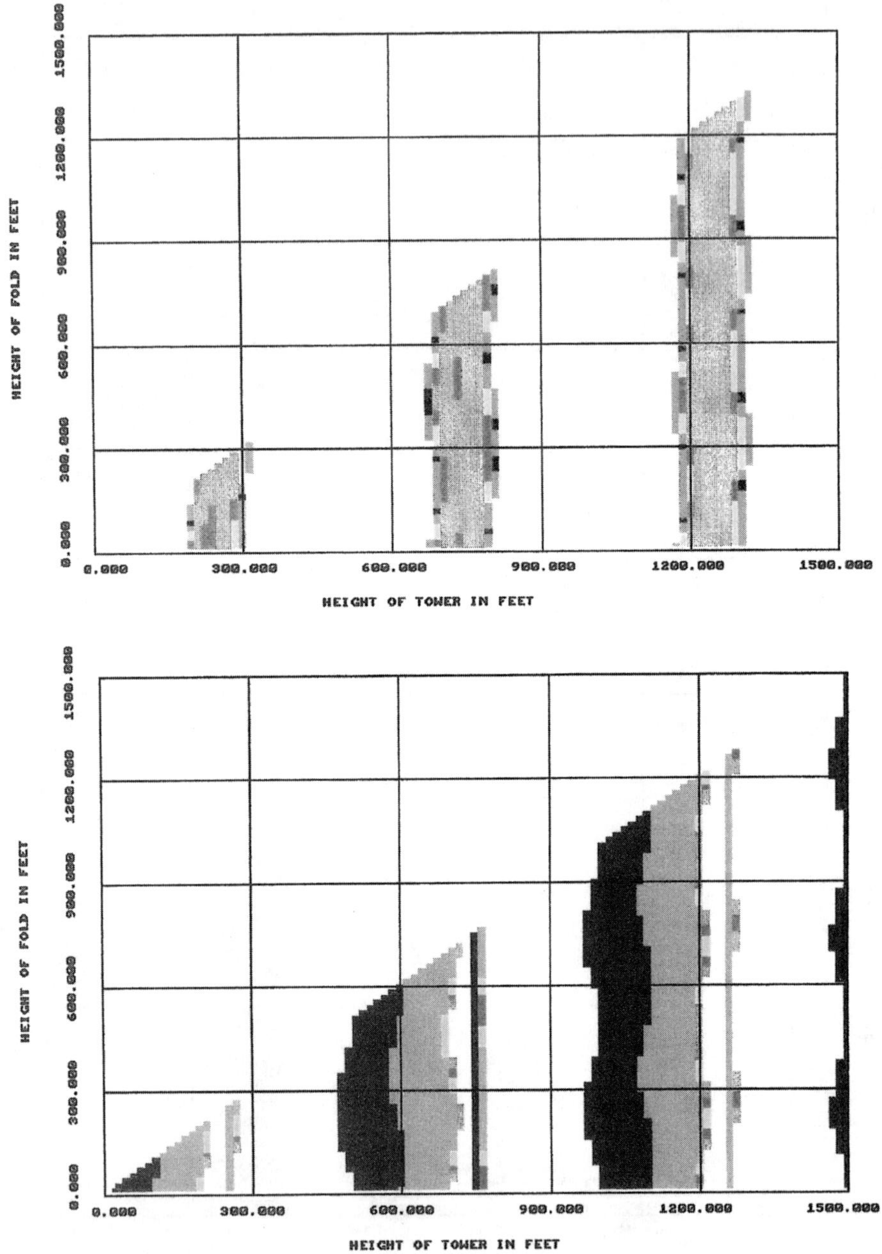

Figure 3-36 Input resistance (upper plot) and reactance (lower plot) for the 2SFU, with a fold terminating reactance of –600 ohms. The two continuous variables are antenna height and fold height.

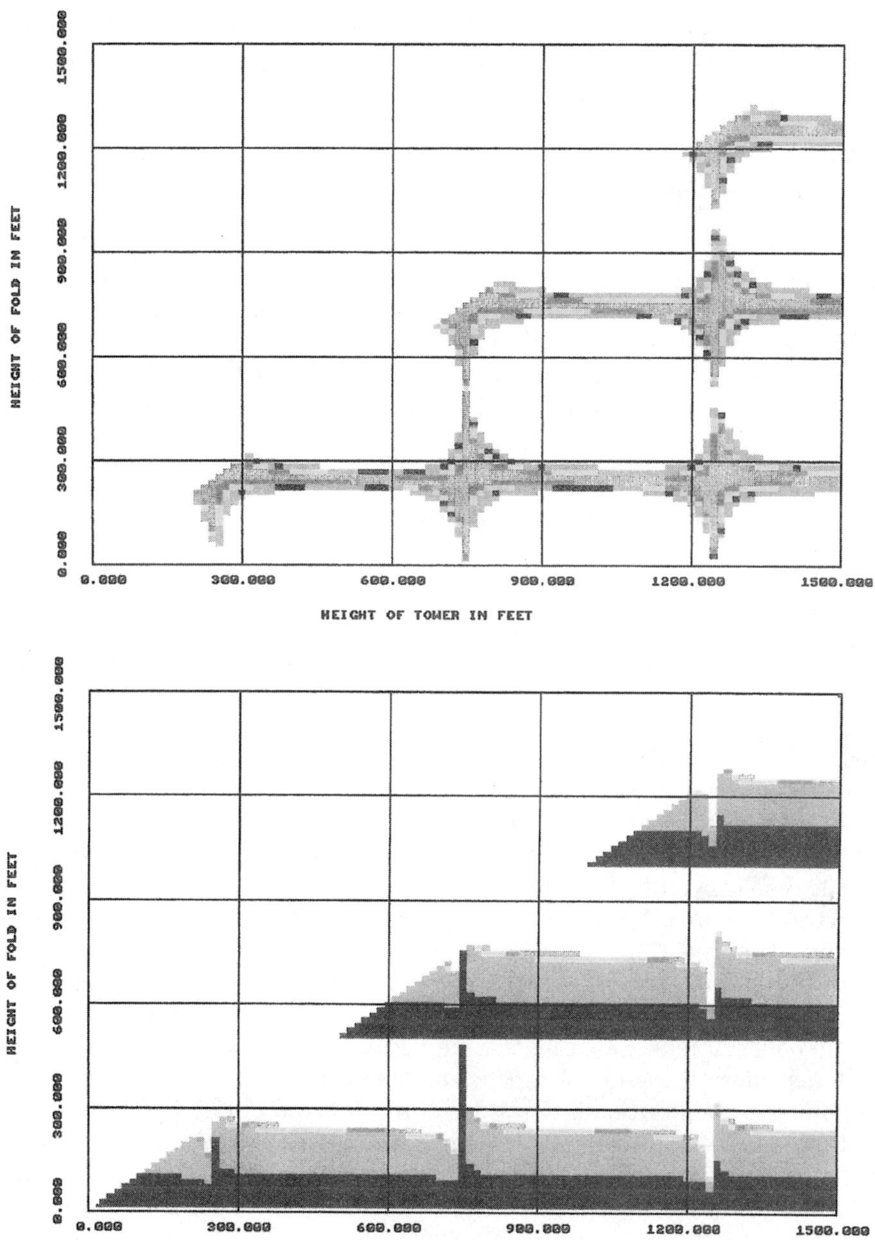

Figure 3-37 Input resistance (upper plot) and reactance (lower plot) for the 2SFU, with a top terminating reactance X_T of 600 ohms. The two continuous variables are antenna height h_2 and fold height h_1. These plots are identical to Fig. 3-32 because the fold terminating reactance X_1 is zero, or a short circuit. Thus, the value of the top terminating reactance has no effect. This is a good sanity check for our mathematical model.

In Fig. 3-36, the fold terminating reactance is set at –600 ohms. The observations are similar to those of the previous figure. So far, it appears that the base terminating reactance has a more dramatic effect on input impedance than the fold terminating reactance; however, that conclusion may be premature because there are still many more combinations of design variables to be explored.

Figure 3-37 is a good sanity check for our mathematical model. Close inspection shows that it is identical to Fig. 3-32, even though the top terminating reactance X_T is set to zero in the former and 600 ohms in the latter. Is there an error in the model? Certainly not! Because the fold terminating reactance X_1 is zero, the differential mode in the upper segment is short-circuited. With no differential mode currents flowing in the upper segment, it does not matter what value X_T has.

3.18 Examples of the 2SFU with Variable Terminating Reactance *and* Fold Height

The examples of the previous section suggested that nonzero terminating reactances were very helpful in obtaining optimal input impedance. We explored only a small sample of discrete reactance values in those examples. In this section, we will use contour plots to explore the effects of a continuum of terminating reactances.

Figures 3-38 and 3-39 show the input impedance for a 375-foot tower with both the height of the lower segment h_1 and a terminating reactance continuously variable. In Fig. 3-38, the base terminating reactance X_B is variable. In Fig. 3-39, the fold terminating reactance X_1 is variable. It is seen that there are many combinations that provide optimal input resistance (dark green) simultaneously with near-optimal input reactance (cyan). This is especially true for Fig. 3-38, with the base terminating reactance continuously varied.

In Figures 3-40 and 3-41, the tower height h_2 has been increased to 500 feet, or about one-half wavelength, or 180 electrical degrees. The same observations continue to apply. In contrast, a 500-foot monopole is much more difficult to impedance match because it exhibits a very high input resistance, on the order of a thousand ohms.

In Figs. 3-42 and 3-43, the tower height is 625 feet, or about 5/8 wavelength, or 225 electrical degrees. The same observations continue to apply. In addition, in Fig. 3-43, the contour for optimal input reactance (dark cyan) occupies a very large area of the plot compared with previous examples.

Figures 3-44 and 3-45, for a 750-foot tower, exhibit remarkable contours compared with previous examples. There are extremely large regions of both optimal input resistance (dark green) and optimal input reactance (dark cyan). This is especially true for Fig. 3-44 in which the

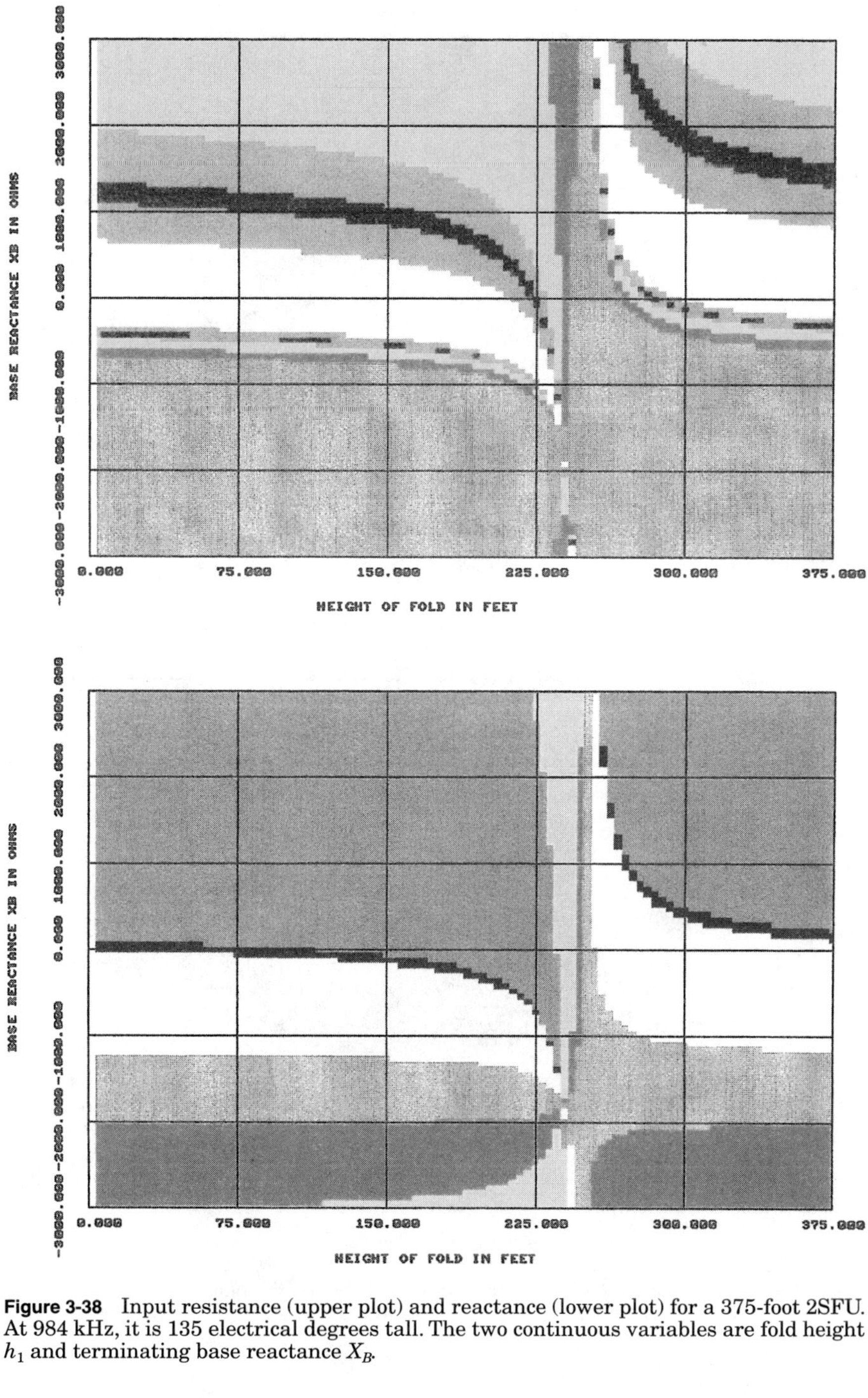

Figure 3-38 Input resistance (upper plot) and reactance (lower plot) for a 375-foot 2SFU. At 984 kHz, it is 135 electrical degrees tall. The two continuous variables are fold height h_1 and terminating base reactance X_B.

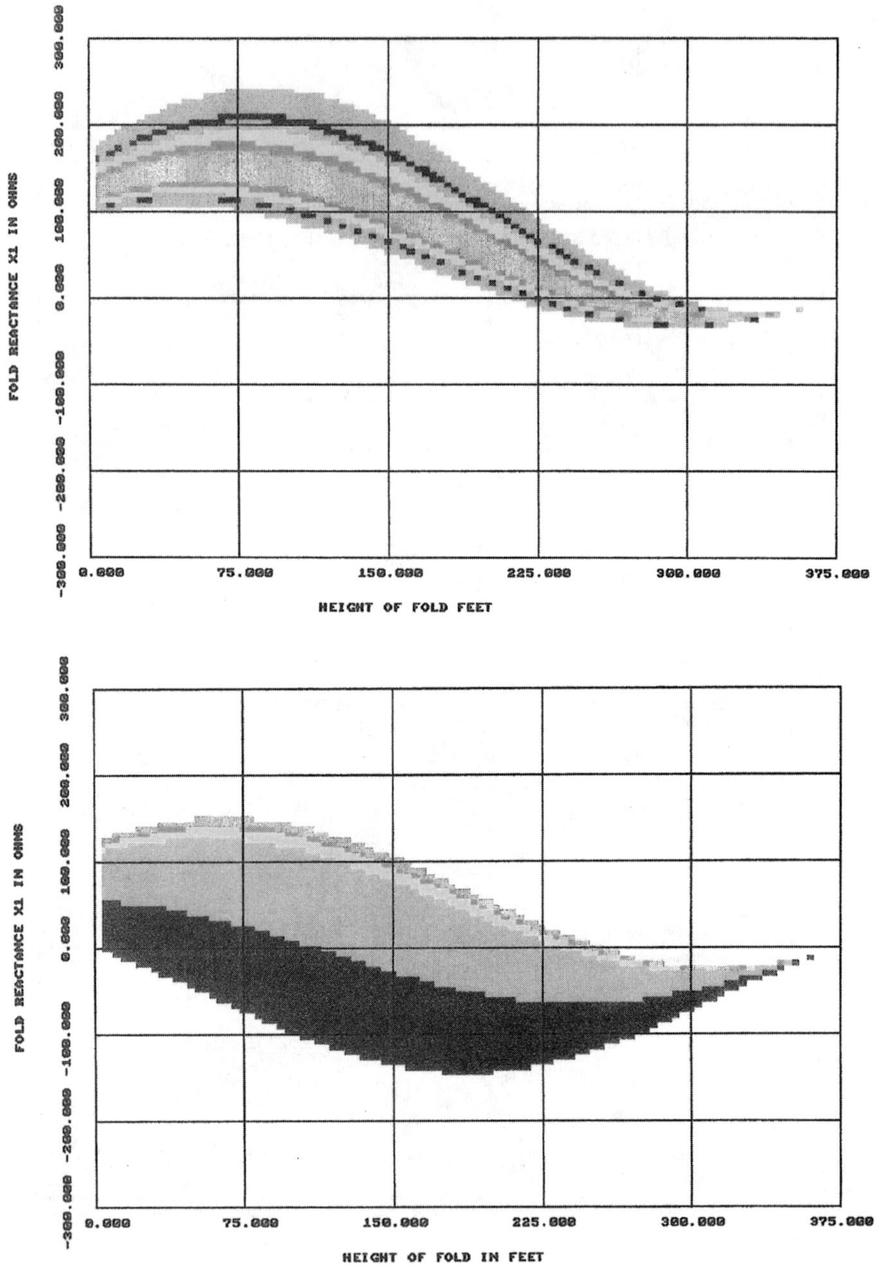

Figure 3-39 Input resistance (upper plot) and reactance (lower plot) for a 375-foot 2SFU. The two continuous variables are fold height h_1 and terminating fold reactance X_1.

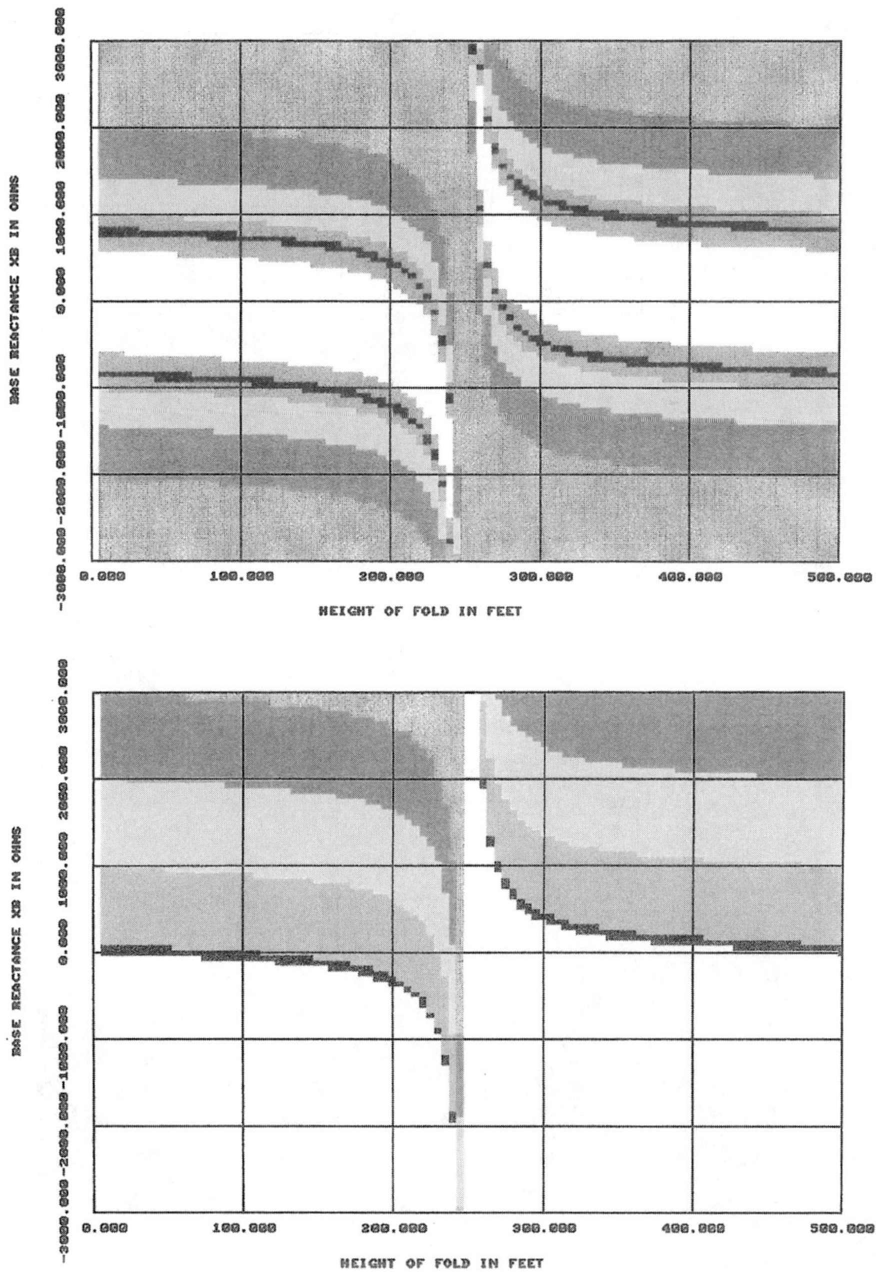

Figure 3-40 Input resistance (upper plot) and reactance (lower plot) for a 500-foot 2SFU. At 984 kHz, it is 180 electrical degrees tall. The two continuous variables are fold height and terminating base reactance.

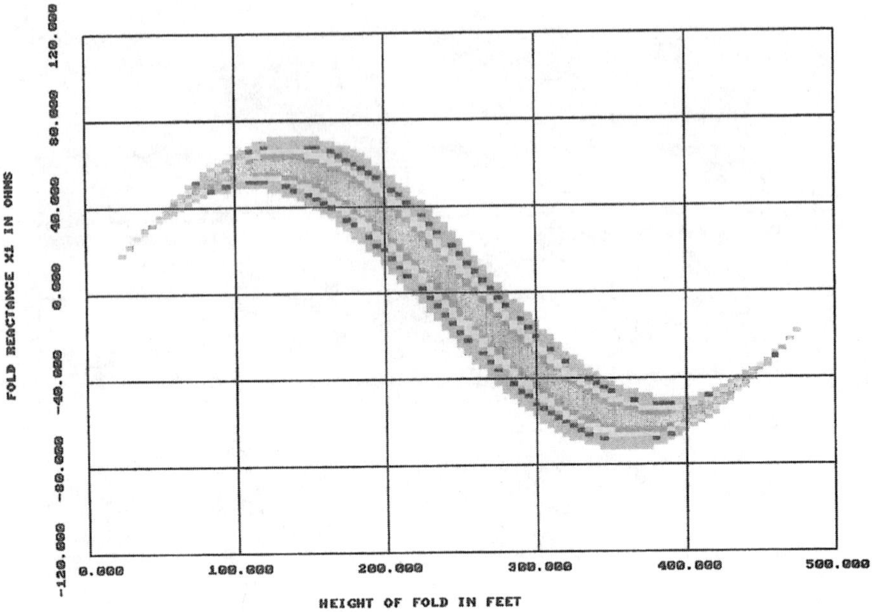

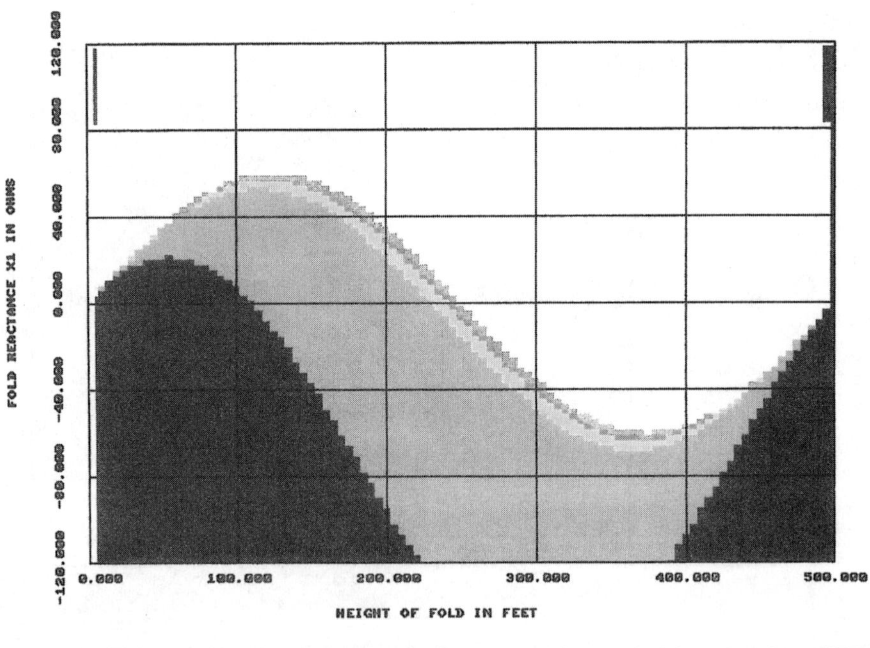

Figure 3-41 Input resistance (upper plot) and reactance (lower plot) for a 500-foot 2SFU. The two continuous variables are fold height and terminating fold reactance.

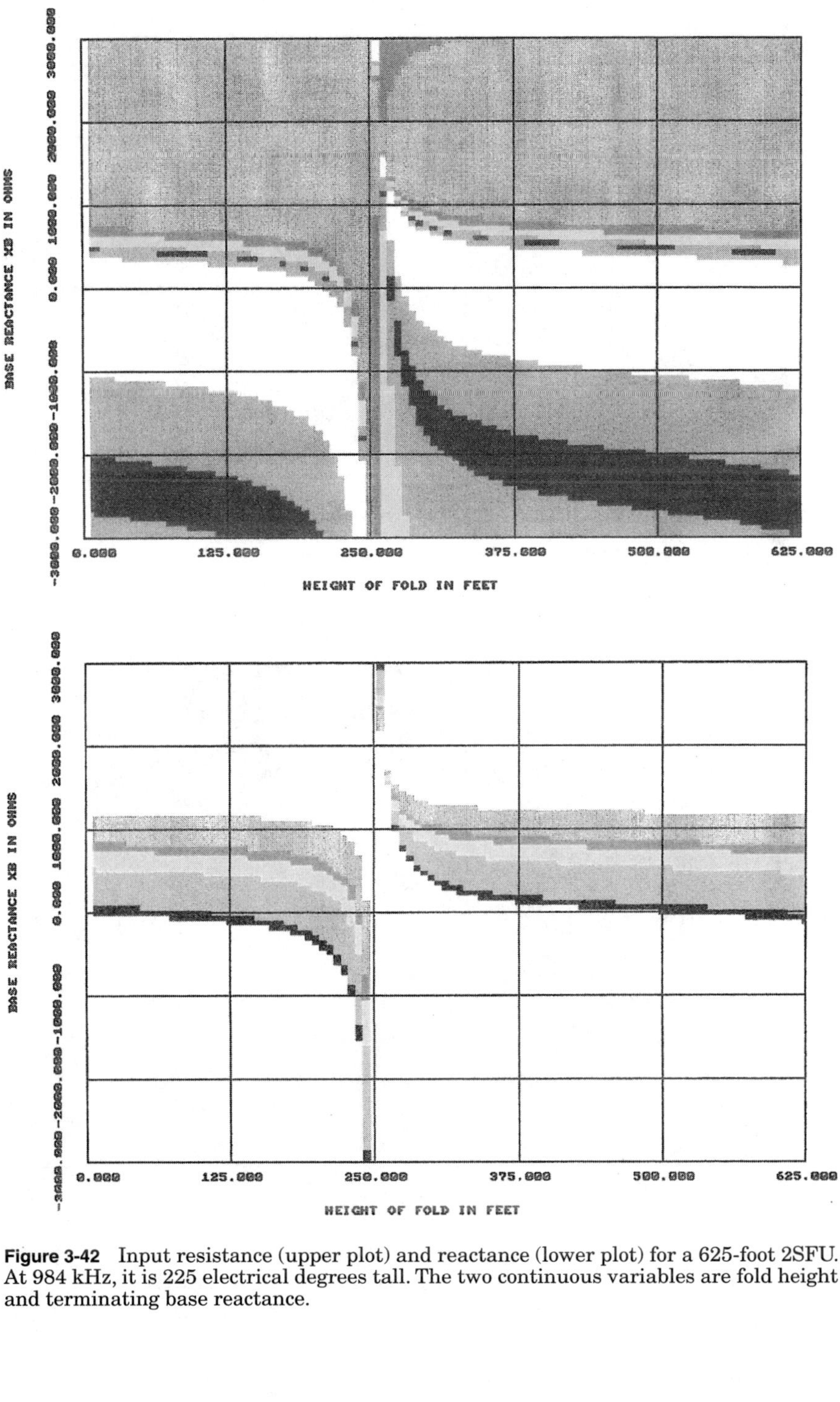

Figure 3-42 Input resistance (upper plot) and reactance (lower plot) for a 625-foot 2SFU. At 984 kHz, it is 225 electrical degrees tall. The two continuous variables are fold height and terminating base reactance.

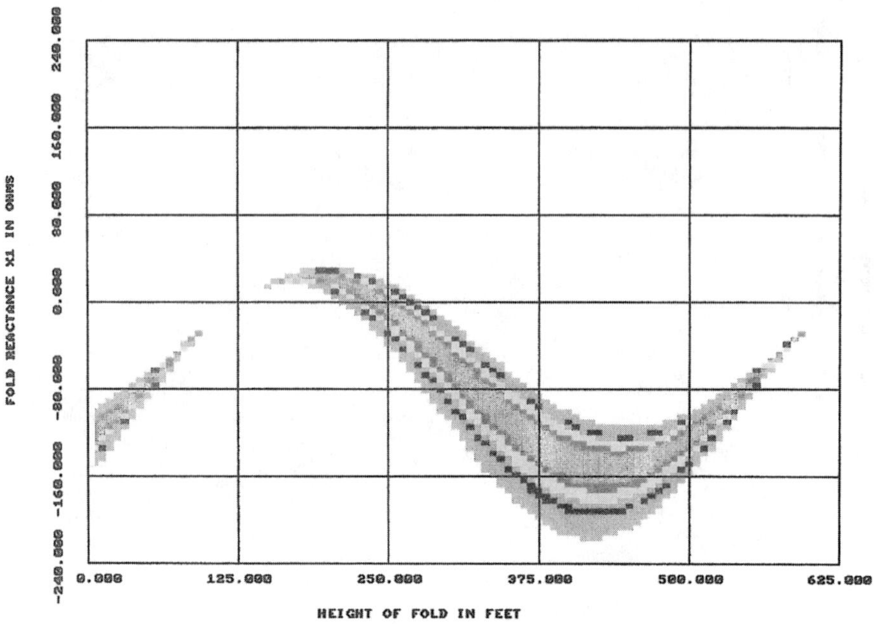

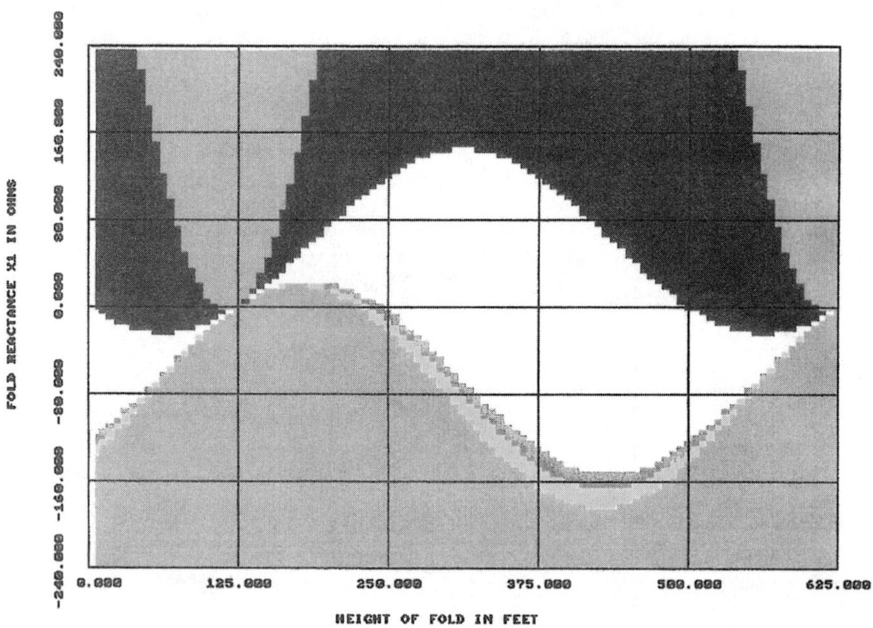

Figure 3-43 Input resistance (upper plot) and reactance (lower plot) for a 625-foot 2SFU. The two continuous variables are fold height and terminating fold reactance.

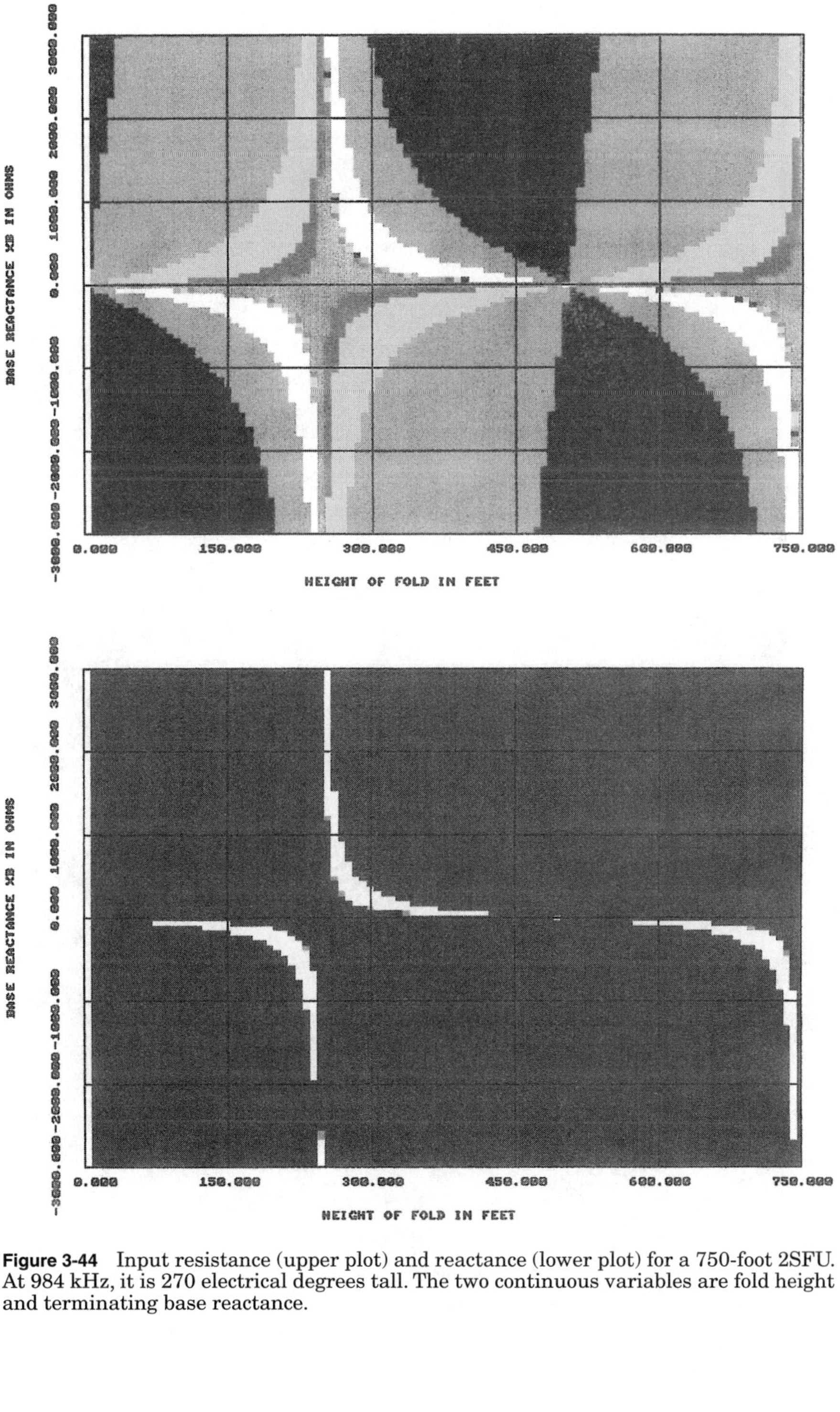

Figure 3-44 Input resistance (upper plot) and reactance (lower plot) for a 750-foot 2SFU. At 984 kHz, it is 270 electrical degrees tall. The two continuous variables are fold height and terminating base reactance.

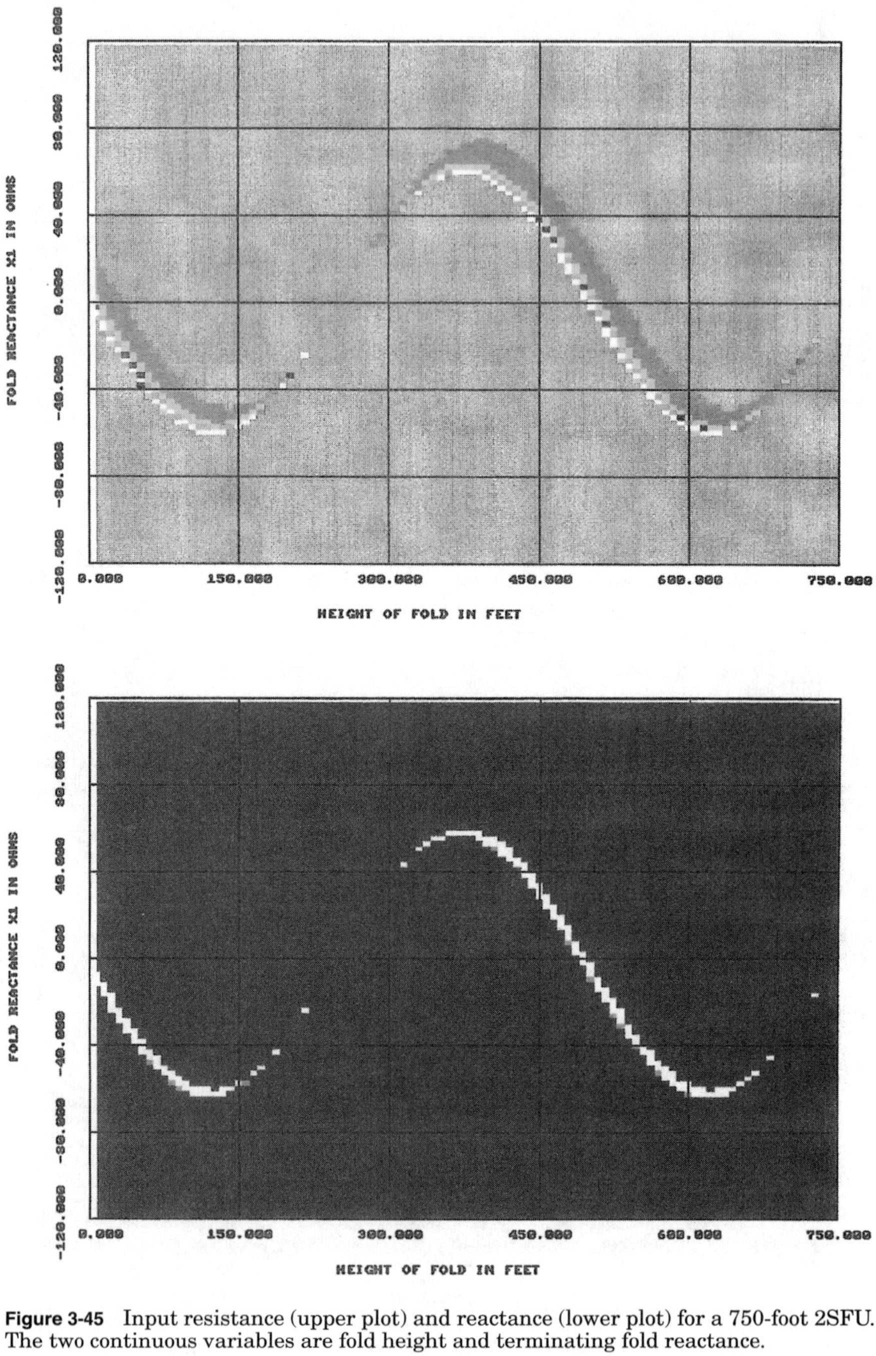

Figure 3-45 Input resistance (upper plot) and reactance (lower plot) for a 750-foot 2SFU. The two continuous variables are fold height and terminating fold reactance.

base terminating reactance is continuously variable. At 984 kHz, the tower is three-quarters wavelength, or 270 electrical degrees. Can we obtain similar plots for other heights? We will explore this possibility in the next section.

Before pursuing that, however, we will round out the set with Figs. 3-46 and 3-47, for a 1,500-foot tower. Periodic behavior is clearly seen. By now it is not surprising to observe that many combinations provide optimal input resistance simultaneously with near-optimal input reactance.

3.19 Examples of the 2SFU with Two Terminating Reactances as Continuous Variables

In the previous section, Fig. 3-44 demonstrated an unusually wide range of variables for which both the input resistance and input reactance were optimal. This was for a particular tower height. We wondered if similar plots might be obtained for other heights. In this section, we will explore that possibility.

To that end, Fig. 3-48 shows the input impedance for a fixed tower height h_2 of 375 feet, and a fixed lower segment length h_1 of 125 feet. These dimensions also fix the upper segment length at 250 feet. The continuous variables are base terminating reactance X_B and fold terminating reactance X_1. It is seen that almost the entire plot is covered by optimal or near-optimal values, that is, green or dark green for resistance, and cyan or dark cyan for reactance.

Figure 3-49 shows the input impedance for h_2 = 750 feet and h_1 = 250 feet. The same observations apply as for the previous example, especially for the input reactance. The contour plot is almost entirely covered by the dark cyan region.

Figure 3-50 shows the input impedance for h_2 = 1,500 feet and h_1 = 500 feet. As observed in the previous examples, the dominant contours are for optimal or near-optimal resistance and reactance.

The implications of Figs. 3-48 through 3-50 are profound, indeed. The dominant regions of green and cyan show that these configurations are very robust. That is, the input impedance will remain optimal or near-optimal even if the design variables are perturbed over a fairly wide range.

Only one logical step is required to conclude that these configurations are very broadband. Perturbing the linear dimensions is equivalent to perturbing the wavelength. Similarly, perturbing the reactance is equivalent to perturbing the frequency. Evidently, a significant benefit from the additional design parameters provided by the 2SFU can be broadband operation.

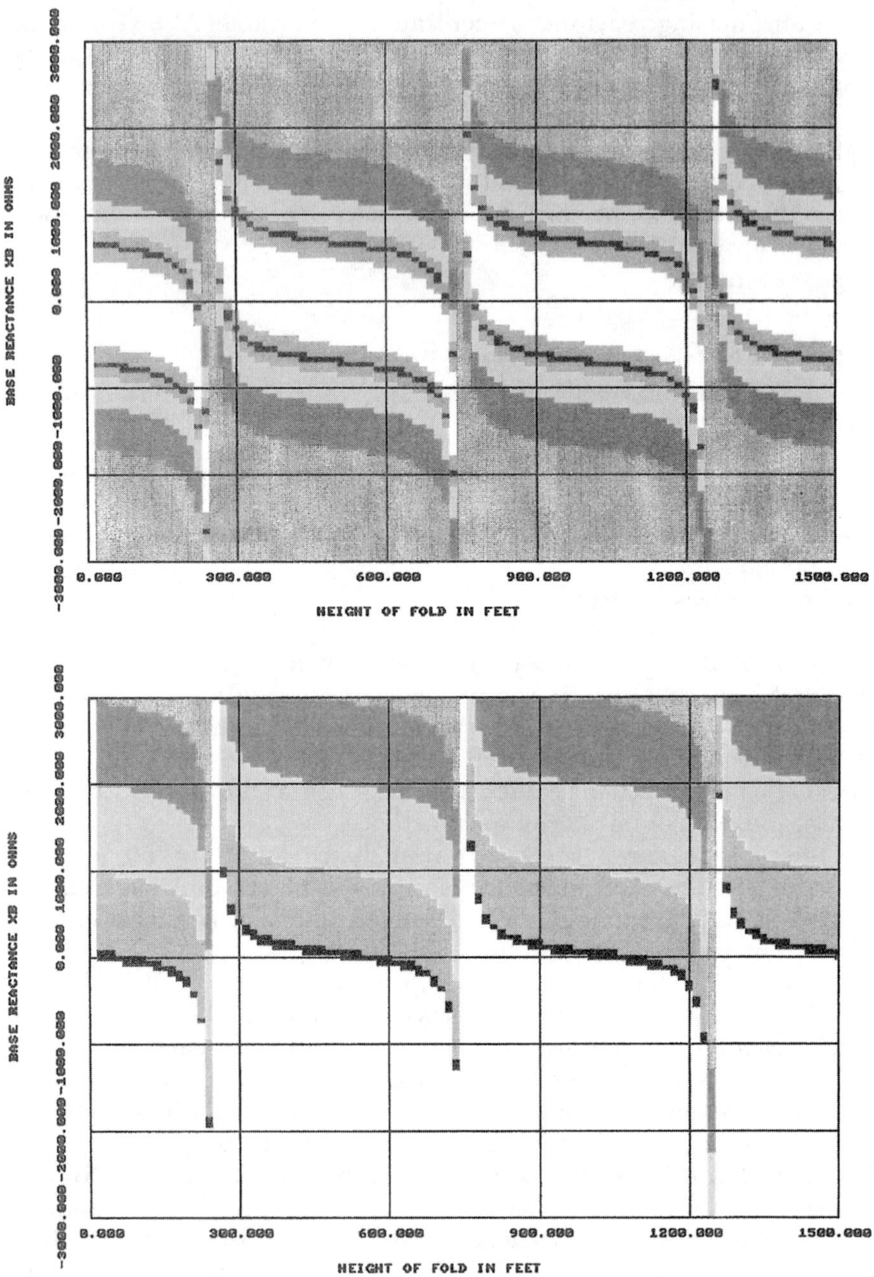

Figure 3-46 Input resistance (upper plot) and reactance (lower plot) for a 1,500-foot 2SFU. At 984 kHz, it is 540 electrical degrees tall. The two continuous variables are fold height and terminating base reactance.

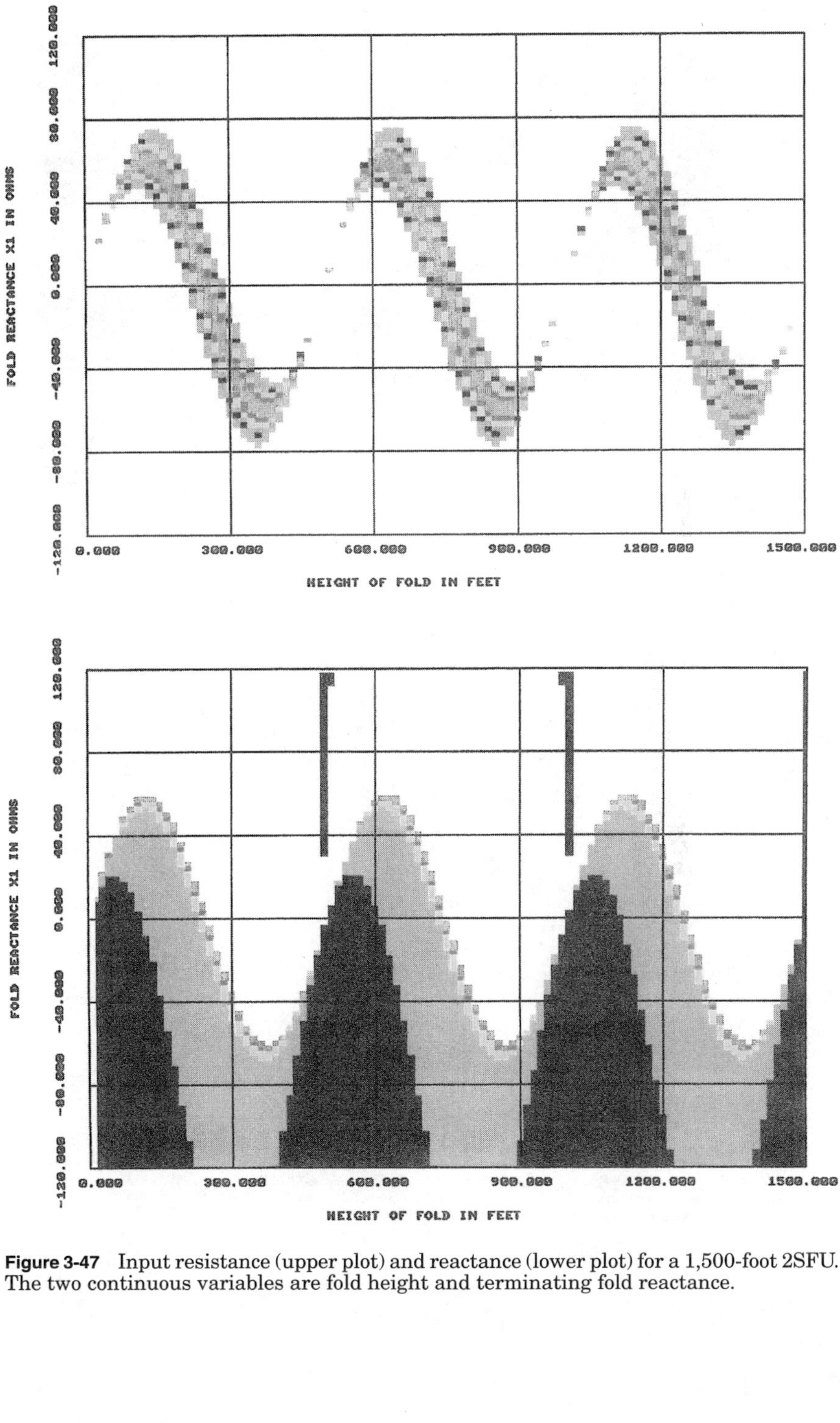

Figure 3-47 Input resistance (upper plot) and reactance (lower plot) for a 1,500-foot 2SFU. The two continuous variables are fold height and terminating fold reactance.

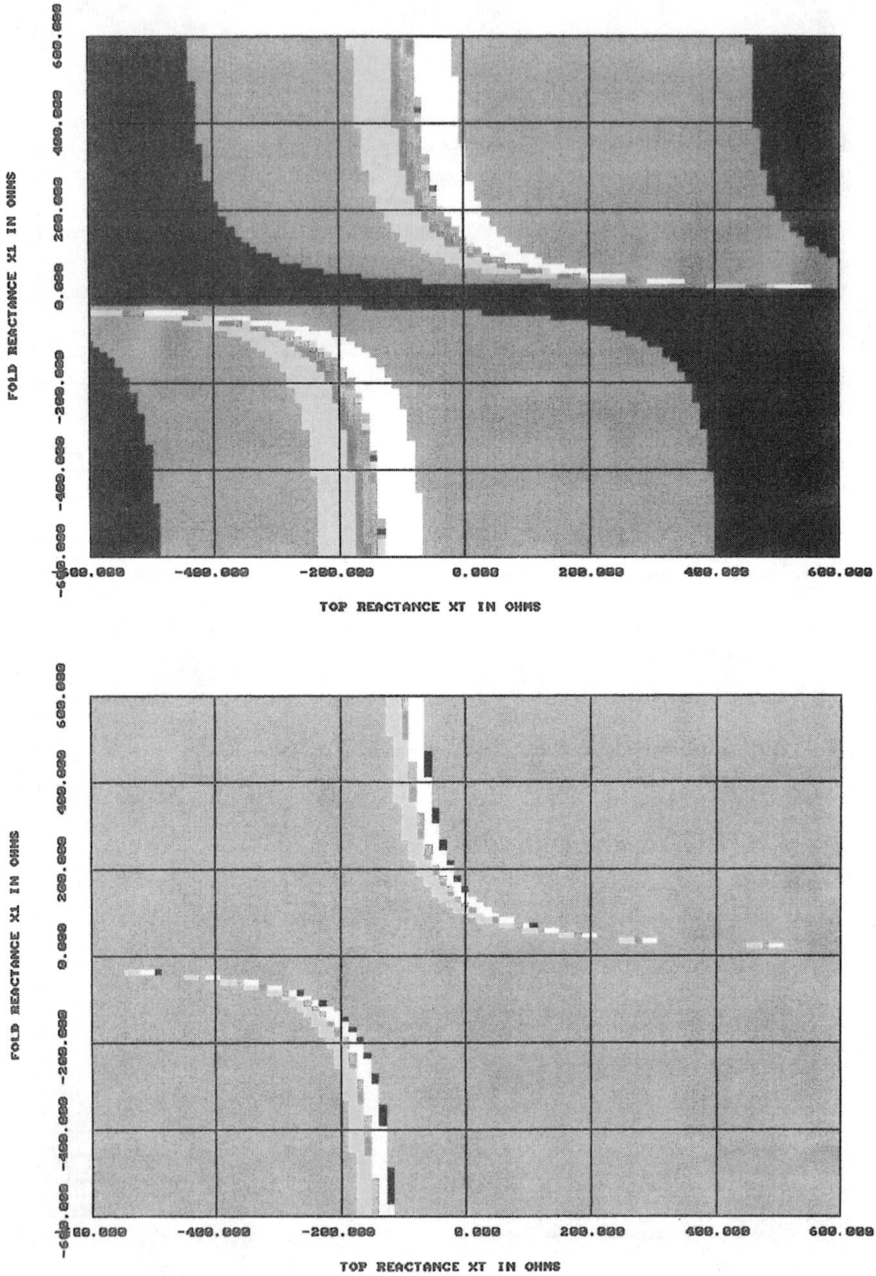

Figure 3-48 Input resistance (upper plot) and reactance (lower plot) for a 375-foot 2SFU with a 125-foot fold, and a terminating base reactance of 1,000 ohms. At 984 kHz, the taller conductor is 135 degrees, and the fold is 45 degrees. The two continuous variables are terminating top reactance X_T and terminating fold reactance X_1.

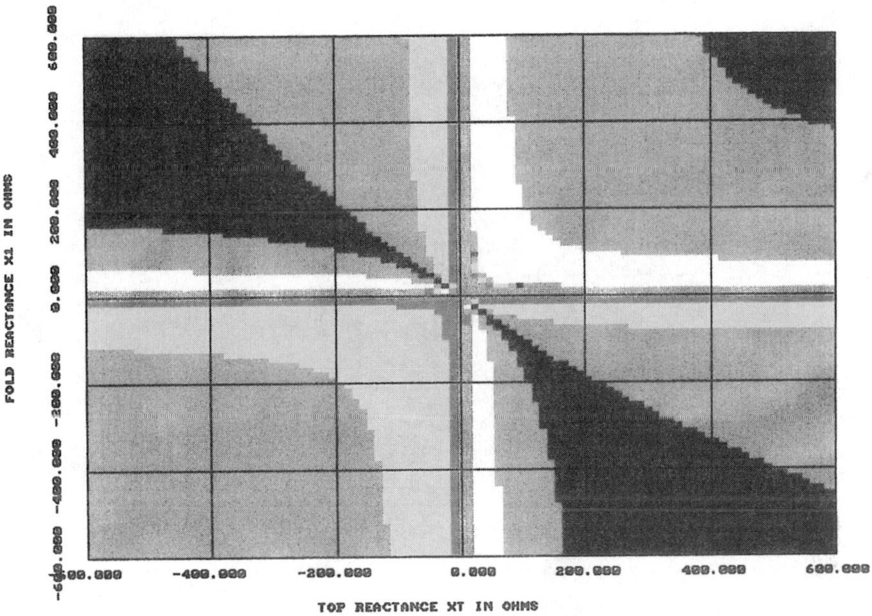

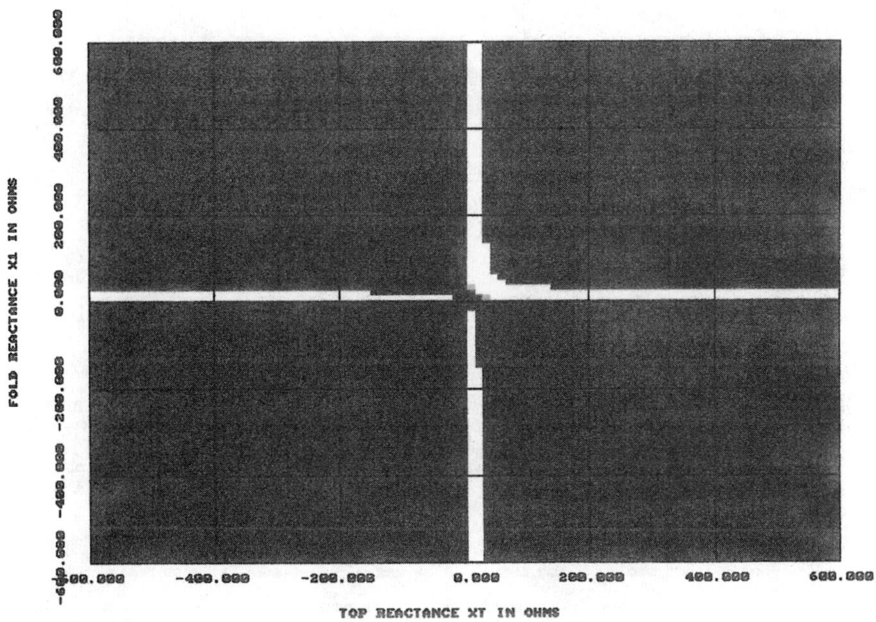

Figure 3-49 Input resistance (upper plot) and reactance (lower plot) for a 750-foot 2SFU with a 250-foot fold, and a terminating base reactance of 500 ohms. At 984 kHz, the taller conductor is 270 degrees, and the fold is 90 degrees. The two continuous variables are terminating top reactance and terminating fold reactance.

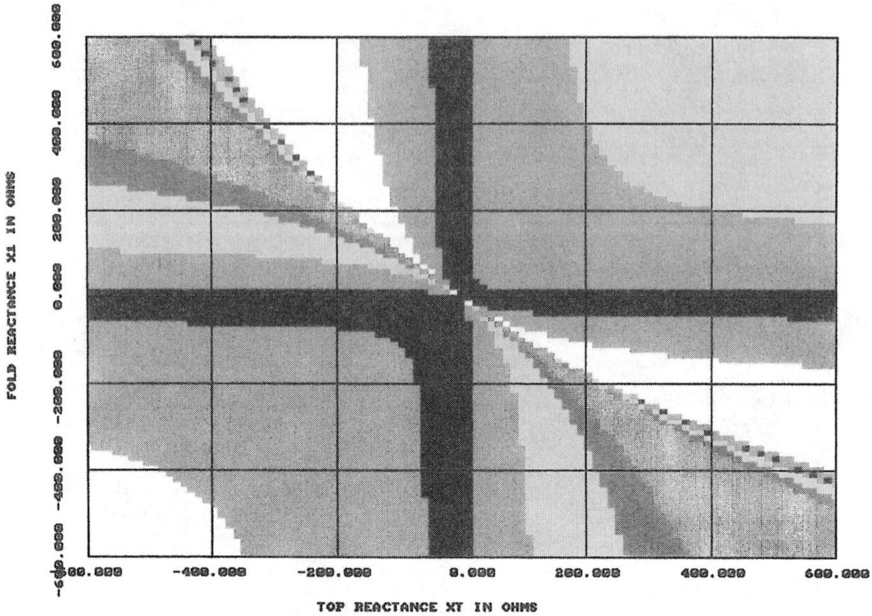

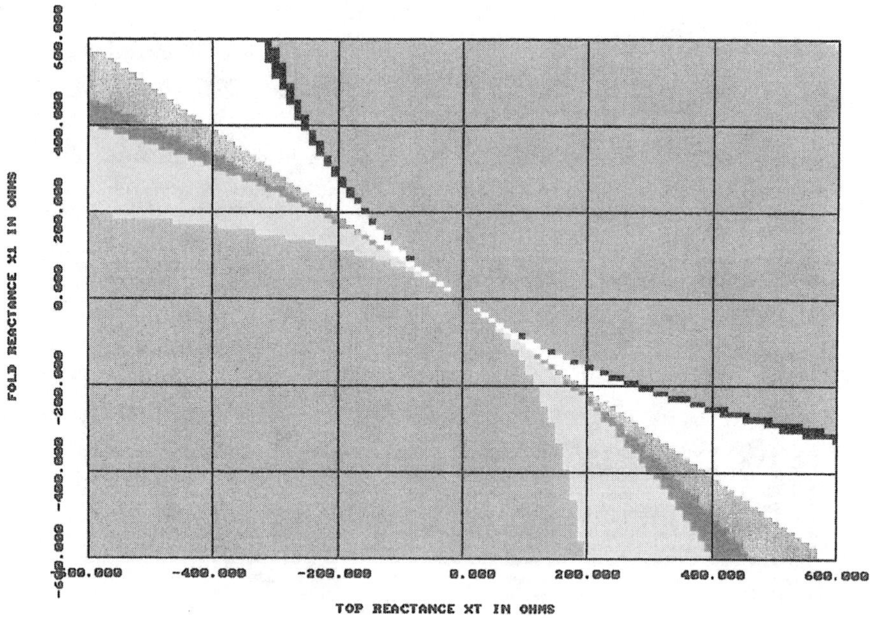

Figure 3-50 Input resistance (upper plot) and reactance (lower plot) for a 1,500-foot 2SFU with a 500-foot fold, and a terminating base reactance of 700 ohms. At 984 kHz, the taller conductor is 540 degrees, and the fold is 180 degrees. The two continuous variables are terminating top reactance and terminating fold reactance.

3.20 Conclusions

We have continued to build the mathematical model which we began in Chap. 2. The underlying physics have remained the same, namely, recognizing the distinction between common mode and differential mode currents and voltages on the antenna. In Chap. 3, we extended the mathematical notation and procedures to include the 2-segment folded unipole (2SFU). The Gamma Match Antenna (GMA) is a special case of the 2SFU. We used the mathematical model to explore the properties of both the GMA and the 2SFU.

The GMA provides no fewer than eight independent design variables, compared with six such variables for the simple configuration of Chap. 2. Accordingly, the GMA offers advantages in performance. Most notably, an optimal or near-optimal impedance match can be obtained with a fold extending only part of the way along the antenna. This saves both space and material. In broadcast applications, it also prevents the fold from interfering with smaller antennas that are attached to the upper part of the tower.

The 2SFU provides no fewer than eleven independent design variables. It offers still more advantages in performance, especially broadband operation.

The above properties of the GMA and 2SFU were demonstrated using our mathematical model with the results displayed as contour plots. Each contour plot displays 10,000 separate iterations of the model. This is the most efficient way to explore the effect of so many independent design variables. Yet each display required less than a minute of computation time on a modest desktop computer. That is one of two chief advantages of our mathematical model. The other advantage, as previously discussed, is low numerical noise.

In contrast, a general purpose, wire antenna analysis computer program would require hours or days to generate the same results. The reason is matrix size. Both approaches use matrix equations. In our model, the matrix for the 2SFU has dimensions 8×8, or 64 individual matrix elements. The general-purpose program requires much larger matrices with on the order of a million individual elements. Thus, the matrix sizes between our model and the general-purpose program ranges between *four and five orders of magnitude.*

Our mathematical model has become more general, with respect to both procedures and notation. So, very little change will be required to explore more complicated configurations. Our basic model building is nearly complete.

Symbols Used in This Chapter

Symbol	Description
$\mathbf{A_{11}}$	Submatrix of **M**
$\mathbf{A_{12}}$	Submatrix of **M**
$\mathbf{A_{21}}$	Submatrix of **M**
$\mathbf{A_{22}}$	Submatrix of **M**
$E(\theta)$	Radiated electric field
$F_n(\theta, a_n, b_n)$	Vertical radiation characteristic, used in computation of $E(\theta)$
h_1	z-coordinate of the top of the first or bottom segment of the folded unipole
h_2	z-coordinate of the top of the second or top segment of the folded unipole
$\mathbf{I}$	Vector of the complex current amplitudes
$\mathbf{I_1}$	Vector of the complex current amplitudes on the first segment of the folded unipole
$\mathbf{I_2}$	Vector of the complex current amplitudes on the second segment of the folded unipole
$\mathbf{I_n}$	Vector of the complex current amplitudes on the nth segment of the folded unipole
$I_{c1}(z)$	Common mode current on the first segment of the folded unipole
$I_{c2}(z)$	Common mode current on the second segment of the folded unipole
$I_{\mathrm{cn}}(z)$	Common mode current on the nth segment of the folded unipole
$I_{d1}(z)$	Differential mode current on the first segment of the folded unipole
$I_{d2}(z)$	Differential mode current on the second segment of the folded unipole
$I_{\mathrm{dn}}(z)$	Differential mode current on the nth segment of the folded unipole
I_{Fcn}	Amplitude of the common mode forward traveling wave current on the nth segment
I_{Fdn}	Amplitude of the differential mode forward traveling wave current on the nth segment
I_{in}	Input current
I_{Rcn}	Amplitude of the common mode reverse traveling wave current on the nth segment
I_{Rdn}	Amplitude of the differential mode reverse traveling wave current on the nth segment
j	Square root of -1
k	Wave number
$\mathbf{M}$	Matrix in the equation for the complex current amplitudes
P_d	Dissipated power
R	Spherical coordinate for distance
R_B	Base terminating resistance
R_n	Terminating resistance for the nth segment of the folded unipole
R_T	Terminating resistance at the top of the folded unipole
V	Input voltage
$V_{c1}(z)$	Common mode voltage on the first or lower segment of the folded unipole
$V_{c2}(z)$	Common mode voltage on the second or upper segment of the folded unipole
$V_{\mathrm{cn}}(z)$	Common mode voltage on the nth segment of the folded unipole
$V_{d1}(z)$	Differential mode voltage on the first or lower segment of the folded unipole
$V_{d2}(z)$	Differential mode voltage on the second or upper segment of the folded unipole
$V_{\mathrm{dn}}(z)$	Differential mode voltage on the nth segment of the folded unipole
X_1	Terminating reactance on the first segment of the folded unipole
X_B	Terminating reactance at the base of the folded unipole
X_T	Terminating reactance at the top of the folded unipole

z	Vertical coordinate
Z_{0c1}	Characteristic impedance for the common mode on the first segment
Z_{0c2}	Characteristic impedance for the common mode on the second segment
Z_{0cn}	Characteristic impedance for the common mode on the nth segment
Z_{0d1}	Characteristic impedance for the differential mode on the first segment
Z_{0d2}	Characteristic impedance for the differential mode on the second segment
Z_{0dn}	Characteristic impedance for the differential mode on the nth segment
Z_1	Terminating impedance for the first segment of the folded unipole
Z_B	Terminating impedance at the base of the folded unipole
Z_T	Terminating impedance at the top of the folded unipole
η	Impedance of free space
ν	Currrent transformation ratio for first segment of Gamma Match Antenna
ν_1	Current transformation ratio for first segment of the folded unipole
ν_2	Current transformation ratio for second segment of the folded unipole
ν_n	Current transformation ratio for the nth segment of the folded unipole
θ	Spherical coordinate for azimuth
$\psi_F(z)$	Forward traveling wave function
$\psi_R(z)$	Reverse traveling wave function

Chapter

4

The Double Gamma Match Antenna

4.1 Introduction

The mathematical procedure for analyzing and designing arbitrarily complicated folded unipole antennas was fully developed by the end of Chap. 3. In particular, Sec. 3.5 and 3.14 introduced boundary conditions between segments of the antenna. Those boundary conditions were the final piece of the mathematical model. From now on, continued practice will serve to demonstrate the flexibility of the mathematical model, and the versatility of the actual folded unipole. To that end, the Double Gamma Match Antenna, or 2GMA, makes a very good example.

Figure 4-1 shows the essential features of the 2GMA. It is seen that it has one more segment than either the GMA or 2SFU, which were the subjects of Chap. 3. The 2GMA has 13 independent design variables. That is five more than the GMA and two more than the 2SFU. Accordingly, we expect more versatility from this antenna. We will explore a few examples that provide a glimpse of the added versatility; however, we will hardly exhaust the possibilities in the limited space of this chapter.

We can generalize the above observations concerning what happens when we add segments to the folded unipole antenna. When we add a nonfolded segment to an nSFU to form an nGMA, as we have done in Fig. 4-1 for $n = 2$, then we add two independent design variables. They are: (1) the z-coordinate h of the top of the extension, and (2) the common mode characteristic impedance of the extension Z_{0c}.

Whenever we add a *folded* segment to an nGMA or nSFU, then we add five independent design variables. They are: (1) the z-coordinate h of the top of the segment, (2) the terminating impedance Z, (3) the current transformation ratio ν; (4) the common mode characteristic impedance Z_{0c}, and (5) the differential mode characteristic impedance Z_{0d}.

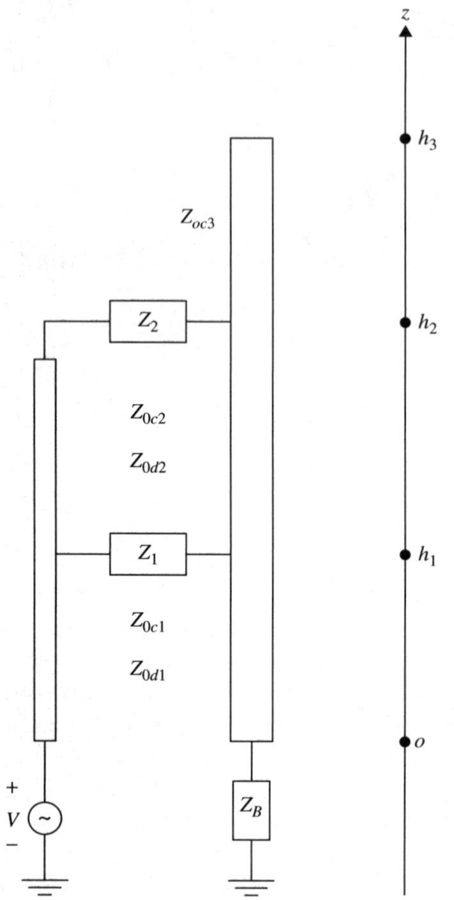

Figure 4-1 Essential features and geometry for the Double Gamma Match Antenna (2GMA). Two parallel conductors are shown only for simplicity in the drawing. In fact, the analysis applies to any arbitrary cross section that can be mathematically described by two characteristic impedances, one for common mode and another for differential mode.

From the above generalizations, we can detect certain patterns with respect to geometry. In the following sections, we will similarly observe patterns with respect to mathematical modeling.

4.2 Equation at the Top of the 2GMA

Figure 4-2 shows the configuration of interest with currents and voltages. As with the folded unipole antennas discussed in previous chapters, the common mode currents $I_{\text{cn}}(z)$ are the sum of forward and reflected traveling waves with amplitudes I_{Fcn} and I_{Rcn}, respectively. This relationship is described by Eq. 3-1, with $n = 1, 2, 3$. The differential mode currents $I_{\text{dn}}(z)$, with forward and reflected traveling wave amplitudes I_{Fdn} and I_{Rdn}, are similarly described by Eq. 3-2, with $n = 1, 2$. The voltages are determined by the currents according to Eqs. 3-6 and 3-7.

To uniquely determine the currents and voltages, we must find ten unknown current amplitudes. They are I_{Fcn} and I_{Rcn} for $n = 1, 2, 3$, and I_{Fdn}

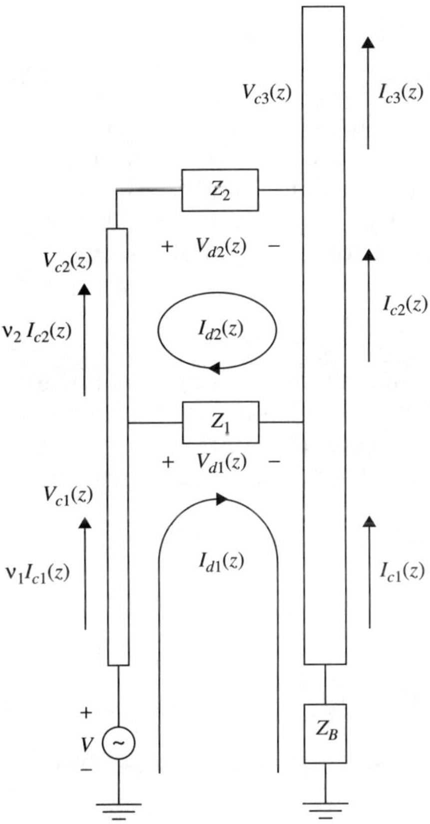

Figure 4-2 Voltages and currents used for analysis of the Double Gamma Match Antenna (2GMA).

and I_{Rdn}, for $n = 1, 2$. We use the same notation as in previous chapters. That is, the subscripts F and R denote forward and reflected waves, respectively. The subscripts c and d denote common and differential modes, respectively. To determine ten unknowns, we will need ten independent equations.

As in Sec. 3.3, one of those equations follows from a boundary condition at the top of the 2GMA. The common mode current must vanish there because it has nowhere to flow:

$$I_{c3}(h_3) = 0 \tag{4-1}$$

4.3 Equations at the Base of the 2GMA

In the very same manner described in Sec. 3.4, two more equations follow from constraints at the base of the 2GMA. Kirchhoff's voltage law (KVL) requires that:

$$Z_B[I_{d1}(0) - I_{c1}(0)] + V_{d1}(0) = V \tag{4-2}$$

Also, from inspection of the fold it is clear that:

$$V_{c1}(0) = V \tag{4-3}$$

Equations 4-2 and 4-3 are identical to Eqs. 3-10 and 3-11, respectively. As anticipated, a pattern in the mathematical model is emerging.

So far, we have formulated three independent equations for the ten unknown current amplitudes. The remaining seven equations must come from the boundaries between the antenna segments.

4.4 Equations at the Boundary $z = h_1$

Four equations describe constraints on the voltages and currents at the boundary between the base and middle segment, that is, at $z = h_1$. The equations are identical to those formulated in Sec. 3.14. This is because the geometry of the boundary there is identical to the one of interest here.

Both differential and common mode voltages must be continuous across the boundary:

$$V_{c1}(h_1) = V_{c2}(h_1) \tag{4-4}$$

$$V_{d2}(h_1) = V_{d1}(h_1) \tag{4-5}$$

KCL requires that the common mode current also be continuous across the boundary:

$$(1 + v_1) I_{c1}(h_1) = (1 + v_2) I_{c2}(h_1) \tag{4-6}$$

The fold impedance Z_1 constrains the differential mode voltage and currents according to the formula,

$$Z_1[I_{d1}(h_1) - I_{d2}(h_1)] = V_{d1}(h_1) \tag{4-7}$$

Equations 4-4 through 4-7 are the same as Eqs. 3-31 through 3-34, demonstrating again the recurring mathematical pattern when it comes to the folded unipole.

4.5 Equations at the Boundary $z = h_2$

So far, we have formulated seven independent equations. The remaining three will follow from constraints at the boundary between the middle segment and the extension, that is, at $z = h_2$.

In Sec. 4.2 through 4.4, the boundary conditions were familiar from our previous examples in Chap. 3. In many cases, they were identical.

That will be the case in this section. Except for changes in subscripts from n to $n + 1$, for $n = 1, 2$, the boundary conditions will be the same as those formulated in Sec. 3.5.

The common mode voltage must be continuous across the boundary:

$$V_{c2}(h_2) = V_{c3}(h_2) \tag{4-8}$$

According to KCL, the common mode currents must also be continuous:

$$(1 + \nu_2)I_{c2}(h_2) = I_{c3}(h_2) \tag{4-9}$$

Finally, for the differential mode, the impedance Z_2 defines the relationship between the voltage and current:

$$Z_2 I_{d2}(h_2) = V_{d2}(h_2) \tag{4-10}$$

Equations 4-1 through 4-10 are the ten independent equations we need to determine the ten unknown current coefficients I_{Fcn} and I_{Rcn}, for $n = 1, 2, 3$, and I_{Fdn} and I_{Rdn}, for $n = 1, 2$. Once these coefficients are determined, then we know the voltages and currents everywhere on the 2GMA.

As observed in each of Secs. 4.2 through 4.5, no new forms of boundary conditions have been introduced. They are all identical, except for a possible change in subscript, to boundary conditions formulated in Chap. 3. So, Eqs. 4-1 through 4-10 are a confirmation that our mathematical modeling tools and procedures are complete.

4.6 Matrix Equations for Currents on the 2GMA

After the examples in Chap. 3, consolidating Eqs. 4-1 through 4-10 into a single matrix equation should be a familiar procedure. The equation is:

$$\mathbf{M} \cdot \mathbf{I} = \begin{pmatrix} V \\ V \\ 0 \\ 0 \\ 0 \\ 0 \\ 0 \\ 0 \\ 0 \\ 0 \end{pmatrix} \tag{4-11}$$

In Eq. 4-11, the current vector is:

$$\mathbf{I} = \begin{pmatrix} \mathbf{I}_1 \\ \mathbf{I}_2 \\ I_{\mathrm{Fc3}} \\ I_{\mathrm{Rc3}} \end{pmatrix} \tag{4-12}$$

In Eq. 4-12, the four-element current subvectors $\mathbf{I}_1$ and $\mathbf{I}_2$ have been introduced previously by Eq. 3-37.

In Eq. 4-11, the matrix **M** has dimensions 10 × 10, and the current **I** is a ten-element vector. As we did in Chap. 3, we must partition **M** into four submatrices just as we did according to Eq. 3-17. Otherwise, the matrix elements would be too small to print legibly on the printed page. Each of the submatrices have dimensions 5 × 5. They are:

$$\mathbf{A}_{11} = \begin{pmatrix} Z_{0c1} & -Z_{0c1} & 0 & 0 & 0 \\ -Z_B & -Z_B & Z_B + Z_{0d1} & Z_B - Z_{0d1} & 0 \\ 0 & 0 & 0 & 0 & 0 \\ Z_{0c1}\psi_F(h_1) & -Z_{0c1}\psi_R(h_1) & 0 & 0 & -Z_{0c2}\psi_F(h_1) \\ 0 & 0 & Z_{0d1}\psi_F(h_1) & -Z_{0d1}\psi_R(h_1) & 0 \end{pmatrix} \tag{4-13}$$

$$\mathbf{A}_{12} = \begin{pmatrix} 0 & 0 & 0 & 0 & 0 \\ 0 & 0 & 0 & 0 & 0 \\ 0 & 0 & 0 & \psi_F(h_3) & \psi_R(h_3) \\ Z_{0c2}\psi_R(h_1) & 0 & 0 & 0 & 0 \\ 0 & -Z_{0d2}\psi_F(h_1) & Z_{0d2}\psi_R(h_1) & 0 & 0 \end{pmatrix} \tag{4-14}$$

$$\mathbf{A}_{21} = \begin{pmatrix} (1+v_1)\,\psi_F(h_1) & (1+v_1)\,\psi_R(h_1) & 0 & 0 & -(1+v_2)\,\psi_F(h_1) \\ 0 & 0 & (Z_1 - Z_{0d1})\,\psi_F(h_1) & (Z_1 + Z_{0d1})\,\psi_R(h_1) & 0 \\ 0 & 0 & 0 & 0 & Z_{0c2}\psi_F(h_2) \\ 0 & 0 & 0 & 0 & (1+v_2)\psi_F(h_2) \\ 0 & 0 & 0 & 0 & 0 \end{pmatrix} \tag{4-15}$$

$$\mathbf{A}_{22} = \begin{pmatrix} -(1+v_2)\,\psi_R(h_1) & 0 & 0 & 0 & 0 \\ 0 & -Z_1\psi_F(h_1) & -Z_1\psi_R(h_1) & 0 & 0 \\ -Z_{0c2}\psi_R(h_2) & 0 & 0 & -Z_{0c3}\psi_F(h_2) & Z_{0c3}\psi_R(h_2) \\ (1+v_2)\,\psi_R(h_2) & 0 & 0 & -\psi_F(h_2) & -\psi_R(h_2) \\ 0 & (Z_2 - Z_{0d2})\,\psi_F(h_2) & (Z_2 + Z_{0d2})\,\psi_R(h_2) & 0 & 0 \end{pmatrix} \tag{4-16}$$

In Eqs. 4-13 through 4-16, the forward and reflected traveling wave functions, $\psi_F(z)$ and $\psi_R(z)$, were previously defined by Eqs. 1-2 and 1-3.

Compared with the analogous equations for submatrices $\mathbf{A}_{11}$, $\mathbf{A}_{12}$, $\mathbf{A}_{21}$, and $\mathbf{A}_{22}$, in Sec. 3.6 and 3.15, it is seen that the matrix elements are perceptibly shrinking on the printed page. With the 10 × 10 matrix, we have reached the limit of legibility. If we were to continue in this manner for more complicated folded unipoles, then it would be necessary to partition the matrix **M** into more than four submatrices. Fortunately, for tutorial purposes, it will no longer be necessary to write down the matrix elements explicitly. The detailed formulation of the 2GMA has completed and confirmed the mathematical model for arbitrarily complicated folded unipole antennas.

It is not necessary to elaborate on the advantages of our matrix formulation in Eq. 4-11. This has already been done multiple times in Secs. 2.8, 3.6, and 3.15 for the matrix equations of less complicated folded unipoles. The chief advantages of computational speed and low numerical noise still apply, perhaps even more than previously.

Further, as observed in Sec. 3.15, we have already derived equations for the radiated electromagnetic field and power, input impedance, and efficiency. Equations 3-22 through 3-25 for field, input current, and dissipated power apply to the 2GMA. Equations 2-41 and 2-42 for efficiency also apply. So, we are ready to use our model to explore the properties and capabilities of the 2GMA.

4.7 Example: 0.3 Wavelength 2GMA with a 0.2 Wavelength Fold

This example straddles the boundary between short (i.e., less than one-quarter wavelength) and tall (greater than one-quarter wavelength) antennas. Other than that, it is purely arbitrary. We cannot exhaust the design possibilities within the space of this chapter or even the entire book. There are too many independent design variables, which, of course, are the main advantage of the folded unipole antenna. The best we can do is present the mathematical model and some examples that are hopefully interesting.

To that end, Fig. 4-3 shows the input impedance as a simultaneous function of the lower segment terminating reactance X_1 and the height of the lower segment h_1. The results are presented as a contour plot. We use the same color code as in Chap. 3. That is, for input resistance,

White	0–25 ohms
Dark green	45–55 ohms
Green	25–75 ohms
Yellow	75–150 ohms
Red	150–300 ohms
Gray	300 ohms and greater

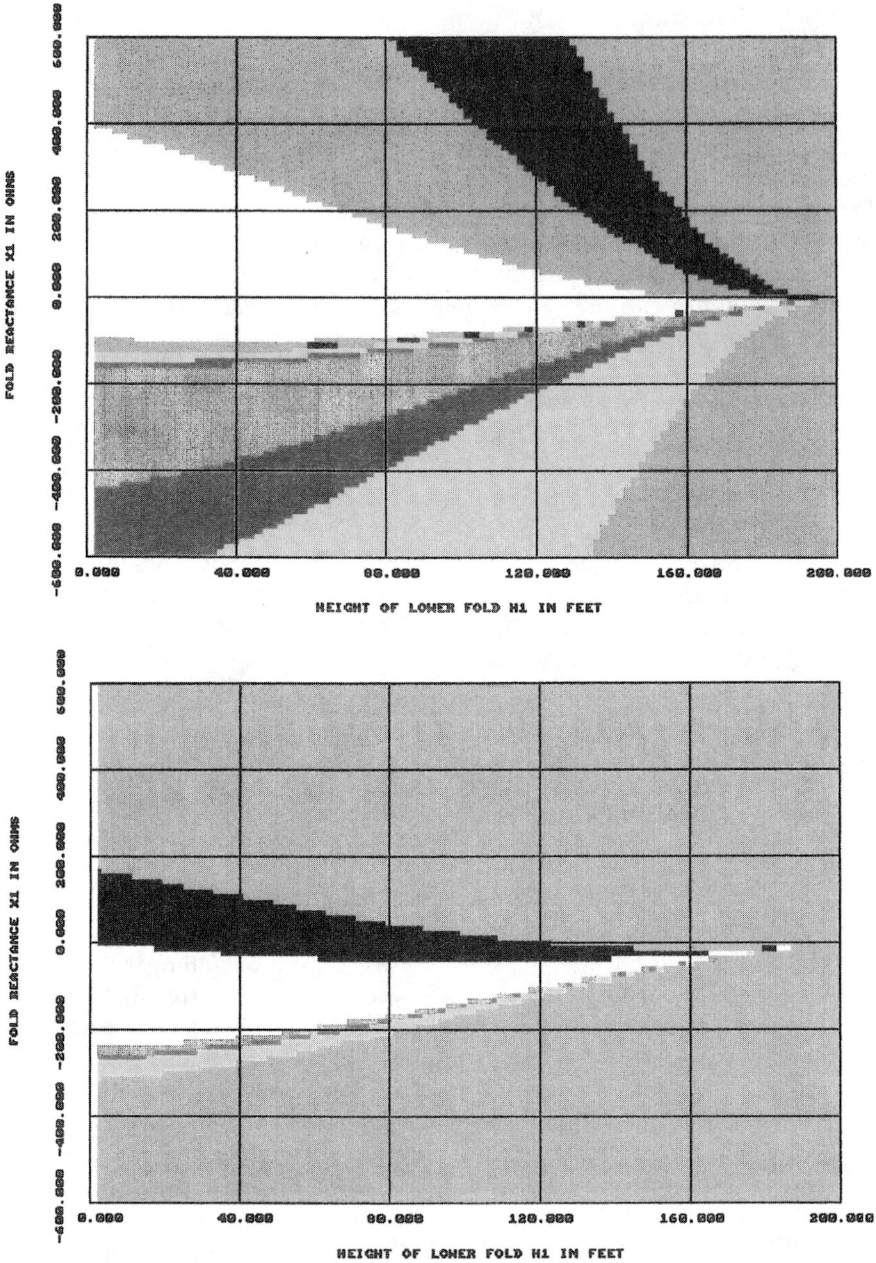

Figure 4-3 Contour plots are an efficient way to explore the Double Gamma Match Antenna (2GMA) when two design parameters are continuously and simultaneously varied. Here the two variables are the height of the lower fold h_1 and the terminating reactance X_1 for the lower fold, for a 300-foot antenna at 984 kHz. The two folds extend for 200 feet from the base of the antenna. The upper plot shows input resistance, and the lower plot shows input reactance. The correspondence between different colors and different ohmic ranges is described in Sec. 4.7. Note: See Preface, Notes concerning the Examples for explanation of the gray scale which has been subsituted for color.

For input reactance, the color scheme is:

White	Less than 0 ohms
Dark cyan	0–100 ohms
Cyan	100–1,000 ohms
Yellow	1,000–2,000 ohms
Magenta	2,000–3,000 ohms
Gray	3,000 ohms and greater

It is seen that over a wide range of the design variables the input resistance is optimal and the input reactance is near optimal.

As in previous examples, many of the design variables, especially those with less dramatic effect, are held constant at "ballpark" values. For the folded segments, the characteristic impedances are $Z_{0c} = 1{,}200$ ohms and $Z_{0d} = 120$ ohms. The current transformation ratio is 0.5. For the extension, the common mode characteristic impedance is $Z_{0c3} = 1{,}600$ ohms. In Chap. 7, we will show how to calculate these design variables exactly for arbitrary antenna cross sections. For now, however, these typical values serve our purpose in the illustrative examples.

Figure 4-4 is similar to Fig. 4-3 except that the base terminating reactance X_B is the continuous variable instead of the fold terminating reactance. It is seen that optimal input resistance is achieved over a much wider range of variables. Evidently, we get more dramatically beneficial results by varying this reactance, at least for this particular geometry.

4.8 More Examples: A Variable Length Antenna

While in the previous section we examined a 0.3 wavelength antenna, in this section we will examine longer ones. We will fix the z-coordinate h_2 of the upper fold at 0.3 wavelength. The resulting geometry is of practical interest. In the AM broadcast band, where dimensions are on the order of hundreds of feet, we are exploring the feasibility of a universal kit for impedance matching towers varying from 0.3 to 1.5 wavelengths, or from 108 to 540 electrical degrees. Such a kit would free the upper part of the tower for mounting auxiliary antennas and other devices.

In the worlds of instrumentation and medicine where dimensions might be on the order of centimeters, the geometry describes a variable length probe. The fixed dimension h_2 allows us to control the coupling of signals in and out of the probe from one end of the antenna.

With the above applications in mind, Figs. 4-5 through 4-9 show the input impedance as a simultaneous function of antenna length h_3 and

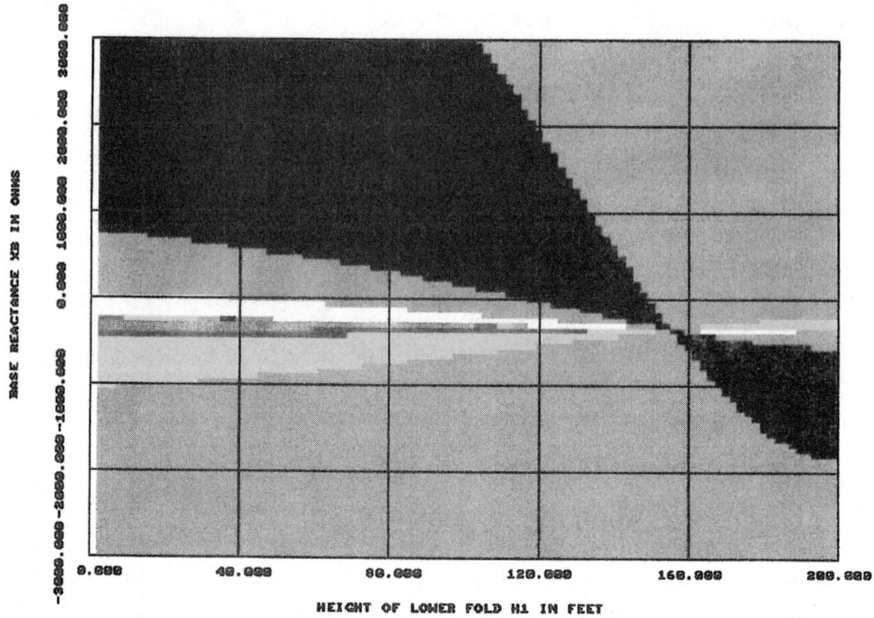

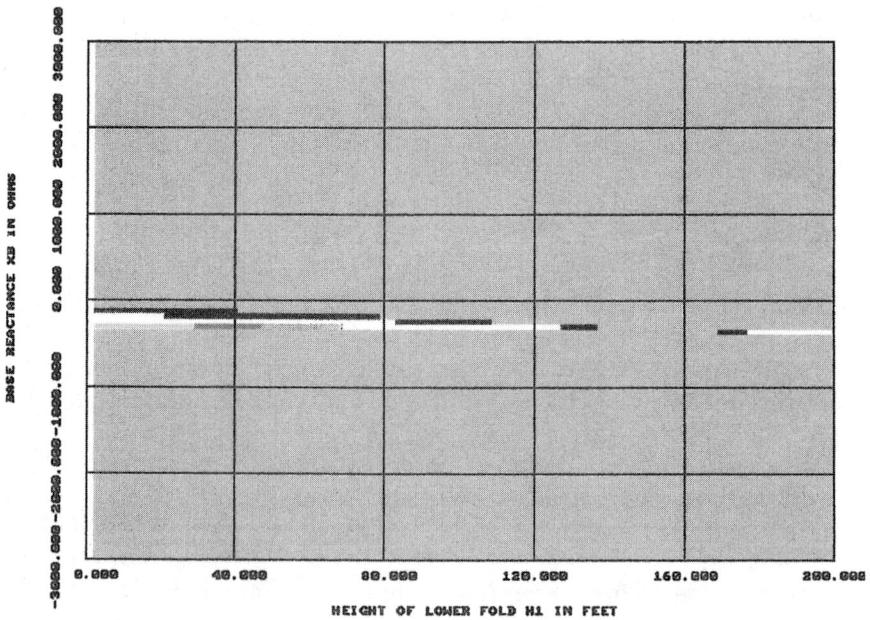

Figure 4-4 Input impedance for a 2GMA when the height of the lower fold h_1 and the base terminating reactance X_B are continuously and simultaneously varied. The terminating reactance for the lower fold is fixed at $X_1 = 300$ ohms. Other dimensions of the antenna are the same as in the previous figure. The large, uninterrupted regions of dark green and cyan indicate the capability for broadband operation.

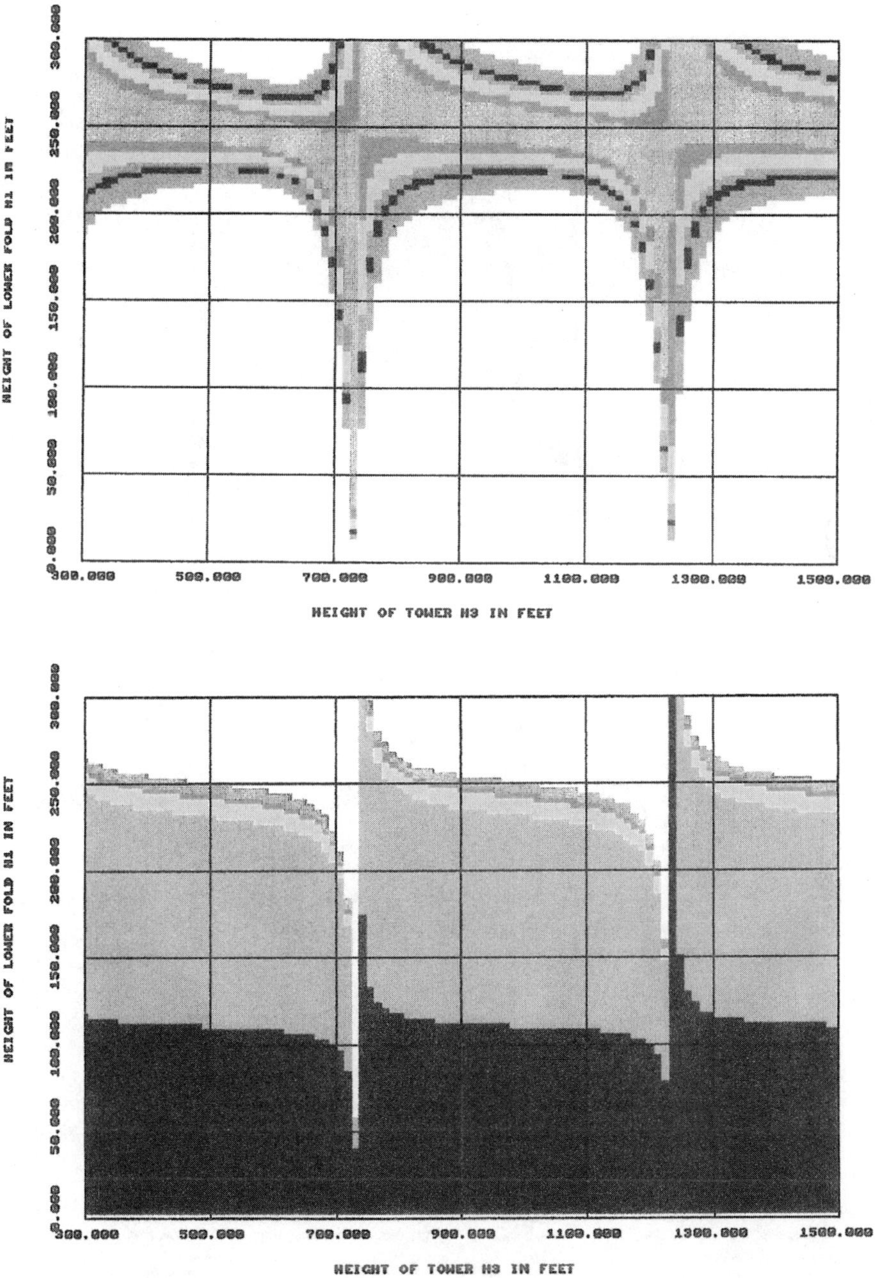

Figure 4-5 Input impedance for a variable length antenna with fixed total fold length of 0.3 wavelength. This could be a standard kit for the AM broadcast band where dimensions are on the order of hundreds of feet, or an extendable probe for instrumentation where dimensions are on the order of centimeters. The two continuous variables are antenna length h_3 and lower fold length h_1. The terminating base reactance is fixed at $X_B = 0$

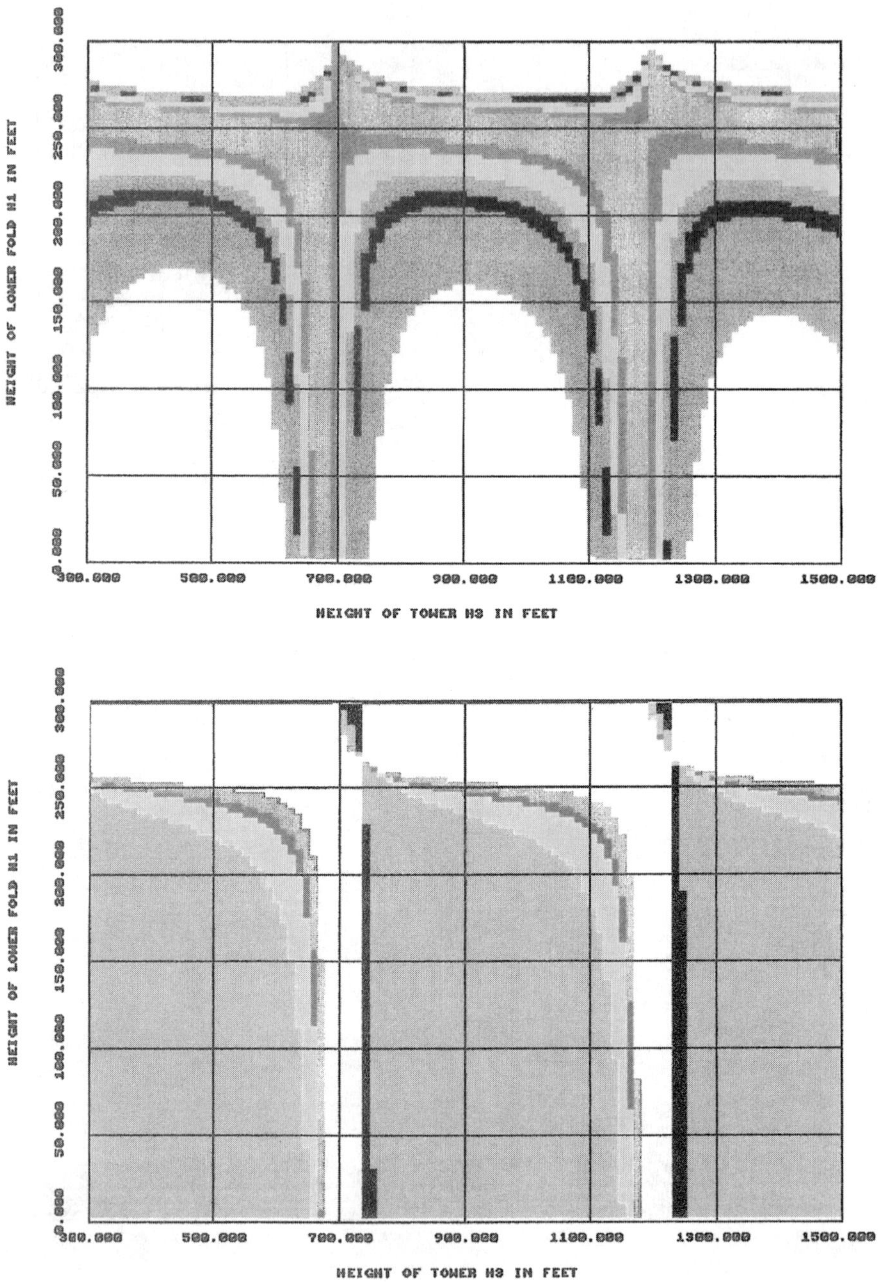

Figure 4-6 Input impedance for a variable length antenna with fixed total fold length of 0.3 wavelength. It is the same antenna as in Fig. 4-5, except that the terminating base reactance is fixed at $X_B = 375$ ohms. For inductive terminating reactances, useful contours extend over large areas of the plot; however, there are significant gaps, for e.g., in the vicinity of $h_3 = 700$ feet and $h_3 = 1{,}200$ feet.

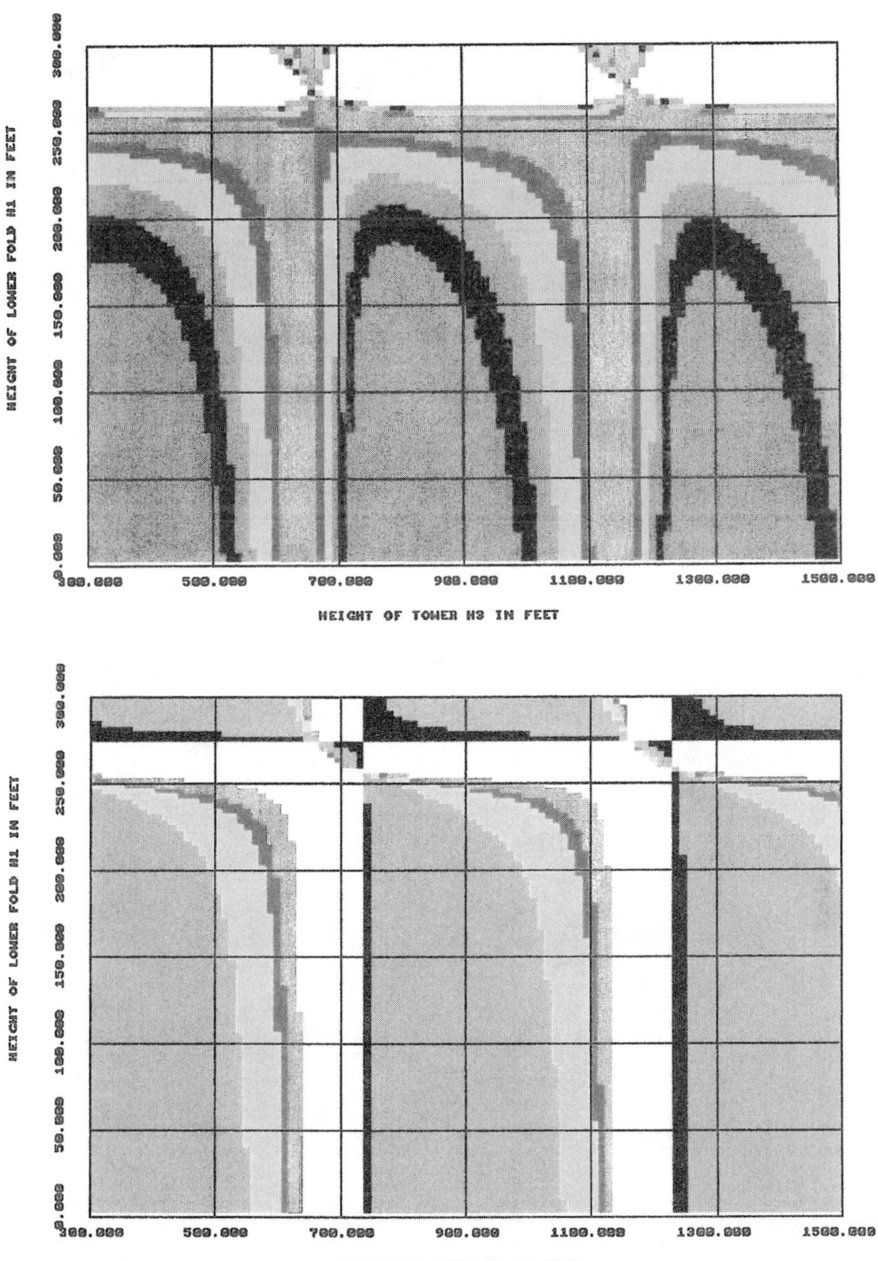

Figure 4-7 Input impedance for a variable length antenna with fixed total fold length of 0.3 wavelength. It is the same antenna as in Fig. 4-5, except that the terminating base reactance is fixed at $X_B = 750$ ohms.

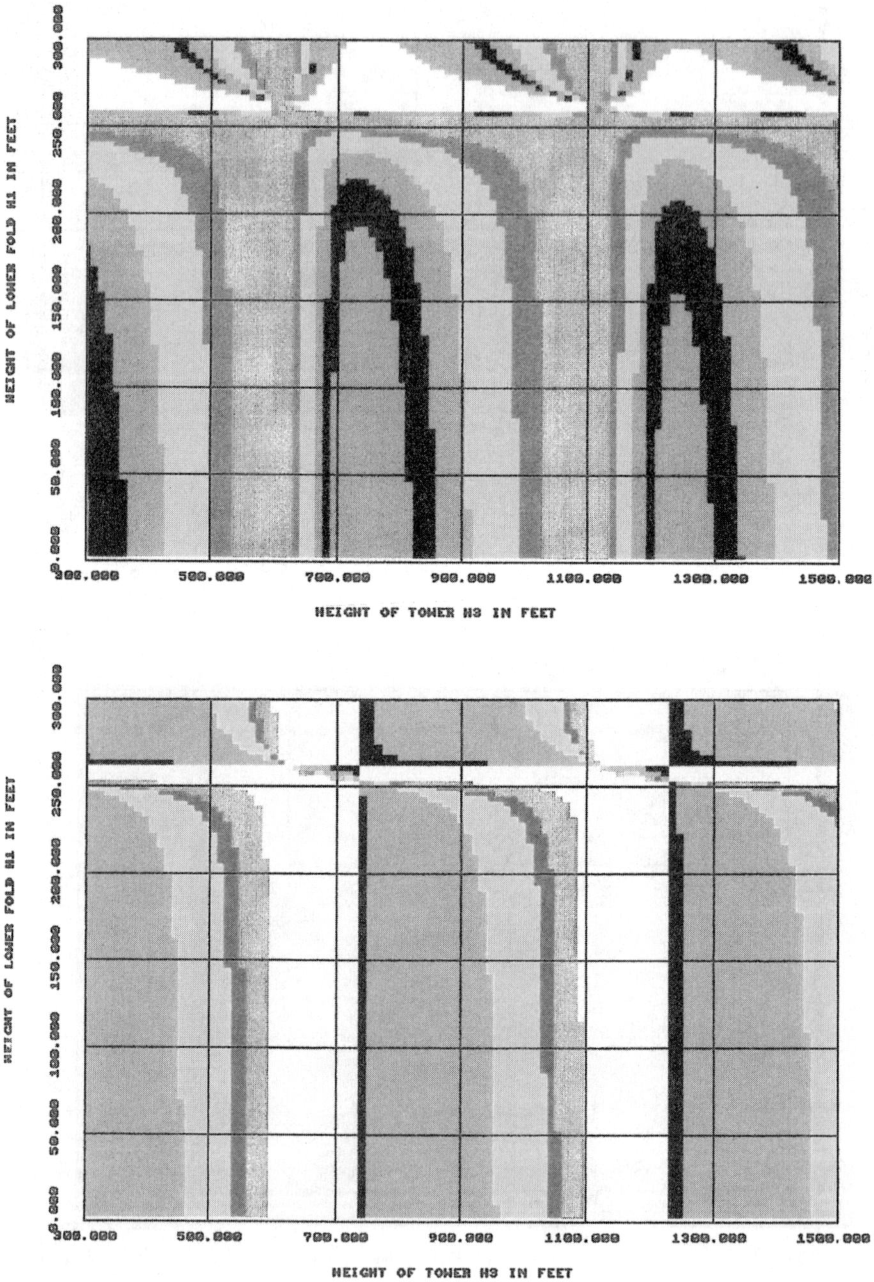

Figure 4-8 Input impedance for a variable length antenna with fixed total fold length of 0.3 wavelength. It is the same antenna as in Fig. 4-5, except that the terminating base reactance is fixed at $X_B = 1{,}500$ ohms.

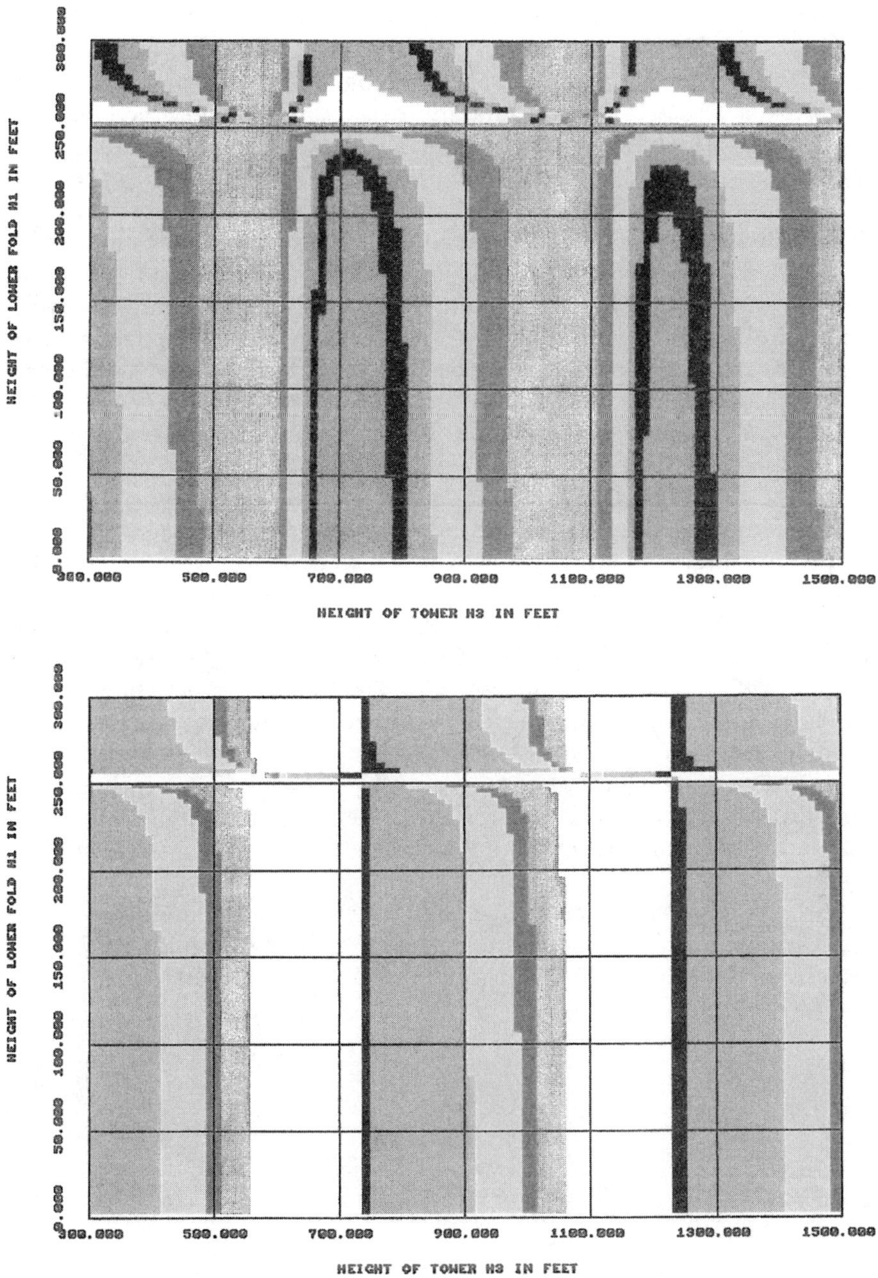

Figure 4-9 Input impedance for a variable length antenna with fixed total fold length of 0.3 wavelength. It is the same antenna as in Fig. 4-5, except that the terminating base reactance is fixed at $X_B = 3{,}000$ ohms.

the length of the lower fold h_1. The only variable change between the figures is the value of the base terminating reactance X_B, which ranges from 0 ohm in Fig. 4-5 to 3,000 ohms in Fig. 4-9. It is seen that optimal input resistance and near-optimal input reactance are obtained for most antenna lengths; however, there are some gaps, especially in the vicinity of 0.65 and 1.2 wavelengths.

In Figs. 4-10 through 4-13, the design variables are the same except that the base terminating reactance is capacitive instead of inductive. X_B is stepped in discrete values from –375 ohms through –3,000 ohms. The gaps in useful input impedance are mostly closed compared with Figs. 4-5 through 4-9; however, the required values for the design variables are more critical. That is, the regions of the contour plots corresponding to optimal and near-optimal values of input impedance are small. With so many independent design variables, there are plenty of possiblities left to explore.

To that end, Figs. 4-14 and 4-15 show the effect of stepping the lower fold terminating reactance X_1. It is seen that there are gaps in useful input impedance as before; however, those gaps are narrower. So we have made some progress.

In Figs. 4-16 through 4-19, both terminating reactances X_B and X_1 are stepped through discrete combinations. Of the four combinations shown, Fig. 4-17, with $X_B = -750$ ohms and $X_1 = 75$ ohms appears to have closed all of the gaps. Fig. 4-20 is a detailed look at that combination in the vicinity of $h_1 = 175$ feet.

Figure 4-21 shows an alternative to the variables used in Fig. 4-20. Instead of varying the length of the lower fold, that length is fixed. Instead, the base terminating reactance X_B is varied, with similar results.

With so many independent design variables available, these preliminary examples strongly support the feasibility for a kit in the AM broadcast band, and for a variable length probe for instrumentation and medicine.

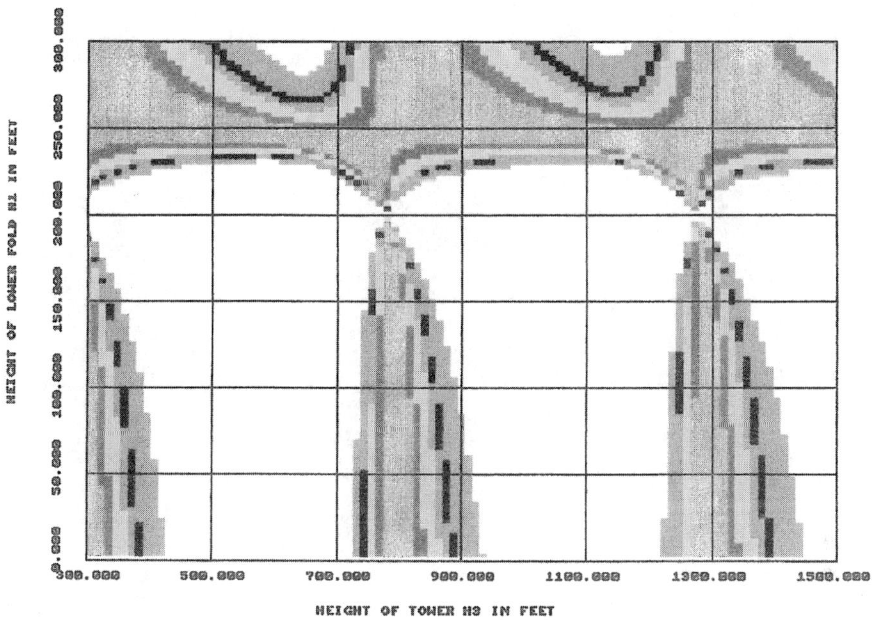

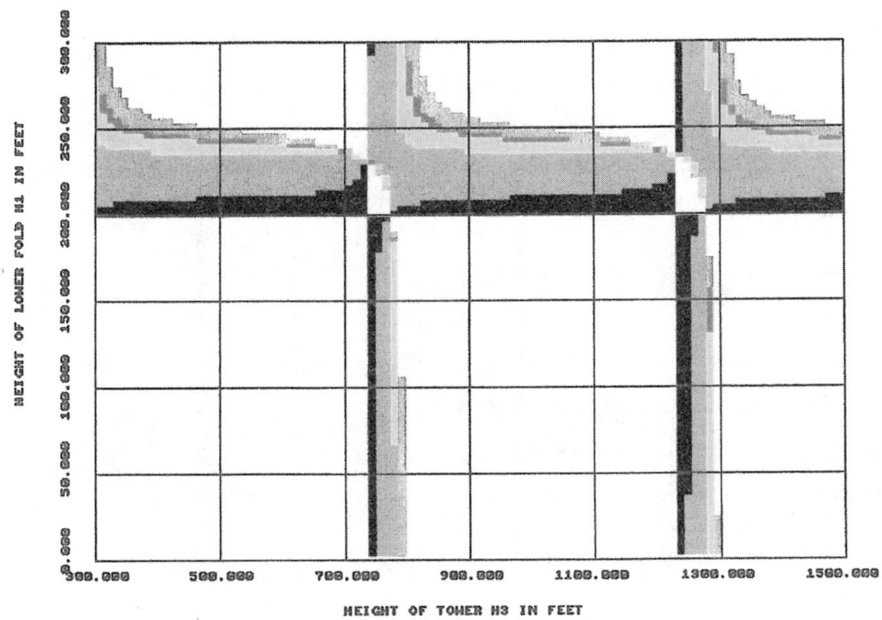

Figure 4-10 Input impedance for a variable length antenna with fixed total fold length of 0.3 wavelength. It is the same antenna as in Fig. 4-5, except that the terminating base reactance is fixed at $X_B = -375$ ohms. Compared with inductive terminating base reactances, the useful contours are not as large; however, they do close most of the gaps evident in Figs. 4-5 through 4-9.

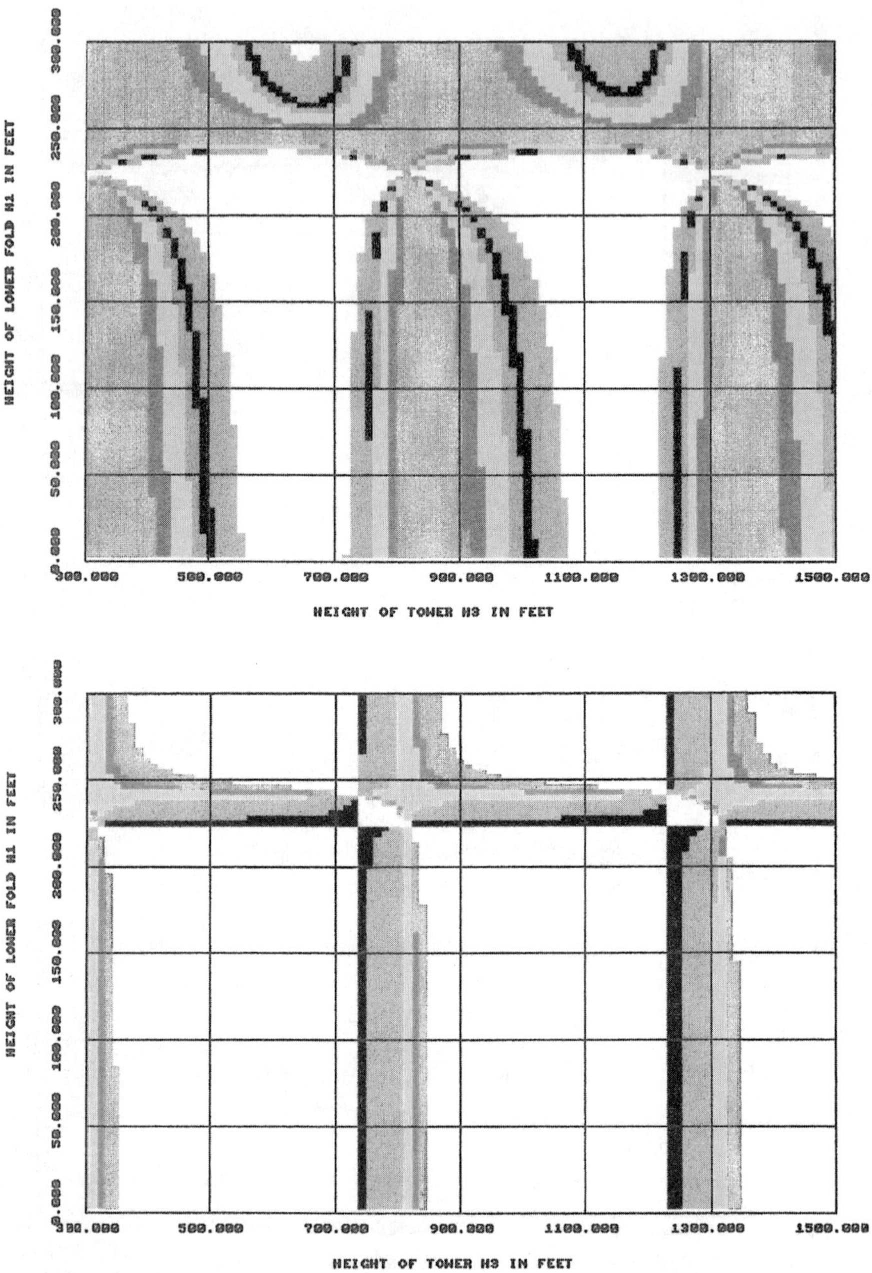

Figure 4-11 Input impedance for a variable length antenna with fixed total fold length of 0.3 wavelength. It is the same antenna as in Fig. 4-5, except that the terminating base reactance is fixed at $X_B = -750$ ohms.

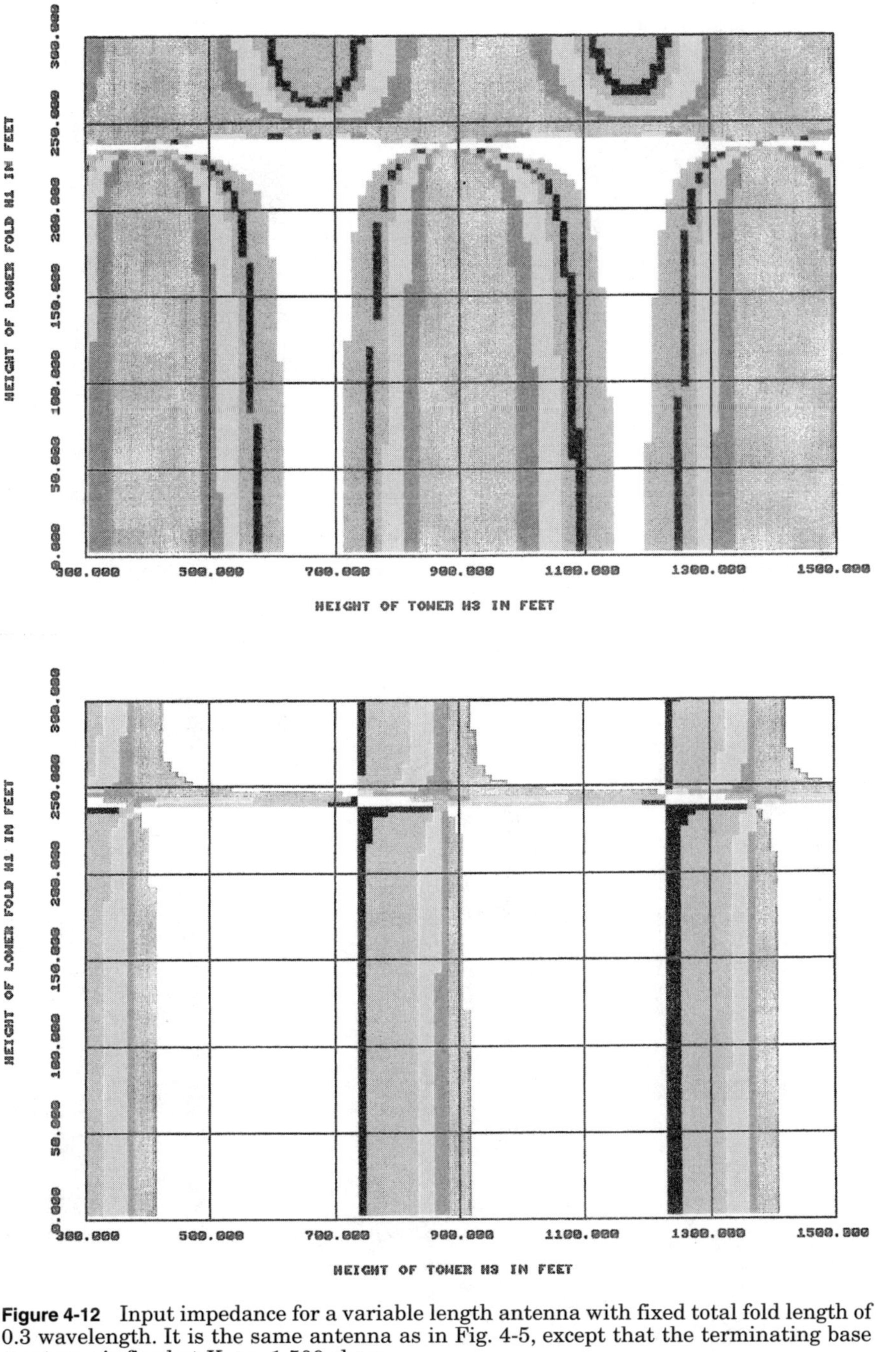

Figure 4-12 Input impedance for a variable length antenna with fixed total fold length of 0.3 wavelength. It is the same antenna as in Fig. 4-5, except that the terminating base reactance is fixed at $X_B = -1{,}500$ ohms.

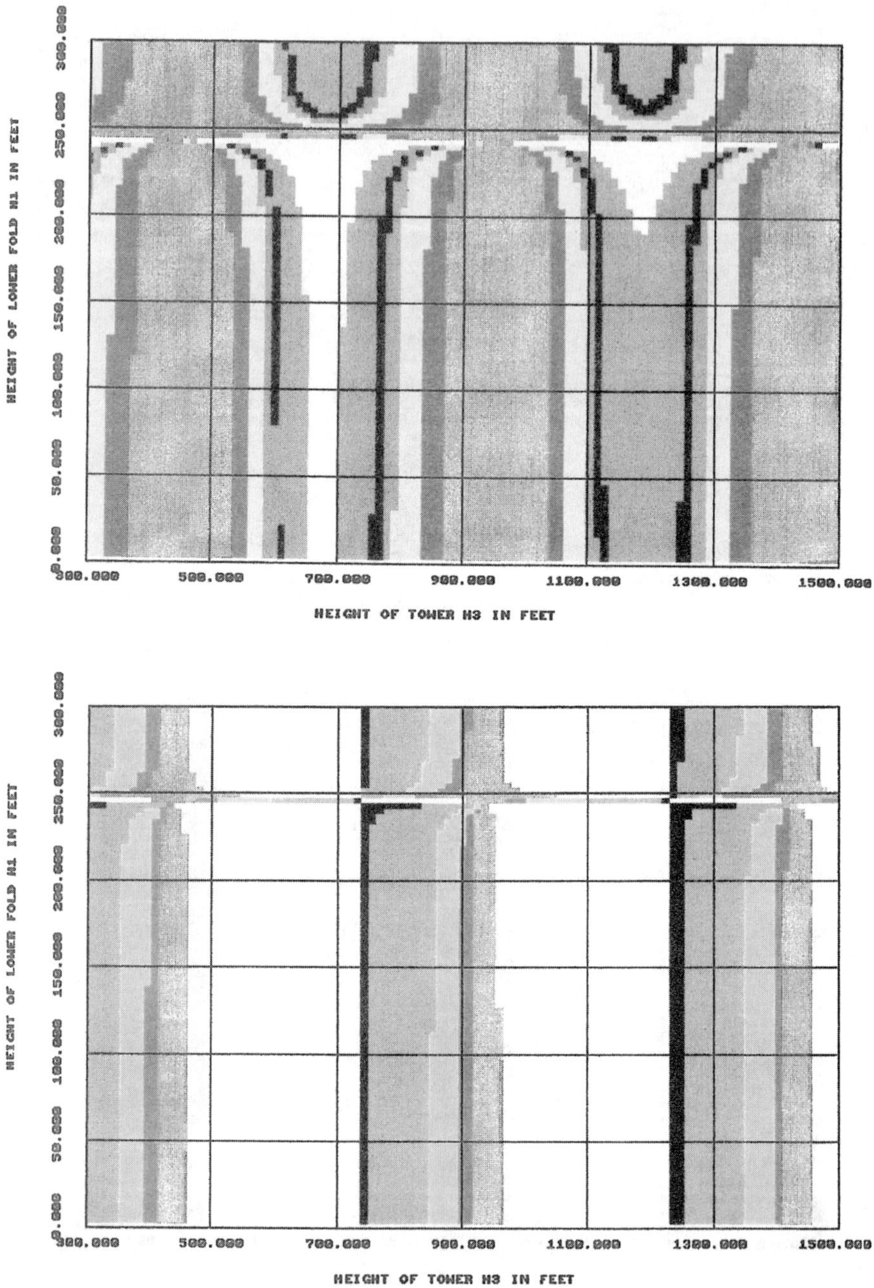

Figure 4-13 Input impedance for a variable length antenna with fixed total fold length of 0.3 wavelength. It is the same antenna as in Fig. 4-5, except that the terminating base reactance is fixed at $X_B = -3{,}000$ ohms.

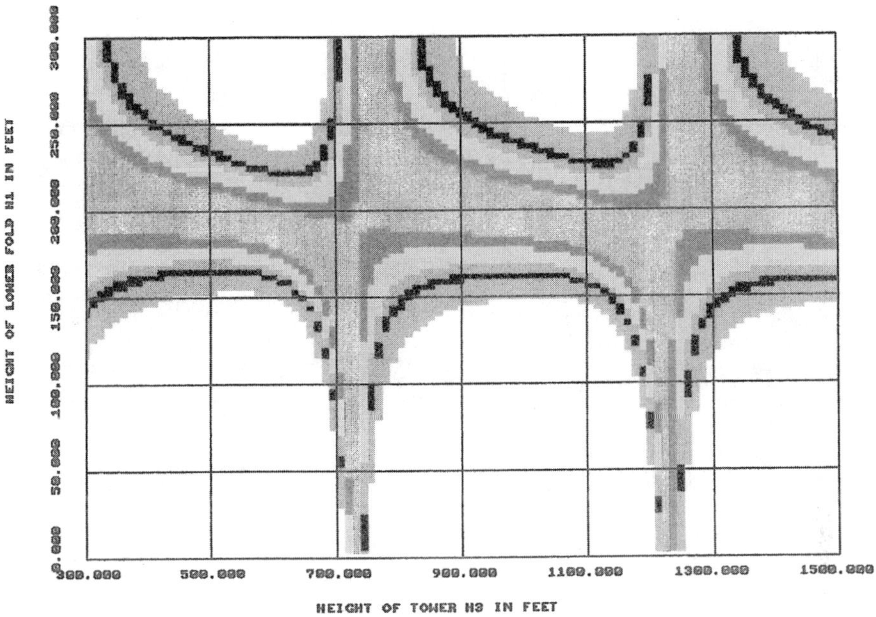

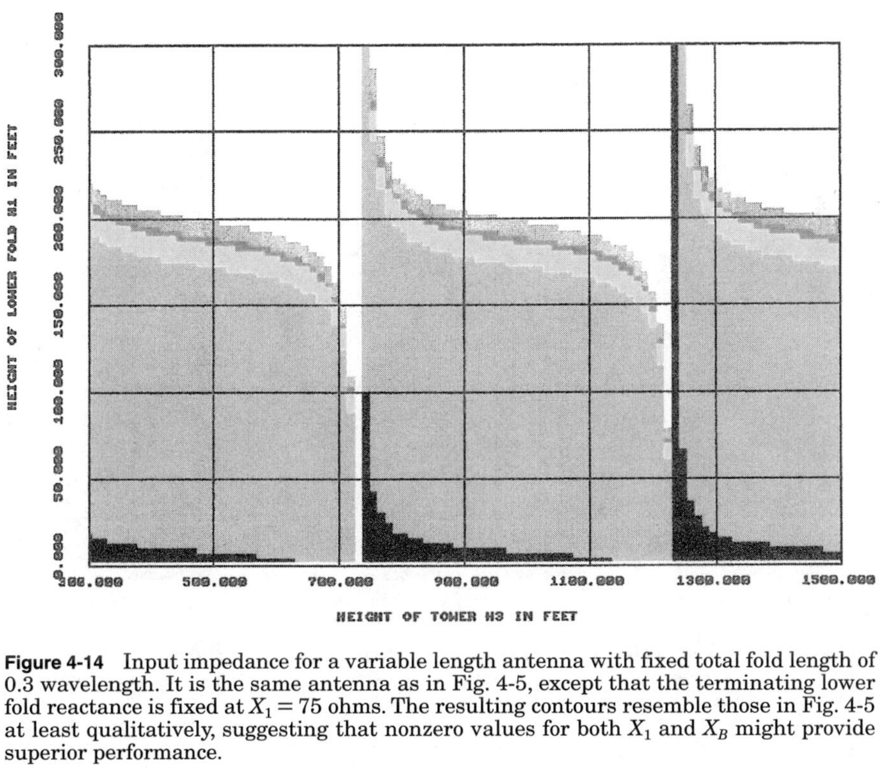

Figure 4-14 Input impedance for a variable length antenna with fixed total fold length of 0.3 wavelength. It is the same antenna as in Fig. 4-5, except that the terminating lower fold reactance is fixed at $X_1 = 75$ ohms. The resulting contours resemble those in Fig. 4-5 at least qualitatively, suggesting that nonzero values for both X_1 and X_B might provide superior performance.

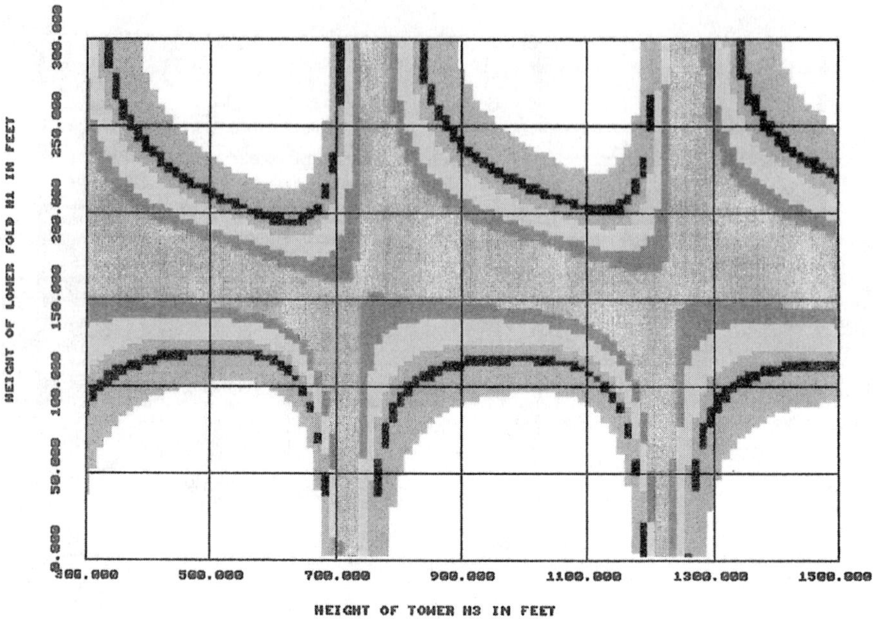

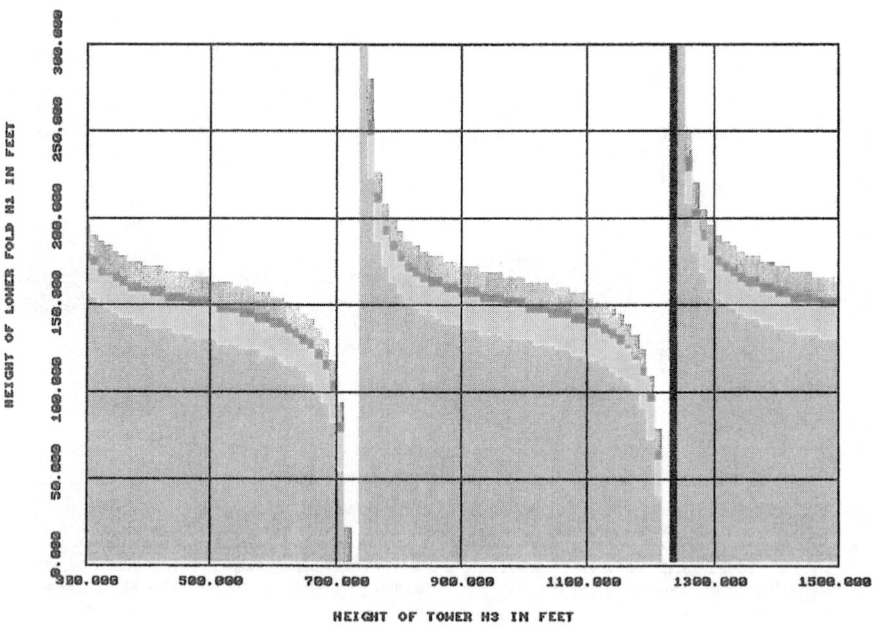

Figure 4-15 Input impedance for a variable length antenna with fixed total fold length of 0.3 wavelength. It is the same antenna as in Fig. 4-5, except that the terminating lower fold reactance is fixed at $X_1 = 150$ ohms.

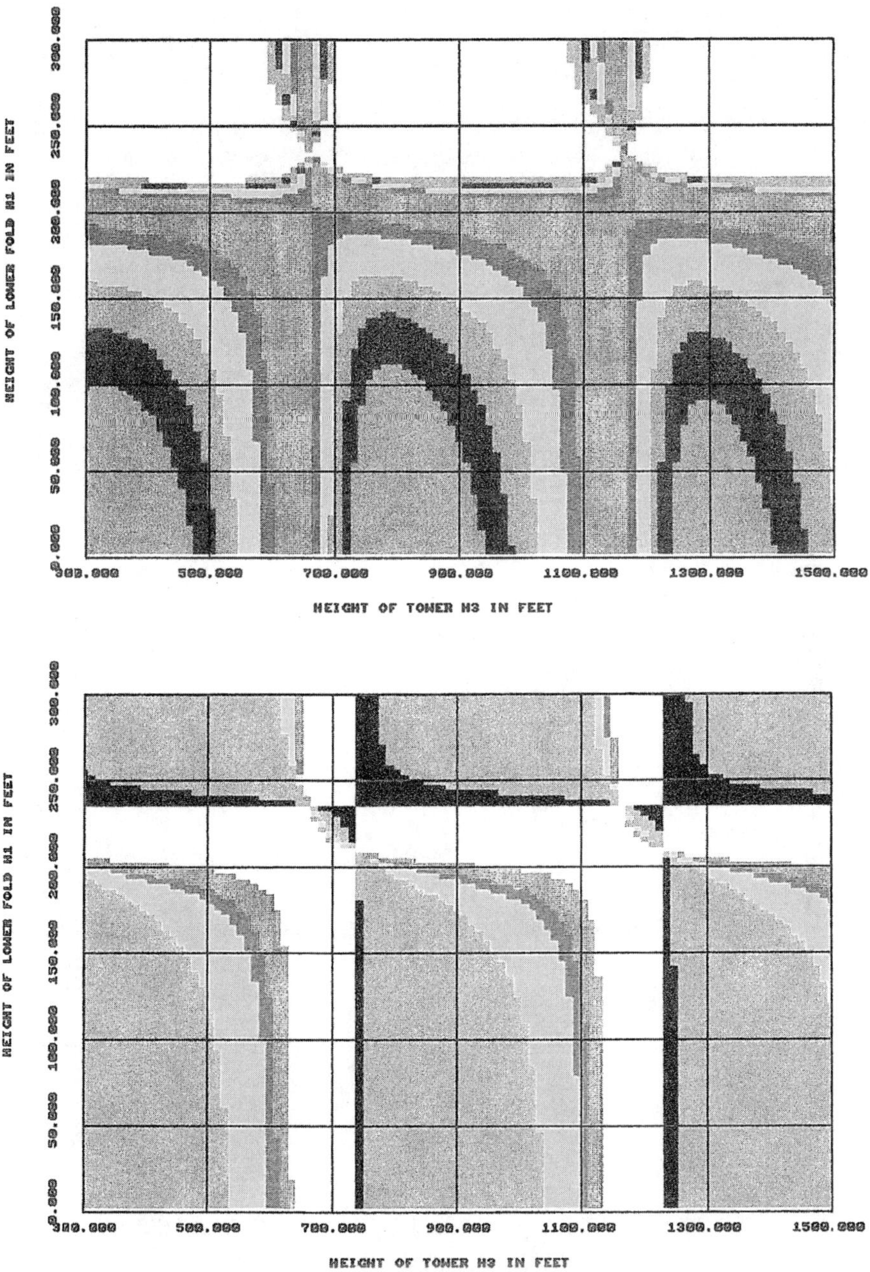

Figure 4-16 Input impedance for a variable length antenna with fixed total fold length of 0.3 wavelength. It is the same antenna as in Fig. 4-5, except that both the terminating base reactance and terminating lower fold reactance are fixed at nonzero values. $X_B = 750$ ohms, and $X_1 = 75$ ohms.

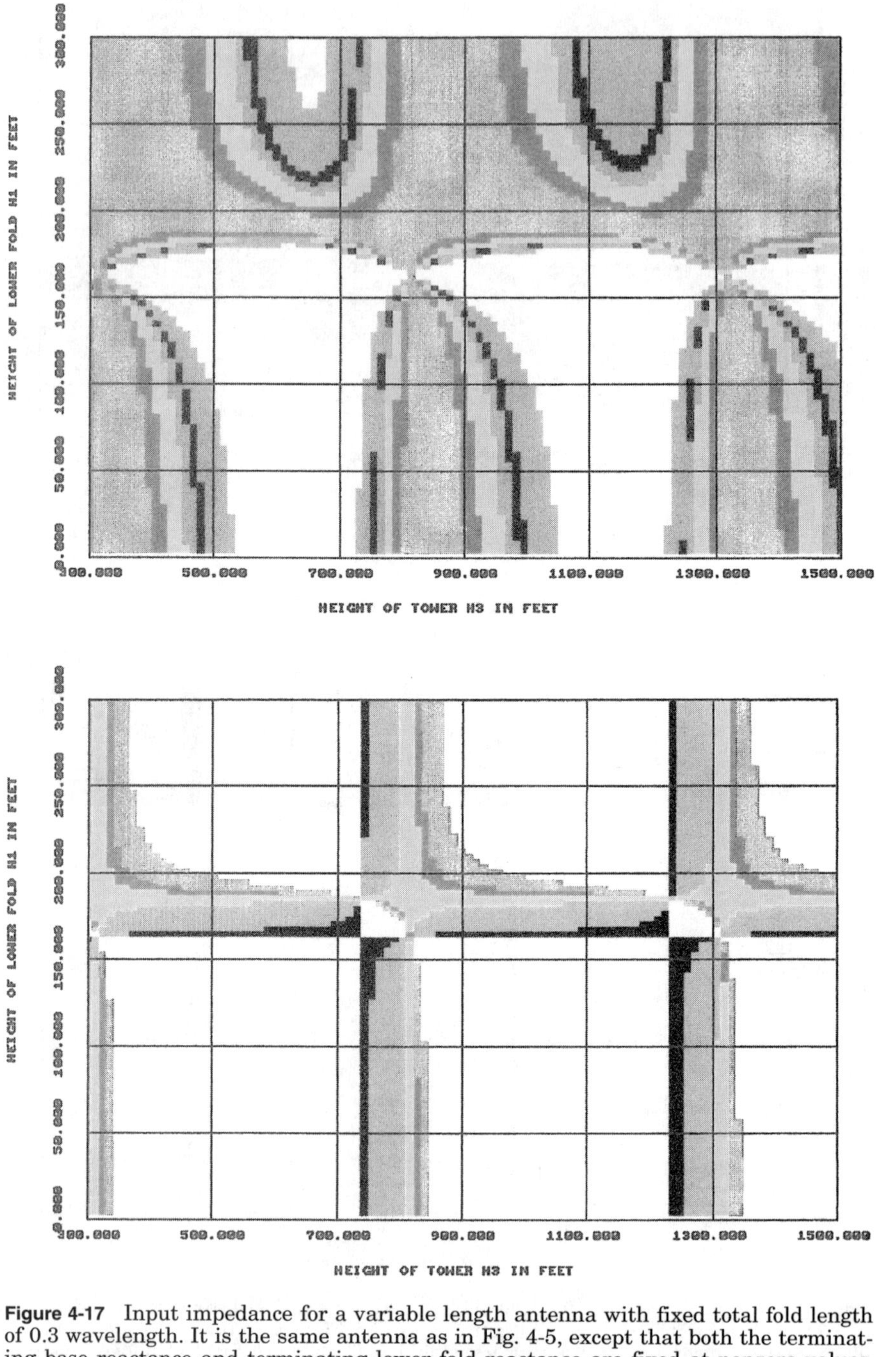

Figure 4-17 Input impedance for a variable length antenna with fixed total fold length of 0.3 wavelength. It is the same antenna as in Fig. 4-5, except that both the terminating base reactance and terminating lower fold reactance are fixed at nonzero values. $X_B = -750$ ohms, and $X_1 = 75$ ohms. Of the four combinations shown in Figs. 4-16 through 4-19, this one appears to close most of the gaps in useful input impedance.

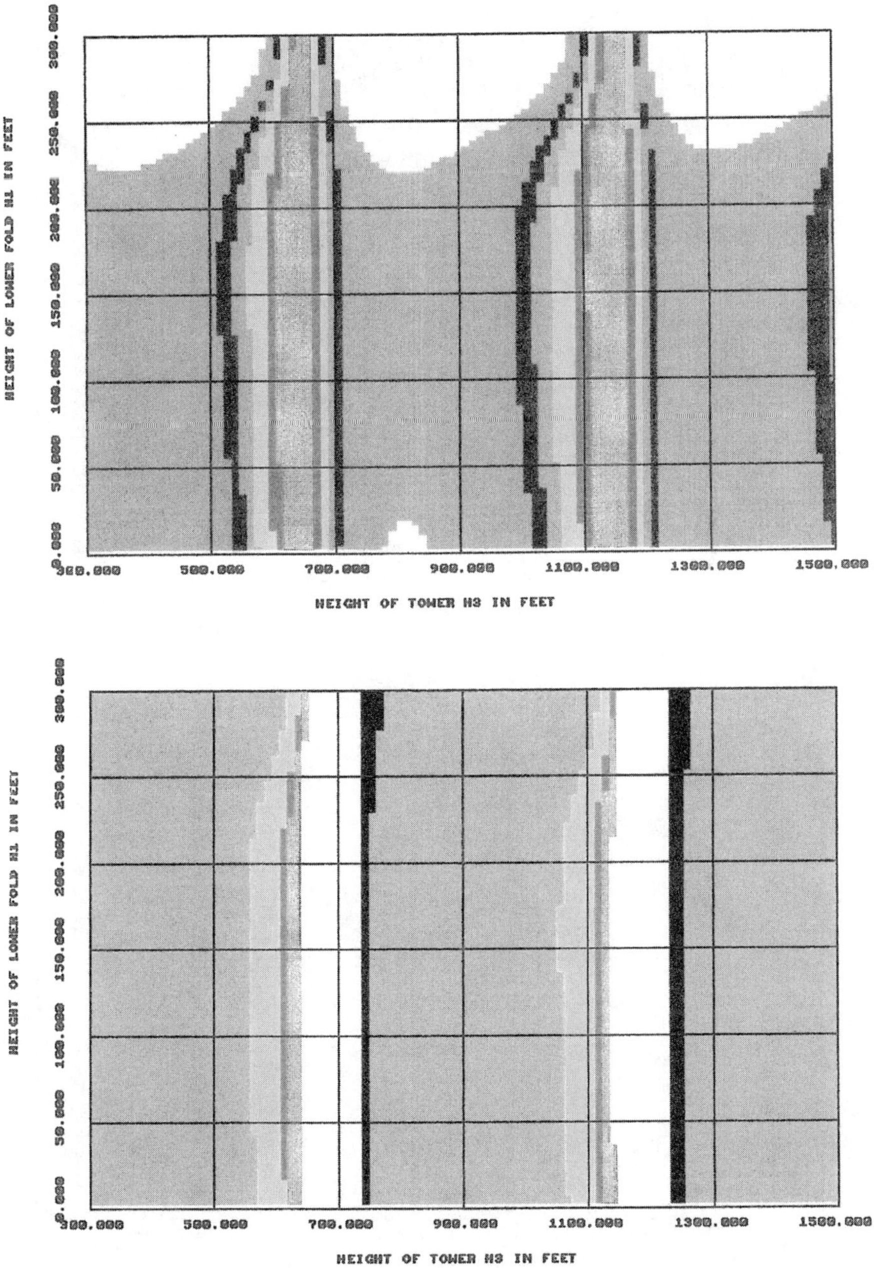

Figure 4-18 Input impedance for a variable length antenna with fixed total fold length of 0.3 wavelength. It is the same antenna as in Fig. 4-5, except that both the terminating base reactance and terminating lower fold reactance are fixed at nonzero values. $X_B = 750$ ohms, and $X_1 = -75$ ohms.

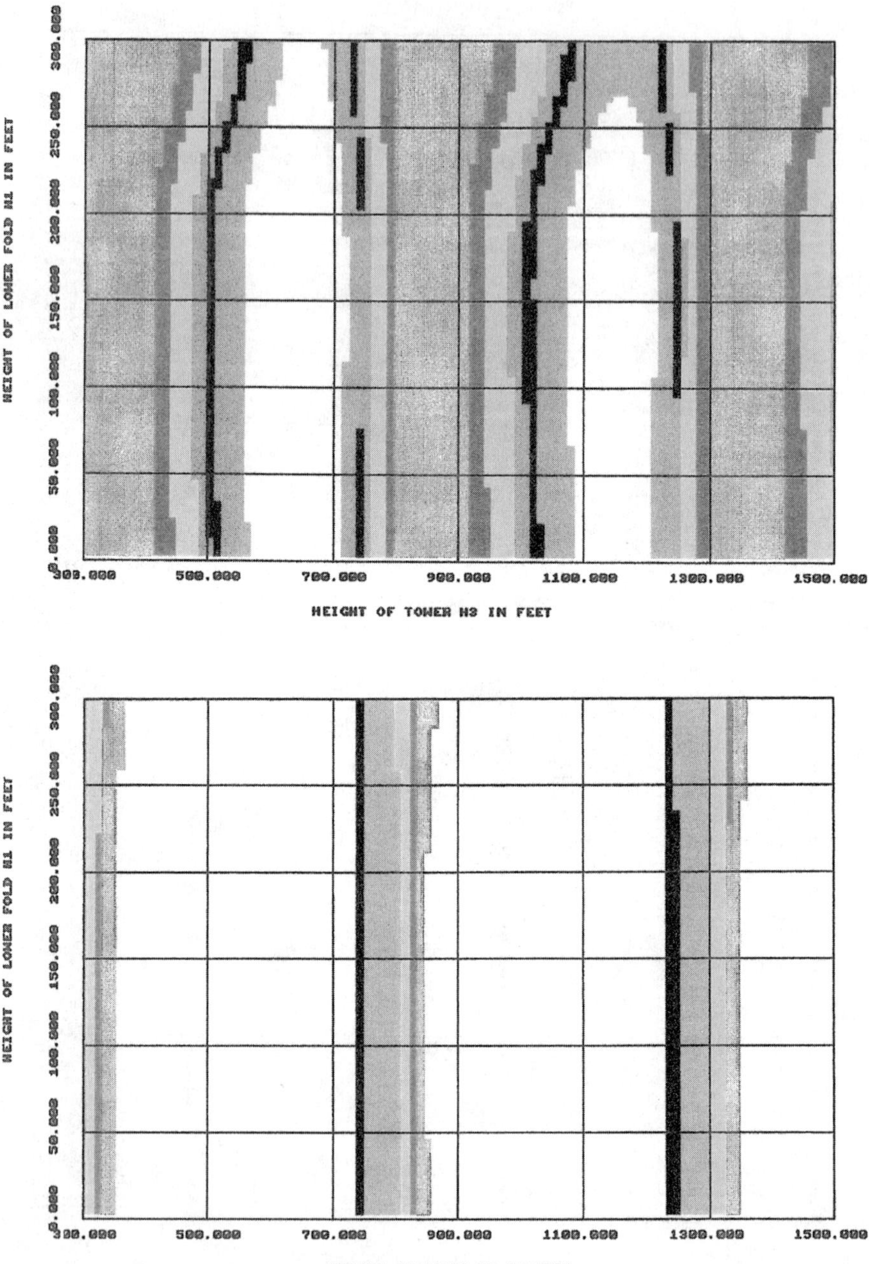

Figure 4-19 Input impedance for a variable length antenna with fixed total fold length of 0.3 wavelength. It is the same antenna as in Fig. 4-5, except that both the terminating base reactance and terminating lower fold reactance are fixed at nonzero values. $X_B = -750$ ohms, and $X_1 = -75$ ohms.

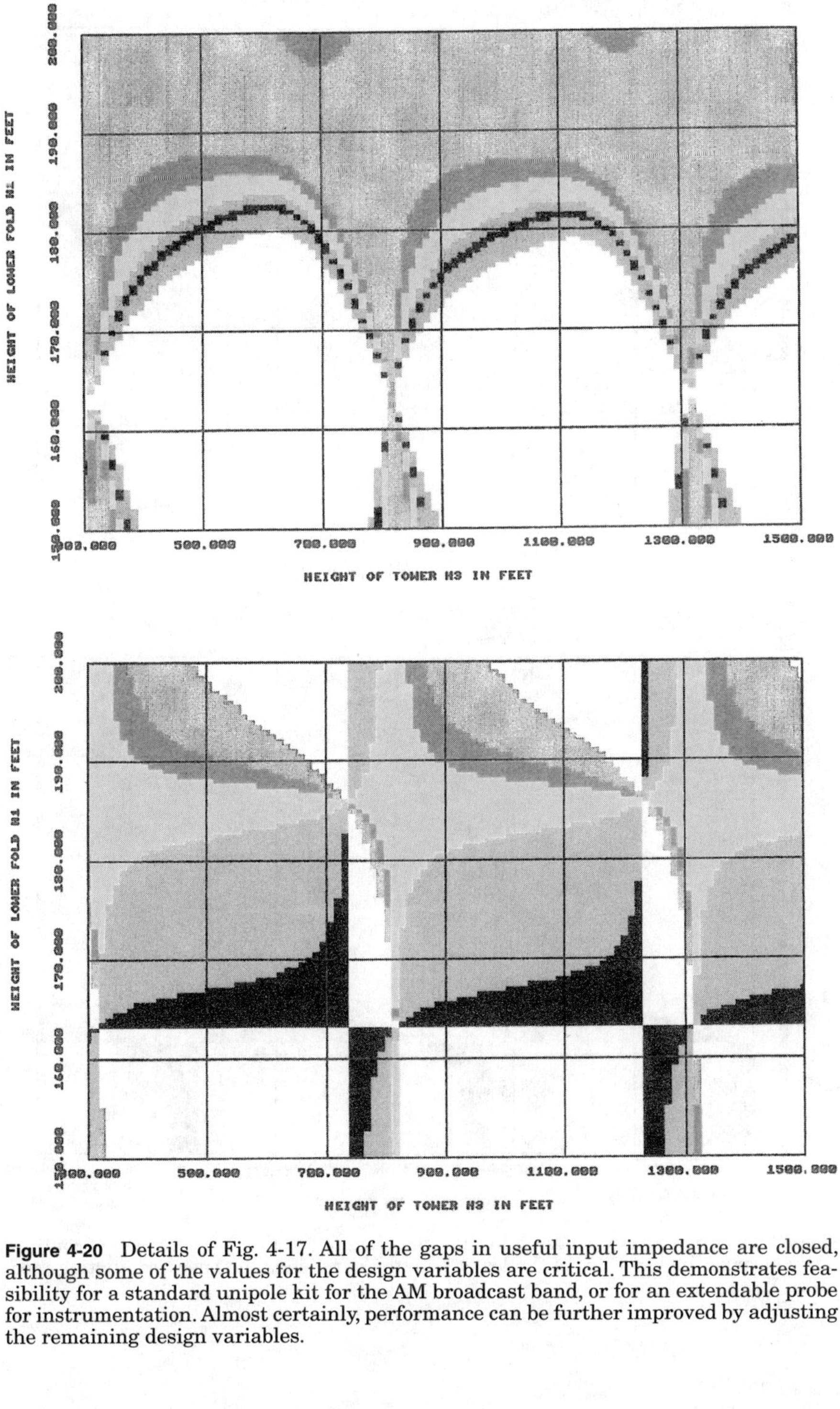

Figure 4-20 Details of Fig. 4-17. All of the gaps in useful input impedance are closed, although some of the values for the design variables are critical. This demonstrates feasibility for a standard unipole kit for the AM broadcast band, or for an extendable probe for instrumentation. Almost certainly, performance can be further improved by adjusting the remaining design variables.

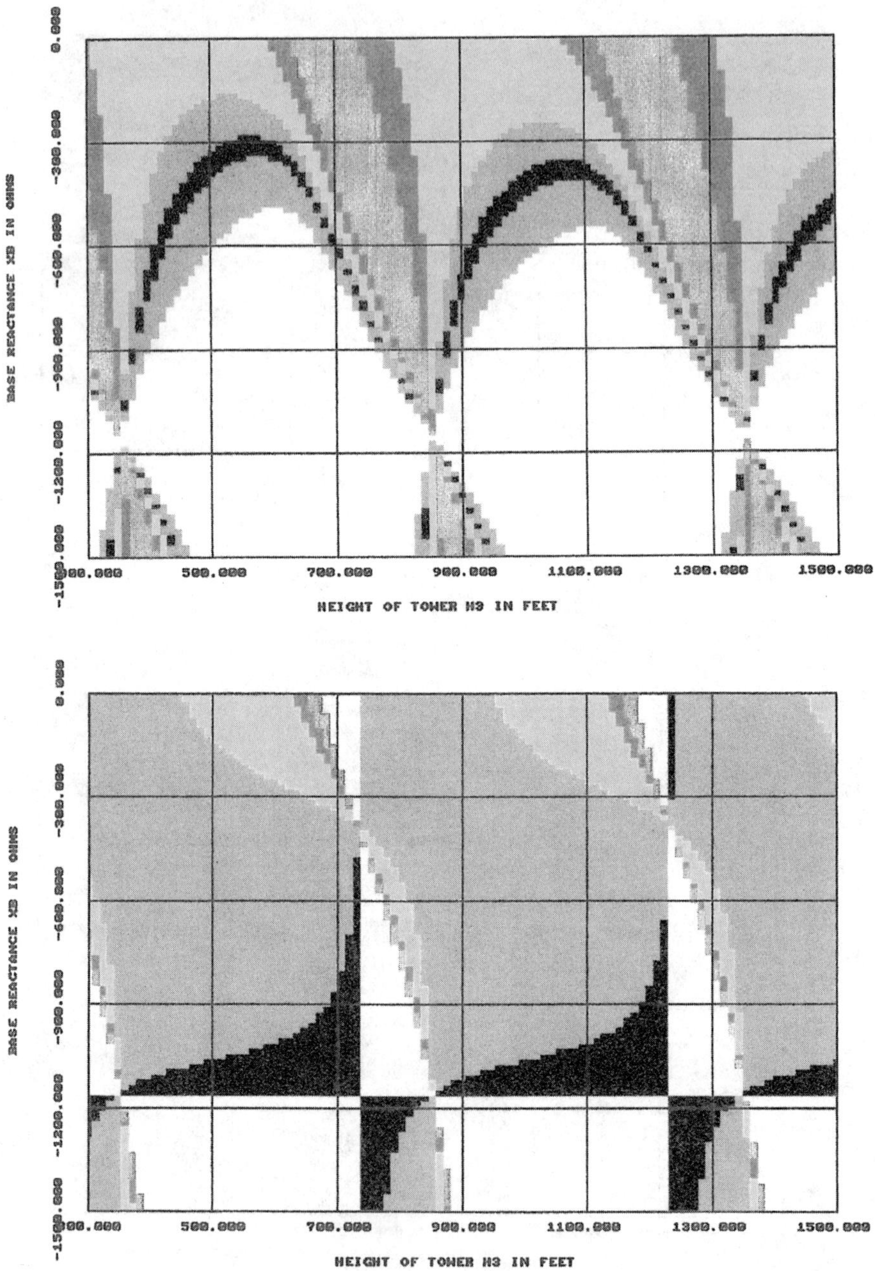

Figure 4-21 An alternative design to the one presented in Fig. 4-20. Instead of continuously varying the length h_1 of the lower fold, the variable is the base terminating reactance X_B.

4.9 Conclusions

In this chapter, we flexed the mathematical model which we began in Chap. 2 and completed in Chap. 3. We found that it is readily extended to analyze and design more complicated configurations of the folded unipole antenna, for example, the 2GMA which was the subject of this chapter.

There are at least two predictable engineering consequences from increasing the complexity of the folded unipole. First, when we add a nonfolded segment, then we add two more independent design variables. Second, when we add a folded segment, then we add five more independent design variables. Each additional variable provides more versatility in antenna design.

The 2GMA has no fewer than 13 independent design variables. We hardly began to explore the possible applications for such a versatile antenna. The two that we did pause to investigate were a universal kit for the AM broadcast band, where dimensions are on the order of hundreds of feet, and a variable length instrumentation probe, where dimensions could be on the order of centimeters. We found that a folded section of fixed length could provide an optimal or near-optimal input impedance to an antenna of variable length.

There are also predictable mathematical consequences from increasing the complexity of the folded unipole. When we add a nonfolded segment, then each dimension of our matrix equation increases by two because we have two more common mode current amplitudes I_{Fc} and I_{Rc} to determine. Thus, adding a nonfolded segment to the 2SFU to create a 2GMA caused the matrix equation to increase from size 8×8 to 10×10. When we add a folded segment, then each dimension increases by four because we have both common and differential mode current amplitudes (I_{Fc}, I_{Rc}, I_{Fd}, and I_{Rd}) to determine. Despite these increases, we are still very much ahead of the game compared with using general purpose, numerical, wire antenna analysis computer programs. The matrices in those programs typically have dimensions around 300×300, and for antennas with dimensions longer than a wavelength and complicated cross sections, they can be much larger. They require orders of magnitude, more execution time, and are more susceptible to numerical noise.

The essential mathematical model of the folded unipole antenna was completed in Chap. 3, and in this chapter we presented a final demonstration using the specific example of the 2GMA. In the next two chapters, we will generalize and formally extend the matrix equation to include arbitrarily complicated folded unipole antennas. After that, we will concentrate on the great variety of applications.

Symbols Used in This Chapter

$\mathbf{A}_{11}$	Submatrix of **M**
$\mathbf{A}_{12}$	Submatrix of **M**
$\mathbf{A}_{21}$	Submatrix of **M**
$\mathbf{A}_{22}$	Submatrix of **M**
h_1	z-coordinate of the top of the first or lower segment of the folded unipole
h_2	z-coordinate of the top of the second or middle segment of the folded unipole
h_3	z-coordinate of the top of the third or upper segment of the folded unipole
$\mathbf{I}$	Vector of the complex current amplitudes
$I_{\mathrm{cn}}(z)$	Common mode current on the nth segment of the folded unipole
$I_{\mathrm{dn}}(z)$	Differential mode current on the nth segment of the folded unipole
I_{Fcn}	Amplitude of the common mode forward traveling wave current on the nth segment
I_{Fdn}	Amplitude of the differential mode forward traveling wave current on the nth segment
$\mathbf{I_n}$	Vector of the complex current amplitudes on the nth segment of the folded unipole
I_{Rcn}	Amplitude of the common mode reverse traveling wave current on the nth segment
I_{Rdn}	Amplitude of the differential mode reverse traveling wave current on the nth segment
$\mathbf{M}$	Matrix in the equation for the complex current amplitudes
V	Input voltage
$V_{\mathrm{cn}}(z)$	Common mode voltage on the nth segment of the folded unipole
$V_{\mathrm{dn}}(z)$	Differential mode voltage on the nth segment of the folded unipole
X_B	Terminating reactance at the base of the folded unipole
X_n	Terminating reactance at the top of the nth segment of the folded unipole
z	Vertical coordinate
$Z_{0\mathrm{c}}$	Characteristic impedance for the common mode
$Z_{0\mathrm{cn}}$	Characteristic impedance for the common mode on the nth segment of the folded unipole
Z_{0d}	Characteristic impedance for the differential mode
$Z_{0\mathrm{dn}}$	Characeristic impedance for the differential mode on the nth segment of the folded unipole
Z_B	Terminating impedance at the base of the folded unipole
Z_n	Terminating impedance at the top of the nth segment of the folded unipole
ν_n	Current transformation ratio along the nth segment
$\psi_F(z)$	Forward traveling wave function
$\psi_R(z)$	Reverse traveling wave function

Chapter

5

The *N*-Segment Folded Unipole

5.1 Introduction

As we observed in Chap. 4, if we add another segment to the folded unipole antenna, then we add five more independent design variables. If we can model an n-segment folded unipole (nSFU), for arbitrary n, then we can work with as many variables as is necessary to satisfy practically any design requirement.

In this chapter we will derive a general matrix equation that applies to the nSFU. The solutions to the matrix equation are the common and differential mode current amplitudes everywhere on the antenna. From these, we can readily obtain input impedance, radiation pattern, and other useful measures of performance. There will not be any new boundary conditions or physical principles introduced here because everything we need for our mathematical model was developed methodically in Chaps. 2 and 3, using one- and two-segment folded unipoles as examples. In Chap. 4, we showed how the model could be readily applied to the Double Gamma Match Antenna, which is a special case of the three-segment folded unipole. In the present chapter, we will show how the model can be applied even more generally to n-segment folded unipoles.

We could have begun with a general matrix equation back in Chap. 2; however, it might have appeared too abstract to be useful at the time. By postponing the generalized version until the present chapter, it will easily follow from our previous developments without any mystery and with minimal confusion.

Our general matrix equation will include many special cases of practical interest. We will show how any number of segments can be "unfolded" as desired. That is, we can concatenate any combination of folded antennas and plain linear antennas, analogous to a slotted waveguide. Of course, the nGMA (n-segment Gamma Match Antenna) is a special case, with only the top segment unfolded.

We will also add a provision to our model to account for both intentional and unintentional top loading. This is due to capacitance at the top of any antenna with finite radius. Because of that capacitance, the common mode current does not quite vanish there. In many cases, that is of little practical importance; however, for radii not too small compared with a wavelength, the effects can be significant. This includes many antennas in the AM broadcast band.

5.2 Essential Features of the *n*SFU

Figure 5-1 shows the *n*SFU. Three segments or partial segments are shown explicitly. They are: the base or bottom segment; the upper portion of the top or *N*th segment; and the boundary between the *n*th and $(n + 1)$st segments. Only two parallel conductors are used to represent each segment but this is only for simplicity in the drawing. In fact, our

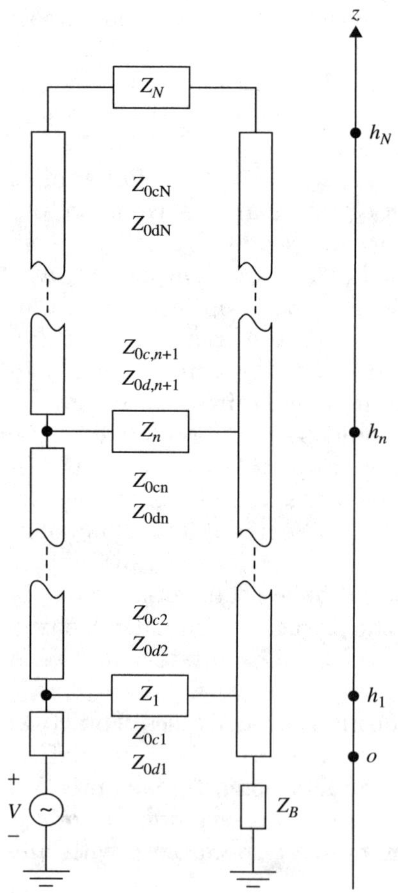

Figure 5-1 Essential features and geometry for the *n*-segment folded unipole antenna (nSFU). Two parallel conductors are shown only for simplicity in the drawing. Each of the *n* segments, in general, has different characteristic impedances and different current transformation ratios.

analysis will apply to antennas with arbitrary cross sections with any number of conductors. This is because any cross section can be represented by a common mode characteristic impedance Z_{0c} and a differential mode characteristic impedance Z_{0d}.

The key to understanding and analyzing the folded unipole is to separate the currents on each segment into common and differential mode traveling waves. The common mode traveling wave currents are described by Eq. 3-1, with $n = 1, 2, \ldots N$. The differential mode traveling wave currents are described by Eqs. 3-1 and 3-2, respectively, with $n = 1, 2, \ldots N$. Common mode and differential mode voltages are related to the currents by the usual transmission line equations. The voltages are described by Eqs. 3-6 and 3-7.

If we count carefully, we will find that there are $5N + 1$ independent design variables. For each segment, five of the variables are: (1) the vertical coordinate h_n, (2) the terminating impedance Z_n, (3) the differential mode characteristic impedance $Z_{0\text{dn}}$, (4) the common mode characteristic impedance $Z_{0\text{cn}}$, and (5) the current transformation ratio ν_n. An additional independent design variable is the base terminating impedance Z_B. By increasing the number of segments, the antenna engineer can obtain as many independent design variables as necessary to satisfy even the most complicated design requirements.

As mentioned in the previous section, any of the segments can be "unfolded", or reduced to a simple linear radiator. For example, if we wish to unfold the nth segment, as shown in Fig. 5-2, then we formally require that,

$$Z_{0\text{dn}} \rightarrow \infty \tag{5-1}$$

in our mathematical model. This automatically makes the folded conductor vanish, leaving only the unbalanced mode current flowing on the remaining conductor. A related consequence is that,

$$\nu_n = 0 \tag{5-2}$$

because there is no current, balanced or otherwise, on a fold that does not exist. Equations 5-1 and 5-2 tell us that for each segment we unfold, we lose the two independent design variables $Z_{0\text{dn}}$ and ν_n.

Depending on what segments we unfold, we also automatically lose terminating impedances as independent design variables. If we unfold two adjacent segments, for example the nth and $(n + 1)$st, then we lose the fold terminating impedance Z_n. We do not have to do anything explicitly in our mathematical model to make this happen. It is simply left hanging with no connections to any folds. So, no current flows through it, and it is effectively useless, no matter what value we may assign to it. Similarly, if we unfold the top or Nth segment, then we lose the top terminating reactance Z_T.

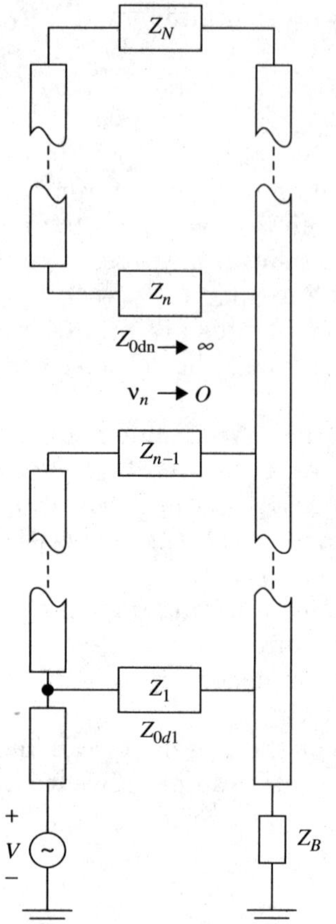

Figure 5-2 Any number of segments can be "unfolded", or converted to simple linear radiators, by mathematically forcing their differential characteristic impedances to become infinite. In that way, we can construct any combination of folded and unfolded segments. If only the top segment is unfolded, then the antenna becomes an $(N-1)$ Gamma Match Antenna. The current transformation ratios automatically vanish on the unfolded segments.

The above remarks tell us much about the versatility of our generalized matrix equation even before we derive it. In the following four sections, that is exactly what we will do.

5.3 Equations at the Top of the *n*SFU

Figure 5-3 shows the top of the *n*SFU with currents and voltages. There are two boundary conditions that must be satisfied there. The common mode current must vanish because it has nowhere to flow beyond the top:

$$(1+v_N)\,I_{\mathrm{cN}}(h_N) = 0 \tag{5-3}$$

The terminating impedance Z_N defines the relationship between the differential mode voltage and current:

$$Z_N I_{\mathrm{dN}}(h_N) = V_{\mathrm{dN}}(h_N) \tag{5-4}$$

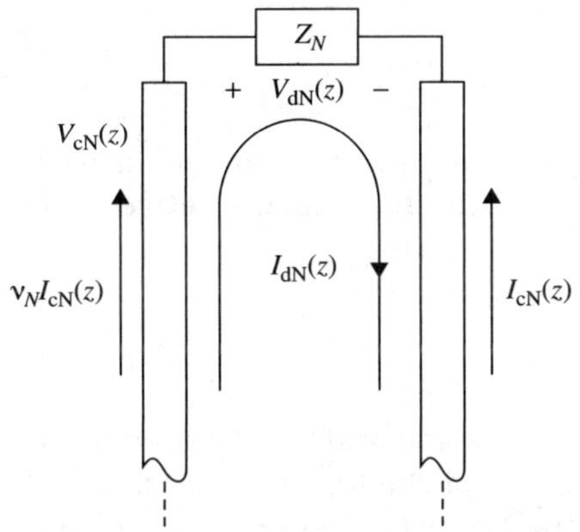

Figure 5-3 Voltages and currents used to formulate boundary conditions at the top of the *n*SFU.

Equations 5-3 and 5-4 are two independent equations out of 4*N* that we require to determine the unknown current amplitudes I_{Fcn}, I_{Rcn}, I_{Fdn}, and I_{Rdn}, for $n = 1, 2, \ldots N$.

5.4 Equations at the Base of the *n*SFU

Figure 5-4 shows the base of the *n*SFU with currents and voltages. We can derive two more independent equations there. Kirchhoff's voltage law (KVL) requires that:

$$Z_B[I_{d1}(0) - I_{c1}(0)] + V_{d1}(0) = V \tag{5-5}$$

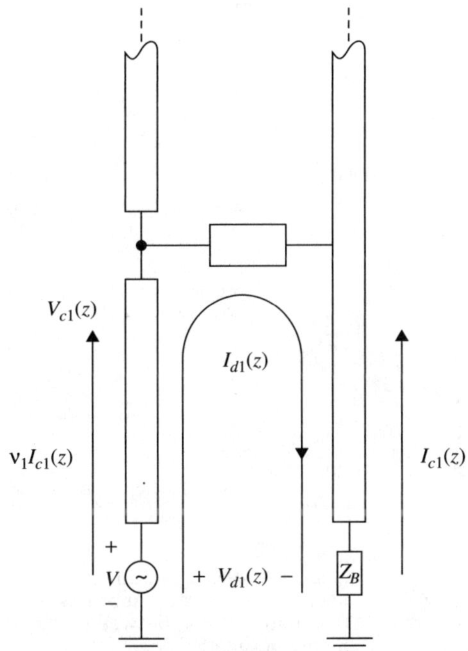

Figure 5-4 Voltages and currents used to formulate boundary conditions at the base of the *n*SFU.

By inspection:

$$V_{c1}(0) = V \tag{5-6}$$

We have picked up two more independent equations at the base. We need $4N-2$ more. Those can be found at the boundaries between the segments for $n = 1, 2, \ldots N-1$.

5.5 Equations at the Boundaries Between Segments

Figure 5-5 shows the boundary between the nth and $(n + 1)$st segments. This geometry applies to all segments except the top one, that is, for all n from 1 through $N-1$. Four constraints apply to the currents and voltages at the boundary. The common mode and differential mode voltages must be continuous across the boundary:

$$V_{cn}(h_n) = V_{c,n+1}(h_n) \tag{5-7}$$

$$V_{dn}(h_n) = V_{d,n+1}(h_n) \tag{5-8}$$

The common mode current must also be continuous across the boundary:

$$(1+v_n)I_{cn}(h_n) = (1+v_{n+1})I_{c,n+1}(h_n) \tag{5-9}$$

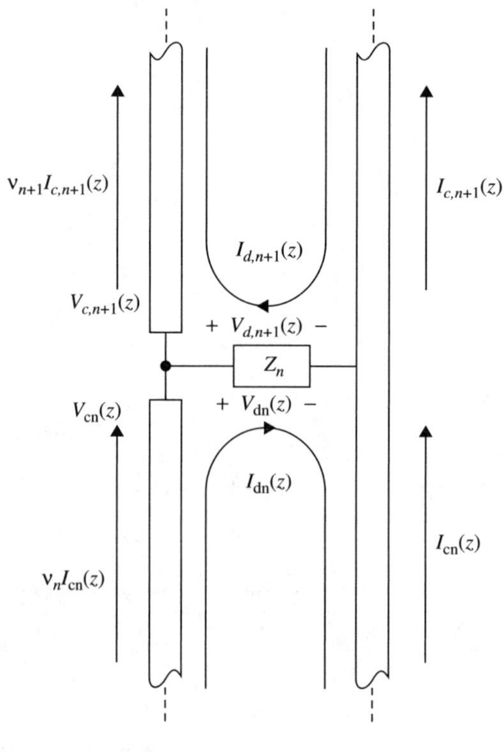

Figure 5-5 Voltages and currents used to formulate equations at the boundaries between segments of the nSFU.

The terminating impedance Z_n defines the relation between the differential mode voltage and currents:

$$Z_n[I_{\text{dn}}(h_n) - I_{d,n+1}(h_n)] = V_{\text{dn}}(h_n) \tag{5-10}$$

Equations 5-7 through 5-10 are called *recursion relations* because once the currents and voltages on the nth segment are specified, then the currents and voltages on the $(n + 1)$st segment are determined. One fortunate consequence is that, when we consolidate these equations into a single matrix equation, most of the matrix elements will be zero.

Since Eqs. 5-7 through 5-10 are a separate set of four equations for each value of n from 1 through $N-1$, we have actually derived $4N-4$ independent equations at the boundaries. Together with the equations at the top and base of the folded unipole, we now have the $4N$ independent equations necessary to uniquely determine the voltages and currents everywhere on the folded unipole antenna.

5.6 Matrix Equation for Currents on the *n*SFU

Equations 5-3 through 5-10 can be consolidated into a single matrix equation, using a procedure similar to that followed in Secs. 2.8, 3.6, 3.15, and 4.6. The equation is:

$$\mathbf{M} \cdot \mathbf{I} = \begin{pmatrix} V \\ V \\ 0 \\ \vdots \\ 0 \end{pmatrix} \tag{5-11}$$

In Eq. 5-11, the dimensions of the matrix **M** are $4N \times 4N$. Both the current vector **I** and the right side of Eq. 5-11 have $4N$ *elements.*

The current vector may be partitioned into vectors for each segment of the folded unipole antenna:

$$\mathbf{I} = \begin{pmatrix} \mathbf{I}_1 \\ \mathbf{I}_2 \\ \vdots \\ \mathbf{I}_N \end{pmatrix} \tag{5-12}$$

In Eq. 5-12 each element In is a four-element vector described by Eq. 3-37.

The matrix **M** may be also be partitioned. In previous chapters we defined four submatrices for that purpose, chiefly so that the matrix elements would remain large enough to be legible on the printed page.

In the case of Eq. 5-11, however, our motivation is different. We will partition **M** with respect to organizational clarity rather than legibility:

$$\mathbf{M} = \left(\begin{array}{c|c|c} \mathbf{A_{11}} & \mathbf{0}\cdots\mathbf{0} & \mathbf{A_{1N}} \\ \hline \mathbf{0} & \mathbf{A_{mn}} & \mathbf{0} \end{array}\right) \tag{5-13}$$

In Eq. 5-13, $\mathbf{A_{11}}$ is a submatrix determined by the boundary conditions at the base of the folded unipole, that is, by Eqs. 5-5 and 5-6, and it has dimensions 4 × 4. Even though these dimensions are modest, for clarity it is useful to partition $\mathbf{A_{11}}$ itself into submatrices:

$$\mathbf{A_{11}} = \left(\begin{array}{c|c} \mathbf{a_{11}} & \mathbf{b_{11}} \\ \hline 0 & 0 \end{array}\right) \tag{5-14}$$

On the right side of Eq. 5-14, each submatrix has dimensions 2 × 2. The nonzero submatrices are:

$$\mathbf{a_{11}} = \left(\begin{array}{c|c} -Z_B & -Z_B \\ \hline Z_{0c1} & -Z_{0c1} \end{array}\right) \tag{5-15}$$

$$\mathbf{b_{11}} = \left(\begin{array}{c|c} Z_B + Z_{0d1} & Z_B - Z_{0d1} \\ \hline 0 & 0 \end{array}\right) \tag{5-16}$$

The submatrix $\mathbf{A_{1N}}$ is determined by the boundary conditions at the top of the folded unipole, that is, by Eqs. 5-3 and 5-4, and like $\mathbf{A_{11}}$ it also has dimensions 4 × 4. Also like $\mathbf{A_{11}}$, it is useful to partition $\mathbf{A_{1N}}$ into submatrices for purposes of clarity:

$$\mathbf{A_{1N}} = \left(\begin{array}{c|c} 0 & 0 \\ \hline \mathbf{c_{1N}} & \mathbf{d_{1N}} \end{array}\right) \tag{5-17}$$

On the right side Eq. 5-17, each submatrix has dimensions 2 × 2. The nonzero submatrices are:

$$\mathbf{c_{1N}} = \left(\begin{array}{c|c} (1+v_N)\psi_F(h_N) & (1+v_N)\psi_R(h_N) \\ \hline 0 & 0 \end{array}\right) \tag{5-18}$$

$$\mathbf{d_{1N}} = \left(\begin{array}{c|c} 0 & 0 \\ \hline (Z_T - Z_{0dN})\psi_F(h_N) & (Z_T + Z_{0dN})\psi_R(h_N) \end{array}\right) \qquad \text{(5-19}$$

In Eqs. 5-18 and 5-19, the forward and reverse traveling wave functions $\psi_F(z)$ and $\psi_R(z)$ are defined by Eqs. 1-2 and 1-3.

It is seen in Eq. 5-13, that the submatrices $\mathbf{A_{11}}$ and $\mathbf{A_{1N}}$ are separated horizontally by a submatrix with dimensions 4 × (4*N*–8), and with all elements equal to zero.

In the bottom portion of matrix **M**, the elements of the submatrix $\mathbf{A_{mn}}$ are defined only for $m = 2, 3,\ldots N-1$, and for $n = 2, 3,\ldots, N-1$.

Note that each element of the submatrix is itself a matrix with dimensions 4 × 4. They are mostly null matrices, that is all the elements are zero. The only matrices with nonzero elements are the diagonal matrices $\mathbf{A_{nn}}$ and the matrices $\mathbf{A_{n,n+1}}$ immediately to the right of the diagonals. They are determined by the recursion relations at the boundaries between the segments of the folded unipole, that is, by Eqs. 5-7 through 5-10. The diagonal matrices are:

$$\mathbf{A_{nn}} = \left(\begin{array}{c|c} \mathbf{a_{nn}} & \mathbf{b_{nn}} \\ \hline \mathbf{c_{nn}} & \mathbf{d_{nn}} \end{array}\right) \qquad n = 2, 3, \ldots N-1 \tag{5-20}$$

In Eq. 5-16, the submatrices are:

$$\mathbf{a_{nn}} = \left(\begin{array}{c|c} (1+\nu_n)\psi_F(h_n) & (1+\nu_n)\psi_R(h_n) \\ \hline 0 & 0 \end{array}\right) \tag{5-21}$$

$$\mathbf{b_{nn}} = \left(\begin{array}{c|c} 0 & 0 \\ \hline Z_{0dn}\psi_F(h_n) & -Z_{0dn}\psi_R(h_n) \end{array}\right) \tag{5-22}$$

$$\mathbf{c_{nn}} = \left(\begin{array}{c|c} Z_{0cn}\psi_F(h_n) & -Z_{0cn}\psi_R(h_n) \\ \hline 0 & 0 \end{array}\right) \tag{5-23}$$

$$\mathbf{d_{nn}} = \left(\begin{array}{c|c} 0 & 0 \\ \hline (Z_n - Z_{0dn})\psi_F(h_n) & (Z_n + Z_{0dn})\psi_R(h_n) \end{array}\right) \tag{5-24}$$

The matrices $\mathbf{A_{n,n+1}}$ immediately to the right of the diagonal matrices are:

$$\mathbf{A_{n,n+1}} = \left(\begin{array}{c|c} \mathbf{a_{n,n+1}} & \mathbf{b_{n,n+1}} \\ \hline \mathbf{c_{n,n+1}} & \mathbf{d_{n,n+1}} \end{array}\right) \qquad n=2,3,\ldots N-1 \tag{5-25}$$

Each element of $\mathbf{A_{n,n+1}}$ is itself a matrix with dimensions 2 × 2:

$$\mathbf{a_{n,n+1}} = \left(\begin{array}{c|c} -(1+\nu_{n+1})\psi_F(h_n) & -(1+\nu_{n+1})\psi_R(h_n) \\ \hline 0 & 0 \end{array}\right) \tag{5-26}$$

$$\mathbf{b_{n,n+1}} = \left(\frac{0}{-Z_{0d,n+1}\psi_F(h_n)} + \frac{0}{-Z_{0d,n+1}\psi_R(h_n)}\right) \tag{5-27}$$

$$\mathbf{c_{n,n+1}} = \left(\begin{array}{c|c} -Z_{0c,n+1}\psi_F(h_n) & Z_{0c,n+1}\psi_R(h_n) \\ \hline 0 & 0 \end{array}\right) \tag{5-28}$$

$$\mathbf{d_{n,n+1}} = \left(\frac{0}{-Z_n\psi_F(h_n)} + \frac{0}{-Z_n\psi_R(h_n)}\right) \tag{5-29}$$

Looking back at Eq. 5-13, it is seen that the submatrix $\mathbf{A}_{mn}$ is bounded horizontally on either side by two matrices with dimensions $(4N-4) \times 4$, and with all elements zero. So, the dimensions of **M**, as determined by its submatrices, are $4N \times 4N$, as expected.

Equation 5-11 is the general matrix equation that we set out to derive. It includes as special cases, all of the matrix equations of the previous chapters. The price we pay for this generality is mathematical notation that is a bit more abstract, as is evident in Eqs. 5-13 through 5-29. We have worked up to that level of abstraction gradually, however, beginning with Chap. 2 for the simplest configuration of folded unipole antenna.

Our general matrix equation offers the same advantages over general purpose, numerical , wire antenna analysis programs that we have discussed in Chaps. 2 through 4. Those advantages are greater computational speed and lower numerical noise. For an N-segment folded unipole, the matrix **M** has $16N^2$ elements. To be sure, as N becomes larger, so does **M**. But, according to Eq. 5-13, most of the elements of **M** are zero, no matter how large N becomes. So, it is still much quicker to compute the relatively few nonzero elements of **M** than it is to fill the matrix of a general purpose, numerical program. Also, **M** is a *sparse matrix*, that is, with many elements equal to zero, and sparse matrices are generally less susceptible to numerical noise.

Now that we have derived a general matrix equation for any combination of folded and unfolded segments, we can concentrate on modeling some of the finer details of the antenna. One that can measurably affect antenna performance is called the *antenna end effect*, and that is the subject of the next section.

5.7 The Antenna End Effect

Until now, we have been asserting that the common mode current vanishes at the top of the antenna. This has been expressed consistently by Eqs. 2-15, 3-9, 3-27, 4-1, and 5-3. This assertion makes a lot of sense. After all, where could the common mode current flow after it reaches the end of the antenna? For many combinations of wavelength and antenna dimensions, the answer is, "Nowhere," and our assertion is correct for all practical purposes.

For certain other combinations, however, there are interesting exceptions. For example, in the AM broadcast band it has often been reported that the radiation patterns of antennas resemble those of significantly taller structures. The difference between the physical height and the apparent height can differ up to about 20 percent. It is as though the antenna had a phantom extension. In fact, the common mode current does not vanish at the top.

The reason is that the top of the antenna has the ability to accumulate electric charge. That is, the top of the antenna is one plate of a capacitor. The other plate is located at the end of the other arm of the dipole, or, in the case of a monopole, at the end of its mirror image in the ground. Often, the value of the capacitor is negligible, on the order of a picofarad or less; however, for the larger dimensions used in the AM broadcast band and at longer wavelengths, it may be tens of picofarads. That is enough so that the common mode current supplies significant amounts of charge to the capacitor rather than simply vanish.

Figure 5-6 shows a disc with radius a, isolated in free space. The classic formula for its capacitance is:

$$C_{\text{disc}} = 8\varepsilon_0 a \tag{5-30}$$

Where ε_0 is the permittivity of free space, about 8.854×10^{-12} farads per meter. Equation 5-30 should be regarded as a lower bound for the capacitance of the top of the antenna. In fact, the sides of a cylinder in the vicinity of the top provide more capacitance than the bottom of a disc.

If Eq. 5-30 is a lower bound on the top capacity, then the upper bound is [1]:

$$C_{\text{cyl}} = \frac{\pi\varepsilon_0 l}{\ln\left(\frac{2l}{a}\right) - 1 - \ln 2} \tag{5-31}$$

Equation 5-31 describes the capacity of a cylinder of length l in free space. Quite clearly, an entire cylindrical antenna should not be regarded as a static capacitor; however, some part of its length l', small compared with a wavelength and near the top, contributes to the end effect.

Some quick mental arithmetic using Eq. 5-30 demonstrates that an antenna with an effective radius of about a meter, has a top capacitance

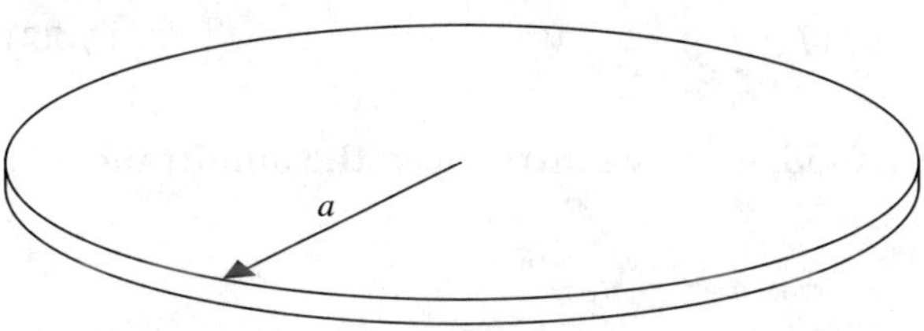

Figure 5-6 A disc isolated in free space has self capacitance that can be described by a simple formula. That capacitance is responsible for antenna end effect which can cause performance to resemble that of a much longer antenna.

of 70.8 picofarads. At a frequency of 1 MHz, that capacitance has a reactance of –2,248 ohms. For some antenna cross sections, this value can be large compared to its common mode characteristic impedance, but not so large that it is an open circuit. In any case, the end of the antenna is not truly an open circuit, and the common mode current does not vanish there.

Schelkunoff & Friis [2] discussed the above phenomenon in detail. They derived a simple formula for the effective lengthening of a dipole antenna due to end effect:

$$\delta_{\text{cap}} = \frac{4aZ_0}{\eta} \tag{5-32}$$

In Eq. 5-32, we have used the original notation of Schelkunoff & Friis, in which Z_0 is the common mode impedance of the dipole and η is the impedance of free space (about 377 ohms). Again resorting to quick mental arithmetic, we see that for an antenna radius of one meter and a common mode characteristic impedance of 1,000 ohms, the lengthening is about 35 feet. This can be 10 percent or more of the total height of an AM broadcast tower.

In Eqs. 5-30 through 5-32, the value of radius a is obvious for antennas with simple circular cross sections. For more complicated cross sections, including that of the nSFU, evaluating C_{cap} requires more tedious and often numerical computations.

Without detailed computations, however, we can observe one important property of the nSFU. When the folds are shortened to the top of the antenna, C_{cap} is obviously greater than when they are not.

With the above discussion as prologue, we can add the end effect to our mathematical model for the n-segment folded unipole antenna. To that end, Fig. 5-7 shows the top of the nSFU, explicitly including the capacitance responsible for end effect. This replaces Fig. 5-3. Equation 5-3, which required the common mode current to vanish at the end, must be replaced by:

$$(1+\nu_N)I_{\text{cN}}(h_N) = Y_{\text{cap}}V_{\text{cN}}(h_N) \tag{5-33}$$

On the right side of Eq. 5-33, we have introduced the admittance,

$$Y_{\text{cap}} = j2\pi f C_{\text{cap}} \tag{5-34}$$

For completeness, it is important to add a small amount of conductance to Y_{cap}. Without it, the reflection coefficient for common mode

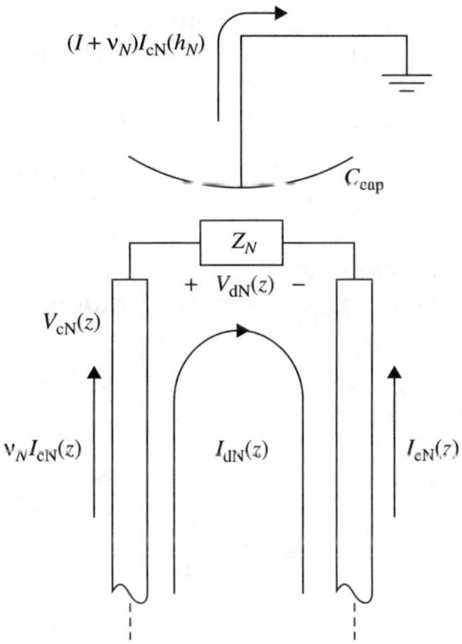

Figure 5-7 Voltages and currents at the top of the nSFU, including the *antenna end effect* due to top capacitance. This replaces the model used in Fig. 5-3.

currents at the top of the antenna has magnitude unity, resulting in perfect standing waves of current along the antenna. For most cases, this is a satisfactory representation; however, for some special cases it is not. These special cases notably include the half-wave antenna, and antennas that are integer multiples of a half wavelength. For those, the standing wave representation results in zero input current and infinite input impedance. For decades, there was heated debate in the literature about the precise value of input current and impedance and how to model them in those cases. We will formulate a simple explanation in the remainder of this section.

The end capacitance may be regarded as an impedor (the origin of the term "impedance") or discontinuity separating two semi-infinite transmission lines. One is the antenna with common mode characteristic impedance Z_{0c}, or common mode characteristic admittance Y_{0c}. The other is free space with characteristic impedance η. According to this model, Eq. 5-34 must be modified to include a series connection between the end capacitance and free space:

$$Y_{cap} = \left(\frac{1}{j2\pi f C_{cap}} + \eta \right)^{-1} \qquad (5\text{-}35)$$

The antenna end termination described by Eq. 5-35 should produce a current reflection coefficient with magnitude close to unity but not quite, according to the formula,

$$\Gamma = \frac{Y_{0c} - Y_{\text{cap}}}{Y_{0c} + Y_{\text{cap}}} \tag{5-36}$$

Let's test the sensibility of Eq. 5-35 with some actual numbers, for example, $f = 1$ MHz, $a = 1$ meter, and $Y_{0c} = 0.001$ mho. Using these numbers, Eq. 5-30 tells us that,

$$C_{\text{cap}} = 8 \times 8.854 \times 10^{-12} \times 1 = 70.8 \text{ pfd}$$

Continuing our example, according to Eq. 5-35,

$$Y_{\text{cap}} = \left(\frac{1}{j2\pi \times 10^6 \times 70.8 \times 10^{-12}} + 377 \right)^{-1} = -j0.4327 + 0.007256 \text{ mmhos}$$

According to Eq. 5-36, the complex reflection coefficient is:

$$\Gamma = \frac{0.9927 + j0.4327}{1.0073 - j0.4327}$$

So, the magnitude of the reflection coefficient is:

$$|\Gamma| = \frac{1.083}{1.096} = 0.988$$

Our model of antenna end effect has produced exactly the result we expected, that is, a reflection coefficient with magnitude not quite unity.

If instead of Eq. 5-30, we had used Eq. 5-31 describing the upper bound on top capacity, for a length of $l = 250$ feet, we would have found that:

$$C_{\text{cap}} = 636 \text{ pfd}$$

and that:

$$|\Gamma| = 0.479$$

So antenna end effect can easily account for the small input currents and large input impedances observed on antennas of a certain length.

According to Eq. 5-36, the reflection coefficient is exactly unity as $Y_{0c} \to 0$, or equivalently, as $Z_{0c} \to \infty$. That is, for an infinitely thin antenna, the reflection coefficient is unity, and the current is a pure standing wave.

In our matrix Eq. 5-11 for the currents on the folded unipole, Eq. 5-35 impacts only the submatrix $\mathbf{c}_{1N}$. Accordingly, Eq. 5-18 must be replaced by:

$$\mathbf{c}_{1N} = \left(\begin{array}{c|c} (1+v_N - Y_{cap} Z_{0cN})\, \psi_F(h_N) & (1+v_N + Y_{cap} Z_{0cN})\, \psi_R(h_N) \\ \hline 0 & 0 \end{array} \right) \tag{5-37}$$

Other than Eq. 5-37, our general matrix equation in Eq. 5-11 remains unchanged.

5.8 Examples of Antenna End Effect

As discussed in the previous section, end effect can profoundly affect antenna performance. As a demonstration, the 2SFU which served as an example of broadbanding in Fig. 3-48 will also serve as a good example of antenna end effect. In that figure, the the top capacitance C_{cap} was zero. In this section, we will vary the top capacitance C_{cap} over a wide range of nonzero values. The other antenna design variables will remain fixed at the same values used for Fig. 3-48. We will see that small values of C_{cap} have little effect; however, eventually larger values change the antenna performance dramatically.

Figures 5-8 and 5-9 show antenna input impedance when C_{cap} is 3 picofarads and 30 picofarads, respectively. The impedance is displayed as a pair of contour plots, in similar fashion to previous chapters. The upper contour plot displays input resistance, using the color scheme:

White	0–25 ohms
Green	25–75 ohms
Dark Green	45–55 ohms
Yellow	75–150 ohms
Red	150–300 ohms
Gray	300 ohms and greater

Dark green regions on the contour plot straddle the optimal input resistance of 50 ohms. The lower contour plot displays input reactance, using the color scheme:

White	Less than 0 ohms
Cyan	0–1,000 ohms
Dark Cyan	0–100 ohms
Yellow	1,000–2,000 ohms
Magenta	2,000–3,000 ohms
Gray	3,000 ohms and greater

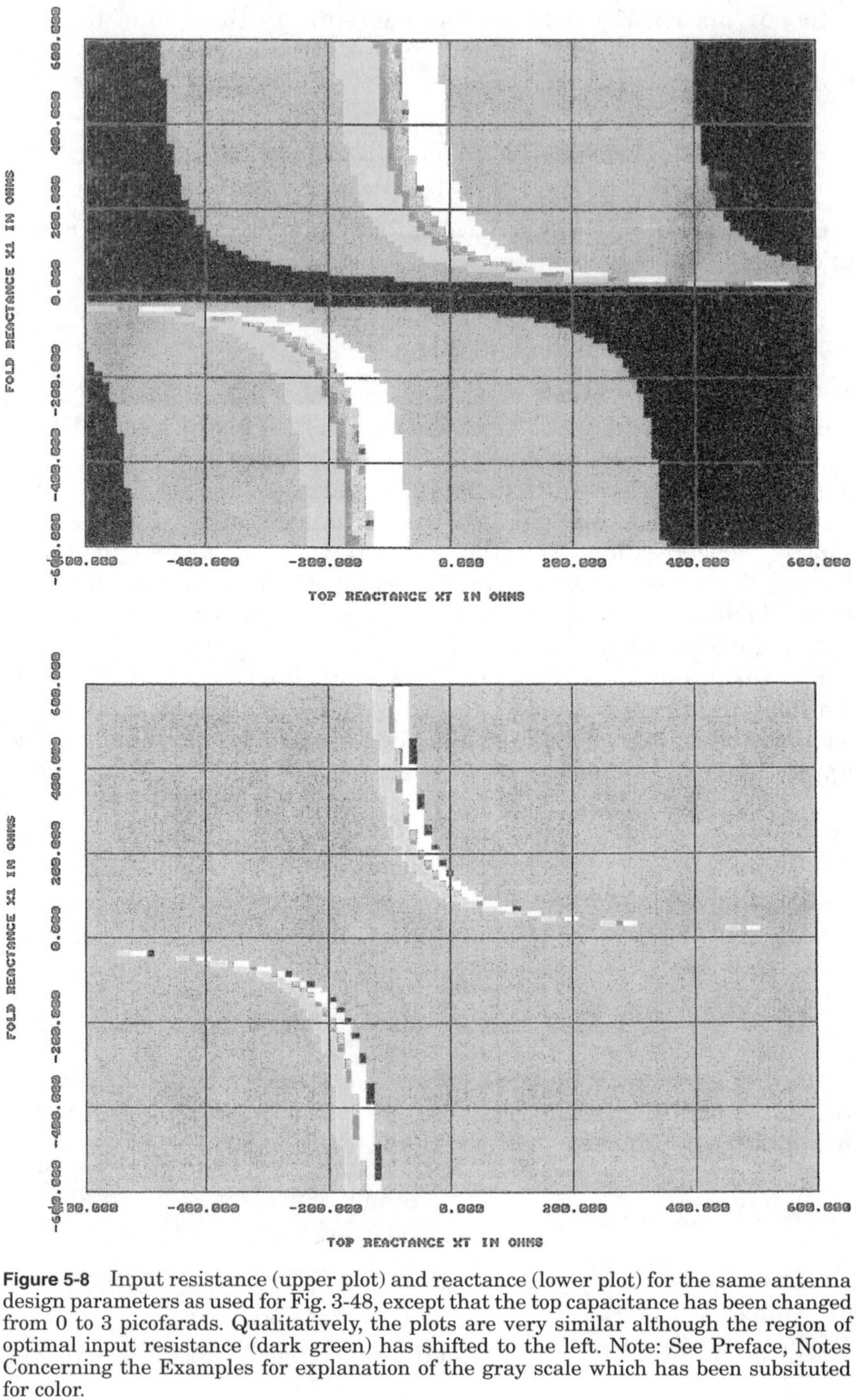

Figure 5-8 Input resistance (upper plot) and reactance (lower plot) for the same antenna design parameters as used for Fig. 3-48, except that the top capacitance has been changed from 0 to 3 picofarads. Qualitatively, the plots are very similar although the region of optimal input resistance (dark green) has shifted to the left. Note: See Preface, Notes Concerning the Examples for explanation of the gray scale which has been subsituted for color.

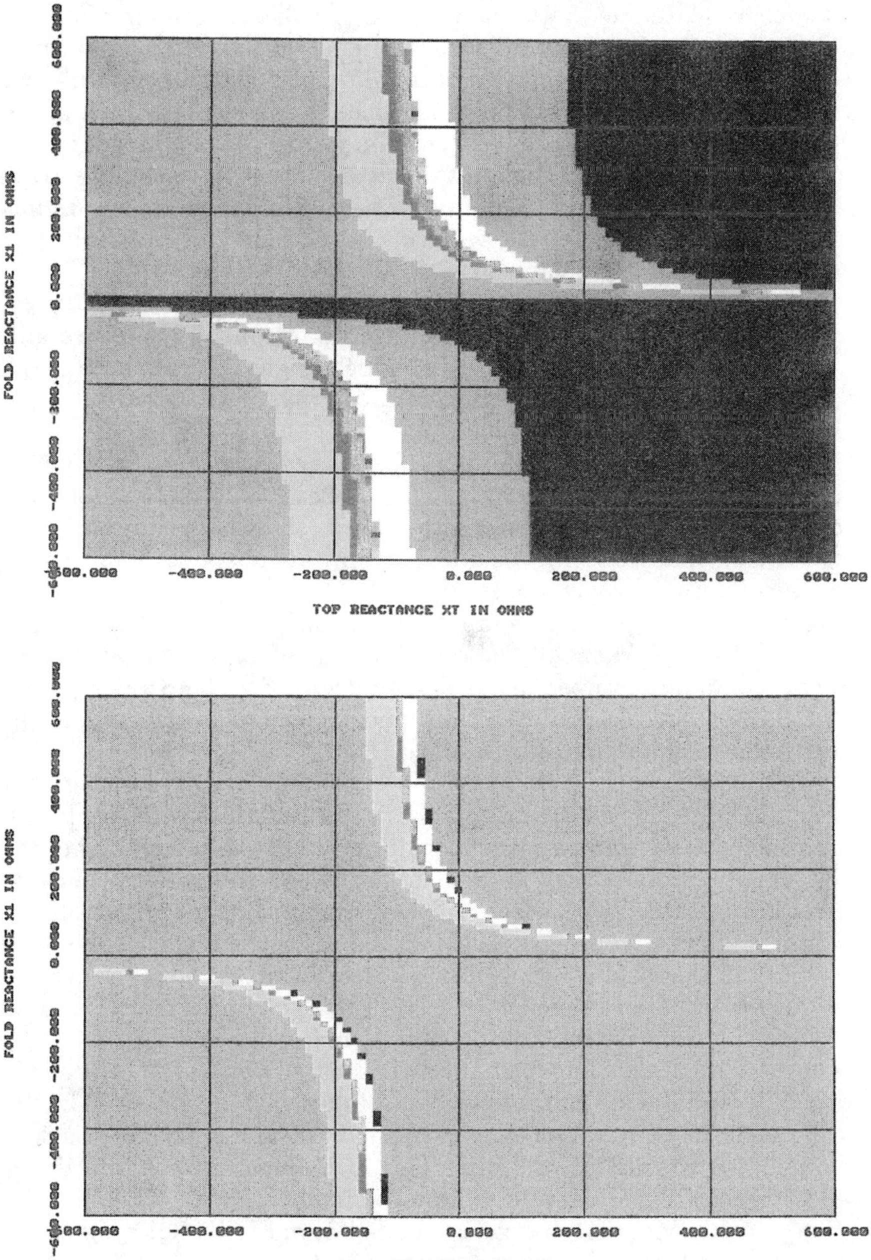

Figure 5-9 Input resistance (upper plot) and reactance (lower plot) for the same antenna design parameters as used for Fig. 3-48, except that the top capacitance has been changed from 0 to 30 picofarads. Qualitatively, the plot has not changed much, although the region of optimal input resistance has shifted further to the left.

Compared with Fig. 3-48, for which $C_{cap} = 0$, the contour plots are qualitatively similar. They all exhibit broad regions of optimal input resistance (dark green) and near-optimal input reactance (cyan). Close examination does reveal some differences, however. The dark green regions shift progressively to the left. The yellow regions progressively broaden and also drift to the left. All three figures, however, illustrate a broadband folded unipole antenna. Note that broadband operation is consistent with perturbing the value of C_{cap}, just as it is consistent with perturbing other design variables.

Figure 5-10 shows antenna input impedance when C_{cap} is 100 picofarads. The contour plots have changed significantly. There are still broad swaths of dark green and cyan which might indicate broadband operation; however, the yellow regions have expanded substantially and have begun to dominate the plots.

Figure 5-11 shows antenna input impedance when C_{cap} is 300 picofarads. Antenna performance can no longer be considered broadband or even desirable. There are few if any combinations of design variables that provide optimal input resistance simultaneously with optimal or near-optimal input reactance.

Figure 5-12 shows antenna input impedance when C_{cap} is 1,000 picofarads. There are practically no combinations of design variables that provide useful input impedance. Compared with Figs. 3-48 and Figs. 5-8 and 5-9, this is a very different antenna, indeed, although C_{cap} is the only design variable that has changed.

Figures 5-10 through 5-12 are examples of deliberate *top loading*, that is, intentionally increasing the top capacitance. Antennas with the dimensions used in these examples would not likely exhibit values of C_{cap} beyond a few tens of picofarads. To achieve larger values, the antenna must be deliberately modified with wires radiating from the top (i.e., umbrella top loading) or alternative structures to increase top capacitance. In Figs. 5-10 through 5-12, the results were clearly counter productive; however, in other cases they may be very beneficial, especially for antennas small compared with a wavelength. We shall return to this subject in Sec. 10.2.

Figure 5-13 shows the effect of continuously varying the top capacitance C_{cap} over a wide range from 0 through 1,000 picofarads. The top terminating reactance X_T has been fixed at 500 ohms. Looking back at Fig. 3-48, for which $C_{cap} = 0$, this value of X_T places input impedance well within an region of optimal input resistance.

Figure 5-13 confirms what we concluded from Figs. 5-8 through 5-12. That is, beyond values of tens of picofarads, C_{cap} causes antenna input impedance to change profoundly.

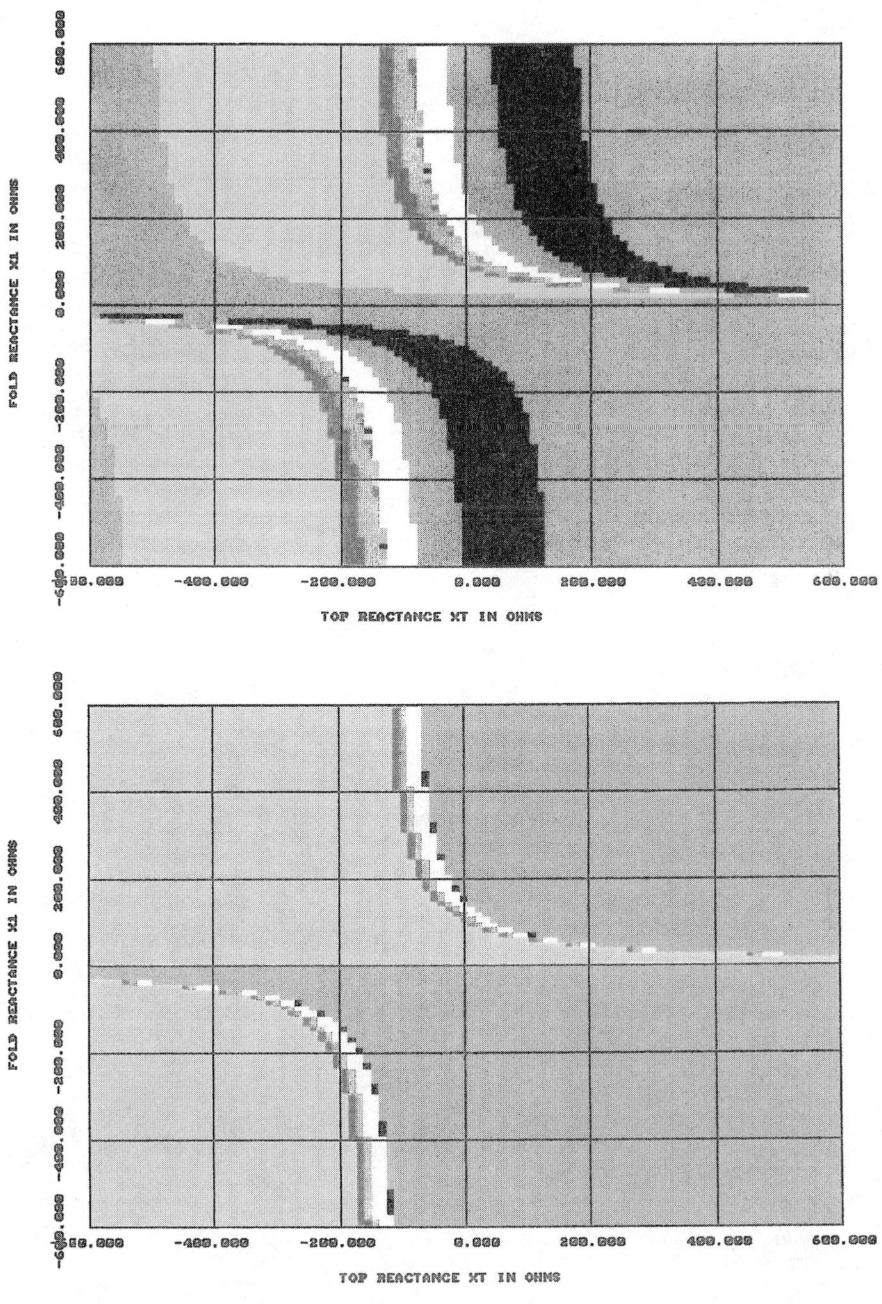

Figure 5-10 Input resistance (upper plot) and reactance (lower plot) for the same antenna design parameters as used for Fig. 3-48, except that the top capacitance has been changed from 0 to 100 picofarads.

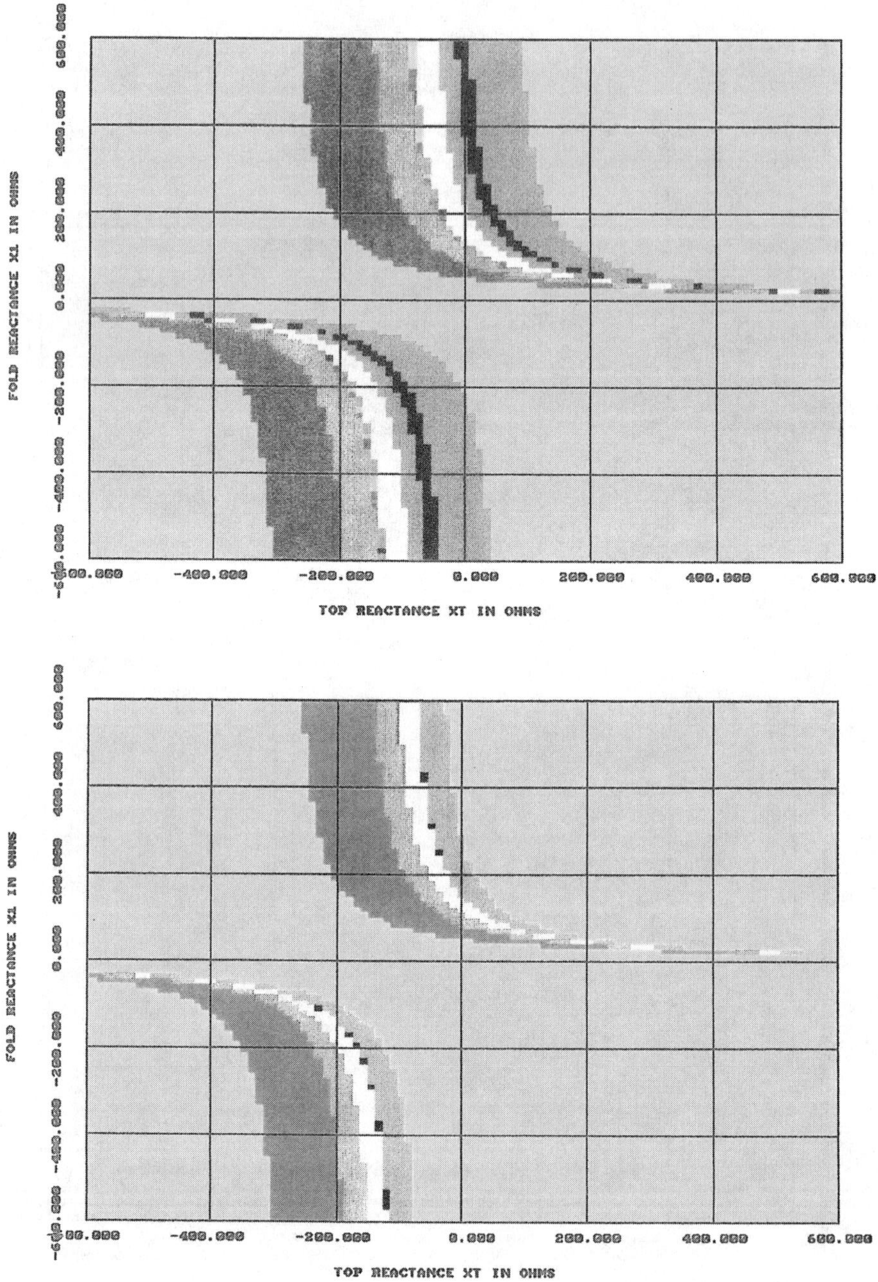

Figure 5-11 Input resistance (upper plot) and reactance (lower plot) for the same antenna design parameters as used for Fig. 3-48, except that the top capacitance has been changed from 0 to 300 picofarads.

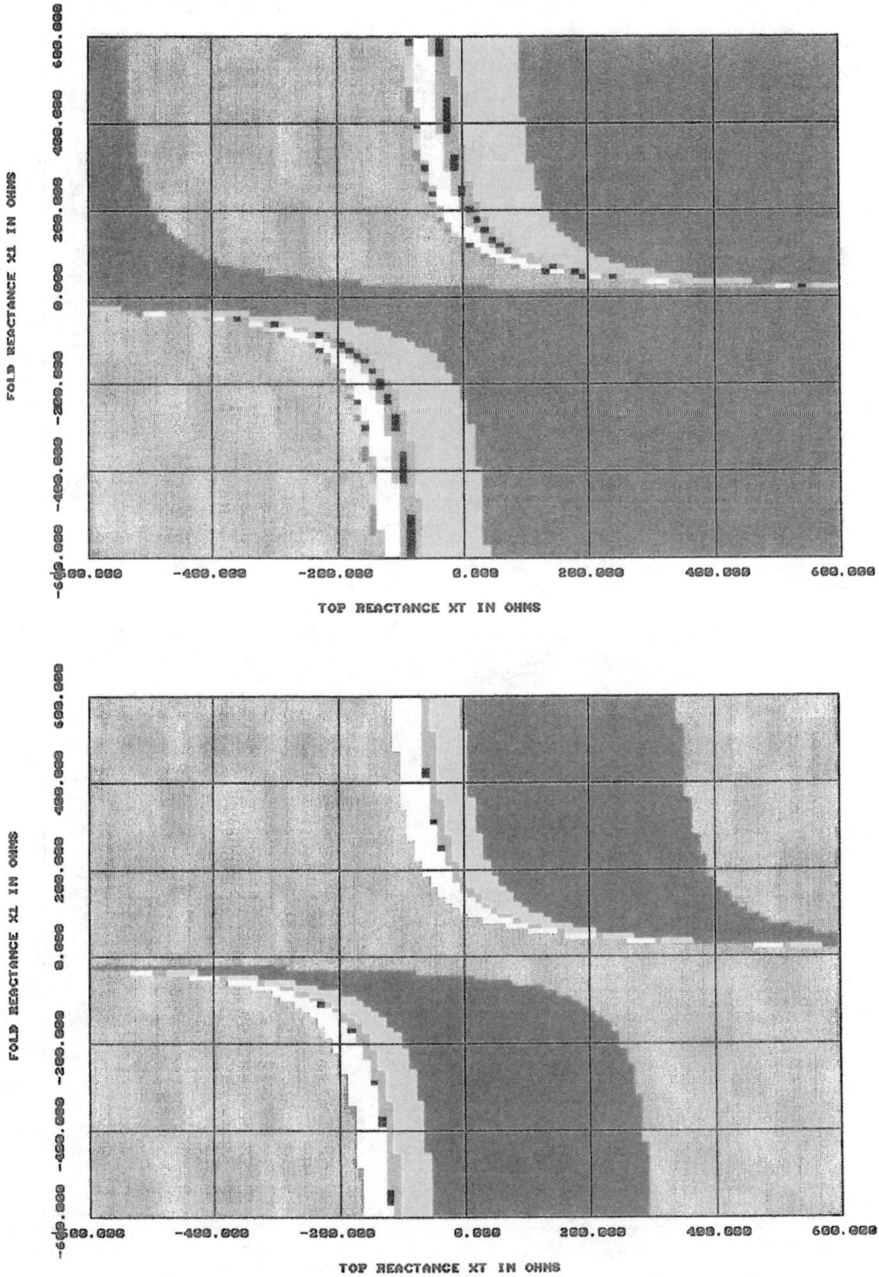

Figure 5-12 Input resistance (upper plot) and reactance (lower plot) for the same antenna design parameters as used for Fig. 3-48, except that the top capacitance has been changed from 0 to 1,000 picofarads.

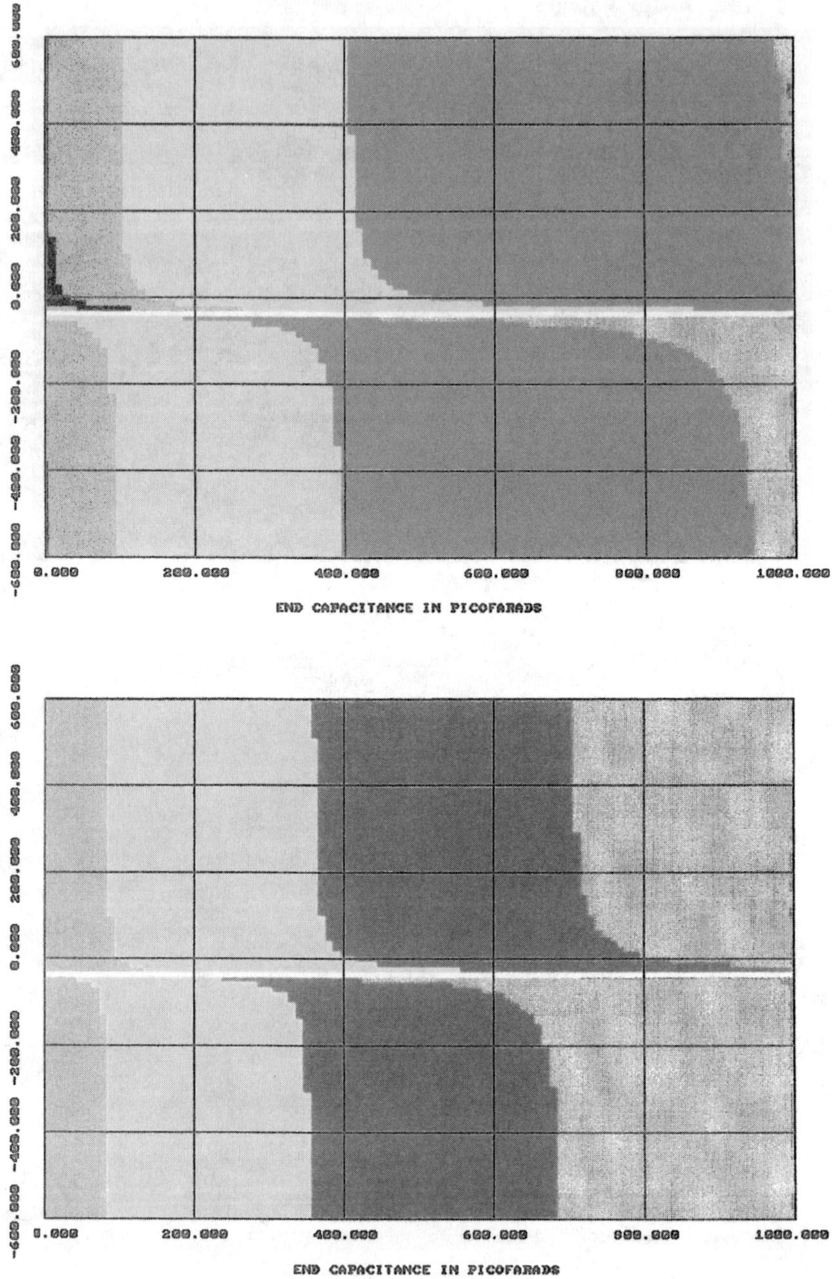

Figure 5-13 Input resistance (upper plot) and reactance (lower plot) for the antenna used for Figs. 5-8 through 5-12, except that the the top terminating reactance X_T is fixed (at 500 ohms), and C_{cap} is varied continuously (from 0 through 1,000 picofarads.) Referring back to Fig. 3-48, when $C_{cap} = 0$ picofarad, a value of $X_T = 500$ ohms puts the antenna input resistance in the midst of an optimal region. Evidently, antenna input impedance changes dramatically for nonzero values of C_{cap} beyond a few tens of picofarads.

5.9 Conclusions

We have good reasons to be pleased with the results of this chapter, both from theoretical and practical points of view.

It is said that a well-engineered project is also a beautiful or aesthetically pleasing one. The converse is often also true. That is, a beautiful project has probably been well engineered. These guidelines most obviously apply to bridges, but they also apply to antennas and other engineering endeavors. Further, they apply to the mathematical models of those endeavors. That is, a well-formulated mathematical model is also an elegant one. The converse, of course, is that an elegant mathematical model is probably a well-formulated one.

With respect to the above philosophy, we can be very pleased with the results of this chapter. We have generalized the mathematical model which began in Chap. 2 and which was essentially complete by the conclusion of Chap. 3. The resulting matrix equation is certainly elegant and beautiful in a mathematical way. To be sure, the price we have paid for elegance is a certain amount of mathematical abstraction; however, we have worked up to that level gradually beginning with the simplest mathematical model of Chap. 2. So, hopefully we have achieved elegance with a minimal amount of confusion.

Aside from mathematical beauty, we have also expanded the practicality and engineering power of our model. We can now predict the performance of an N-segment folded unipole antenna, for an arbitrary integer value of N. Because each segment introduces five new independent variables, the antenna engineer is now able to meet practically any design specifications.

We can also design arbitrary combinations of folded and unfolded segments. This is accomplished simply by letting the differential mode characteristic impedance become unbounded (i.e., go to infinity) for the latter.

Finally, we have explicitly introduced the *antenna end effect* into the model. For certain combinations of dimensions and wavelength, this phenomenon causes the antenna to perform as though it were significantly longer, on the order of 20 percent. This effective lengthening is often observed in the AM broadcast band.

Our mathematical model offers at least three substantial advantages over general purpose, numerical, wire antenna analysis programs. The two advantages discussed in previous chapters continue to apply. They are faster computation time and lower numerical noise. Even though our matrix becomes larger as N^2, most of the elements are zero. So, it is a *sparse matrix* with relatively few nonzero elements, and they can be computed rapidly. Sparse matrices are also usually robust matrices which are resistant to numerical noise.

Explicitly including the *antenna end effect,* as described in the above paragraph, is an important third advantage. General purpose, numerical programs do not automatically or implicitly model this phenomenon. Instead, they require currents to vanish at the ends of wire segments that are open circuited. This can lead to significant errors.

Even though we could be satisfied with the elegance and generality of the results of this chapter, we can still advance our mathematical model to a higher level of versatility. That is the subject of the next chapter.

References

1. Schelkunoff, S.A., and H. T. Friis, Antennas Theory and Practice, New York: John Wiley & Sons, 1952, p. 306.
2. *Ibid.*, pp. 244–246.

Symbols Used in This Chapter

Symbol	Description
a	Radius of a disc in free space or of the top of an antenna
$\mathbf{A}_{11}$	Upper left submatrix of $\mathbf{M}$
$\mathbf{a}_{11}$	Submatrix of $\mathbf{A}_{11}$
$\mathbf{A}_{1N}$	Upper right submatrix of $\mathbf{M}$
$\mathbf{A}_{mn}$	Lower submatrices of $\mathbf{M}$
$\mathbf{A}_{nn}$	Diagonal submatrices of $\mathbf{A}_{mn}$
$\mathbf{a}_{nn}$	Submatrix of $\mathbf{A}_{nn}$
$\mathbf{A}_{n,n+1}$	Submatrices of $\mathbf{A}_{mn}$ immediately to the right of $\mathbf{A}_{nn}$
$\mathbf{a}_{n,n+1}$	Submatrices of $\mathbf{A}_{n,n+1}$
$\mathbf{b}_{11}$	Submatrix of $\mathbf{A}_{11}$
$\mathbf{b}_{nn}$	Submatrix of $\mathbf{A}_{nn}$
$\mathbf{b}_{n,n+1}$	Submatix of $\mathbf{A}_{n,n+1}$
$\mathbf{c}_{1N}$	Submatrix of $\mathbf{A}_{1N}$
C_{cap}	Capacitance of antenna top
C_{disc}	Capacitance of a disc in free space
C_{cyl}	Capacitance of a cylinder in free space
$\mathbf{c}_{nn}$	Submatrix of $\mathbf{A}_{nn}$
$\mathbf{c}_{n,n+1}$	Submatrix of $\mathbf{A}_{n,n+1}$
$\mathbf{d}_{1N}$	Submatrix of $\mathbf{A}_{1N}$
$\mathbf{d}_{nn}$	Submatrix of $\mathbf{A}_{nn}$
$\mathbf{d}_{n,n+1}$	Submatirx of $\mathbf{A}_{n,n+1}$
f	Frequency
h_N	Vertical coordinate of the top of the Nth or top segment of the folded unipole
h_n	Vertical coordinate of the top of nth segment of the folded unipole
$\mathbf{I}$	Column vector of current amplitudes on the folded unipole
$I_{c1}(z)$	Common mode current on the first or bottom segment of the folded unipole
$I_{cN}(z)$	Common mode current on the Nth or top segment of the folded unipole
$I_{cn}(z)$	Common mode current on the nth segment of the folded unipole
$I_{c,n+1}(z)$	Common mode current on the $(n + 1)$st segment of the folded unipole
$I_{d1}(z)$	Differential mode current on the first or bottom segment of the folded unipole
$I_{dN}(z)$	Differential mode current on the Nth or top segment of the folded unipole
$I_{dn}(z)$	Differential mode current on the nth segment of the folded unipole
$I_{d,n+1}(z)$	Differential mode current on the $(n + 1)$st segment of the folded unipole
$\mathbf{I}_n$	Four-element vector of the current amplitudes on the nth segment of the folded unipole; an element of $\mathbf{I}$

j	Square root of -1
$\mathbf{M}$	Matrix in the equation for current amplitudes everywhere on the folded unipole
V	Input voltage to the folded unipole
$V_{c1}(z)$	Common mode voltage on the first or bottom segment of the folded unipole
$V_{cN}(z)$	Common mode voltage on the Nth or top segment of the folded unipole
$V_{cn}(z)$	Common mode voltage on the nth segment of the folded unipole
$V_{c,n+1}(z)$	Common mode voltage on the $(n + 1)$st segment of the folded unipole
$V_{d1}(z)$	Differential mode voltage on the first or bottom segment of the folded unipole
$V_{dN}(z)$	Differential mode voltage on the top or Nth segment of the folded unipole
$V_{dn}(z)$	Differential mode voltage on the nth segment of the folded unipole
$V_{d,n+1}(z)$	Differential mode voltage on the $(n + 1)$st segment of the folded unipole
Y_{cap}	Admittance of the antenna top capacitance
Y_{0c}	Common mode characteristic admittance.
z	Vertical coordinate
Z_0	Common mode characteristic impedance in Schelkunoff's formula for end effect
Z_{0c}	Common mode characteristic impedance
Z_{0c1}	Common mode characteristic impedance for first or bottom segment of the folded unipole
Z_{0cN}	Common mode characteristic impedance for the Nth or top segment of the folded unipole
Z_{0cn}	Common mode characteristic impedance for the nth segment of the folded unipole
$Z_{0c,n+1}$	Common mode characteristic impedance for the $(n + 1)$st segment of the folded unipole
Z_{0d1}	Differential mode characteristic impedance for the first or bottom segment of the folded unipole
Z_{0dN}	Differential mode characteristic impedance for the Nth or top segment of the folded unipole
Z_{0dn}	Differential mode characteristic impedance for the nth segment of the folded unipole
$Z_{0d,n+1}$	Differential mode characteristic impedance for the $(n + 1)$st segment of the folded unipole
Z_B	Base terminating impedance
Z_N	Terminating impedance for the Nth or top segment of the folded unipole
Z_n	Terminating impedance for the nth segment of the folded unipole
Z_T	Same as Z_N
Γ	Complex reflection coefficient at termination of transmission line
δ_{cap}	Effective lengthening of antenna due to antenna end effect using Schelkunoff's formula
ε_0	Permittivity of free space
η	Impedance of free space
ν_N	Current transformation ratio for common mode current on Nth or top segment of the folded unipole
ν_n	Current transformation ratio for common mode current on nth segment of the folded unipole
ν_{n+1}	Current transformation ratio for common mode current on $(n+1)$st segment of the folded unipole
$\psi_F(z)$	Forward traveling wave function
$\psi_R(z)$	Reverse traveling wave function

Chapter

6

The Most General Configuration

6.1 Introduction

In the previous chapter, we showed how our mathematical model included the folded unipole antenna for any number of segments, and for any combination of folded and unfolded segments. There are still practical restrictions on the model; however, and they concern the terminating impedances. First, those impedances are all two-terminal devices, or one-port networks. That is, they are simple combinations of inductors, capacitors, and resistors. Second, they are all connected as jumpers, that is, between the fold and the main conductor or tower. There are no components connected in series between segments. In this chapter, we will remove these restrictions.

There are important design advantages to advancing from one-port to two-port terminating networks. First, the latter will add two more independent design variables to each segment of the antenna. Earlier there were five such variables for each folded segment. Now there will be seven. So, there are even more degrees of design freedom available to the antenna engineer.

Second, we will see that the inherent topology of a two-port network provides greater control over the amplitude and phase shift of the common mode current as we transition between antenna segments. For structures comparable to a wavelength and longer, this means the ability to design a vertical array of antennas. Such an array can shape the vertical radiation pattern as desired. In the case of AM broadcast, we can reduce sky wave interference. In the case of "smart antennas" for cellular telephone and other applications, we can steer or scan the main lobe as desired. So, the two-port terminating network is a feature worth the extra effort required to incorporate it into our mathematical model.

Mathematically, we will find that we have already done most of the work. The resulting matrix equation and matrix elements will be very similar to what we derived in the previous chapter.

6.2 Essential Features of the Most General Configuration

Figure 6-1 shows the configuration of interest. It is similar to Fig. 5-1 except that the terminating impedances at the boundaries between segments have been replaced by two-port (*i.e.*, four-terminal) networks.

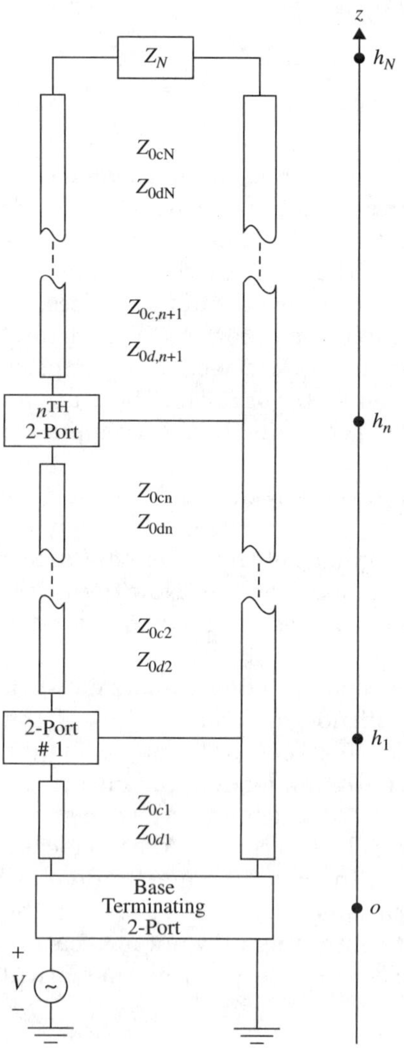

Figure 6-1 Essential features and geometry for the most general configuration (MGC) of the folded unipole antenna. They are similar to those shown in Fig. 5-1 except that the two-terminal terminating impedances have been replaced by the more general two-port (i.e., four-terminal) network. This adds two independent design variables for each segment. It also provides greater capability for designing phase and amplitude changes between segments. As usual, only two parallel conductors are shown for simplicity in the drawing. In fact, the mathematical model of the MGC applies to arbitrary cross sections.

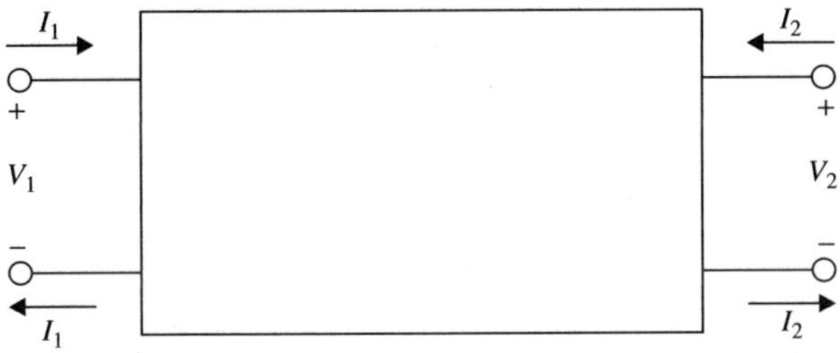

Figure 6-2 In the MGC, the terminating impedances of the nSFU are replaced by two-port networks. Such networks are defined by the relations between the input voltages and currents at the ports.

Most of the latter are drawn with only three terminals because it is neither practical nor desirable to break up the main conductor or tower. Nonetheless, the terminating networks are two-port networks, as we shall shortly see.

The same remarks made concerning Fig. 5-1 also apply here. The antenna can be an arbitrary combination of folded or unfolded segments. To unfold a particular segment, its differential mode characteristic impedance is forced to a very large value approximating infinity, and the common mode current transformation ratio is set to zero.

We can say a lot more about the antenna by being more specific about the circuit components inside the two-port terminating networks. To that end, Fig. 6-2 shows an isolated two-port network in its most general form. It is completely defined by the relations between the currents and voltages at the terminals. In particular, for a linear, reciprocal network, those relations are:

$$V_1 = Z_{11} I_1 + Z_{21} I_2 \tag{6-1}$$

$$V_2 = Z_{22} I_2 + Z_{12} I_1 \tag{6-2}$$

In Eqs. 6-1 and 6-2, Z_{11} and Z_{22} are called *self impedances*. Z_{21} and Z_{12} are called *mutual impedances.* A fundamental property of reciprocal networks is that the mutual impedances are equal:

$$V_{12} = Z_{21} \equiv Z_M \tag{6-3}$$

There are two simple, three-component circuits that satisfy Eqs. 6-1 through 6-3, and we can use either of them in our mathematical model of the folded unipole antenna. They are the Tee and the Pi networks, shown in Figs. 6-3 and 6-4, respectively. In the Tee network, it is seen that the left hand series impedance is:

$$Z_L = Z_{11} - Z_M \tag{6-4}$$

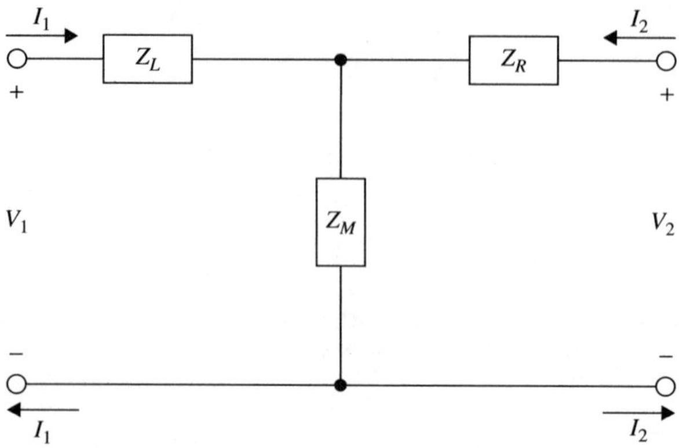

Figure 6-3 A linear, reciprocal, time-invariant two-port network can be constructed as a Tee circuit from three impedances.

The right hand series impedance is:

$$Z_R = Z_{22} - Z_M \tag{6-5}$$

According to Eqs. 6-4 and 6-5, specifying Z_L, Z_R, and Z_M uniquely determines the two-port network. For antenna design purposes, each one is an independent variable. So, each two-port provides three variables, in contrast to the single variable provided by terminating impedance in previous models.

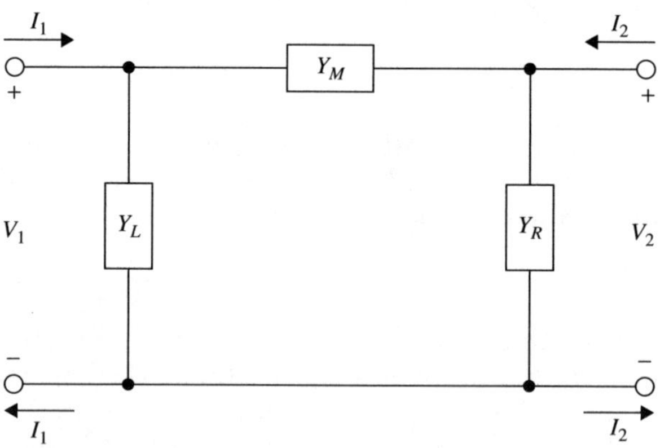

Figure 6-4 A linear, reciprocal, time-invariant two-port network can also be constructed as a Pi circuit from three admittances. The Pi circuit is the *dual* of the Tee circuit.

The Pi network shown in Fig. 6-4 is the *dual* of the Tee network. The relations analogous to Eqs. 6-1 through 6-3 are:

$$I_1 = Y_{11}V_1 + Y_{21}V_2 \tag{6-6}$$

$$I_2 = Y_{22}V_2 + Y_{12}V_1 \tag{6-7}$$

$$Y_{21} = Y_{12} \equiv Y_M \tag{6-8}$$

The admittances analogous to the impedances of Eqs. 6-4 and 6-5 are:

$$Y_L = Y_{11} - Y_M \tag{6-9}$$

$$Y_R = Y_{22} - Y_M \tag{6-10}$$

In similar fashion to the Tee network, specifying the admittances Y_L, Y_R, and Y_M completely determines the two-port network. We can use either the Tee or Pi network, or both, in our model of the folded unipole antenna. They are completely equivalent, or *duals* of each other. In the rest of this chapter, the mathematics will prove less awkward if we stick with the Tee network; however, the Pi network could prove useful in other models of the folded unipole antenna. For example, if we wish to separate adjacent folds with an open circuit, then we might insert a Pi network between them, and then let $Y_M = 0$.

In Figs. 6-3 and 6-4, it is readily seen that if the lower two terminals are connected together so as to form a single terminal, then the result is a three-terminal network like the ones shown in Fig. 6-1. So, the three-terminal network is, in fact, a two-port.

The two-port network is the distinguishing feature of the folded unipole antenna of interest in this chapter. Now that we understand its basic properties, it is time to include it in the mathematical model of the most general configuration (MGC).

6.3 Equations at the Top of the MGC

Figure 6-5 shows the top of the MGC with currents and voltages. It also shows the top capacitance due to antenna end effect, a phenomenon that was discussed in Sec. 5.7. The equations are the very same as the ones derived in Chap. 5. The common mode current flows into the top capacitor:

$$(1 + \nu_N) I_{\text{CN}}(h_N) = Y_{\text{cap}} V_{\text{CN}}(h_N) \tag{6-11}$$

In Eq. 6-11, Y_{cap} is defined by Eq. 5-34.

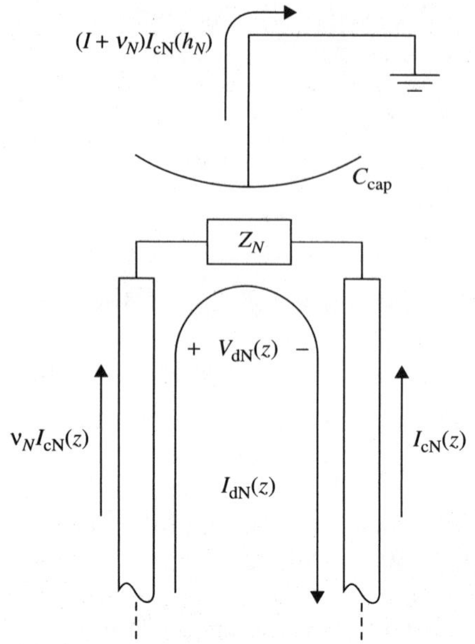

Figure 6-5 Voltages and currents at the top of the MGC, including the *antenna end effect* due to top capacitance.

The differential mode current and voltage are related by the terminating impedance Z_N:

$$Z_N I_{dN}(h_N) = V_{dN}(h_N) \tag{6-12}$$

Equations 6-11 and 6-12 are identical to Eqs. 5-33 and 5-4, respectively, because the model of the top of the MGC is identical to the model for the *n*SFU. Below the top, however, the similarity ends, and the equations will be different.

For an *N*-segment MGC, we will need *4N* independent equations to determine *4N* current amplitudes, I_{Fcn}, I_{Rcn}, I_{Fdn}, and I_{Rdn}. Equations 6-11 and 6-12 are two of those. We will derive the rest at the base of the MGC and at the boundaries between segments.

6.4 Equations at the Base of the MGC

Figure 6-6 shows the base of the MGC with currents and voltages. It is seen that we have introduced an extra current I_B in addition to the common mode and differential mode currents $I_{c1}(z)$ and $I_{d1}(z)$. This is necessary because of the current path provided by the circuit component Z_{MB}. As a result, we can derive three independent equations at the base; however, we will reduce them to only two of the sort we need for our final tally of *4N*.

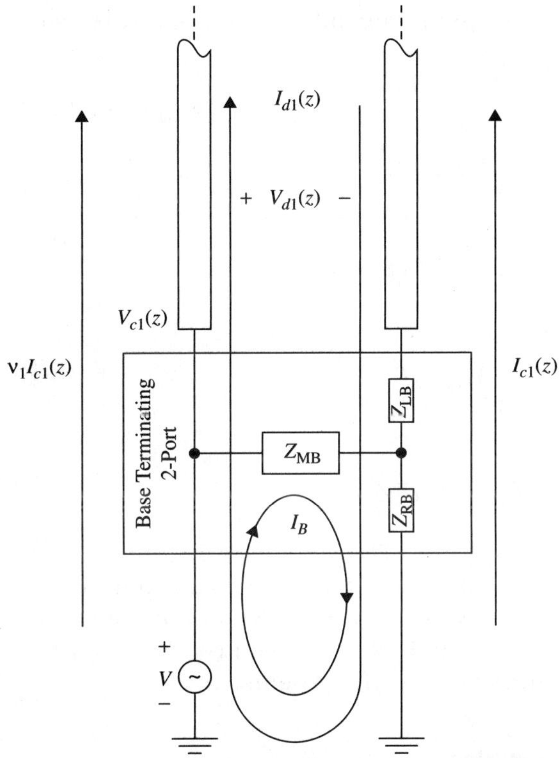

Figure 6-6 Voltages and currents at the base of the MGC. Compared with the nSFU shown in Fig. 5-4, it is necessary to define the additional current I_B because of the extra loop created by the mutual impedance Z_{MB}.

Kirchhoff's voltage law (KVL) provides us with two independent equations:

$$Z_{LB}[I_{d1}(0) - I_{c1}(0) + I_B] + Z_{MB} I_B = V \tag{6-13}$$

and

$$-Z_{MB} I_B + Z_{RB}[I_{d1}(0) - I_{c1}(0)] + V_{d1}(0) = 0 \tag{6-14}$$

By inspection, we obtain a third equation,

$$V_{c1}(0) = V \tag{6-15}$$

Equations 6-13 and 6-14 are independent; however, they contain an intermediate unknown I_B that is not of interest with respect to formulating our mathematical model. So, we can eliminate I_B from Eqs. 6-13 and 6-14 and end up with a single independent equation. After some algebra, that equation is:

$$(Z_{\text{LB}} + Z_{\text{RB}} + Y_{\text{MB}} Z_{\text{LB}} Z_{\text{RB}})[I_{d1}(0) - I_{c1}(0)] + (1 + Y_{\text{MB}} Z_{\text{LB}}) V_{d1}(0) = V(1 + Y_{\text{MB}} Z_{\text{LB}}) \quad (6\text{-}16)$$

In Eq. 6-16, the mutual admittance is:

$$Y_{\text{MB}} = \frac{1}{Z_{\text{MB}}} \quad (6\text{-}17)$$

We see that in the limit as,

$$Y_{\text{MB}} \to 0 \quad (6\text{-}18)$$

and if we set

$$Z_{\text{LB}} + Z_{\text{RB}} \equiv Z_B \quad (6\text{-}19)$$

then, Eq. 6-16 is identical to Eq. 5-5. This is a good sanity check for Eqs. 6-13 and 6-14, and for the subsequent algebraic manipulations.

Equations 6-15 and 6-16 are two independent equations we can use in our mathematical model of the MGC. We still require $4N-4$ more. We will find them at the boundaries between segments.

6.5 Equations at the Boundaries Between Segments

Figure 6-7 shows the boundary between the nth and $(n + 1)$st segments. This geometry applies to all segments except the top one. KVL provides us with three independent equations:

$$V_{\text{cn}}(h_n) - Z_{\text{Rn}}[I_{\text{dn}}(h_n) + v_n I_{\text{cn}}(h_n)] - Z_{\text{Ln}}[I_{d,n+1}(h_n) + v_{n+1} I_{c,n+1}(h_n)] = V_{c,n+1}(h_n) \quad (6\text{-}20)$$

$$V_{\text{dn}}(h_n) - Z_{\text{Rn}}[I_{\text{dn}}(h_n) + v_n I_{\text{cn}}(h_n)] - Z_{\text{Ln}}[I_{d,n+1}(h_n) + v_{n+1} I_{c,n+1}(h_n)] = V_{d,n+1}(h_n) \quad (6\text{-}21)$$

$$V_{\text{dn}}(h_n) - Z_{\text{Rn}}[I_{\text{dn}}(h_n) + v_n I_{\text{cn}}(h_n)] + Z_{\text{Mn}}[I_{d,n+1}(h_n) - I_{\text{dn}}(h_n)] = 0 \quad (6\text{-}22)$$

There are easy sanity checks for Eqs. 6-20 through 6-22, just as there was for Eq. 6-16. In the limit as

$$Z_{\text{Ln}} = Z_{\text{Rn}} = 0 \quad (6\text{-}23)$$

and if we define,

$$Z_n \equiv Z_{\text{Mn}} \quad (6\text{-}24)$$

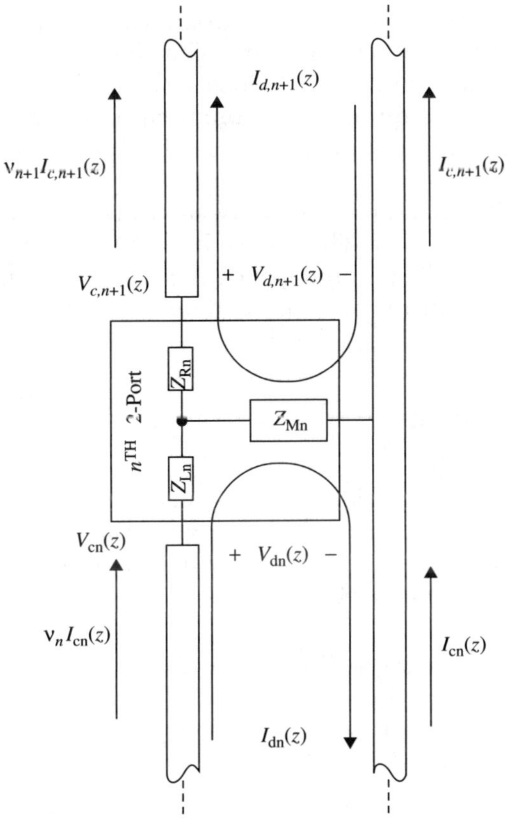

Figure 6-7 Voltages and currents at the boundaries between segments of the MGC. In contrast to the *n*SFU shown in Fig. 5-5, the differential and common mode voltages are not necessarily continuous across the boundaries.

then Eqs. 6-20 through 6-22 become identical to equations in Sec. 5.5. Equation 6-20 becomes Eq. 5-7. Equation 6-21 becomes Eq. 5-8, and Eq. 6-22 becomes Eq. 5-10.

We have one more independent equation from the boundary. The common mode current must be continuous across it:

$$(1+v_n)I_{cn}(h_n) = (1+v_{n+1})I_{c,n+1}(h_n) \tag{6-25}$$

Equation 6-25 is identical to Eq. 5-9.

Equations 6-20 through 6-22, and Eq. 6-25, fall into the category of mathematical formulas called *recursion relations*. They state that once the currents and voltages in the *n*th segment are known, then the currents and voltages in the $(n + 1)$st segment are automatically determined. We shall see in the next section that a practical and beneficial consequence of the recursion relation is a matrix equation with a *sparse matrix*. Most elements of a *sparse matrix* are zero. The time needed to compute the elements of such a matrix is obviously reduced. Also, a matrix element with value zero is quite clearly immune to numerical noise. So, there is less cumulative numerical noise in the matrix equation.

Equations 6-20 through 6-22, and Eq. 6-25, apply separately to all boundaries from $n = 1$ through $n = N-1$. That is, we have $4N-4$ independent equations. Combined with the four independent equations from Secs. 6.3 and 6.4, we have the necessary number to determine the $4N$ unknown current amplitudes $I_{\mathrm{Fc}n}$, $I_{\mathrm{Rc}n}$, $I_{\mathrm{Fd}n}$, and $I_{\mathrm{Rd}n}$.

6.6 Matrix Equation for the Currents on the MGC

The equations of the previous three sections can be consolidated into a single matrix equation, and that equation has many similarities to Eq. 5-11 for the *n*SFU. Many of the matrix elements will be identical to those in Eq. 5-11. The matrix equation for the MGC is:

$$\mathbf{M} \cdot \mathbf{I} = \begin{pmatrix} (1+Y_{\mathrm{MB}} Z_{\mathrm{LB}})V \\ V \\ 0 \\ \vdots \\ 0 \end{pmatrix} \tag{6-26}$$

It is seen that in the limit given by Eq. 6-18, Eq. 6-26 reduces to Eq. 5-11 exactly.

In Eq. 6-26 both the matrix **M** and the current vector **I** may be partitioned exactly as they were in Chap. 5, according to Eqs. 5-13 and 5-12, respectively. Further, the submatrices of **M** may be partitioned as they were in Chap. 5. In particular, $\mathbf{A_{11}}$, $\mathbf{A_{1N}}$, $\mathbf{A_{nn}}$, and $\mathbf{A_{n,n+1}}$ may be partitioned precisely as described by Eqs. 5-14, 5-17, 5-20, and 5-25, respectively.

Some of the individual elements of the submatrices will be different from those of Chap. 5 as a result of the different equations derived in Secs. 6.3, 6.4, and 6.5. For example, the nonzero submatrices of $\mathbf{A_{11}}$ are:

$$\mathbf{a}_{11} = \left(\begin{array}{c|c} -Z_{\mathrm{LB}} - Z_{\mathrm{RB}} - Y_{\mathrm{MB}} Z_{\mathrm{LB}} Z_{\mathrm{RB}} & -Z_{\mathrm{LB}} - Z_{\mathrm{RB}} - Y_{\mathrm{MB}} Z_{\mathrm{LB}} Z_{\mathrm{RB}} Z \\ \hdashline Z_{0c1} & -Z_{0c1} \end{array} \right) \tag{6-27}$$

$$\mathbf{b}_{11} = \left(\begin{array}{c|c} Z_{\mathrm{LB}} + Z_{\mathrm{RB}} + Y_{\mathrm{MB}} Z_{\mathrm{LB}} Z_{\mathrm{RB}} + (1 + Y_{\mathrm{MB}} Z_{\mathrm{LB}}) Z_{0d1} & Z_{\mathrm{LB}} + Z_{\mathrm{RB}} + Y_{\mathrm{MB}} Z_{\mathrm{LB}} Z_{\mathrm{RB}} - (1 + Y_{\mathrm{MB}} Z_{\mathrm{LB}}) Z_{0d1} \\ \hdashline 0 & 0 \end{array} \right) \tag{6-28}$$

It is seen that in the limits given by Eqs. 6-18 and 6-19, Eqs. 6-27 and 6-28 reduce to Eqs. 5-15 and 5-16 exactly. This is a good sanity check on our matrix equation for the MGC.

The submatrix $\mathbf{A_{1N}}$ is unchanged from Chap. 5. Its nonzero elements are given by Eqs. 5-35 and 5-19.

The nonzero elements for the submatrix $\mathbf{A_{nn}}$ are:

$$\mathbf{a}_{\mathbf{nn}} = \left(\begin{array}{c|c} (1+\nu_n)\psi_F(h_n) & (1+\nu_n)\psi_R(h_n) \\ \hline -\nu_n Z_{\mathrm{Rn}}\psi_F(h_n) & -\nu_n Z_{\mathrm{Rn}}\psi_R(h_n) \end{array}\right) \tag{6-29}$$

$$\mathbf{b}_{\mathbf{nn}} = \left(\begin{array}{c|c} 0 & 0 \\ \hline (Z_{\mathrm{0dn}} - Z_{\mathrm{Rn}})\psi_F(h_n) & (-Z_{\mathrm{0dn}} - Z_{\mathrm{Rn}})\psi_R(h_n) \end{array}\right) \tag{6-30}$$

$$\mathbf{c}_{\mathbf{nn}} = \left(\begin{array}{c|c} (Z_{\mathrm{0cn}} - \nu_n Z_{\mathrm{Rn}})\psi_F(h_n) & (-Z_{\mathrm{0cn}} - \nu_n Z_{\mathrm{Rn}})\psi_R(h_n) \\ \hline \nu_n Z_{\mathrm{Rn}}\psi_F(h_n) & \nu_n Z_{\mathrm{Rn}}\psi_R(h_n) \end{array}\right) \tag{6-31}$$

$$\mathbf{d}_{\mathbf{nn}} = \left(\begin{array}{c|c} -Z_{\mathrm{Rn}}\psi_F(h_n) & -Z_{\mathrm{Rn}}\psi_R(h_n) \\ \hline (Z_{\mathrm{Rn}} + Z_{\mathrm{Mn}} - Z_{\mathrm{0dn}})\psi_F(h_n) & (Z_{\mathrm{Rn}} + Z_{\mathrm{Mn}} + Z_{\mathrm{0dn}})\psi_R(h_n) \end{array}\right) \tag{6-32}$$

In the limit described by Eq. 6-23, it is seen that Eqs. 6-29 through 6-32 reduce exactly to Eqs. 5-21 through 5-24. This is yet another confirmation of the matrix equation for MGC, Eq. 6-26.

The nonzero elements of the submatrix $\mathbf{A_{n,n+1}}$ are:

$$\mathbf{a}_{\mathbf{n,n+1}} = \left(\begin{array}{c|c} -(1+\nu_{n+1})\psi_F(h_n) & -(1+\nu_{n+1})\psi_R(h_n) \\ \hline -\nu_{n+1} Z_{\mathrm{Ln}}\psi_F(h_n) & -\nu_{n+1} Z_{\mathrm{Ln}}\psi_R(h_n) \end{array}\right) \tag{6-33}$$

$$\mathbf{b}_{\mathbf{n,n+1}} = \left(\begin{array}{c|c} 0 & 0 \\ \hline (-Z_{\mathrm{Ln}} - Z_{0d,n+1})\psi_F(h_n) & (-Z_{\mathrm{Ln}} + Z_{0d,n+1})\psi_R(h_n) \end{array}\right) \tag{6-34}$$

$$\mathbf{c}_{\mathbf{n,n+1}} = \left(\begin{array}{c|c} (-Z_{0c,n+1} - \nu_{n+1} Z_{\mathrm{Ln}})\psi_F(h_n) & (Z_{0c,n+1} - \nu_{n+1} Z_{\mathrm{Ln}})\psi_R(h_n) \\ \hline 0 & 0 \end{array}\right) \tag{6-35}$$

$$\mathbf{d}_{\mathbf{n,n+1}} = \left(\begin{array}{c|c} -Z_{\mathrm{Ln}}\psi_F(h_n) & -Z_{\mathrm{Ln}}\psi_R(h_n) \\ \hline -Z_{\mathrm{Mn}}\psi_F(h_n) & -Z_{\mathrm{Mn}}\psi_R(h_n) \end{array}\right) \tag{6-36}$$

In a similar fashion to the elements of $\mathbf{A_{nn}}$, in the limit described by Eq. 6-23, Eqs. 6-33 through 6-36 reduce exactly to Eqs. 5-26 through 5-29.

6.7 Conclusions

We have generalized our mathematical model of a single folded unipole antenna about as much as possible. In Chap. 5, we generalized the length of the antenna by extending it to include an arbitrary number of folded and unfolded segments, in any combination. In Chap. 6, we continued the generalization by expanding the terminating impedance of each segment, from a one-port (i.e., two-terminal) to a two-port (i.e., four-terminal) network.

There are at least two advantages to generalizing from a one-port to a two-port terminating network. First, it increases the number of independent design variables from five to seven for each folded segment. For unfolded segments, the number increases from three to five.

The second advantage is a fundamental property of two-port networks, namely, the capability for easily designing phase and amplitude changes between ports. For the folded unipole antenna, this means that each segment can now be regarded as an element in a vertical antenna array. Such an array provides control over the shape of its vertical radiation pattern. For AM radio arrays, this can mean sky wave reduction. For cellular radio and other "smart antenna" applications, it means that the vertical radiation pattern can be steered or scanned.

A two-port network can be constructed from either of two simple circuits. They are the Tee and Pi networks. Each network has only three components, and they can be specified independently. The Tee network has a right side impedance Z_R, a left side impedance Z_L, and a mutual impedance Z_M. The Pi network has a right side admittance Y_R, a left side admittance Y_L, and a mutual admittance Y_M. The Tee and Pi networks are *duals* of one another.

The extra mathematical effort to achieve this final stage of generalization was minimal. We already did most of the work in Chap. 5. The elements of the matrix equations in Chap. 5 and 6 are remarkably similar, and the latter readily reduce to the former as the two-port network degenerates to the special case of a one-port.

Symbols Used in This Chapter

$\mathbf{A}_{11}$	Upper left submatrix of $\mathbf{M}$
$\mathbf{a}_{11}$	Submatrix of $\mathbf{A}_{11}$
$\mathbf{A}_{1N}$	Upper right submatrix of $\mathbf{M}$
$\mathbf{A}_{mn}$	Lower submatrices of $\mathbf{M}$
$\mathbf{A}_{nn}$	Diagonal submatrices of $\mathbf{A}_{mn}$
$\mathbf{a}_{nn}$	Submatrix of $\mathbf{A}_{nn}$
$\mathbf{A}_{n,n+1}$	Submatrices of $\mathbf{A}_{mn}$ immediately to the right of $\mathbf{A}_{nn}$
$\mathbf{a}_{n,n+1}$	Submatrices of $\mathbf{A}_{n,n+1}$
$\mathbf{b}_{11}$	Submatrix of $\mathbf{A}_{11}$
$\mathbf{b}_{nn}$	Submatrix of $\mathbf{A}_{nn}$

$\mathbf{b}_{\mathbf{n,n+1}}$ Submatix of $\mathbf{A}_{\mathbf{n,n+1}}$
$\mathbf{c}_{\mathbf{1N}}$ Submatrix of $\mathbf{A}_{\mathbf{1N}}$
C_{cap} Capacitance of antenna top
$\mathbf{c}_{\mathbf{nn}}$ Submatrix of $\mathbf{A}_{\mathbf{nn}}$
$\mathbf{c}_{\mathbf{n,n+1}}$ Submatrix of $\mathbf{A}_{\mathbf{n,n+1}}$
$\mathbf{d}_{\mathbf{1N}}$ Submatrix of $\mathbf{A}_{\mathbf{1N}}$
$\mathbf{d}_{\mathbf{nn}}$ Submatrix of $\mathbf{A}_{\mathbf{nn}}$
$\mathbf{d}_{\mathbf{n,n+1}}$ Submatrix of $\mathbf{A}_{\mathbf{n,n+1}}$
h_N Vertical coordinate of the top of the Nth or top segment of the folded unipole
h_n Vertical coordinate of the top of nth segment of the folded unipole
$\mathbf{I}$ Column vector of current amplitudes on the folded unipole
I_1 Input current to the left side of a two-port network
I_2 Input current to the right side of a two-port network
I_B Loop current in the base terminating two-port network
$I_{c1}(z)$ Common mode current on the first or bottom segment of the folded unipole
$I_{\mathrm{cN}}(z)$ Common mode current on the Nth or top segment of the folded unipole
$I_{\mathrm{cn}}(z)$ Common mode current on the nth segment of the folded unipole
$I_{c,n+1}(z)$ Common mode current on the $(n + 1)$st segment of the folded unipole
$I_{d1}(z)$ Differential mode current on the first or bottom segment of the folded unipole
$I_{\mathrm{dN}}(z)$ Differential mode current on the Nth or top segment of the folded unipole
$I_{\mathrm{dn}}(z)$ Differential mode current on the nth segment of the folded unipole
$I_{d,n+1}(z)$ Differential mode current on the $(n + 1)$st segment of the folded unipole
$\mathbf{M}$ Matrix in the equation for current amplitudes everywhere on the folded unipole
V input voltage to the folded unipole
V_1 Input voltage to the left side of a two-port network
V_2 Input voltage to right side of a two-port network
$V_{c1}(z)$ Common mode voltage on the first or bottom segment of the folded unipole
$V_{\mathrm{cN}}(z)$ Common mode voltage on the Nth or top segment of the folded unipole
$V_{\mathrm{cn}}(z)$ Common mode voltage on the nth segment of the folded unipole
$V_{c,n+1}(z)$ Common mode voltage on the $(n + 1)$st segment of the folded unipole
$V_{d1}(z)$ Differential mode voltage on the first or bottom segment of the folded unipole
$V_{\mathrm{dN}}(z)$ Differential mode voltage on the top or Nth segment of the folded unipole
$V_{\mathrm{dn}}(z)$ Differential mode voltage on the nth segment of the folded unipole
$V_{d,n+1}(z)$ Differential mode voltage on the $(n + 1)$st segment of the folded unipole
Y_{11} Self admittance of the left side of a two-port network
Y_{12} Mutual admittance from the left port to the right port of a two-port network
Y_{21} Mutual admittance from the right port to the left port of a two-port network
Y_{22} Self-admittance of the right port of a two-port network
Y_{cap} Admittance of the antenna top capacitance
Y_L Left side admittance of a Pi network
Y_M Mutual admittance of a Pi network
Y_{MB} Reciprocal of Z_{MB}
Y_R Right side admittance of a Pi network
z Vertical coordinate
Z_{0c1} Common mode characteristic impedance for first or bottom segment of the folded unipole
$Z_{0\mathrm{cN}}$ Common mode characteristic impedance for the Nth or top segment of the folded unipole
$Z_{0\mathrm{cn}}$ Common mode characteristic impedance for the nth segment of the folded unipole
$Z_{0c,n+1}$ Common mode characteristic impedance for the $(n + 1)$st segment of the folded unipole

Z_{0d1}	Differential mode characteristic impedance for the first or bottom segment of the folded unipole
$Z_{0\mathrm{dN}}$	Differential mode characteristic impedance for the Nth or top segment of the folded unipole
$Z_{0\mathrm{dn}}$	Differential mode characteristic impedance for the nth segment of the folded unipole
$Z_{0d,n+1}$	Differential mode characteristic impedance for the $(n + 1)$st segment of the folded unipole
Z_{11}	Self-impedance of the left side of a two-port network
Z_{12}	Mutual impedance from left to right side of a two-port network
Z_{21}	Mutual impedance from right to left side of a two-port network
Z_{22}	Self-impedance of the right side of a two-port network
Z_B	Base terminating impedance
Z_L	Left side impedance of a Tee network
Z_{LB}	Left side impedance of base terminating Tee network
Z_{Ln}	Left side impedance of terminating Tee network for nth segment of the folded unipole
Z_M	Mutual impedance of a Tee network
Z_{MB}	Mutual impedance of base terminating Tee network
Z_{Mn}	Mutual impedance of terminating Tee network for nth segment of the folded unipole
Z_N	Terminating impedance for the Nth or top segment of the folded unipole
Z_n	Terminating impedance for the nth segment of the folded unipole
Z_R	Right side impedance of Tee network
Z_{RB}	Right side impedance of base terminating Tee network
Z_{Rn}	Right side impedance of terminating Tee network for nth segment of the folded unipole
ν_N	Current transformation ratio for common mode current on Nth or top segment of the folded unipole
ν_n	Current transformation ratio for common mode current on nth segment of the folded unipole
ν_{n+1}	Current transformation ratio for common mode current on $(n + 1)$st segment of the folded unipole
$\psi_F(z)$	Forward traveling wave function
$\psi_R(z)$	Reverse traveling wave function

Chapter

7

Characteristic Impedance

7.1 Introduction

In this chapter, we will derive formulas for the three independent design variables Z_{0d} (differential mode characteristic impedance), Z_{0c} (common mode characteristic impedance), and ν (current transformation ratio). They are functions of the antenna cross-sectional geometry and wavelength. In other chapters, they permit us to conveniently avoid the details of that geometry, which would overwhelm our ability to consider other design variables. In 1943, S. A. Schelkunoff eloquently expressed the advantages of this approach when he wrote [1]:

> "All physical fields are three dimensional; but in some circumstances either two or all three dimensions are unimportant; then they may be 'integrated out' and thus 'concealed'; in the first case the problem belongs to 'transmission line theory' and in the second to 'network theory.' This suppression of some or all physical dimensions is analogous to the method of 'ignoration of coordinates' in mechanics; and it may or may not involve approximations. It is a mistake to say that the circuit and line theories are approximate while only the field theory is exact. In fact in many important cases a three-dimensional problem is rigorously reducible to a set of one-dimensional problems. Once the one-dimensional problem has been solved in sufficiently general terms, the results can be used repeatedly in the solution of more general problems."

In the case of characteristic impedance and current transformation ratio, we are integrating out the transverse or cross-sectional dimensions. After we do this, we no longer need to worry about the number, sizes, shapes, and positions of the conductors that make up the folds and other parts of the folded unipole. For purposes of input impedance and radiation pattern calculations, those details are completely contained within the three parameters Z_{0d}, Z_{0c}, and ν.

We will find that both characteristic impedances can vary over a wide range of values, from the order of an ohm to well over 1,000 ohms. The common mode characteristic impedance Z_{0c} can be less than or greater than Z_{0d}.

We will also find that the current transformation ratio ν can vary from a tiny fraction to 100 or more. The larger values will lead us to consider a configuration called the *inverted unipole*, with the driven section and passive sections interchanged.

In any case, these three parameters will provide us with a large "ball-park" of design values. We simply must design the antenna cross section to obtain the desired results. In this chapter, we will examine how the radius, spacing, and number of wires come into play.

Curiously, of the three design parameters, the common mode characteristic impedance Z_{0c} has been a controversial quantity at least since the 1930s. As we will see in the following two sections, there are competing formulas for it. Also, the formulas depend upon the length of the antenna. This contradicts the fundamental definition of characteristic impedance, that is, the ratio of voltage to current for a traveling wave launched on a semi-infinitely long transmission line. In this chapter, we hope that our approach will resolve both the controversy and the contradiction.

7.2 Hallén's Definition of Common Mode Characteristic Impedance

As discussed in Sec. 1.3, in 1938, Erik Hallén inspired a major school of thought concerning antenna analysis when he derived and solved an integral equation for the current along monopole and dipole antennas [2]. The approximate solution for the current is [3]:

$$I(z) = \frac{jV}{60\Omega}\left[\frac{\sin\beta(l-|z|)+\left(b_1/\Omega\right)}{\cos\beta l+\left(d_1/\Omega\right)}\right] \tag{7-1}$$

In Eq. 7-1, V is the input voltage, β is the wavenumber (same as k), and the constants b_1 and d_1 are complicated combinations of the sine, cosine, and exponential integrals. Also,

$$\Omega = 2\ln\left(\frac{2l}{a}\right) \tag{7-2}$$

In Eqs. 7-1 and 7-2, l is the length of the monopole, or half-length of the dipole, and a is its radius.

Equation 7-1 can be readily rearranged to obtain a formula for the input impedance of the antenna:

$$Z_{in} = \frac{V}{I(0)} = -j60\Omega\left[\frac{\cos\beta l + \left(d_1/\Omega\right)}{\sin\beta l + \left(b_1/\Omega\right)}\right] \tag{7-3}$$

Equation 7-3 resembles the formula for the input impedance of an open circuited transmission line, except for the higher order terms d_1/Ω and b_1/Ω. Following through on that resemblance, it is very tempting to identify the factor outside of the brackets as the characteristic impedance of the antenna:

$$Z_{0c} = 60\Omega = 120\ln\left(\frac{2l}{a}\right) \tag{7-4}$$

As mentioned in the previous section, Eq.7-4 contains a conceptual contradiction because it depends upon the length of the antenna. The characteristic impedance of electromagnetic devices such as waveguides and transmission lines certainly depend on radius and other cross-sectional dimensions. They may also depend upon wavelength and mode numbers; however, they never depend upon the longitudinal dimension, or length. As we will show in Sec. 7.4, this contradicts the fundamental definition of characteristic impedance.

So, while Eq. 7-3 is a remarkable achievement and very accurate for modeling purposes, the extrapolation of Eq. 7-4 to the general definition of antenna characteristic impedance is risky. In 1952, Schelkunoff discussed the risk when he wrote [4]:

> The input impedance of an antenna contains one important parameter analogous to the characteristic impedance of a transmission line. The value of this parameter, however, depends on the particular method of analysis and, as in the case of integral equations, even on how the analysis is carried out. Theoretically these differences are compensated by the corresponding differences in associated functions involved in the input impedance formula.

Schelkunoff's message is clear. We are free to arbitrarily define terms that resemble characteristic impedance *as* characteristic impedance; however, that definition is useful only within the particular formula from which the terms originate.

7.3 Schelkunoff's Definition of Common Mode Characteristic Impedance

As discussed in Sec. 1.5, Sergei A. Schelkunoff pursued antenna modeling using analogies to transmission lines. He obtained many useful and elegant results and inspired a school of thought that competed vigorously with the integral equation approach initiated by Hallén. Not surprisingly, Schelkunoff's transmission analogy included an expression interpreted as characteristic impedance.

Initially, Schelkunoff's version of characteristic impedance did not depend on the length of the antenna. The reason is that his model of antennas began with a mathematically convenient biconical geometry, illustrated in Fig. 1-5. For this geometry, the characteristic impedance is [5]:

$$Z_{0c} = 120\ln\left[\cot\left(\psi/_2\right)\right] \tag{7-5}$$

In Eq. 7-5, ψ is the angle of the cone. It is seen that the characteristic impedance does not depend on the length of the antenna.

One can contrive to introduce antenna length into Eq. 7-5 by noting that the half angle of a truncated cone is:

$$\psi/_2 = \tan^{-1}\left(a/_l\right) \tag{7-6}$$

Using Eq. 7-6, Eq. 7-5 becomes:

$$Z_{0c} = 120\ln\left(\frac{2l}{a}\right) \tag{7-7}$$

Equation 7-7 is identical to Eq. 7-4, except that the former is for biconical antennas and the latter is for cylindrical antennas. For any antenna different from the biconical one, Schelkunoff argued that each incremental segment at position z along its length is a section of a biconical antenna with characteristic impedance,

$$Z_{0c}(z) = 120\ln\left[\frac{2z}{a(z)}\right] \tag{7-8}$$

He then asserted that this function could be averaged to obtain the characteristic impedance for the entire structure:

$$Z_{0c} = \frac{120}{l}\int_0^l \ln\left[\frac{2z}{a(z)}\right]dz \tag{7-9}$$

For cylindrical antennas, the radius $a(z)$ is a constant a, and Eq. 7-9 may be evaluated exactly to obtain:

$$Z_{0c} = 120\left[\ln\left(\frac{2l}{a}\right) - 1\right] \tag{7-10}$$

Equation 7-10 is Schelkunoff's version of characteristic impedance for cylindrical antennas. It is similar to Hallén's Eq. 7-4; however, the two do not agree. Further, both Eqs. 7-10 and 7-4 have a disturbing dependence upon the length of the antenna. At least Schelkunoff began his derivation with a formula that was independent of length, but in subsequent steps that independence was lost.

So far, we have reviewed two very popular schools of thought, and each has a different formula for the characteristic impedance of a cylindrical antenna. Both formulas depend upon the length of the antenna, which contradicts formulas for the characteristic impedance of transmission lines and waveguides. In the next section, we will return to basic transmission line theory and attempt to resolve these dilemmas.

7.4 Characteristic Impedance According to Fundamental Transmission Line Theory

In the previous two sections, we saw how integral equations led to one formula for characteristic impedance and biconical geometry led to a different formula. In this section, we will use a simpler and more fundamental approach for determining characteristic impedance.

To that end, Fig. 7-1 shows a small segment of a transmission line. The length of the segment is Δz. It can have any cross-sectional geometry.

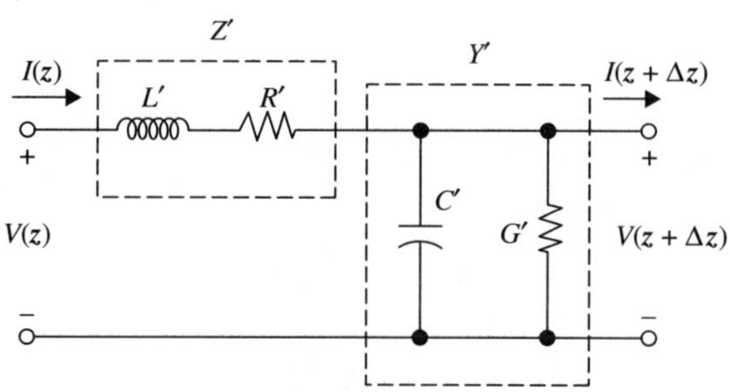

Figure 7-1 An incremental section of transmission line of length Δz is completely represented by an impedance per unit length Z' and an admittance per unit length Y'. These determine the characteristic impedance and wave number for currents and voltages along the line. The components of Z' and Y' depend upon the line geometry and the environment. They may include losses in the form of resistance per unit length R' and conductance per unit length G'.

Whatever that geometry might be, it can be modeled as an impedance per unit length Z' and an admittance per unit length Y'. Typically, Z' may be further resolved into a resistance R' and an inductance L' connected in series; however, this is not a strict requirement for our derivation. Similarly, Y' is typically resolved into a conductance G' and a capacitance C' connected in parallel. For all of these quantities the prime superscript is a reminder that they are incremental, that is, they include the dimension $(length)^{-1}$. For example, R' has dimensions ohms per meter, not simply ohms.

Applying Kirchhoff's voltage law to the transmission line segment,

$$V(z+\Delta z)+I(z)Z'\,\Delta z = V(z) \tag{7-11}$$

Equation 7-11 is easily rearranged to obtain:

$$I(z)Z' = -\frac{V(z+\Delta z)-V(z)}{\Delta z} \tag{7-12}$$

Passing to the limit as $\Delta z \to 0$, Eq. 7-12 becomes:

$$I(z)Z' = -\frac{\partial V}{\partial z} \tag{7-13}$$

Equation (7-13) is one of the two Telegrapher's Equations. The second equation can be similarly derived by applying Kirchhoff's current law to the transmission line segment:

$$I(z+\Delta z) = I(z) - Y'\,\Delta z\, V(z) \tag{7-14}$$

Equation 7-14 can be rearranged as:

$$Y'V(z) = -\frac{I(z+\Delta z)-I(z)}{\Delta z} \tag{7-15}$$

Passing to the limit as $\Delta z \to 0$, Eq. 7-15 becomes:

$$Y'V(z) = -\frac{\partial I}{\partial z} \tag{7-16}$$

Equations 7-13 and 7-16 are the two Telegrapher's Equations.

From the Telegrapher's Equations, we can derive a wave equation for the current and voltage. Differentiating Eq. 7-16,

$$Y'\frac{\partial V}{\partial z} = -\frac{\partial^2 I}{\partial z^2} \tag{7-17}$$

Combining Eqs. 7-13 and 7-17, we obtain a wave equation for the current:

$$\frac{\partial^2 I}{\partial z^2} - Z'Y'I = 0 \tag{7-18}$$

We can similarly obtain a wave equation for the voltage:

$$\frac{\partial^2 V}{\partial z^2} - Z'Y'\,V = 0 \tag{7-19}$$

The solution to Eq. 7-18 is:

$$I(z) = I_F\, e^{-\Gamma z} + I_R e^{\Gamma z} \tag{7-20}$$

Equation 7-20 shows that the current is the linear combination of forward and reverse traveling waves. The wave number is:

$$\Gamma = \sqrt{Z'Y'} \tag{7-21}$$

In general, Γ is a complex number because both Z' and Y' may be complex. The real part of Γ is an attenuation constant, and the imaginary part is a propagation constant. When the transmission line is immersed in, or in close proximity to, a lossy medium, neither part will be negligible. In the remainder of this chapter, we will not be concerned with that situation; however, we will revisit it in Chap. 10.

From Eqs. 7-20 and 7-16, we readily obtain the voltage:

$$V(z) = \sqrt{\frac{Z'}{Y'}}\left(I_F e^{-\Gamma z} - I_R e^{\Gamma z}\right) \tag{7-22}$$

Equation 7-22 contains the fundamental definition of characteristic impedance:

$$Z_0 = \sqrt{\frac{Z'}{Y'}} \tag{7-23}$$

Comparing Eqs. 7-20 and 7-22, we find a simple interpretation for Z_0. It is the ratio of voltage amplitude to current amplitude for a traveling wave on the transmission line.

We are interested in low loss transmission lines, for which R' and G' contribute negligibly to Z' and Y', respectively:

$$Z' = j\omega L' \tag{7-24}$$

$$Y' = j\omega C' \tag{7-25}$$

Using Eqs. 7-24 and 7-25, the wave number in Eq. 7-21 becomes:

$$\Gamma = \sqrt{-\omega^2 L'C'} = j\omega\sqrt{L'C'} \tag{7-26}$$

On the right side of Eq.7-26, we recognize the wave number k from previous chapters:

$$k = \omega\sqrt{L'C'} \tag{7-27}$$

The wave number is also,

$$k = \frac{\omega}{c} \tag{7-28}$$

where c is the phase velocity of the traveling wave. Comparing Eqs.7-27 and 7-28, the phase velocity is evidently,

$$c = \frac{1}{\sqrt{L'C'}} \tag{7-29}$$

Returning to Eqs. 7-24 and 7-25, and applying them toEq. 7-23, the characteristic impedance becomes:

$$Z_0 = \sqrt{\frac{L'}{C'}} \tag{7-30}$$

Combining Eqs. 7-29 and 7-30, the characteristic impedance can be written entirely in terms of C' or L':

$$Z_0 = cL' = \frac{1}{cC'} \tag{7-31}$$

If the propagation medium is free space, then,

$$c = \frac{1}{\sqrt{\varepsilon_0 \mu_0}} \tag{7-32}$$

and Eq. 7-31 becomes:

$$Z_0 = \frac{L'}{\sqrt{\varepsilon_0 \mu_0}} = \frac{\sqrt{\varepsilon_0 \mu_0}}{C'} \tag{7-33}$$

Equation 7-33 is an especially useful and practical result. It tells us that if we can somehow compute either the inductance per unit length L' or the capacitance per unit length C' of a transmission line, then we can readily obtain the characteristic impedance. That is exactly what we will do in later sections of this chapter.

Throughout the derivations of this section, the length of the transmission line has not come into play. In particular, it has no bearing on the calculation of characteristic impedance.

7.5 Quasi Static Log Potential Theory

In this section we will develop the mathematical tools necessary for calculating the capacitance per unit length C' for both balanced and unbalanced transmission lines with arbitrary cross sections. Then we will be able to use Eq. 7-33 to obtain the characteristic impedances Z_{0d} and Z_{0c}.

Our approach is an extension of a method described in 1952, by Edmund A. LaPorte [6]. He called the method "logarithmic potential theory." Essentially, it used electrostatics to find the potential V of a point charge Q using the formula,

$$V = 138cQ\log_{10}\left(\frac{1}{R}\right) + x \tag{7-34}$$

In Eq.7-34, we have replaced the symbol ρ, used in LaPorte's original notation, with R to avoid confusing distance with the charge density $\rho(\boldsymbol{r})$.

Using Eq. 7-34 and similar systems of equations, LaPorte interpreted the ratio Q/V as the capacitance per unit length C' of a transmission line.

In Eq. 7-34, x is an unknown constant. It is important because without it, the equation causes problems with dimensions. That is, the argument $1/R$ of the logarithm is not dimensionless, and the answer varies depending upon whether inches, feet, meters, or some other unit of distance is used.

LaPorte's examples were limited to balanced transmission lines, and so the unknown constant x did not matter. It always "cancelled out." For unbalanced transmission lines, however, we are not so fortunate.

If we go back to fundamental electromagnetic theory, we can reformulate LaPorte's method and avoid the dimensional problems caused by the unknown constant x. To that end, we can write the electromagnetic field exactly as:

$$\mathbf{E} = -\nabla V - j\omega\mathbf{A} \tag{7-35}$$

$$\mathbf{H} = \frac{1}{\mu_0}\nabla\times\mathbf{A} \tag{7-36}$$

In Eqs. 7-35 and 7-36 the potential functions are:

$$\mathbf{A}(\mathbf{r}) = \mu_0 \iint \mathbf{J}(\mathbf{r}')G(\mathbf{r},\mathbf{r}')d^2r' \tag{7-37}$$

$$V(\mathbf{r}) = \frac{1}{\varepsilon_0} \iiint \rho(\mathbf{r}')G(\mathbf{r},\mathbf{r}')d^3r' \tag{7-38}$$

In Eqs. 7-37 and 7-38, **r** is the vector position of the observer, and **r′** is the vector position of the current density **J**, or charge density ρ. The integrations are carried out over all regions of space where those densities are nonzero.

Within the integrals, the weighting function $G(\mathbf{r}, \mathbf{r}')$ is called the *Green's Function*. In three dimensions, it is:

$$G(\mathbf{r},\mathbf{r}') = \frac{e^{-jkR}}{4\pi R} \tag{7-39}$$

where:

$$R = |\mathbf{r} - \mathbf{r}'| \tag{7-40}$$

Equations 7-35 through 7-40 are exact at all frequencies. As the frequency ω goes to zero, the wavenumber k also vanishes, and the equations transition smoothly to the electrostatic case:

$$\lim_{k \to 0} G(\mathbf{r},\mathbf{r}') = \frac{1}{4\pi R} \tag{7-41}$$

Equations 7-35 and 7-38 become:

$$\mathbf{E} = -\nabla V \tag{7-42}$$

$$V = \frac{1}{4\pi\varepsilon_0} \iiint \frac{\rho(\mathbf{r}')}{R} d^3r' \tag{7-43}$$

Equations 7-42 and 7-43 are the classic formulas for the electrostatic field in three dimensions.

We wish to derive the two-dimensional version of Eq. 7-43. As in LaPorte's logarithmic potential theory, our transmission lines are infinite in extent, and we may confine our computations to the cross-sectional dimensions only. The steps leading to the two-dimensional version of Eq. 7-43 are similar but involve some Bessel functions. In two dimensions, the Green's function is:

$$G(\mathbf{r},\mathbf{r}') = \frac{H_0^{(2)}(kR)}{j4} \tag{7-44}$$

On the right side of Eq. 7-44, the function in the numerator is called the *Hankel function of the second kind*. It is one of the Bessel functions. Fortunately, we can replace it with simpler functions if our attention is restricted to a region small compared with a wavelength. The cross section of a transmission line is just such a region, and so,

$$\lim_{kR \to 0} \quad H_0^{(2)}(kR) = -j\frac{2}{\pi}\ln\left(\frac{e^{\gamma}}{2}kR\right) \tag{7-45}$$

In Eq. 7-45, γ is called *Euler's constant*:

$$\gamma = 0.5772156649... \tag{7-46}$$

We can simplify the notation of Eq. 7-45, by defining a *quasi static wave number* k':

$$k' = \frac{e^{\gamma}}{2}k \approx 0.891\,k = \frac{k}{1.123} \tag{7-47}$$

Using Eqs. 7-45 and 7-47, the two-dimensional Green's function of Eq. 7-44 becomes:

$$G(\mathbf{r},\mathbf{r}') = -\frac{1}{2\pi}\ln(k'R) \tag{7-48}$$

Using Eq. 7-48, the two dimensional version of Eq. 7-43 is:

$$V(\mathbf{r}) = -\frac{1}{2\pi\varepsilon_0}\iint \rho(\mathbf{r}')\ln(k'R)\,d^2r' \tag{7-49}$$

For point charges Q, or for charge distributions that can be approximated by point charges, Eq. 7-49 becomes:

$$V(\mathbf{r}) = -\frac{1}{2\pi\varepsilon_0}\sum_{\text{all charges}} Q(\mathbf{r}')\ln(k'R) \tag{7-50}$$

For a single charge, Eq. 7-50 simplifies to:

$$V = -\frac{Q}{2\pi\varepsilon_0}\ln(k'R) \tag{7-51}$$

Equation 7-51 may be compared directly with Eq. 7-34 due to LaPorte. It is seen that $k'R$ is dimensionless because we have retained frequency dependence in our derivation. As a result, we do not have the dimensional problems associated with the unknown constant x

in LaPorte's formula. So, we should be able to apply Eq. 7-50 to both balanced and unbalanced transmission lines. That is exactly what we will do in the following sections.

7.6 Characteristic Impedance of a Single Circular Cylinder

Equations 7-33 and 7-50, derived in the previous two sections, are powerful tools that will help us calculate the characteristic impedance of practically any antenna cross section. As an example, let's begin with the simplest case, a single circular, conducting cylinder, as shown in Fig. 7-2. The cylinder carries a total charge per unit length Q coulombs/meter. Because it is an isolated conductor, by symmetry the charge exhibits fields as though it were concentrated at its center. Therefore, according to Eq. 7-50, the potential at the surface of the cylinder is:

$$V(a) = -\frac{Q}{2\pi\varepsilon_0}\ln(k'a) \tag{7-52}$$

The capacitance per unit length of the cylinder is:

$$C' = \frac{Q}{V(a)} = -\frac{2\pi\varepsilon_0}{\ln(k'a)} \tag{7-53}$$

Using Eq. 7-53, Eq. 7-33 gives us the characteristic impedance:

$$Z_{0c} = -\frac{\eta}{2\pi}\ln(k'a) \tag{7-54}$$

where the impedance of free space is,

$$\eta = \sqrt{\frac{\mu_0}{\varepsilon_0}} \tag{7-55}$$

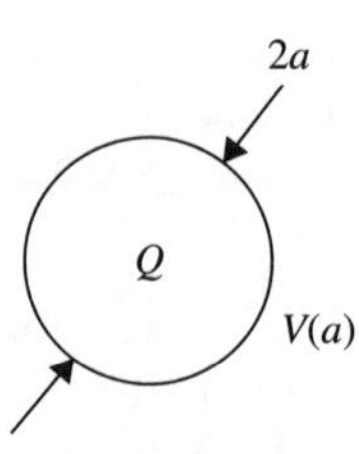

Figure 7-2 Cross section of a single, conducting cylinder or wire. The charge density Q in coulombs per meter is directly proportional to the current flowing on the wire due to the continuity relation, which is derived from Maxwell's Equations. There will be only a common mode characteristic impedance Z_{0c} because, by definition, a single cylinder is unbalanced. There is no path for a current with equal magnitude and opposite sign to flow.

In Eq. 7-55, we have written the characteristic impedance Z_{0c} as the common mode characteristic impedance because, by definition, a single isolated cylinder is unbalanced.

Also, in Eq. 7-55, k' is the quasi static wave number derived in the previous section, and defined in Eq. 7-47. So, the argument of the logarithm in Eq. 7-54 is:

$$k'a = 0.891\,ka = 5.6\frac{a}{\lambda} \tag{7-56}$$

We now have three different formulas for the characteristic impedance of a cylindrical antenna. They are Hallén's Eq. 7-4, Schelkunoff's Eq. 7-10, and Eq. 7-54. Unlike the first two, Eq. 7-54 does not depend on the length of the antenna. Also unlike the first two, Eq. 7-54 does depend upon wavelength. This is most consistent with formulas for the characteristic impedance of other electromagnetic devices. As we have previously discussed, the characteristic impedance for transmission lines and waveguides never depends upon their lengths, and it may depend upon wavelength.

Clearly, there is controversy surrounding which of the three formulas is most general and accurate; however, that controversy has merely simmered rather than erupted into a hot debate. Why? Perhaps the reason is that the formulas all give comparable answers for a monopole in the AM broadcast band. It is the AM broadcast antenna engineers who have made most use of these formulas since the 1930s.

In support of the above, let's compare some actual numbers using the three different formulas. For example, take the wavelength $\lambda = 1000$ feet (984 kHz); length $l = 250$ feet (one-quarter wavelength); and radius $a = 1$ foot. With these numbers, Hallén's formula predicts $Z_{0c} = 373$ ohms. Schelkunoff's formula predicts $Z_{0c} = 313$ ohms. Equation 7-54 predicts $Z_{0c} = 311$ ohms. (Note that Hallén's and Schelkunoff's formulas are for dipoles, and we must halve the answers to obtain values for a monopole). So, the answers are all in the same "ballpark," at least for an example in the AM broadcast band. In particular, Schelkunoff's formula and Eq. 7-54 agree with each other to within less than 1 percent.

Depending upon radius and wavelength, the characteristic impedance according to Eq. 7-54 can exhibit a wide range of values, far different from the 311 ohms in the earlier example. Figure 7-3 shows that range of values for the electromagnetic spectrum from ELF through SHF, and for extreme wire sizes, from 0000(radius 0.230 inches) to 40 Gage (radius 0.0015725 inches). It is seen that Z_{0c} ranges from over 1,500 ohms to less than one ohm.

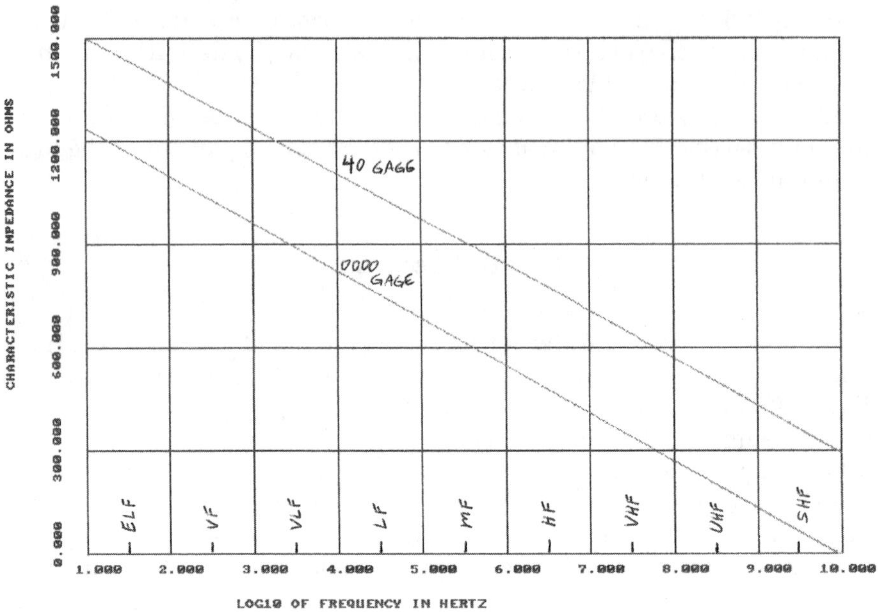

Figure 7-3 Characteristic impedance of single, conducting wires as a function of frequency, from the ELF (30–300 Hertz) through SHF (3–30 GHz) bands. The bounds for two extreme wire gages are explicitly shown. They are for 0000(with radius 0.230 inch) and 40 Gage (with radius 0.0015725 inch).

Figure 7-4 is similar to Fig. 7-3, except that the cylinder radii range from 0.1 to 100 inches. Z_{0c} ranges from over 1,200 ohms to less than one ohm.

Figures 7-5 and 7-6 are a different way of illustrating how Z_{0c} can vary over a wide range. The continuous independent variable is the radius of the cylinder. In Fig. 7-5, two fixed frequencies are from the ELF and VHF bands. In Fig. 7-6, two fixed frequencies are from the VLF and SHF bands. As in the previous figures, values for Z_{0c} run the gamut from over 1,200 ohms to less than one ohm.

Our results for the circular cylinder are readily extended to certain noncircular cylinders. The equivalent radii for ellipses, ribbons, triangles, and squares are well documented [7, 8], and they are summarized in Fig. 7-7.

This is a logical place to insert a note of caution. Figures 7-3 through 7-6, and Eq. 7-54 suggest that the characteristic impedance can vanish; however, this never actually happens. When the radius a is not very small compared with a wavelength, then the quasi static approximation as expressed in Eq. 7-45 is no longer valid, and the electromagnetic field in the vicinity of the cylinder is no longer TEM-like.

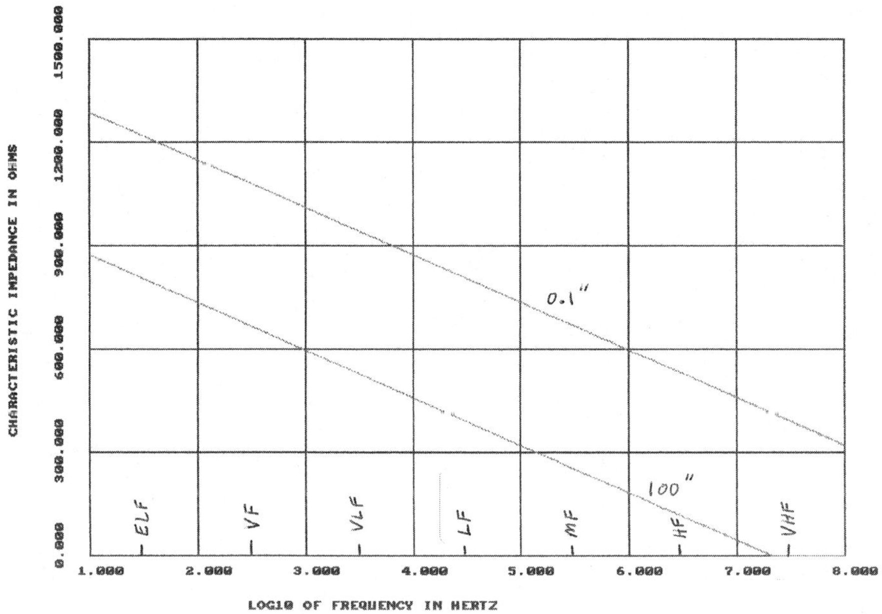

Figure 7-4 Characteristic impedance of a single, conducting wire or cylinder as a function of frequency, from the ELF through VHF bands. The bounds for a 1,000-to-1 range of radii are shown, covering most practical cases.

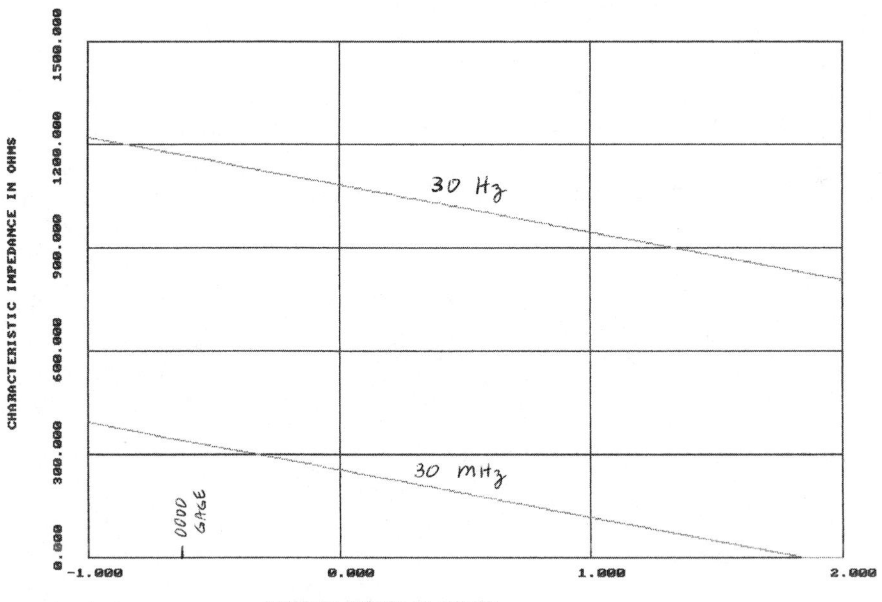

Figure 7-5 Charcteristic impedance of a single, conducting wire or cylinder as a function of radius, from 0.1 to 100 inches. The bounds for a million-to-one frequency range are shown, from ELF (30 Hz) through VHF (30 MHz).

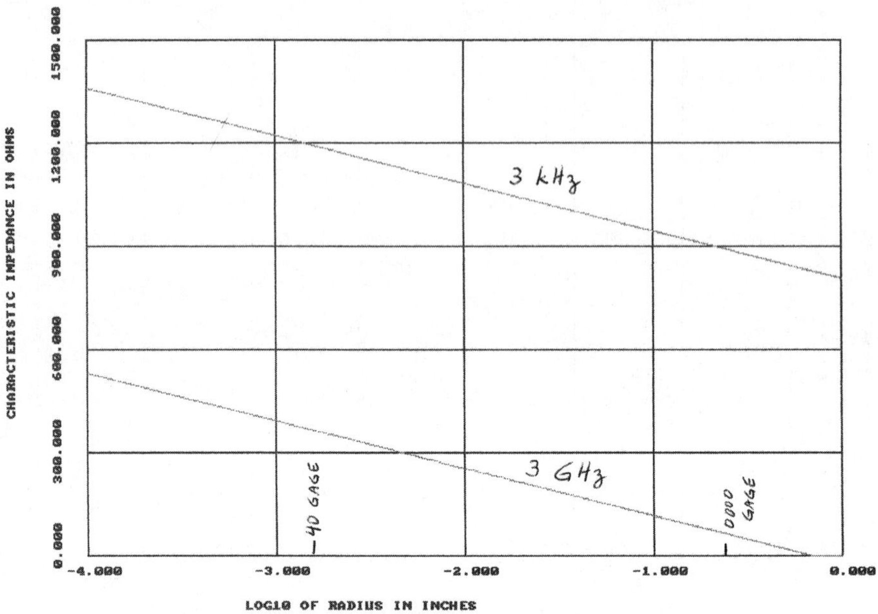

Figure 7-6 Characteristic impedance of a single, conducting wire or cylinder as a function of radius, from 0.0001 to 1 inch. The bounds for a million-to-one frequency range are shown, from VLF (3 kHz) through SHF (3 GHz).

With the above limitation in mind, we are well on our way to calculating characteristic impedances for arbitrary cross sections. An informative intermediate step will be a cross section composed of two parallel cylinders. That will be the subject of the next section.

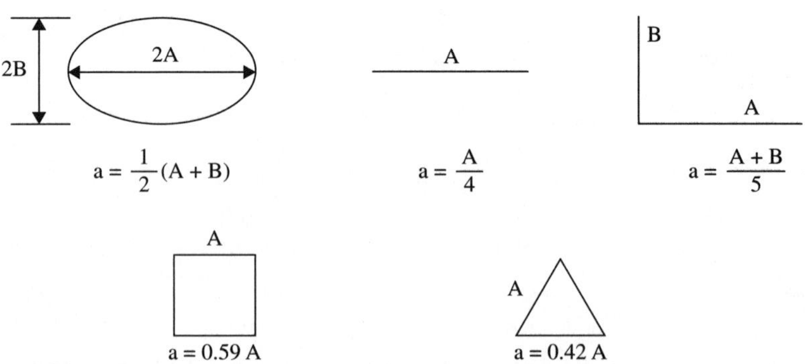

Figure 7-7 Equivalent radii for some cylinders with noncircular cross sections.

7.7 Characteristic Impedances of Two Parallel Cylinders

Advancing from a single cylinder to two parallel cylinders is not a trivial step, either mathematically or conceptually. A lot happens. For example, while the single cylinder could support only common mode charges and currents, two cylinders will exhibit both common and differential modes. Also, we will encounter the concept of the *current transformation ratio* because, in general, the charge densities and currents on one cylinder will not equal those on the other.

Figure 7-8 shows the geometry of interest. The radii of the cylinders are not necessarily equal. The separation between the cylinders is arbitrary, provided the cylinders do not touch. The charge densities (i.e., coulombs per meter) on each are not necessarily equal.

We can readily apply the Quasi Static Log Potential Theory developed in Sec. 7.5 to the geometry of interest here. According to Eq. 7-50, the potentials at the surfaces of the cylinders are:

$$V_a = -\frac{1}{2\pi\varepsilon_0}\{Q_a \ln(k'a) + Q_b \ln[k'(s-a)]\} \tag{7-57}$$

$$V_b = -\frac{1}{2\pi\varepsilon_0}\{Q_a \ln[k'(s-b)] + Q_b \ln(k'b)\} \tag{7-58}$$

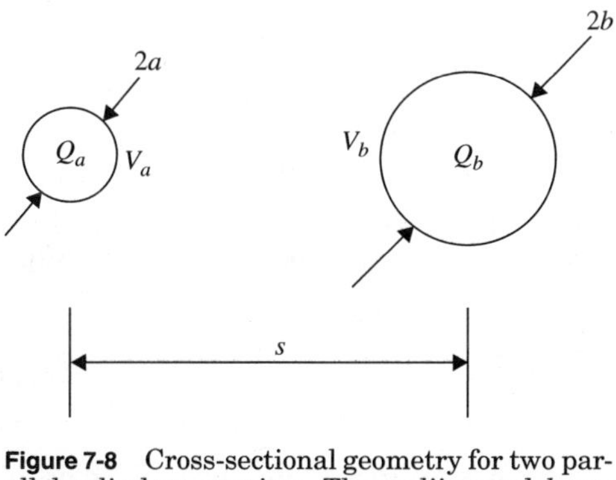

Figure 7-8 Cross-sectional geometry for two parallel cylinders or wires. The radii a and b are not necessarily equal, and the separation s is arbitrary. The fold is wire a, and the tower or main conductor is wire b. Both common mode and differential mode characteristic impedances, Z_{0c} and Z_{0d}, will exist. There will also be a current transformation ratio ν between the two wires that, in general, has a value other than one.

In Eqs. 7-57 and 7-58 we have simplified the notation by writing V_a for $V(a)$ and V_b for $V(b)$.

Using straightforward algebra, we can solve Eqs. 7-57 and 7-58 for the charge densities:

$$Q_a = 2\pi\varepsilon_0 \frac{V_a \ln(k'b) - V_b \ln[k'(s-a)]}{\ln[k'(s-a)]\ln[k'(s-b)] - \ln(k'a)\ln(k'b)} \tag{7-59}$$

$$Q_b = -2\pi\varepsilon_0 \frac{V_a \ln[k'(s-b)] - V_b \ln(k'a)}{\ln[k'(s-a)]\ln[k'(s-b)] - \ln(k'a)\ln(k'b)} \tag{7-60}$$

From Eqs. 7-59 and 7-60, the ratio ν of the charge densities is:

$$\nu = \frac{Q_a}{Q_b} = \frac{V_a \ln(k'b) - V_b \ln[k'(s-a)]}{V_b \ln(k'a) - V_a \ln[k'(s-b)]} \tag{7-61}$$

Equation 7-61 is also the ratio of the currents on the cylinders. This is because of the continuity relation, which is derived from the Maxwell Equations, and which relates the charge density and currents on each cylinder:

$$j\omega Q - jkI = 0 \tag{7-62}$$

Using Eq. 7-28, Eq. 7-62 tells us that:

$$Q = \frac{I}{c} = \sqrt{\varepsilon_0 \mu_0}\, I \tag{7-63}$$

Equations 7-57 through 7-61 are correct for both common and differential modes, or for any linear combination of them. Actually, we have not specialized to any particular mode yet. Now is the logical point in our derivation to do just that. Let's begin with the common, or unbalanced, mode. In that mode, both cylinders are maintained at the same potential, that is:

$$V_a = V_b, \qquad \text{common mode} \tag{7-64}$$

For mathematical simplicity and convenience, and without loss of generality, set $V_a = 1$. Then, the capacitance per unit length is:

$$C' = \frac{Q_a + Q_b}{1} \tag{7-65}$$

Using Eq. 7-64 to simplify Eqs. 7-59 and 7-60, Eq. 7-65 becomes:

$$C' = 2\pi\varepsilon_0 \frac{\ln\left[\left(\frac{b}{s-a}\right)\left(\frac{a}{s-b}\right)\right]}{\ln[k'(s-a)]\ln[k'(s-b)] - \ln(k'a)\ln(k'b)} \tag{7-66}$$

Using Eq. 7-33, the common mode characteristic impedance is:

$$Z_{0c} = \frac{\eta}{2\pi} \frac{\ln[k'(s-a)]\ln[k'(s-b)] - \ln(k'a)\ln(k'b)}{\ln\left[\frac{ab}{(s-a)(s-b)}\right]} \tag{7-67}$$

Using Eq. 7-64, Eq. 7-61 provides us with the *current transformation ratio*:

$$\nu = \frac{\ln\left(\frac{b}{s-a}\right)}{\ln\left(\frac{a}{s-b}\right)} \tag{7-68}$$

For the special case of cylinders with equal radii, $b = a$. From Eq. 7-68, $\nu = 1$, and from Eq. 7-67, the common mode characteristic impedance is:

$$Z_{0c} = -\frac{\eta}{4\pi} \ln[(k')^2 a(s-a)] \qquad b = a \tag{7-69}$$

In Eqs. 7-67 and 7-69, it is seen that the quasi static wave number k' is an important factor. We could not obtain these results using the logarithmic potential theory described by LaPorte.

Let's move on to the differential, or balanced, mode. In that mode, we require that the currents on the cylinders be equal in magnitude but opposite in phase. According to Eq. 7-63, that is equivalent to placing the same requirement on the charge densities:

$$Q_b = -Q_a, \qquad \text{differential mode} \tag{7-70}$$

For mathematical simplicity and convenience, and without loss of generality, set $Q_a = 1$. Then, the capacitance per unit length is:

$$C' = \frac{1}{V_a - V_b} \tag{7-71}$$

Using Eq. 7-70 to simplify Eqs. 7-57 and 7-58, Eq. 7-71 becomes:

$$C' = \frac{2\pi\varepsilon_0}{\ln\left[\frac{(s-a)(s-b)}{ab}\right]} \tag{7-72}$$

Then, according to Eq. 7-33, the differential mode characteristic impedance is:

$$Z_{0d} = \frac{\eta}{2\pi} \ln\left[\frac{(s-a)(s-b)}{ab}\right] \tag{7-73}$$

For the special case of cylinders with equal radii, Eq. 7-73 becomes:

$$Z_{0d} = \frac{\eta}{\pi} \ln\left(\frac{s-a}{a}\right) \qquad b = a \tag{7-74}$$

In Eqs. 7-73 and 7-74, it is seen that the quasi static wave number k' is nowhere to be found. Evidently, it "cancelled out" during the derivation of those formulas. We could have used the logarithmic potential theory of LaPorte to obtain the same results. In that case, the unknown constant x in Eq. 7-34 would also have "cancelled out."

Equations 7-67, 7-68, and 7-73 are general formulas for two parallel cylinders. They are explicit prescriptions for evaluating Z_{0c}, ν, and Z_{0d}, respectively, as functions of cross-sectional dimensions and wavelength. They do not depend upon the length of the cylinders.

Figures 7-9 through 7-12 are examples that result from using those formulas over a wide range of frequencies. Figure 7-9 is for the VLF band. Figures 7-10 and 7-11 are for the MF band. Figure 7-12 is for the VHF band. We arbitrarily picked one cylinder with a radius ten times greater than the radius of the other, that is, an order of magnitude difference. Thus, the various radii span a range from 0.023 to 50 inches. The examples are not exhaustive, but cover a broad range of values and result in some interesting presentations.

From the examples, it is seen that Z_{0d} may be greater than Z_{0c} or less than Z_{0c}, depending most dramatically upon the spacing between the cylinders. The current transformation ratio ν can range from 0.1 to over 0.6. The lower value is not surprising. It occurs when the cylinders are most closely spaced. Since we picked the cylinders to have a 10:1 ratio in radii, they act as simple resistance dividers when they are close, and the current divides in the same ratio. As the spacing increases, other electromagnetic phenomena come into play, changing the current transformation ratio.

By expanding our theory from a single cylinder to two cylinders, we have introduced the differential, or balanced mode, characteristic impedance, and the current transformation ratio. That completes the trio of design parameters of interest in this chapter. If we increase the number of cylinders, no more parameters will be encountered; however, that does not mean that the values will not be substantially different or otherwise interesting. So, in the next section, we will generalize the theory to include N cylinders.

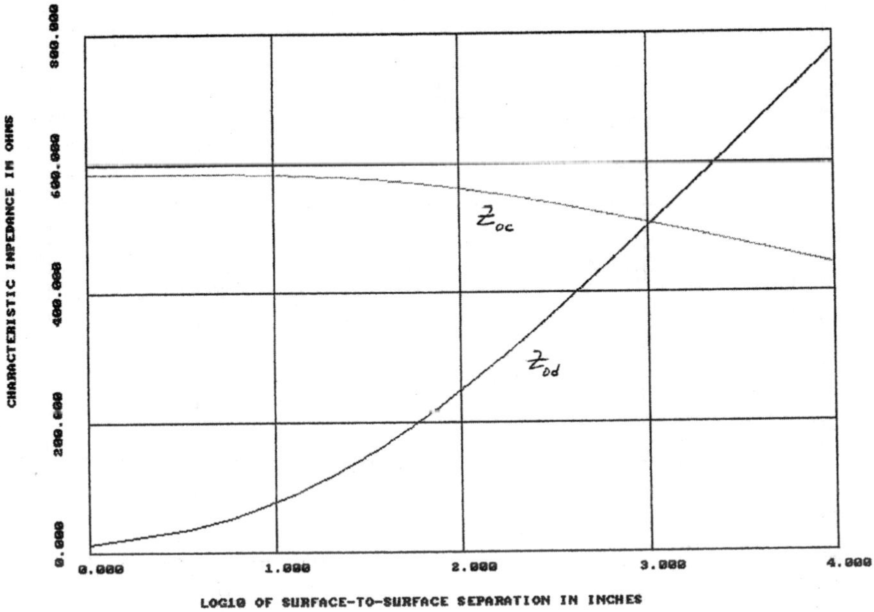

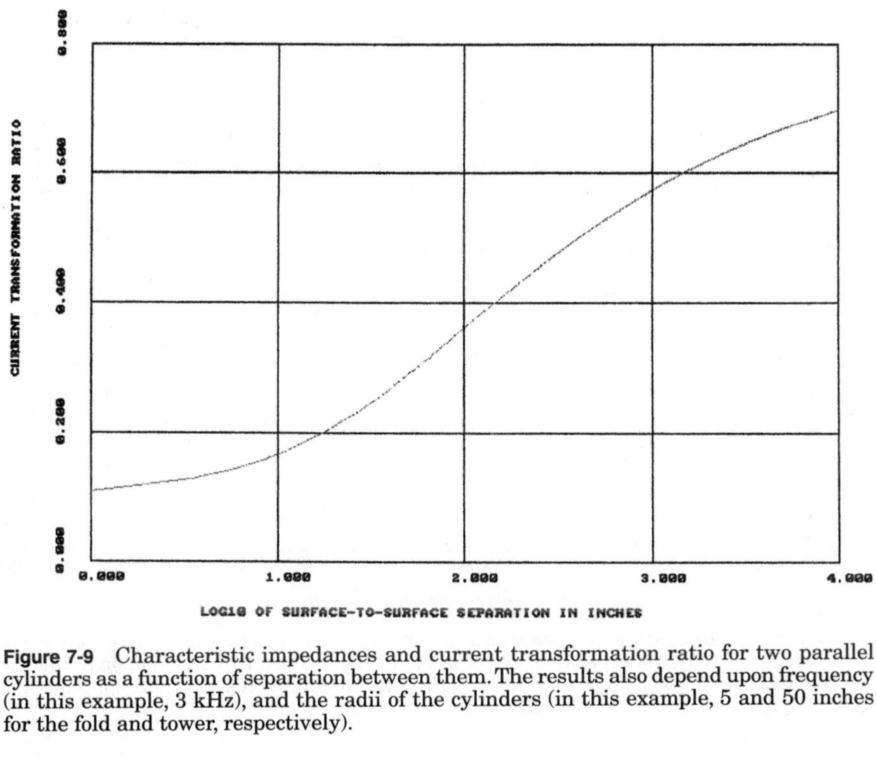

Figure 7-9 Characteristic impedances and current transformation ratio for two parallel cylinders as a function of separation between them. The results also depend upon frequency (in this example, 3 kHz), and the radii of the cylinders (in this example, 5 and 50 inches for the fold and tower, respectively).

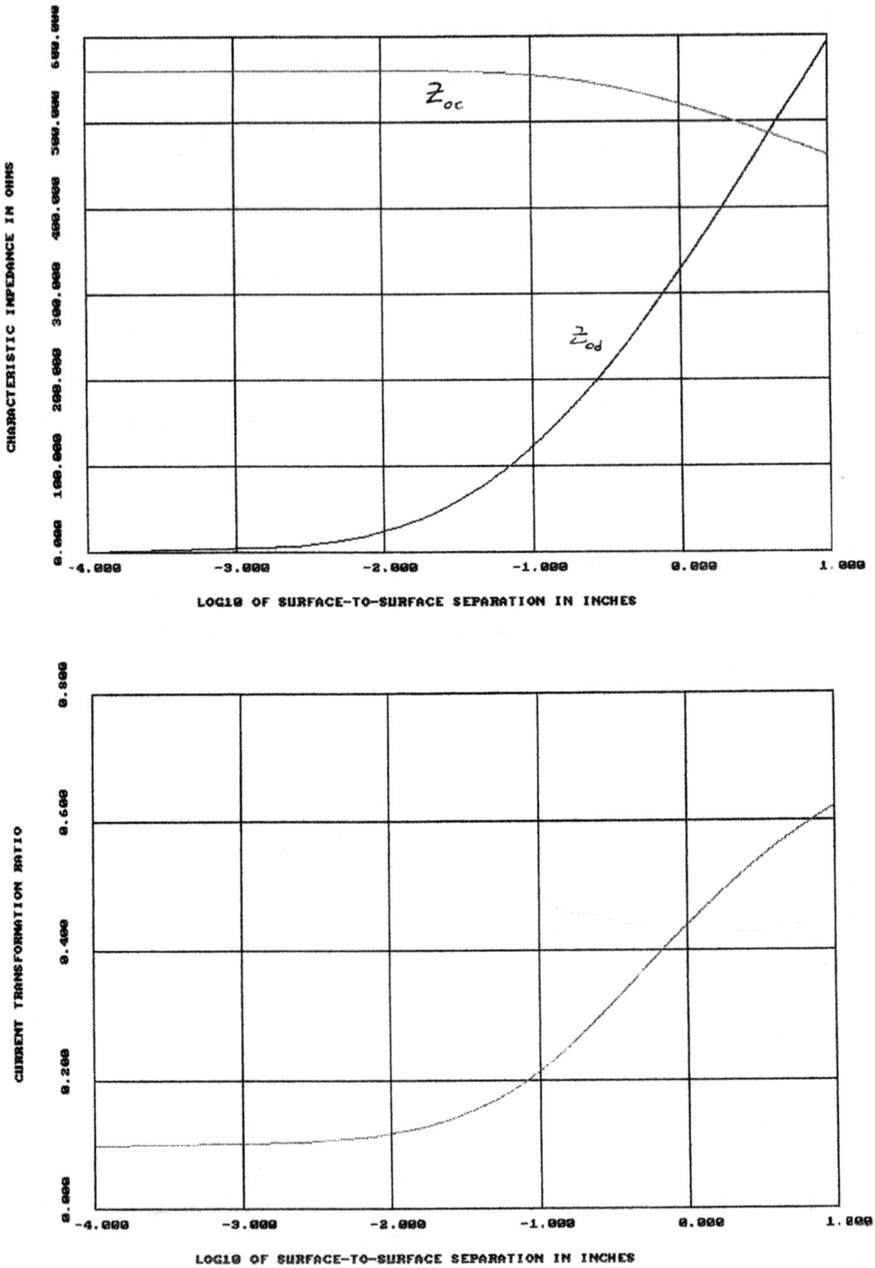

Figure 7-10 Characteristic impedances and current transformation ratio for two parallel cylinders as a function of separation between them. The results also depend upon frequency (in this example, 984 kHz), and the radii of the cylinders (in this example, 17 Gage and 0000 Gage for the fold and tower, respectively).

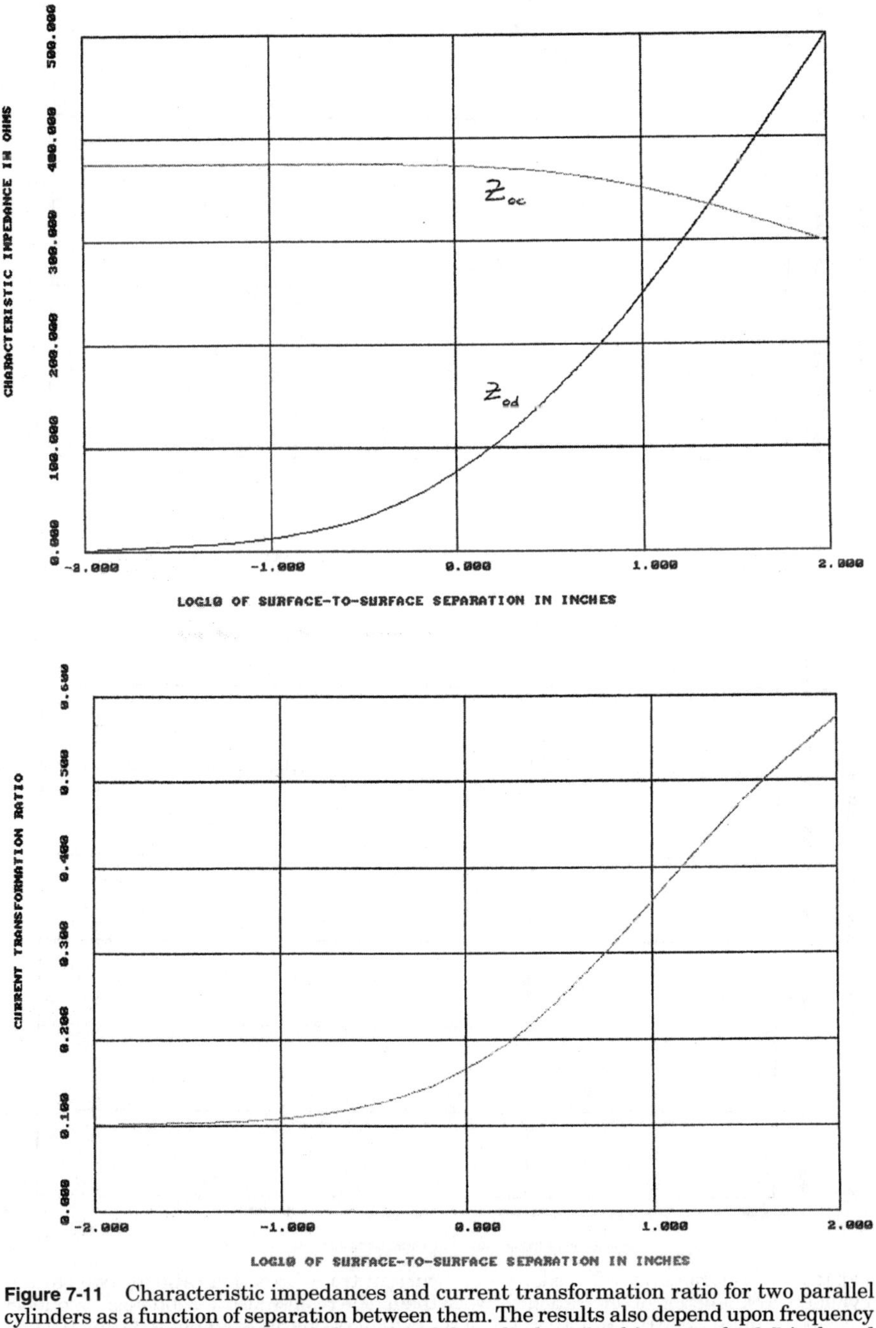

Figure 7-11 Characteristic impedances and current transformation ratio for two parallel cylinders as a function of separation between them. The results also depend upon frequency (in this example, 984 kHz), and the radii of the cylinders (in this example, 0.5 inch and 5 inches for the fold and tower, respectively).

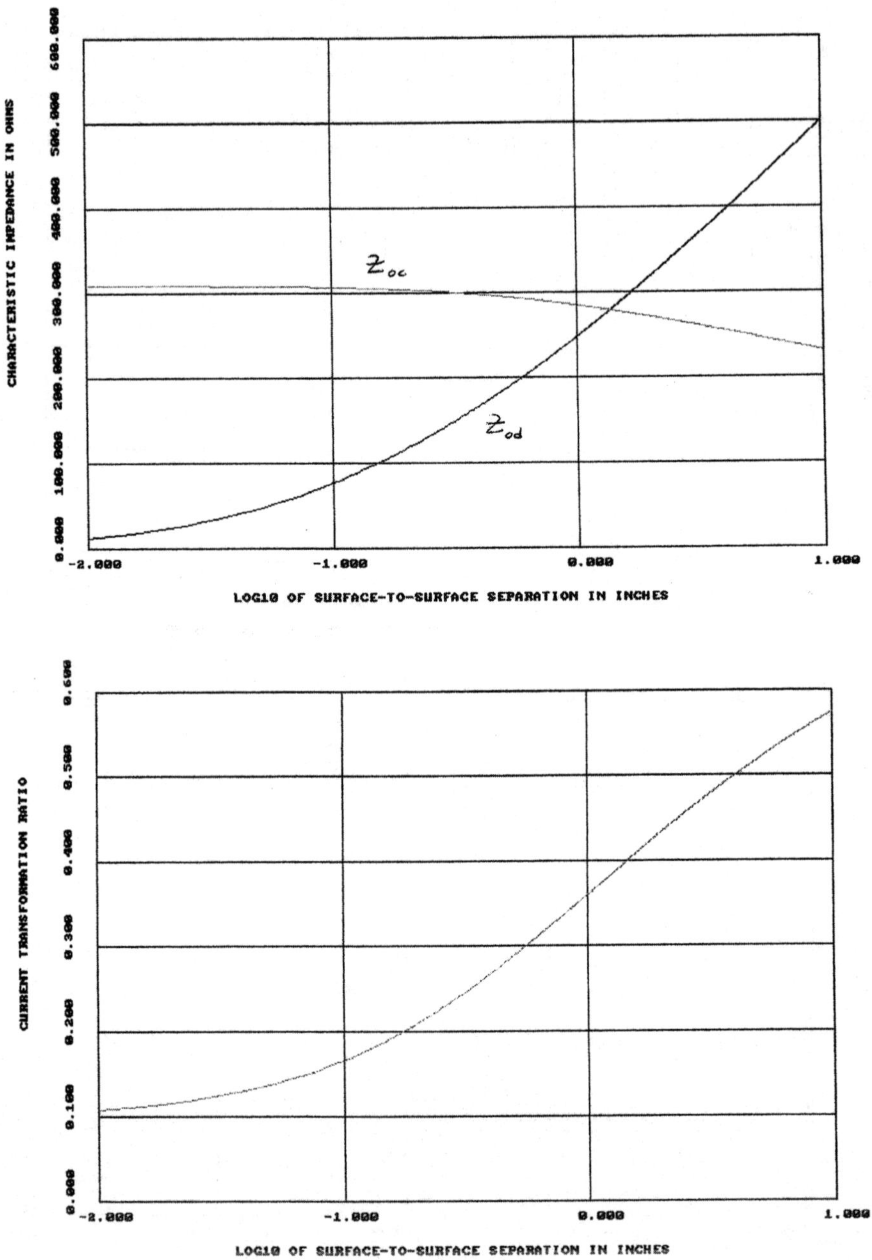

Figure 7-12 Characteristic impedances and current transformation ratio for two parallel cylinders as a function of separation between them. The results also depend upon frequency (in this example, 30 MHz), and the radii of the cylinders (in this example, 0.05 and 0.5 inch for the fold and tower, respectively).

7.8 Characteristic Impedances of *N* Parallel Cylinders

In this section, the careful mathematical developments of this chapter finally pay off. In Sections 7.5 and 7.6, we did not compress the number of design variables. We simply exchanged them for different ones. For the single cylinder of Sec. 7.5, we simply exchanged the radius a for the common mode characteristic impedance Z_{0c}. For the two parallel cylinders of Sec. 7.6, we exchanged the two radii a and b, and the separation s, for two characteristic impedances Z_{0c} and Z_{0d}, and for the current transformation ratio ν. In this section, however, we will compress arbitrary numbers of radii and spacings into the same three design variables. As was discussed in Sec. 7.1, this permits us to pursue the design of folded unipole antennas in a much more efficient manner.

To that end, Fig. 7-13 shows the geometry of interest, although the representation is a bit abstract. The fold is constructed from an arbitrary number of parallel cylinders N_a. Similarly, an arbitrary number N_b comprise the "main" conductor or "tower." The fold is maintained at a potential V_a. The tower is maintained at V_b. The total charge densities on the fold and tower are Q_a and Q_b, respectively. As discussed in the previous section, the charge densities are directly proportional to the currents flowing on the cylinders.

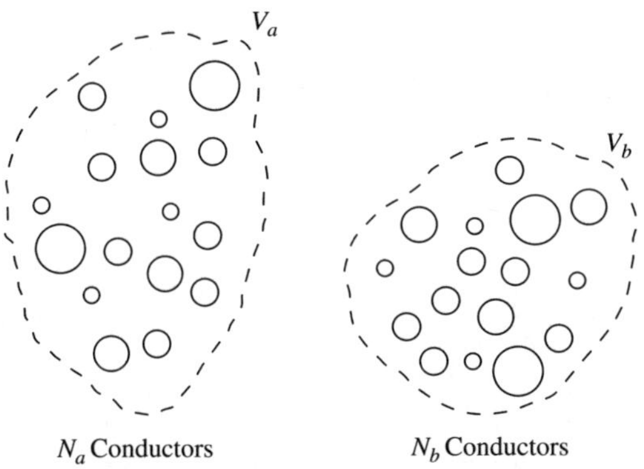

Figure 7-13 Cross-sectional geometry for an antenna with any cross section that can be modeled either exactly or approximately by an arbitrary number of wires $N_a + N_b$. The wires may have any radius and any position within the cross section. N_a wires make up the fold, and N_b wires make up the tower or main conductor. They do not have to be in separate regions as shown. The regions may overlap. The fold wires are distinguished by being maintained at a voltage V_a. The tower wires are distinguished by being maintained at a voltage V_b.

Our mathematical procedure will be very similar to that in the previous section for two parallel cylinders. Using Eq. 7-50 from quasi static log potential theory, the potential at the surface of each fold wire is:

$$V_a = -\frac{1}{2\pi\varepsilon_0} \sum_{n=1}^{N_a+N_b} Q_n \ln(k' s_{\text{mn}}) \qquad m = 1,2,3\ldots N_a \tag{7-75}$$

Similarly, the potential of each wire making up the tower is:

$$V_b = -\frac{1}{2\pi\varepsilon_0} \sum_{n=1}^{N_a+N_b} Q_n \ln(k' s_{\text{mn}}) \qquad m = N_a+1, N_a+2, \ldots N_a+N_b \tag{7-76}$$

In Eqs. 7-75 and 7-76, s_{mn} is the separation between wires m and n. When the two wires are one and the same, that separation is the radius of the wire:

$$s_{\text{mm}} = a_m \tag{7-77}$$

Equations 7-75 and 7-76 may be consolidated into a single matrix equation for the unknown charge densities:

$$\mathbf{M} \cdot \mathbf{Q} = -2\pi\varepsilon_0 \mathbf{V} \tag{7-78}$$

In Eq. 7-78, the matrix M is:

$$M_{\text{mn}} = \ln(k' s_{\text{mn}}) \tag{7-79}$$

The vector of current densities is:

$$\mathbf{Q} = \begin{pmatrix} \mathbf{Q_a} \\ \mathbf{Q_b} \end{pmatrix} \tag{7-80}$$

In Eq. 7-80, the subvectors are:

$$\mathbf{Q_a} = \begin{pmatrix} Q_1 \\ Q_2 \\ \vdots \\ Q_{N_a} \end{pmatrix} \tag{7-81}$$

$$\mathbf{Q}_b = \begin{pmatrix} Q_{N_a+1} \\ Q_{N_a+2} \\ \vdots \\ Q_{N_b+N_b} \end{pmatrix} \tag{7-82}$$

The voltage column vector in Eq. 7-78 is:

$$\mathbf{V} = \begin{pmatrix} \mathbf{V_a} \\ \mathbf{V_b} \end{pmatrix} \tag{7-83}$$

In Eq. 7-83, V_a is a constant vector with N_a elements, and V_b is a constant vector with N_b elements.

Equations 7-78 through 7-83 apply to the common mode, differential mode, or to any linear combination of those modes. To find the common mode charge densities, we follow the same logic as in the previous section for two parallel cylinders. That is, in the common mode, both the fold and the tower are maintained at the same potential:

$$V_a - V_b, \qquad \text{common mode} \tag{7-84}$$

Equation 7-84 is identical to Eq. 7-64, which described the common mode for two parallel cylinders. For mathematical convenience, and without loss of generality, we can arbitrarily set

$$V_a = V_b = -\frac{1}{2\pi\varepsilon_0} \tag{7-85}$$

Using Eq. 7-85, the matrix Eq. 7-78 becomes:

$$\mathbf{M} \cdot \mathbf{Q} = \mathbf{1} \tag{7-86}$$

The solutions to Eq. 7-86 are the charge densities on each wire Q_n, and so the total charge density in the antenna cross section is:

$$Q_{\text{total}} = \sum_{n=1}^{N_a+N_b} Q_n \tag{7-87}$$

The capacitance per unit length is:

$$C' = \frac{Q_{\text{total}}}{V_a} = \frac{Q_{\text{total}}}{V_b} = -2\pi\varepsilon_0 Q_{\text{total}} \tag{7-88}$$

According to Eq. 7-33, the common mode characteristic impedance is:

$$Z_{0c} = -\frac{\eta}{2\pi Q_{\text{total}}} \tag{7-89}$$

The current transformation ratio is:

$$\nu = \frac{\sum_{n=1}^{N_a} Q_n}{\sum_{n=N_a+1}^{N_a+N_b} Q_n} \tag{7-90}$$

Equations 7-86, 7-87, 7-89, and 7-90 are a complete prescription for determining the common mode characteristic impedance and current transformation ratio. That takes care of two of the three design parameters of interest in this chapter.

The remaining parameter is the differential mode characteristic impedance. In that mode, we impose a requirement analogous to Eq. 7-70. That is, the currents on the fold and tower must be equal in magnitude but opposite in sign. According to the continuity relation discussed in connection with Eq. 7-70, that is equivalent to imposing the same requirement on the charge densities:

$$\sum_{n=1}^{N_a} Q_n = -\sum_{n=N_a+1}^{N_a+N_b} Q_n, \qquad \text{differential mode} \tag{7-91}$$

For mathematical convenience, and without any loss of generality, we can set,

$$\sum_{n=1}^{N_a} Q_n = 1 \tag{7-92}$$

and

$$\sum_{n=N_a+1}^{N_a+N_b} Q_n = -1 \tag{7-93}$$

Equations 7-92 and 7-93 are two constraints on the charge densities in addition to the matrix Eq. 7-78. Actually, although Eqs 7-78 through 7-83 still apply, they are not very useful for determining Z_{0d}. This is because the potentials V_a and V_b are unknown. In the common mode, they were equal to each other and constant; however, in the differential mode they will adjust themselves to whatever values are necessary such that Eq. 7-91 is satisfied.

In view of the reasoning in the previous paragraph, it is best to abandon Eq. 7-78 and derive an alternate matrix equation. To that end, we may rewrite Eqs. 7-75 and 7-76 as,

$$\sum_{n=1}^{N_a+N_b} Q_n \ln(k's_{\mathrm{mn}}) + 2\pi\varepsilon_0 V_a = 0 \qquad m = 1,2,3\ldots N_a \tag{7-94}$$

$$\sum_{n=1}^{N_a+N_b} Q_n \ln(k's_{\mathrm{mn}}) + 2\pi\varepsilon_0 V_b = 0 \qquad m = N_a+1, N_a+2, \ldots N_a+N_b \tag{7-95}$$

Equations 7-92 through 7-95 can be consolidated into a single matrix equation:

$$\mathbf{M}\cdot\begin{pmatrix}\mathbf{Q_a}\\ \mathbf{Q_b}\\ V_a\\ V_b\end{pmatrix}=\begin{pmatrix}1\\ -1\\ \mathbf{0}\end{pmatrix} \tag{7-96}$$

In Eq. 7-96 the unknown charge density subvectors are defined by Eqs. 7-81 and 7-82. The elements V_a and V_b are individual scalars. On the right side of Eq. 7-96, the first two elements are individual scalars. The third element is a subvector with all elements zero.

The matrix M in Eq. 7-96 is logically partitioned into four submatrices:

$$\mathbf{M}=\begin{pmatrix}\leftarrow & \mathbf{A} & \rightarrow\\ \leftarrow & \mathbf{B} & \rightarrow\\ \leftarrow & \mathbf{C} & \rightarrow\\ \leftarrow & \mathbf{D} & \rightarrow\end{pmatrix} \tag{7-97}$$

In Eq. 7-97, the submatrices **A** and **B** are single row vectors with N_a+N_b+2 elements. **A** is:

$$A_n=\begin{cases}1 & n=1,2,\ldots,N_a\\ 0 & n>N_a\end{cases} \tag{7-98}$$

The submatrix **B** is:

$$B_n=\begin{cases}0 & n=1,2,\ldots,N_a\\ 1 & n=N_a+1,N_a+2,\ldots,N_a+N_b\\ 0 & n>N_a+N_b\end{cases} \tag{7-99}$$

The submatrix **C** has N_a rows and N_a+N_b+2 columns:

$$C_{\mathrm{mn}}=\begin{cases}\ln(k's_{\mathrm{mn}}) & n=1,2,\ldots,N_a,N_a+1,\ldots,N_a+N_b\\ 2\pi\varepsilon_0 & n=N_a+N_b+1\\ 0 & n=N_a+N_b+2\end{cases} \tag{7-100}$$

The submatrix **D** has N_b rows and N_a+N_b+2 columns:

$$D_{\mathrm{mn}}=\begin{cases}\ln(k's_{\mathrm{mn}}) & n=1,2,\ldots N_a,N_a+1,\ldots,N_a+N_b\\ 0 & n=N_a+N_b+1\\ 2\pi\varepsilon_0 & n=N_a+N_b+2\end{cases} \tag{7-101}$$

After Eq. 7-96 is solved for the charge densities and potentials, the capacitance per unit length is:

$$C' = \frac{\sum_{n=1}^{N_a} Q_n}{V_a - V_b} \tag{7-102}$$

Using Eq. 7-92, Eq. 7-102 becomes:

$$C' = \frac{1}{V_a - V_b} \tag{7-103}$$

According to Eq. 7-33, the differential mode characteristic impedance is:

$$Z_{0d} = \sqrt{\varepsilon_0 \mu_0}\,(V_a - V_b) \tag{7-104}$$

The solutions to the matrix Eq. 7-96, combined with Eq. 7-104, are a complete prescription for the differential mode characteristic impedance of any collection of parallel cylinders.

Equations 7-89, 7-90, and 7-104 are formulas for the three parameters Z_{0c}, ν, and Z_{0d}, respectively, for any collection of parallel cylinders. It is not necessary to keep separate track of multiple radii a_m and spacings s_{mn} in the antenna cross section during the design process. They have all been assimilated into Z_{0c}, ν, and Z_{0d}. These three parameters are all we need for two cylinders, three cylinders, or a hundred cylinders.

The formulas also apply to cylinders that are not circular, that is, to cross sections that are triangular, square, any closed polygonal shape, or any open polygon. We can readily simulate those cross sections with multiple, closely spaced, circular cylinders that trace an arbitrary perimeter.

Figure 7-14 shows some examples of triangular and square cross sections, with three, four, six, and eight separate fold wires. It is seen that the current transformation ratios in some examples are greater than one. This raises an interesting design possibility which is the subject of the next section.

7.9 The Inverted Unipole

The examples and formulas derived in the previous section suggest a useful type of reciprocity. Referring back to Fig. 7-13, suppose the fold wires and tower wires were interchanged. That is, what if the subscript a denoted tower wires and the subscript b denoted fold wires?

Geometry		Z_{0d}	Z_{0c}	V
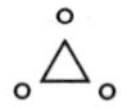		206.1	305.8	1.137
		197.0	310.6	1.090
	Spacing Measured from Corner	151.0	284.4	2.662
	Spacing Measured from Face	142.6	290.4	2.565
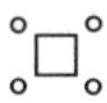		62.8	291.0	1.347
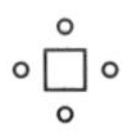		61.9	294.4	1.300
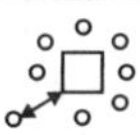	Corner	116.6	270.6	3.863
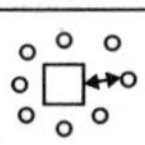	Face	158.2	362.7	0.785

Figure 7-14 Characteristic impedances and current transformation ratios for some antenna cross sections with more than two parallel cylinders, and with cylinders that are noncircular. The examples shown include some commonly found in the AM broadcast band. For all examples, the frequency is 984 kHz; the face of the tower is 12 inches; the fold wires 6 Gage; and the separation of the fold wires from the tower is 24 inches. It is seen that the current transformation ratio can exceed unity, suggesting the desirability of an inverted unipole configuration.

How would the results of the previous section change? We derived formulas for Z_{0c}, v, and Z_{0d}. If we interchanged the subscripts a and b in all of our derivations, what would the resulting formulas for Z'_{0c}, v', and Z'_{0d} be?

Reviewing the derivations for Eqs. 7-89 and 7-104, it is clear that the characteristic impedances would be unchanged. That is,

$$Z'_{0c} = Z_{0c} \tag{7-105}$$

and

$$Z'_{0d} = Z_{0d} \tag{7-106}$$

The current transformation ratio is a different story, however. If the subscripts a and b now denote the tower and fold wires, respectively, then Eq. 7-90 must be replaced by,

$$\nu' = \frac{\sum_{n=N_a+1}^{N_a+N_b} Q_n}{\sum_{n=1}^{N_a} Q_n} \tag{7-107}$$

Comparing Eq. 7-107 with Eq. 7-90,

$$\nu' = 1/\nu \tag{7-108}$$

Equations 7-105, 7-106, and 7-108 describe properties of the *inverted unipole*. By interchanging the connections which define the fold and tower wires, we can retain the characteristic impedances while *inverting* the current transformation ratio. This is an obvious advantage for design purposes.

Figure 7-15 demonstrates why we might want to configure an inverted unipole. It shows how the characteristic impedances and current transformation ratio change as the number of fold wires is increased. The radii of the tower and fold wires are the same as in Fig. 7-11. It is seen that the current transformation ratio varies over a wide range of values from less than one to almost 100. The value is about one when the number of folds equals three. So, if we desire more than three fold wires (for example, to satisfy the FCC requirement for geometric symmetry in AM broadcast band antennas) combined with a current transformation ratio less than one, then the inverted unipole configuration is indicated.

In practice, the inverted unipole has motivated some innovative geometries. For example, in the AM broadcast band, a single wire is sometimes positioned inside the tower. That is, the latticework of the tower is external to the wire. The latticework is clearly in the conventional position of the fold. The wire inside is in the conventional position of the tower. Figure 7-16 shows the cross section of the configuration. It works only because the external conductor is not solid. The latticework permits electromagnetic fields to penetrate inside so that the fold wire can function.

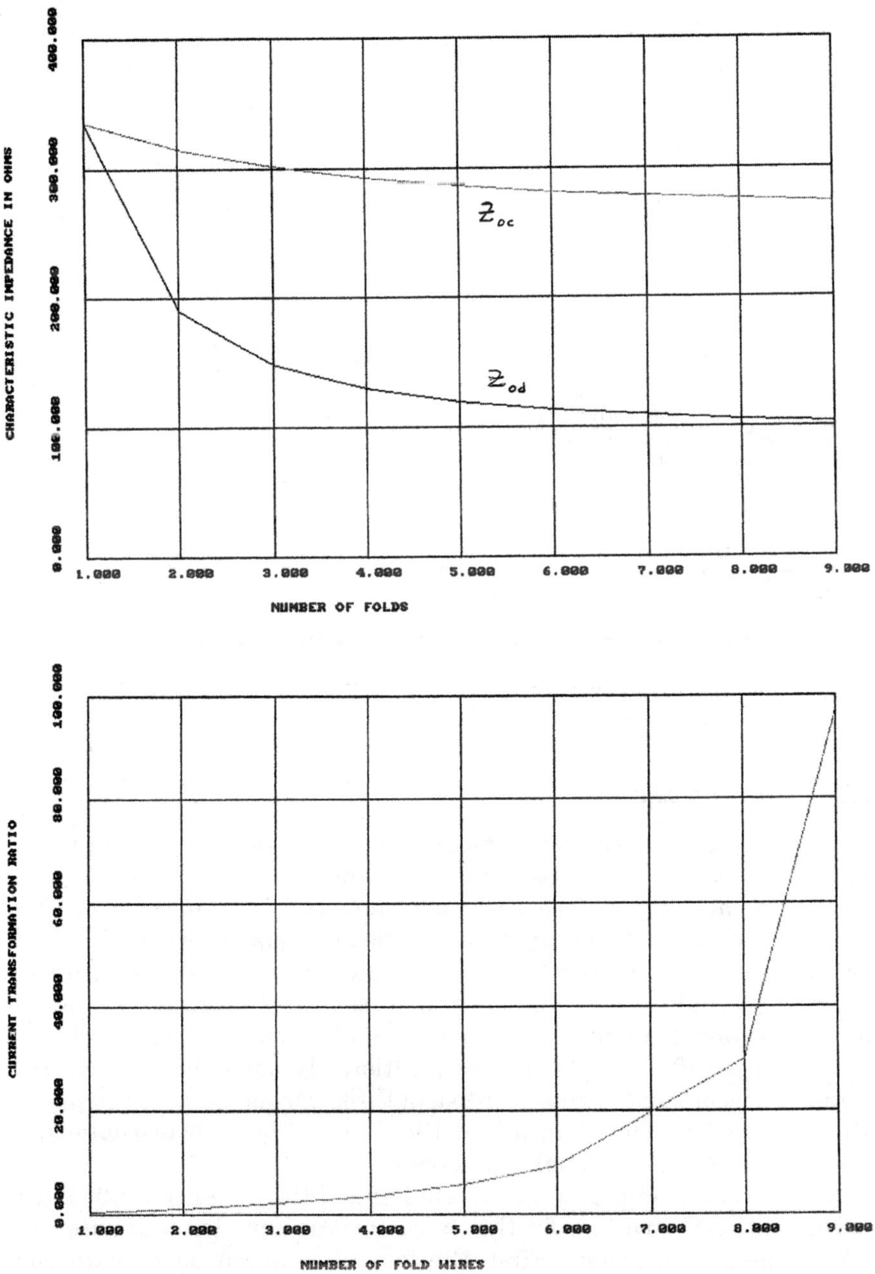

Figure 7-15 Characteristic impedances and current transformation ratio as a function of the number of fold wires. The frequency is 984 kHz; the radius of the tower is 5 inches; the radius of each fold wire is 0.5 inch; and the separation between each fold wire and the tower is 27.8 inches. It is seen that beyond three folds, the current transformation ratio exceeds unity, suggesting the desirability of an inverted unipole configuration. In that configuration, the characteristic impedances would remain the same; however, the current transformation ratio would be *inverted*, that is, become the reciprocal of its former value.

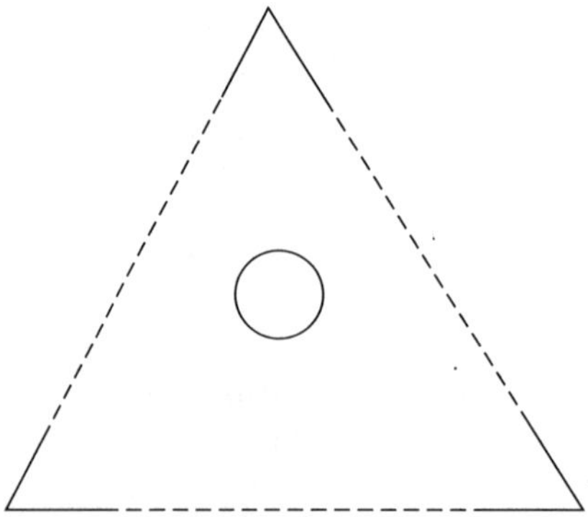

Figure 7-16 One innovative configuration of the inverted unipole, with a single cylinder or wire inside a cylinder with triangular cross section. The triangle actually functions as the fold, and the inside wire functions as the tower or main conductor. It is essential that the sides of the triangle not be solid conductors. They must allow electromagnetic fields to leak or penetrate inside so that the wire is not completely shielded. Commercial broadcast and communication towers made from latticework satisfy this requirement.

7.10 Conclusions

In this chapter, we explored in great detail the three design parameters Z_{0c} (common mode characteristic impedance), ν (current transformation ratio), and Z_{0d} (differential mode characteristic impedance). They efficiently express the effects on antenna performance of every possible cross-sectional geometry, from the simplest to the most complicated. There is no limit to the number of parallel wires that make up the fold and the tower, or main conductor, of the folded unipole, and each wire may have a different radius and position. By assimilating the great number of geometric variables into just three parameters, we can pursue the design of the folded unipole without fear of being overwhelmed by complicated cross-sectional geometries.

As part of our exploration, we challenged the general applicability of conventional formulas for the characteristic impedance of antennas, for at least two reasons. First, the formulas do not agree with each other. Second, they depend upon the length of the antenna, and that contradicts the fundamental definition of characteristic impedance.

That definition is the ratio of voltage to current (or transverse electric field to transverse magnetic field) for a traveling wave launched on a transmission line (or waveguide) of infinite length.

We derived new formulas for characteristic impedance, starting with fundamental transmission line theory. We extended the Logarithmic Potential Theory, based upon electrostatics and described by LaPorte in 1952, into a Quasi Static Log Potential Theory. Using the latter theory, we developed general formulas for Z_{0c}, ν, and Z_{0d}. The formulas relate characteristic impedance and current transformation ratio to the detailed cross-sectional geometry of any antenna, and to wavelength.

These formulas are widely applicable, and they should prove of general interest beyond the design of the folded unipole antenna.

We found that the values for characteristic impedance vary over a wide range, from over 1,000 ohms to less than one ohm. We also found that the current transformation ratio can also vary over a wide range, from a small fraction to the order of 100.

The wide range of values for current transformation ratio suggested an innovative configuration called the *inverted unipole*. For that configuration, the fold wires and tower, or main conductor, wires are interchanged. As a result, the characteristic impedances remain unchanged; however, the current transformation ratio inverts, or becomes the reciprocal of its former value.

In today's age of affordable and powerful computers, our efforts in this chapter have nevertheless shown that formulas and analytic mathematics still offer certain advantages over general purpose, numerical, wire antenna analysis programs. We have reduced complicated antenna cross sections, consisting of an arbitrary number of wires, to only three independent design variables. This is something the general purpose, wire analysis programs simply can not do!

References

1. Schelkunoff, S. A., Electromagnetic Waves, New York: D. Van Nostand Company, 1943, p. vii.
2. Hallén, E., Theoretical Investigations into Transmitting and Receiving Antennas, Upsala, Sweden: *Nova Acta Regiae Soc. Sci. Upaliensis*, **11**, 1938, pp. 1–44.
3. Kraus, J. D., Antennas, New York: McGraw-Hill, 1950, p. 237.
4. Schelkunoff, S. A., and H. T. Friis, Antennas Theory and Practice, New York: John Wiley & Sons, 1952, p. 425.
5. *Ibid.*, p. 426.
6. LaPorte, E. A., Radio Antenna Engineering, New York: McGraw-Hill, 1952, pp. 513–526.
7. Bowman, J. J., T. B.A Senior, and P. L. E.Uslenghi, Electromagnetic and Acoustic Scattering by Simple Shapes, New York: Hemisphere Publishing Corporation, 1987, p. 473.
8. Wolff, E. A., Antenna Analysis, New York: John Wiley & Sons, 1966, p. 61.

Symbols Used in This Chapter

A	1. Magnetic vector potential. 2. Submatrix of M
A_n	Element of submatrix **A**
a	Radius of cylinder or wire
a_n	Radius of nth cylinder or wire
B	Submatrix of **M**
B_n	Element of submatrix **B**
b_1	Higher order term in Hallén's formula for antenna current and input impedance
C	Submatrix of **M**
C'	Capacitance per unit length of transmission lines and antennas
C_{mn}	Element of submatrix **C**
$\boldsymbol{c}$	Phase velocity or speed of light
D	Submatrix of **M**
D_{mn}	Element of submatrix **D**
d_1	Higher order term in Hallén's formula for antenna current and input impedance
d^2r'	Incremental area, used in integration formulas
d^3r'	Incremental volume, used in integration formulas
E	Electric field
G	Green's function
H	Magnetic field
$\mathrm{H}_0^{(2)}$	Hankel function of the second kind, a form of Bessel function
I(z)	Current along an antenna or transmission line
I_F	Complex amplitude of forward traveling wave of current
I_R	Complex amplitude of reverse traveling wave of current
J	Current density in amps per square meter
j	Square root of −1
k	Wave number
k'	Quasi static wave number
L'	Inductance per unit length along antennas or transmission lines
l	Length of monopole, or half length of dipole antenna
M	Matrix in the equation for characteristic impedance
N_a	Number of parallel cylinders or wires from which the fold is constructed
N_b	Number of parallel cylinders or wires from which the tower, or main conductor, is constructed
$\mathbf{Q}$	Vector of unknown charge densities in the equation for characteristic impedance
$\mathbf{Q_a}$	Subvector of **Q** for unknown charge densities on the fold wires
$\mathbf{Q_b}$	Subvector of **Q** for unknown charge densities on the tower wires
Q	Charge density along an antenna or transmission line in coulombs per meter
Q_a	Charge density along cylinder or wire a
Q_b	Charge density along cylinder or wire b
Q_n	Charge density along the nth cylinder or wire
Q_{total}	Sum of charge densities along all wires or cylinders in the antenna cross section
R	Distance between r and r′
r	Vector position of observer
r′	Vector position of charge or current density
s	Separation between wires or cylinders a and b
s_{mn}	Separation between wires or cylinders m and n

$\mathbf{V}$	Vector of potentials at the surfaces of wires and cylinders in the equation for characteristic impedance
$\mathbf{V_a}$	Subvector of V for potentials at the surfaces of cylinders from which the folds are constructed
$\mathbf{V_b}$	Subvector of V for potentials at the surfaces of cylinders from which the tower is constructued
V	1. Electrostatic potential. 2. Input voltage in Hallén's formulas for antenna current and input impedance
$V(z)$	Voltage along antenna or transmission line
V_a	Voltage at the surface of cylinder or wire a
V_b	Voltage at the surface of cylinder or wire b
x	Unknown constant in LaPorte's formula for electrostatic potential
Y'	Admittance per unit length of an antenna or transmission line
Z'	Impedance per unit length of an antenna or transmission line
Z_0	Characteristic impedance
Z_{0c}	Common mode characteristic impedance
Z'_{0c}	Common mode characteristic impedance of an inverted unipole antenna
Z_{0d}	Differential mode characteristic impedance
Z'_{0d}	Differential mode characteristic impedance of an inverted unipole antenna
Z_{in}	Input impedance
z	Position or coordinate along an antenna or transmission line
Δz	Incremental position or coordinate along a transmission line
Γ	Complex wave number for antennas or transmission lines
Ω	Parameter containing length and radius in Hallén's equation for antenna current and input impedance
β	Wave number in Hallén's formulas for antenna current and input impedance, same as k
ε_0	Permittivity of free space
γ	Euler's constant
η	Impedance of free space
λ	Wavelength
μ_0	Permeability of free space
ν	Current transformation ratio
ν'	Current transformation ratio for inverted unipole antenna
ρ	Volume charge density in coulombs per cubic meter
ω	Frequency in radians per second
ψ	Angle of biconical antenna in Schelkunoff's formula for characteristic impedance

Chapter

8

Reradiation and Detuning

8.1 Introduction

The folded unipole antenna is not limited to use as a transmitting or broadcast device. In fact, one of its most prolific deployments is in detuning. That is, a great many structures such as communication towers, water tanks, and utility poles, can be made electronically invisible by reconfiguring them as folded unipole antennas. This does not imply dismantling them. Rather, folds can be added to them. In that way, interference and electromagnetic scattering by these structures can be practically eliminated.

There are also stealth applications of detuning. Leading edges of airplane wings can be detuned by embedding folded unipoles within them. Similarly, other prominent features of military vehicles and apparatus can be detuned at their resonant frequencies. In that way, their radar signatures can be substantially reduced.

Whatever the application, we will find that electromagnetic interference and scattering can almost always be reduced by 20 to 40 dB, and sometimes by over 60 dB!

The mathematical model that we have developed in Chaps. 2 through 7 for the transmitting folded unipole antenna is readily extended to include the detuning phenomenon. Even so, there is so much that could be written about this application that we have space for only a brief sampling in this chapter. From the selection of mathematical models and examples, it will be seen that an entire book could easily be dedicated to this one application.

8.2 Current Induced by an Incident Electromagnetic Wave

An electromagnetic wave incident from a distant source causes electric current to flow on the folded unipole antenna. That current, in turn, radiates more electromagnetic waves. That is the mechanism of reradiation or scattering. In this section, we will derive a simple formula relating the induced current to the incident electromagnetic field.

As Sergei A. Schelkunoff explained in his landmark 1941 paper, concerning linear antennas of arbitrary size and shape [1], the biconical dipole readily lends itself to visualizing electromagnetic waves in close proximity to radiators. Figure 8-1 shows a wave incident upon a biconical dipole. It is seen that the wave propagates inwardly toward the terminals of the antenna.

Figure 8-2 shows the same phenomenon as the biconical dipole is deformed into a cylindrical monopole. The incident wave still propagates inwardly toward the antenna terminals. Along the surface of the monopole, the tangential (i.e., z-component) of the electric field can be described mathematically as:

$$E_z = E_0 \psi_R(z) \tag{8-1}$$

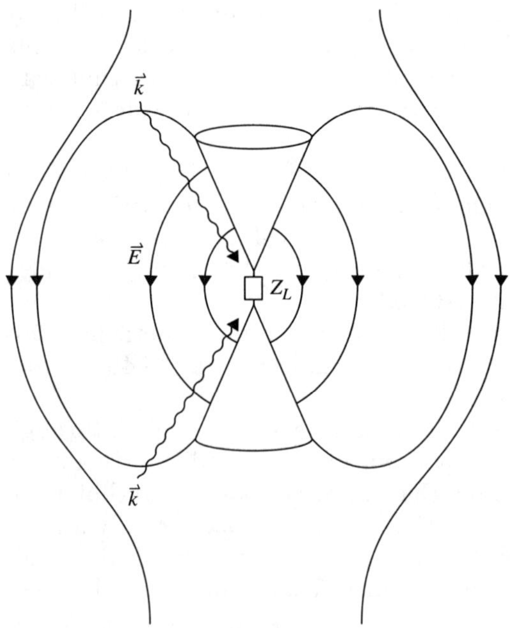

Figure 8-1 The biconical antenna, introduced conceptually by Schelkunoff in 1941, provides the clearest visualization of how an incident or incoming electromagnetic wave induces a reverse traveling wave current.

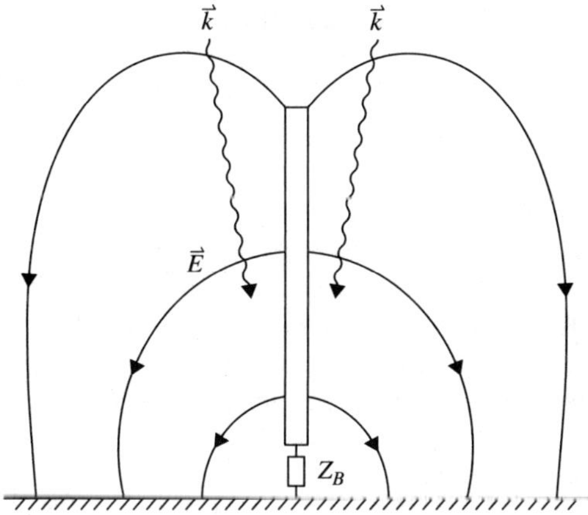

Figure 8-2 The currents induced along a linear monopole by an incident electromagnetic wave are similar to the ones more easily visualized on the biconical antenna.

In Eq. 8-1, the reverse or reflected wave function is the same one we have been using in previous chapters. It is defined by Eq. 1-3.

The tangential electric field and the common mode voltage along the monopole are related:

$$E_z = -\frac{\partial V_c(z)}{\partial z} \tag{8-2}$$

Equation 8-2 tells us that the incident electromagnetic wave is equivalent to a common mode voltage with only a reflected wave component:

$$V_c(z) = -Z_{0c} I_{\text{Rc}} \psi_R(z) \tag{8-3}$$

Substituting Eq. 8-3 into Eq. 8-2, we obtain:

$$E_z = Z_{0c} I_{\text{Rc}} \frac{\partial \psi_R(z)}{\partial z} \tag{8-4}$$

From the definition of the reverse wave function, Eq. 1-3, it is seen that:

$$\frac{\partial \psi_R(z)}{\partial z} = jk\psi_R(z) \tag{8-5}$$

Combining Eqs. 8-1, 8-4, and 8-5, we derive a simple relation between I_{Rc} and E_0:

$$I_{\text{Rc}} = \frac{E_0}{jkZ_{0c}} \tag{8-6}$$

The common mode voltage in Eq. 8-3 cannot exist along the monopole without a common mode current:

$$I_c(z) = I_{Rc}\psi_R(z) \tag{8-7}$$

In Eq. 8-7, I_{Rc} is described by Eq. 8-6.

There is one remaining logical step from electromagnetic theory to complete our description of the induced current. At the surface of a good conductor, the total tangential component of the electric field must vanish. For this to happen, a current equal in magnitude but opposite in sign to $I_c(z)$, as described by Eq. 8-7, must flow. Therefore, the induced current is:

$$I_i(z) = -I_c(z) = \frac{jE_0}{kZ_{0c}}\psi_R(z) \tag{8-8}$$

For simplicity of notation in the remaining sections of this chapter, we will simply write the induced current as:

$$I_i(z) = I_{i0}\psi_R(z) \tag{8-9}$$

Comparing Eqs. 8-8 and 8-9, the amplitude of the induced current must be:

$$I_{i0} = \frac{jE_0}{kZ_{0c}} \tag{8-10}$$

Equations 8-9 and 8-10 describe a current that functions as a distributed source for additional currents and voltages on the antenna. In the remaining sections of this chapter, we will develop mathematical models for those currents and voltages.

8.3 Full Length Detuning

In the previous section, we developed a simple formula for the current induced on an antenna by an incident electromagnetic wave. In this section, we will see how that induced current acts as a distributed source. It excites both common and differential mode voltages and currents, similar to those we have mathematically modeled in previous chapters.

By "full length" we mean that an entire structure is reconfigured as a folded unipole antenna, not just part of it. Figure 8-3 shows the essential features of such a reconfigured structure. The original structure is depicted as the conductor on the right. The added folds are depicted as the conductor on the left. It is seen that there are terminating impedances at both the base and top of the folds.

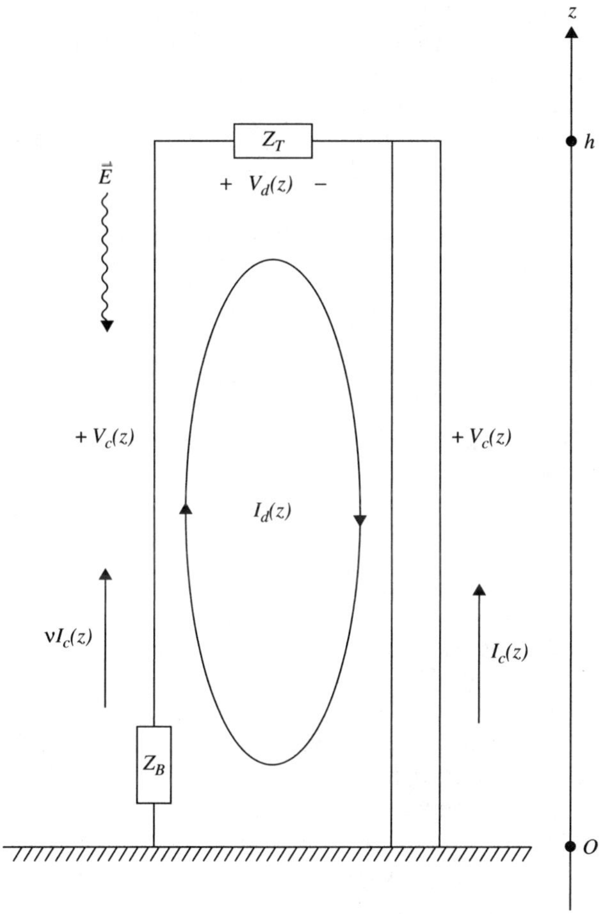

Figure 8-3 The essential features of full length detuning. The reradiating structure is reconfigured as a one-segment folded unipole antenna, except that the base terminating impedance is connected to the fold, not the main structure.

In the previous section, we found that the incident electromagnetic wave induces a common mode current on the structure:

$$I_i(z)=I_{i0}\psi_R(z) \tag{8-11}$$

In Eq. 8-11, the amplitude I_{i0} is given by Eq. 8-10. The total common mode current is the superposition of the induced current and whatever other common mode currents it may excite:

$$I_c(z)=I_{\mathrm{Fc}}\psi_F(z)+(I_{\mathrm{Rc}}+I_{i0})\psi_R(z) \tag{8-12}$$

Equation 8-12 is similar or identical to formulas for the common mode current that we have derived in previous chapters, except for the term I_{i0}. In the terminology of differential equations, we may regard I_{i0} as the *particular solution*, and the remaining terms as the *homogeneous solution*. In any case, I_{i0} is the only amplitude that has been specified so far by the incident electromagnetic wave. The other amplitudes remain to be determined.

In addition to the common mode, there may also be a differential mode current:

$$I_d(z) = I_{\text{Fd}}\psi_F(z) + I_{\text{Rd}}\psi_R(z) \tag{8-13}$$

In Eqs. 8-12 and 8-13, the forward traveling wave function $\psi_F(z)$ is defined by Eq. 1-2.

In both common and differential modes, there are voltages that are simply related to the currents:

$$V_c(z) = Z_{0c}[I_{\text{Fc}}\psi_F(z) - I_{\text{Rc}}\psi_R(z)] \tag{8-14}$$

$$V_d(z) = Z_{0d}[I_{\text{Fd}}\psi_{\text{Fd}}(z) - I_{\text{Rd}}\psi_R(z)] \tag{8-15}$$

In Eqs. 8-14 and 8-15, Z_{0c} and Z_{0d} are the characteristic impedances of the common and differential modes, respectively.

There are four unknown amplitudes I_{Fc}, I_{Rc}, I_{Fd}, and I_{Rd}, in Eqs. 8-12 through 8-15. To determine them, we will need four independent equations. In similar fashion to previous chapters, we can formulate those equations at the boundaries on the antenna.

At the top of the antenna, if we neglect the antenna end effect discussed in Sec. 5.7, the common mode current must vanish because it has nowhere to flow:

$$(1 + \nu)I_c(h) = 0 \tag{8-16}$$

Also at the top, the differential mode voltage and current are related by the terminating impedance:

$$I_d(h)Z_T = V_d(h) \tag{8-17}$$

Equations 8-16 and 8-17 are two independent equations. We need two more, and we will formulate those at the base of the antenna. There, Kirchhoff's voltage law (KVL) requires that:

$$[\nu I_c(0) + I_d(0)]Z_B + V_c(0) = 0 \tag{8-18}$$

KVL also requires that:

$$V_d(0) = V_c(0) \tag{8-19}$$

Equations 8-16 through 8-19 are four independent equations. We need to determine the four unknown complex amplitudes I_{Fc}, I_{Rc}, I_{Fd}, and I_{Rd}. We proceed in exactly the same fashion as in previous chapters to consolidate the equations into a single matrix equation.

$$\mathbf{M}\cdot\mathbf{I} = \begin{pmatrix} -I_{i0}\psi_R(h) \\ -\nu Z_B I_{i0} \\ 0 \\ 0 \end{pmatrix} \tag{8-20}$$

In Eq. 8-20, the unknown vector to be determined is:

$$\mathbf{I} = \begin{pmatrix} I_{\mathrm{Fc}} \\ I_{\mathrm{Rc}} \\ I_{\mathrm{Fd}} \\ I_{\mathrm{Rd}} \end{pmatrix} \tag{8-21}$$

The matrix **M** has dimensions 4×4:

$$\mathbf{M} = \begin{pmatrix} \psi_F(h) & \psi_R(h) & 0 & 0 \\ \nu Z_B + Z_{0c} & \nu Z_B - Z_{0c} & Z_B & Z_B \\ 0 & 0 & (Z_T - Z_{0d})\psi_F(h) & (Z_T + Z_{od})\psi_R(h) \\ -Z_{0c} & Z_{0c} & Z_{0d} & -Z_{0d} \end{pmatrix} \tag{8-22}$$

Comparing Eqs. 8-20 through 8-22 with Eqs. 2-19 through 2-21, we observe some remarkable similarities. These should not come as a surprise, however, because the two configurations are the same. Only the excitations are different. In the case of Chap. 2, the excitation is a voltage source connected to the base of the fold. In this section, the excitation is a distributed induced current.

Consistent with the above, we see that Eqs. 8-20 and 2-19 are almost identical in form. Only the driving terms on the right sides of the equations are different. In Eq. 2-19 the driving terms are from the voltage source with amplitude V. In Eq. 8-20, they are from the current with amplitude I_{i0} induced by the incident electromagnetic wave.

After Eq. 8-20 is solved for the vector **I** of unknown current amplitudes, then the power reradiated from the antenna can be readily calculated. Equation 8-12 for the total common mode current may be used in the procedure described in Sec. 2.9 for calculating radiated electromagnetic field and power. In Eq. 2-33, the expression $I_{Rc} + I_{i0}$ must be used instead of the term I_{Rc}.

When the reradiated power is evaluated as described above, pronounced minima will be observed for certain combinations of antenna design parameters. That is, the antenna is *detuned* for those values of design parameters. Evidently, the common mode currents, excited by the induced current, exhibit just the right amplitude and phase so that their respective reradiated fields cancel each other. In the next section, we will explore some examples that demonstrate the detuning phenomenon.

8.4 Examples of Full Length Detuning

There are very few discussions about detuning in the literature, and those discussions unanimously recommend that the folds be a quarter-wavelength long. They further recommend that that the base terminating reactance be an open circuit and that the top terminating reactance be a short circuit. In this section, we will go beyond those recommendations. The lengths of the folds may be shorter or longer than a quarter-wavelength. Further, the base terminating reactance is not necessarily an open circuit, nor is the top terminating reactance necessarily a short circuit.

Figure 8-4 verifies the recommendations of the existing literature, and it is also a good sanity check for our mathematical model. It is seen

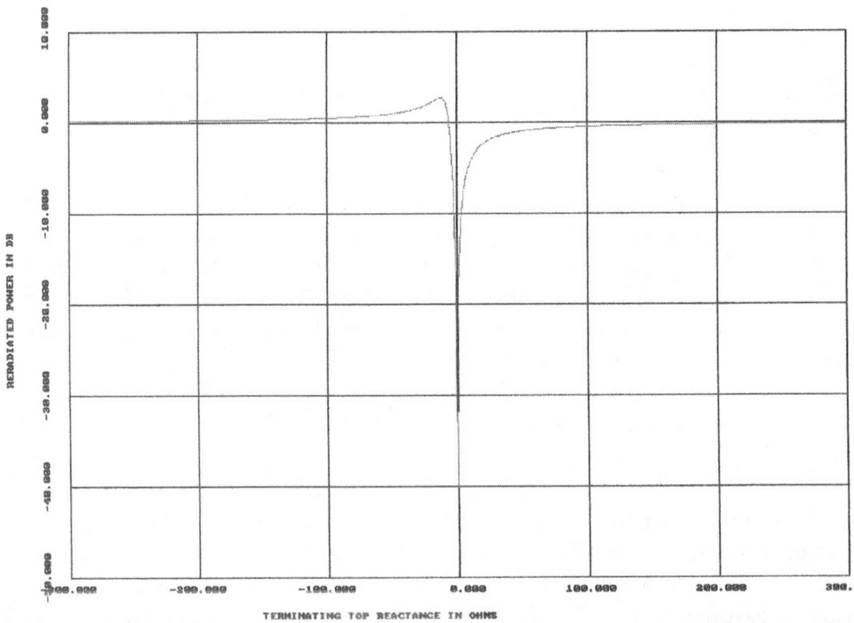

Figure 8-4 Reradiation from the full length detuning (FLD) configuration is reduced by about 40 dB in this example. The configuration conforms to the conventional wisdom, that is, the structure is a quarter-wave long. It is terminated in a short circuit at the top and an open circuit at the base.

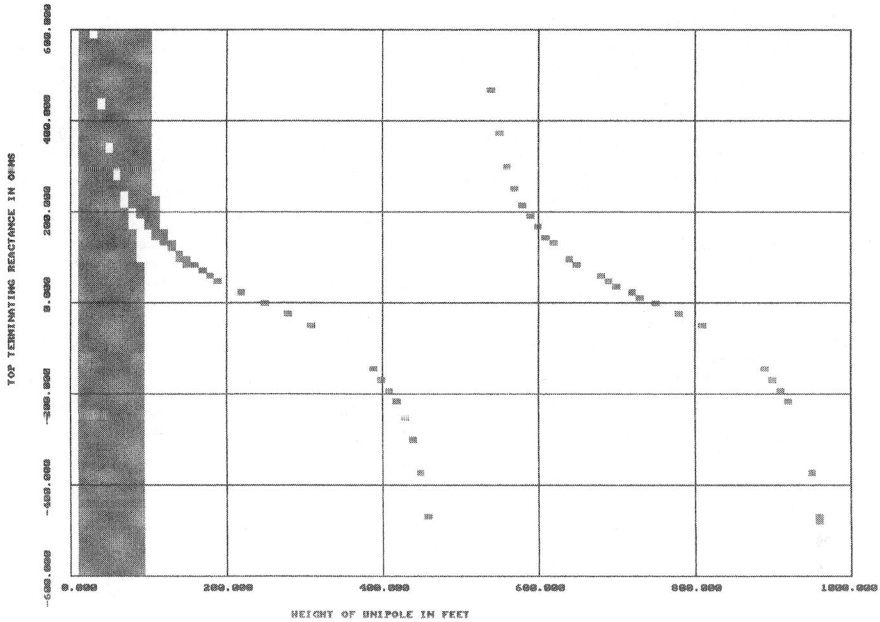

Figure 8-5 This contour plot shows reradiation reduction as a function of two variables for the FLD configuration. The variables are the length of the structure and the top terminating reactance. At 984 kHz, 1,000 feet is a full wavelength. It is seen that with a suitable terminating reactance, reradiation can be reduced by at least 20 dB for lengths shorter and longer than one-quarter wavelength.

that a quarter-wave fold terminated by an open circuit at the base, and by a short circuit at the top, exhibits a 40 dB reduction in reradiated power. In this example, certain design parameters are held fixed at "ballpark" values. The common mode and differential mode characteristic impedances are 1,200 ohms and 120 ohms, respectively. The current transformation ratio is 0.5. These values will also be used in the remaining examples.

Figure 8-5 shows that detuning is not limited to quarter wavelength folds. It is a contour plot with reradiated power as a function of two antenna design variables. The variables are the height of the fold (and also the entire structure) and terminating top reactance. The dark regions represent reductions of 20 dB or greater in reradiated power. It is seen that there are many combinations that produce those reductions, not just quarter wavelength heights.

Figure 8-6 shows an example of a detuned structure that is shorter than a quarter wavelength. It is seen that an inductive terminating top reactance reduces reradiated power by about 37 dB.

Figures 8-7 and 8-8 show detuned structures that are longer than a quarter wavelength. It is seen that nonzero terminating reactances reduce reradiated power by up to 54 dB.

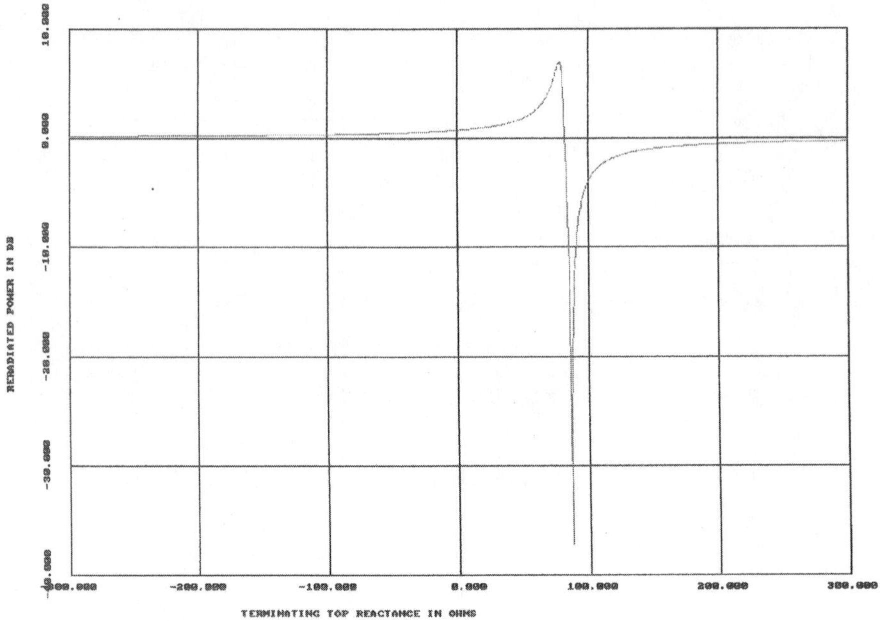

Figure 8-6 Reradiation reduction for a FLD configuration, as a function of terminating top reactance. The base is terminated in an open circuit. At 984 kHz, the length of the structure is 54 electrical degrees.

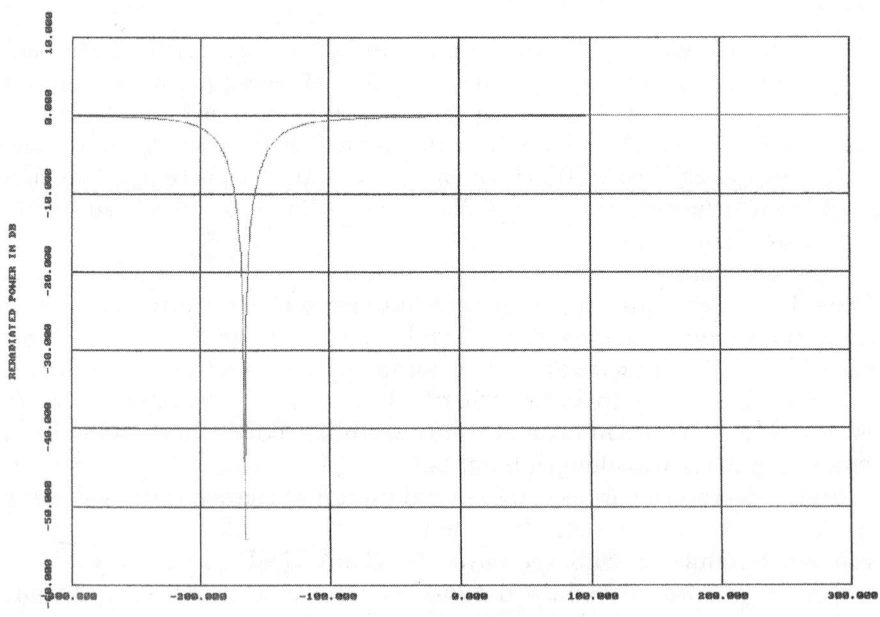

Figure 8-7 Reradiation reduction for a 144 degree FLD configuration, as a function of terminating top reactance. The base is terminated in an open circuit.

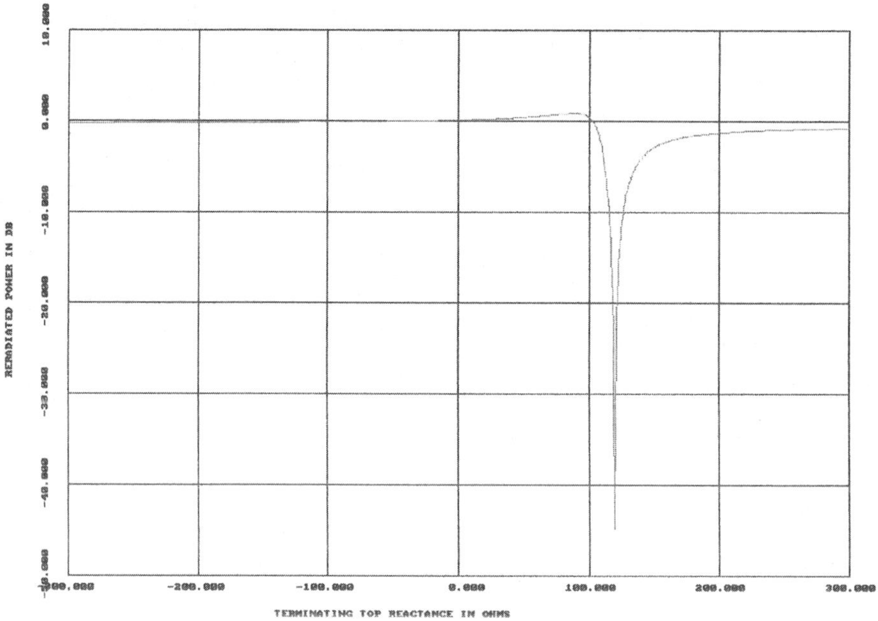

Figure 8-8 Reradiation reduction for a 225 degree FLD configuration, as a function of terminating top reactance. The base is terminated in an open circuit.

Perhaps the most challenging structures to detune are one-half wavelength long. Figures 8-9 and 8-10 show that even these may be detuned. In Figs. 8-9, the terminating base reactance has magnitude 500 ohms. It is seen that a top terminating reactance of nearly the same magnitude but opposite sign reduces reradiated power by over 50 dB. Figure 8-10 is similar except that the terminating base reactance has magnitude 4,000 ohms. As before, if the terminating top reactance has a comparable magnitude and opposite sign, then the reradiated power is reduced by over 35 dB. One interpretation of these results is that the structure has been converted into a resonant tank circuit where incident electromagnetic energy is trapped and not reradiated.

8.5 Lower Section Detuning

The detuning theory is not limited to an entire structure, that is, to full length detuning, as described in Secs. 8.3 and 8.4. If desired, we can detune only part of a structure. In this section, we will develop a mathematical model that will guide us in detuning the lower section.

Figure 8-11 shows the essential features of the structure reconfigured to detune its lower section. The height of that section h_1 will be arbitrary in our mathematical model. It is seen that the fold is terminated by both base and top impedances. The incident electromagnetic wave **E**, in general, will induce different currents in the top and bottom sections.

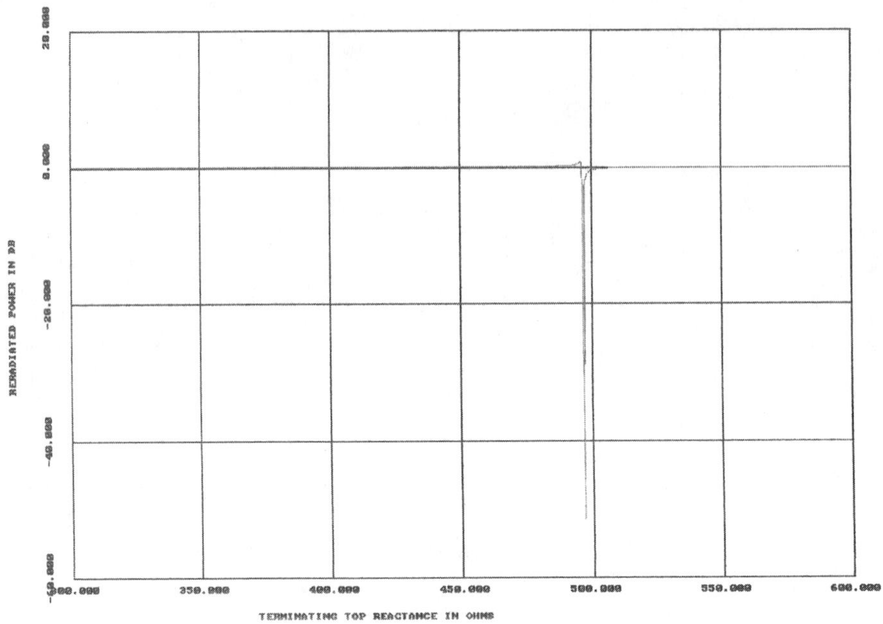

Figure 8-9 Reradiation reduction for a 180 degree FLD configuration, as a function of terminating top reactance. The base is terminated with a capacitive reactance of –500 ohms.

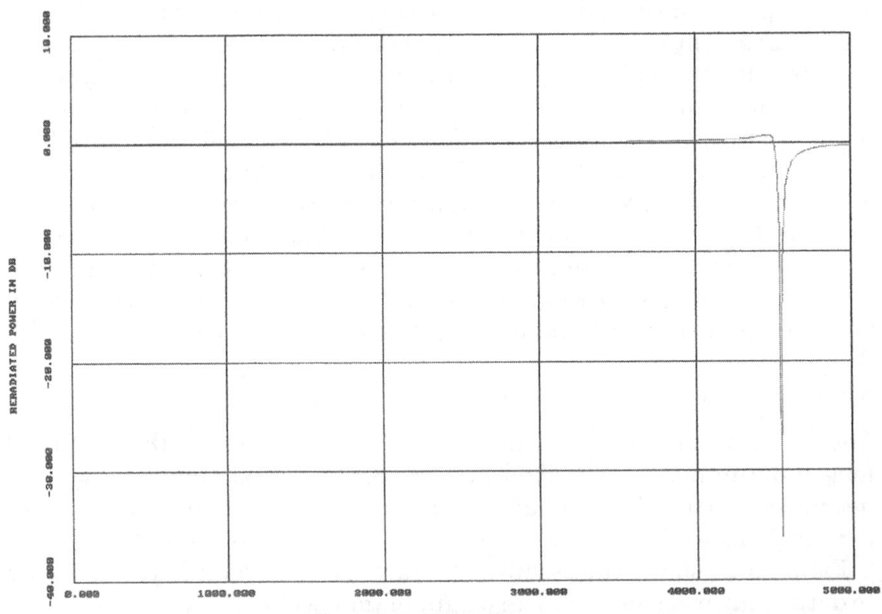

Figure 8-10 Reradiation reduction for a 180 degree FLD configuration, as a function of terminating top reactance. The base is terminated with a capacitive reactance of –4,000 ohms.

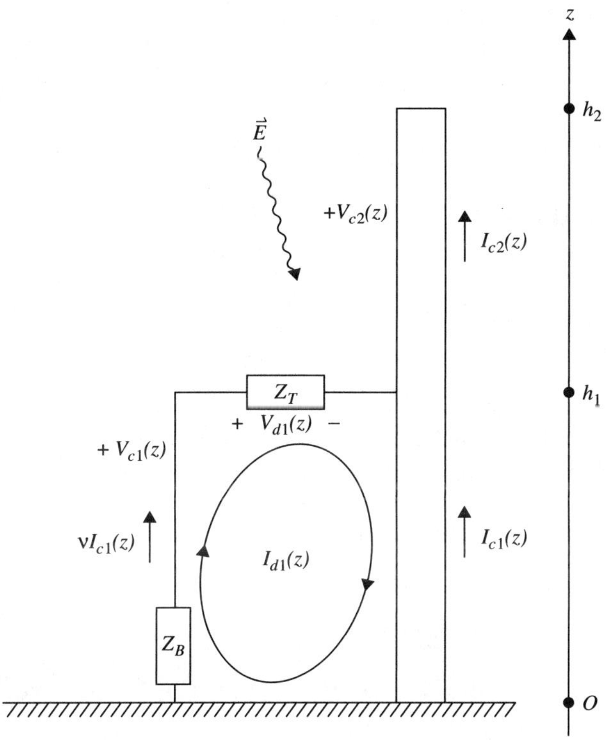

Figure 8-11 The essential features of lower section detuning (LSD) configuration. The reradiating structure is reconfigured as a Gamma Match Antenna (GMA), except that the base terminating impedance is connected to the fold, not the main structure.

We will develop a mathematical model for lower section detuning by following steps very similar to those used in Sec. 8.3. The induced currents are:

$$I_{\text{in}}(z) = I_{i0n}\psi_R(z) \qquad n=1,2 \tag{8-23}$$

In Eq. 8-23, the complex amplitudes of the induced currents are:

$$I_{i0n} = \frac{jE_0}{kZ_{0\text{cn}}} \qquad n=1,2 \tag{8-24}$$

Equations 8-23 and 8-24 are the same as Eqs. 8-9 and 8-10, except for the addition of the subscript n. That reminds us that, in general, the induced currents will be different on the top and bottom sections because the common mode characteristic impedances are different.

The common mode currents and voltages are:

$$I_{cn}(z) = I_{Fcn}\psi_F(z) + (I_{Rcn} + I_{i0n})\psi_R(z) \qquad n=1,2 \tag{8-25}$$

$$V_{cn}(z) = Z_{0cn}[I_{Fcn}\psi_F(z) - I_{Rcn}\psi_R(z)] \qquad n=1,2 \tag{8-26}$$

On the lower section, there are also a differential mode voltage and current:

$$I_{d1}(z) = I_{Fd1}\psi_F(z) + I_{Rd1}\psi_R(z) \tag{8-27}$$

$$V_{d1}(z) = Z_{0d1}[I_{Fd1}\psi_F(z) - I_{Rd1}\psi_R(z)] \tag{8-28}$$

In Eqs. 8-25 through 8-28, there are six unknown current amplitudes. They are I_{Fcn} and I_{Rcn} for $n = 1$ and $n = 2$, and I_{Fd1} and I_{Rd1}. To determine them, we must formulate six independent boundary conditions. One of those will be at the top of the antenna. There, the common mode current must vanish:

$$I_{c2}(h_2) = 0 \tag{8-29}$$

KVL provides two more boundary conditions at the base of the antenna:

$$[v_1 I_{c1}(0) + I_{d1}(0)]Z_B + V_{c1}(0) = 0 \tag{8-30}$$

$$V_{d1}(0) = V_{c1}(0) \tag{8-31}$$

The remaining three independent equations are formulated at the boundary between segments, at $z = h_1$. There, the common mode current must be continuous across the boundary:

$$(v_1 + 1)I_{c1}(h_1) = I_{c2}(h_1) \tag{8-32}$$

KVL requires that,

$$[v_1 I_{c1}(h_1) + I_{d1}(h_1)]Z_T = V_{d1}(h_1) \tag{8-33}$$

and that,

$$V_{d1}(h_1) = V_{c1}(h_1) - V_{c2}(h_1) \tag{8-34}$$

Equations 8-29 through 8-34 are six independent equations. We need to determine the six unknown current amplitudes. As we have done many times in previous chapters, these equations can be consolidated into a single matrix equation:

$$\mathbf{M}\cdot\mathbf{I}=\begin{pmatrix} -I_{i02}\psi_R(h_2) \\ -v_1 I_{i01}Z_B \\ I_{i02}\psi_R(h_1)-(v_1+1)I_{i01}\psi_R(h_1) \\ -v_1 I_{i01}Z_T\psi_R(h_1) \\ 0 \\ 0 \end{pmatrix} \tag{8-35}$$

In Eq. 8-35, the vector of unknown current amplitudes is:

$$\mathbf{I}=\begin{pmatrix} I_{\mathrm{Fc1}} \\ I_{\mathrm{Rd1}} \\ I_{\mathrm{Fd1}} \\ I_{\mathrm{Rd1}} \\ I_{\mathrm{Fc2}} \\ I_{\mathrm{Rc2}} \end{pmatrix} \tag{8-36}$$

The matrix **M** has dimensions 6 × 6. So that the individual matrix elements are legible on the printed page, it is necesssary to partition **M** into four submatrices:

$$\mathbf{M}=\begin{pmatrix} \mathbf{A}_{11} & \mathbf{A}_{12} \\ \mathbf{A}_{21} & \mathbf{A}_{22} \end{pmatrix} \tag{8-37}$$

In Eq. 8-37, each submatrix has dimensions 3 × 3. They are:

$$\mathbf{A}_{11}=\begin{pmatrix} 0 & 0 & 0 \\ v_1 Z_B + Z_{0c1} & v_1 Z_B - Z_{0c1} & Z_B \\ (v_1+1)\psi_F(h_1) & (v_1+1)\psi_R(h_1) & 0 \end{pmatrix} \tag{8-38}$$

$$\mathbf{A}_{12}=\begin{pmatrix} 0 & \psi_F(h_2) & \psi_R(h_2) \\ Z_B & 0 & 0 \\ 0 & -\psi_F(h_1) & -\psi_R(h_1) \end{pmatrix} \tag{8-39}$$

$$\mathbf{A}_{21}=\begin{pmatrix} v_1 Z_T\psi_F(h_1) & v_1 Z_T\psi_R(h_1) & (Z_T-Z_{0d1})\psi_F(h_1) \\ Z_{0c1}\psi_F(h_1) & -Z_{0c1}\psi_R(h_1) & -Z_{0d1}\psi_F(h_1) \\ -Z_{0c1} & Z_{0c1} & Z_{0d1} \end{pmatrix} \tag{8-40}$$

$$\mathbf{A}_{22}=\begin{pmatrix} (Z_T+Z_{0d1})\psi_R(h_1) & 0 & 0 \\ Z_{0d1}\psi_R(h_1) & -Z_{0c2}\psi_F(h_1) & Z_{0c2}\psi_R(h_1) \\ -Z_{0d1} & 0 & 0 \end{pmatrix} \tag{8-41}$$

The geometry of lower section detuning is similar to the geometry of the Gamma Match Antenna (GMA) discussed in Chap. 3. Therefore, we might expect that Eqs. 8-35 through 8-41 would be similar to Eqs. 3-15 through 3-21. In fact, however, there are differences as well as similarities. Comparing the matrix equations 8-35 and 3-15, we see that there are four nonzero terms on the right side of the former, and only two nonzero terms on the right side of the latter. We can interpret this as the difference between a lumped voltage source in the case of the GMA, and distributed current sources in the case of the detuned structure. Comparing the individual matrix elements in **M**, we find that some but not all are identical. This is not surprising because, even though the two structures are geometrically similar, the terminating base impedances are in different locations.

8.6 Examples of Lower Section Detuning

Using the model developed in the previous section, we will explore a variety of detuned lower sections, ranging in length from less than a quarter wavelength to greater than a half wavelength. We will see that all can be detuned; however, the degree of success depends upon whether the boundary between the lower and upper sections coincides with a minimum in the common mode current. If it does, then the two sections are mostly decoupled, and a greater reduction in reradiation is possible.

In Sec. 8.4, some design parameters were fixed at "ballpark" values, and those values will be used here also. In addition, the overall length of the structure will be fixed at a full wavelength. The common mode characteristic impedance of the unfolded part of the structure is 1,600 ohms.

Figure 8-12 shows what results are possible when a detuned lower section is shorter than a quarter wavelength. It is seen that reradiation is reduced by almost 30 dB.

Figure 8-13 shows a detuned lower section that is exactly one-quarter wavelength. Curiously, reradiation is reduced by only about 22 dB. This may contradict intuitive expectations until we realize that the upper section of the tower in this particular example is 0.75 wavelengths. So, the lower and upper sections are strongly coupled to each other, and it is difficult to isolate the former.

Figure 8-14 shows a detuned lower section that is greater than one-quarter wavelength. It is seen that reradiation is reduced by about 35 dB.

Figure 8-15 shows a detuned lower section that is one-half wavelength. Historically, this is a most difficult structure to detune; however, we are remarkably successful for the same reasons discussed in Sec. 8.4. The terminating base and top impedances are comparable in magnitude but opposite in sign. Evidently, we have converted the lower section into a resonant tank that traps incident electromagnetic energy

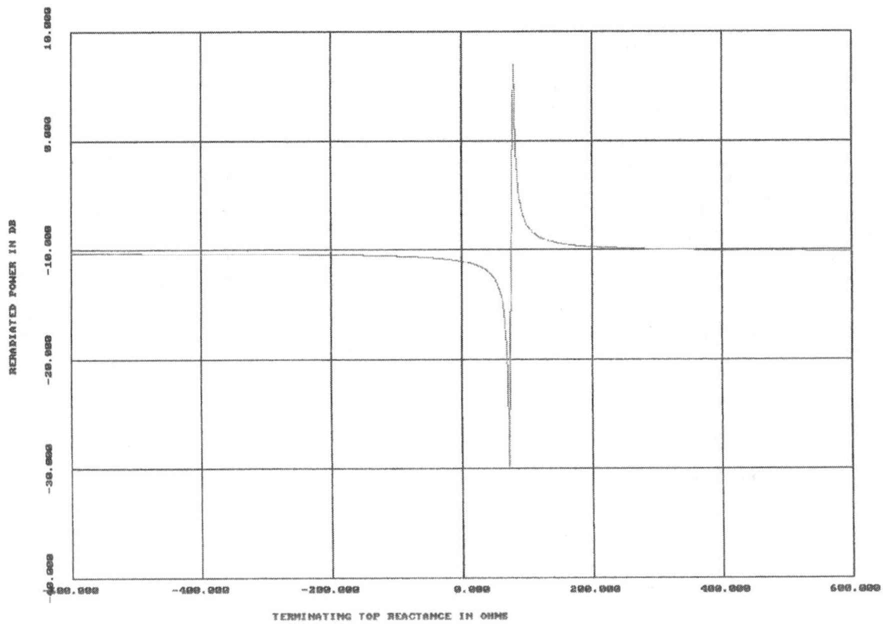

Figure 8-12 Reradiation from the LSD configuration as a function of top terminating reactance. The folded section is 54 electrical degrees. The base terminating reactance is an open circuit.

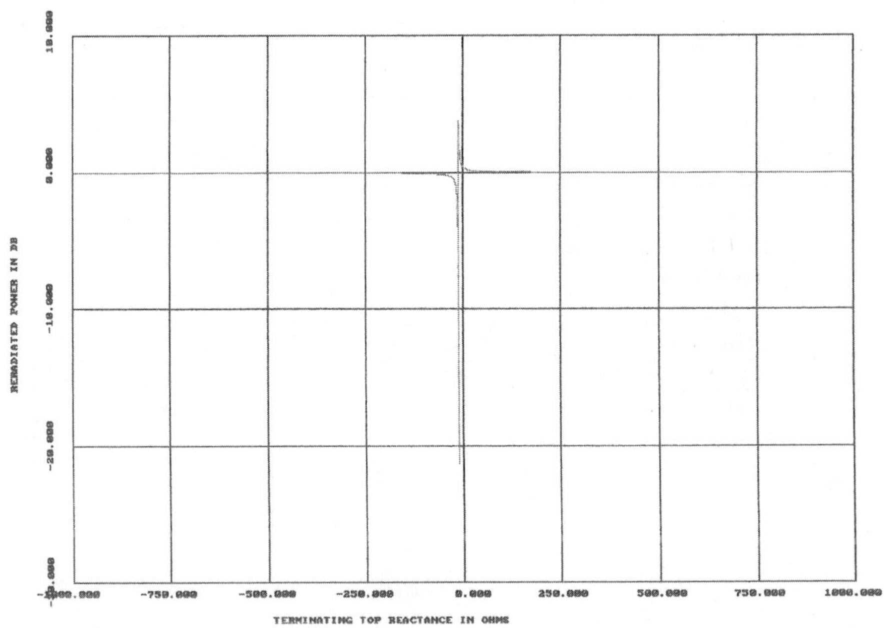

Figure 8-13 Reradiation from the LSD configuration as a function of top terminating reactance. The folded section is 90 electrical degrees, or one-quarter wavelength. The base is terminated with a capacitive reactance of –1,200 ohms.

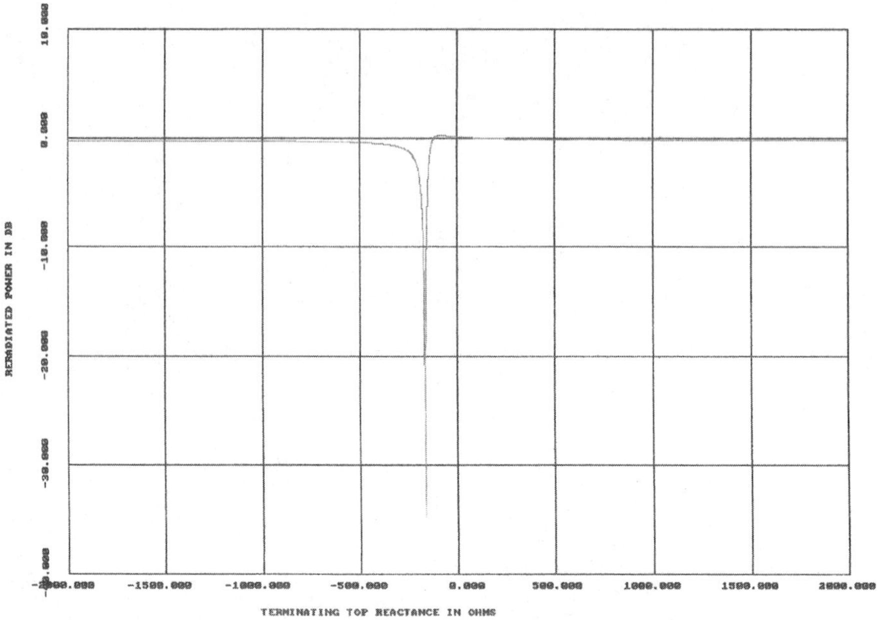

Figure 8-14 Reradiation from the LSD configuration as a function of top terminating reactance. The folded section is 144 electrical degrees. The base terminating reactance is an open circuit.

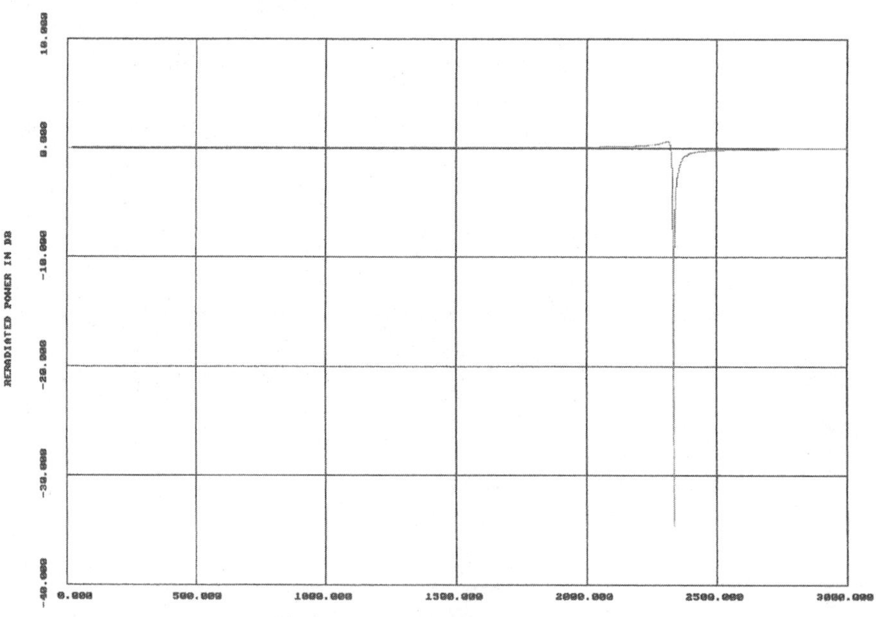

Figure 8-15 Reradiation from the LSD configuration as a function of top terminating reactance. The folded section is 180 electrical degrees, or one-half wavelength. The base is terminated with a capacitive reactance of –2,400 ohms.

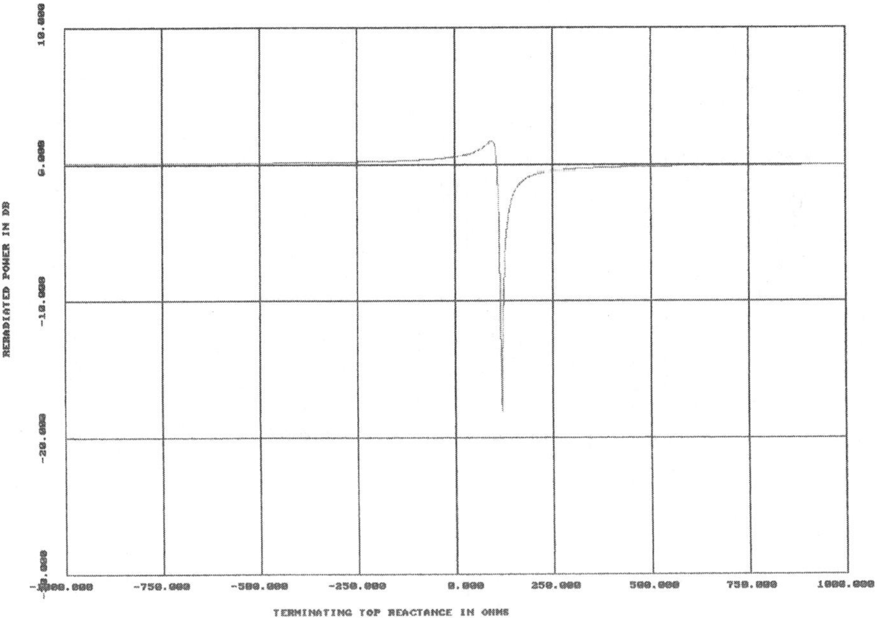

Figure 8-16 Reradiation from the LSD configuration as a function of top terminating reactance. The folded section is 225 electrical degrees. The base is terminated with an open circuit.

rather than reradiate it. The top section of this particular example is also one-half wavelength. So the boundary between the two sections coincides approximately with a minimum in common mode current. Therefore, the two sections are not tightly coupled and this contributes significantly to our success in detuning, with a reduction in reradiation by almost 35 dB.

Figure 8-16 shows a detuned lower section that is greater than one-half wavelength. The reduction in reradiation is less than expected, only about 18 dB. This is consistent with the example in Fig. 8-13. The boundary between upper and lower sections does not coincide with a common mode current minimum, so that two sections are strongly coupled. This makes it more difficult to detune the lower section.

8.7 Upper Section Detuning

The mathematical model and examples for upper section detuning are similar in most ways to those for lower section detuning; however, there are some differences in the details. We would have similar expectations if the geometry of a Gamma Match Antenna were inverted.

Figure 8-17 shows the essential features of a structure reconfigured for upper section detuning. It is seen that there are terminating impedances at both top and bottom of the detuned section.

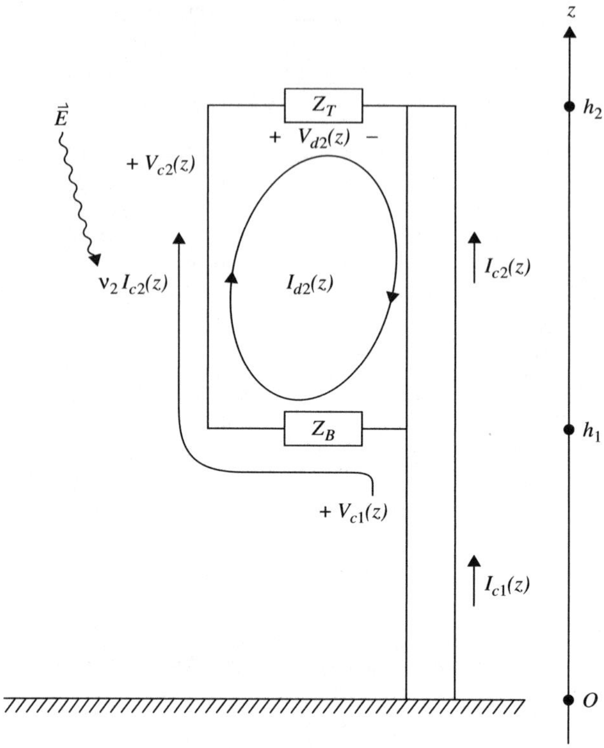

Figure 8-17 Essential features of the upper section detuning (USD) configuration. For the first time in our mathematical modeling, we will require the common mode voltage to vanish at the base.

The currents induced by the incident electromagnetic wave **E** are described by the same formulas as for lower section detuning, that is, by Eqs. 8-23 and 8-24. The common mode currents and voltages are also described by formulas which are the same as for lower section detuning, that is, by Eqs. 8-25 and 8-26.

A differential mode current and voltage exist on the upper section. They are described by

$$I_{d2}(z) = I_{\text{Fd2}}\psi_F(z) + I_{\text{Rd2}}\psi_R(z) \tag{8-42}$$

Eqs. 8-27 and 8-28, with a simple change in subscript:

$$V_{d2}(z) = Z_{0d2}[I_{\text{Fd2}}\psi_F(z) - I_{\text{Rd2}}\psi_R(z)] \tag{8-43}$$

Equations 8-25 and 8-26, and Eqs. 8-42, and 8-43 completely describe the currents and voltages on the detuned structure. To evaluate those equations, we must first determine the six unknown current amplitudes they contain. They are I_{Fcn} and I_{Rcn} for $n = 1$ and $n = 2$, and I_{Fd2} and I_{Rd2}. To determine them, we must formulate six independent boundary conditions. Two of those will be at the top of the antenna. There, the common mode current must vanish:

$$(1+v_2)I_{c2}(h_2) = 0 \tag{8-44}$$

The differential mode voltage and current are constrained by the terminating top impedance:

$$I_{d2}(h_2)Z_T = V_{d2}(h_2) \tag{8-45}$$

There is only one boundary condition at the base, and this is the first time we have used it. The common mode voltage must vanish there:

$$V_{c1}(0) = 0 \tag{8-46}$$

The remaining three independent equations can be formulated at the boundary between the folded and unfolded sections, at $z = h_1$. The common mode current must be continuous across the boundary:

$$I_{c1}(h_1) = (1+v_2)I_{c2}(h_1) \tag{8-47}$$

KVL requires that:

$$[v_2 I_{c2}(h_1) + I_{d2}(h_1)]Z_B = -V_{d2}(h_1) \tag{8-48}$$

and that:

$$V_{d2}(h_1) = V_{c2}(h_1) - V_{c1}(h_1) \tag{8-49}$$

Equations 8-44 through 8-49 are six independent equations. We need to determine the unknown current amplitudes. The equations can be consolidated into a single matrix equation:

$$\mathbf{M} \cdot \mathbf{I} = \begin{pmatrix} -(1+v_2)I_{i02}\psi_R(h_2) \\ [-I_{i01} + (1+v_2)I_{i02}]\psi_R(h_1) \\ -v_2 I_{i02} Z_B \psi_R(h_1) \\ 0 \\ 0 \\ 0 \end{pmatrix} \tag{8-50}$$

In Eq. 8-50, **I** is a vector of the unknown current amplitudes:

$$\mathbf{I} = \begin{pmatrix} I_{\mathrm{Fc1}} \\ I_{\mathrm{Rc1}} \\ I_{\mathrm{Fc2}} \\ I_{\mathrm{Rc2}} \\ I_{\mathrm{Fd2}} \\ I_{\mathrm{Rd2}} \end{pmatrix} \tag{8-51}$$

The matrix **M** has dimensions 6 × 6. So that individual elements are legible on the printed page, it must be partitioned exactly as we partitioned the matrix for lower section detuning, according to Eq. 8-37. The resulting submatrices are:

$$\mathbf{A}_{11} = \begin{pmatrix} 0 & 0 & (1+\nu_2)\psi_F(h_2) \\ \psi_F(h_1) & \psi_R(h_1) & -(1+\nu_2)\psi_F(h_1) \\ 0 & 0 & \nu_2 Z_B \psi_F(h_1) \end{pmatrix} \tag{8-52}$$

$$\mathbf{A}_{12} = \begin{pmatrix} (1+\nu_2)\psi_R(h_2) & 0 & 0 \\ -(1+\nu_2)\psi_R(h_1) & 0 & 0 \\ \nu_2 Z_B \psi_R(h_1) & (Z_B + Z_{0d2})\psi_F(h_1) & (Z_B - Z_{0d2})\psi_R(h_1) \end{pmatrix} \tag{8-53}$$

$$\mathbf{A}_{21} = \begin{pmatrix} 0 & 0 & 0 \\ Z_{0c1} & -Z_{0c1} & 0 \\ Z_{0c1}\psi_F(h_1) & -Z_{0c1}\psi_R(h_1) & -Z_{0c2}\psi_F(h_1) \end{pmatrix} \tag{8-54}$$

$$\mathbf{A}_{22} = \begin{pmatrix} 0 & (Z_T - Z_{0d2})\psi_F(h_2) & (Z_T + Z_{0d2})\psi_R(h_2) \\ 0 & 0 & 0 \\ Z_{0c2}\psi_R(h_1) & Z_{0d2}\psi_F(h_1) & -Z_{0d2}\psi_R(h_1) \end{pmatrix} \tag{8-55}$$

Comparison of Eqs. 8-52 through 8-55 with Eqs. 8-38 through 8-41 does not reveal any striking similarities. Some individual elements are identical but others are different. Evidently, inverting the configuration from lower section detuning to upper section detuning results in significant differences, at least mathematically. This is due to the presence of the ground plane. Had the configurations been in empty space, then at most a simple transformation would have been required to make the equations identical.

8.8 Examples of Upper Section Detuning

Despite differences in the mathematical models for lower and upper detuning configurations, we will observe similar general trends in reradiation reduction. We will examine upper sections ranging in length from less than a quarter-wave to greater than a half-wave. As with lower section detuning, we will find that reradiation reduction is most successful when the boundary between the upper and lower sections does not coincide with a common mode current maximum. A current maximum at the boundary means that the two sections are tightly coupled, and this makes it very difficult to isolate the upper section and reduce reradiation from it.

The design parameters that were fixed at "ballpark" values in Secs. 8.4 and 8.6 will have the same values in this section.

Figure 8-18 shows results for an upper section shorter than a quarter wavelength. It is seen that reradiation is reduced by almost 30 dB for a certain value of terminating top reactance. The top reactance is inductive, and that serves to make the upper section appear electrically longer so that it is similar to an open circuited, quarter wavelength of transmission line.

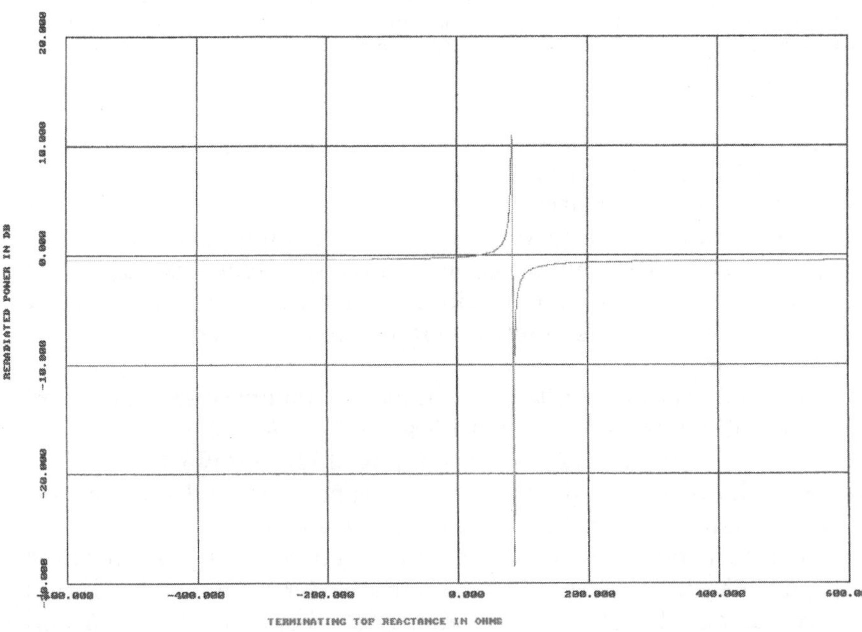

Figure 8-18 Reradiation from the USD configuration as a function of top terminating reactance. The folded section is 54 electrical degrees. The base is terminated with an open circuit.

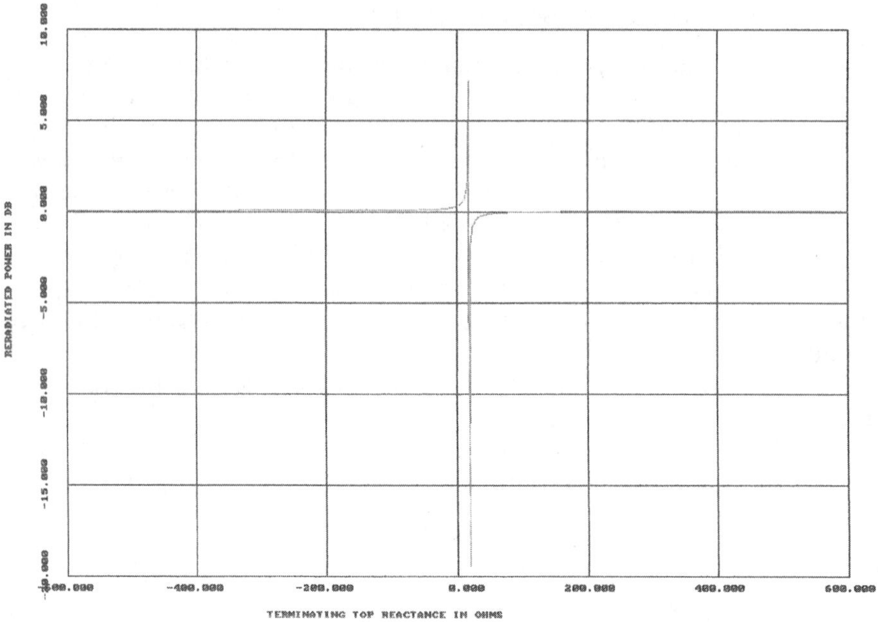

Figure 8-19 Reradiation from the USD configuration as a function of top terminating reactance. The folded section is 117 electrical degrees. The base is terminated with an open circuit. This is as close as we can get to a quarter wavelength folded section with a useable reradiation reduction for reasons explained in Sec. 8.8.

Figure 8-19 shows results for an upper section slightly shorter than a quarter wavelength. Reradiation is reduced by only about 18 dB. Actually, for an upper section of exactly one-quarter wavelength, we would have failed to obtain any reduction in reradiation. For that length of upper section, the lower section is three-quarters wavelength, and the boundary is at a current maximum. Therefore, the two sections are tightly coupled, and it is practically impossible to reduce reradiation in the upper one.

Figure 8-20 shows results for an upper section between one-quarter and one-half wavelength. As with Fig. 8-18, reradiation is reduced by almost 30 dB for a certain value of top terminating reactance. Unlike Fig. 8-18, however, the top reactance is capacitive, and that serves to make the upper section appear electrically shorter.

Figure 8-21 shows results for an upper section slightly shorter than a half wavelength. This example is remarkable for the amount of reradiation reduction, almost 65 dB! Evidently, the boundary between upper and lower sections of the antenna very nearly coincide with a current minima.

Figure 8-22 shows results for an upper section exactly one-half wavelength. The base terminating reactance is not an open circuit, as it was for shorter upper sections. It is a finite value, and the top terminating

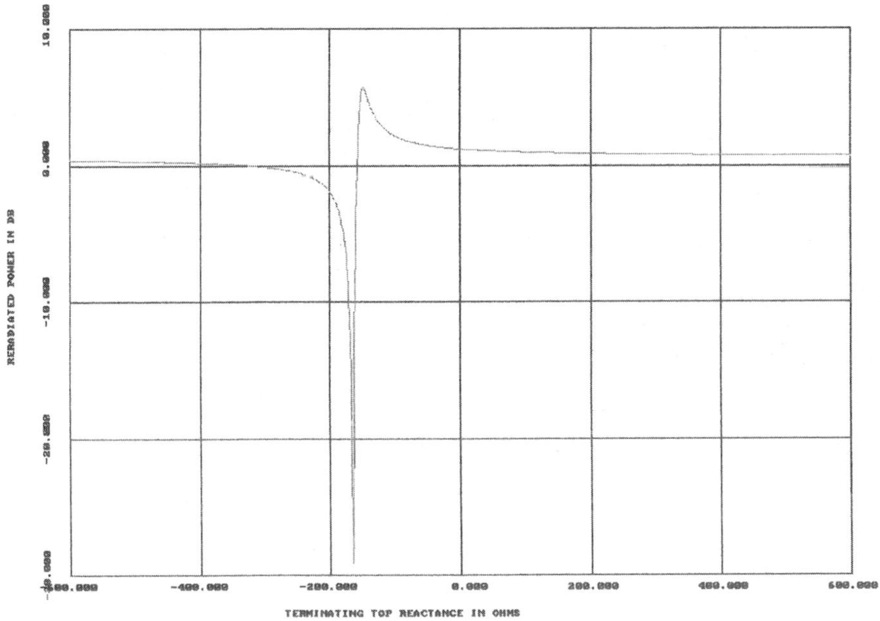

Figure 8-20 Reradiation from the USD configuration as a function of top terminating reactance. The folded section is 144 electrical degrees. The base is terminated with an open circuit.

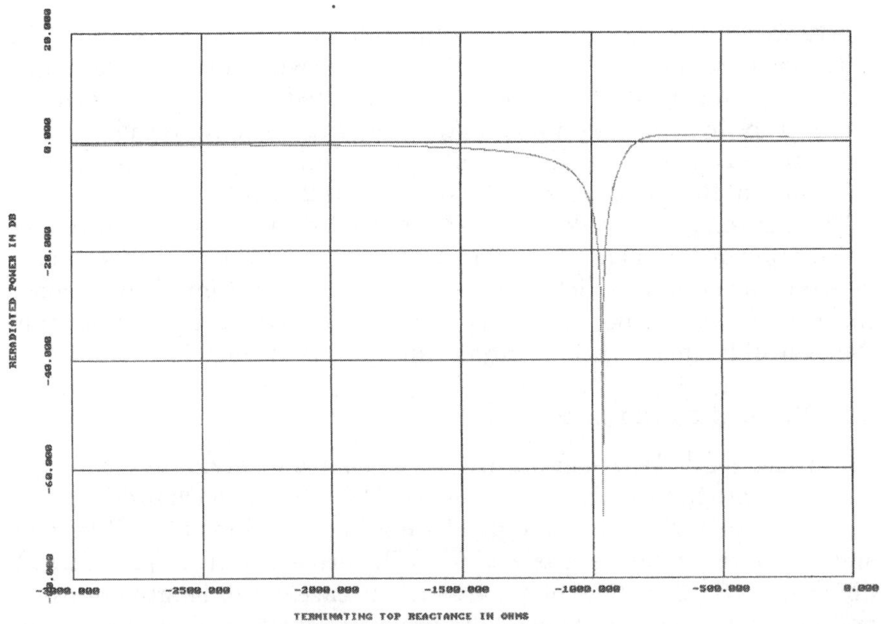

Figure 8-21 Reradiation from the USD configuration as a function of top terminating reactance. The folded section is 172.8 electrical degrees, not quite one-half wavelength. The base is terminated with an open circuit. This example is remarkable because the reradiation reduction is almost 65 decibels!

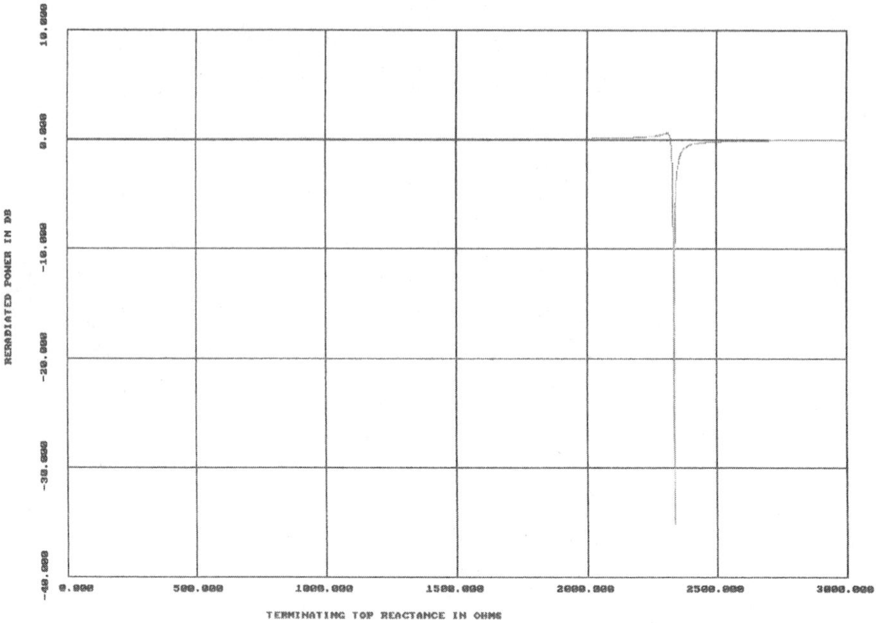

Figure 8-22 Reradiation from the USD configuration as a function of top terminating reactance. The folded section is 180 electrical degrees, or one-half wavelength. The base is terminated with a capacitive reactance of –2,400 ohms.

reactance has nearly the same magnitude but the opposite sign. The reduction in reradiation is over 35 dB. As we did with full section detuning and with lower section detuning, we have created a resonant tank circuit which traps incident electromagnetic energy rather than reradiates it.

Figure 8-23 shows results for an upper section longer than one-half wavelength. Reradiation is reduced by about 24 dB.

The above examples demonstrate that in most cases we can reduce reradiation by tens of decibels. The surprising exception was exactly one-quarter wavelength, which may seem counter to intuition. The reason is the strong coupling between upper and lower sections, a reminder that the overall length of the antenna is an important consideration.

8.9 MidSection Detuning

In this section we will complicate the geometry by dividing the overall structure into three sections, as shown in Fig. 8-24. We wish to detune the middle section. Not surprisingly, the mathematical model will become slightly more complicated as well. We will derive a matrix equation with dimensions 8×8 instead of 4×4, as in the case of full length detuning, or 6×6, as in the cases of lower and upper section detuning. Despite the increased complexity, much of the mathematical modeling has already been accomplished in the simpler detuning configurations.

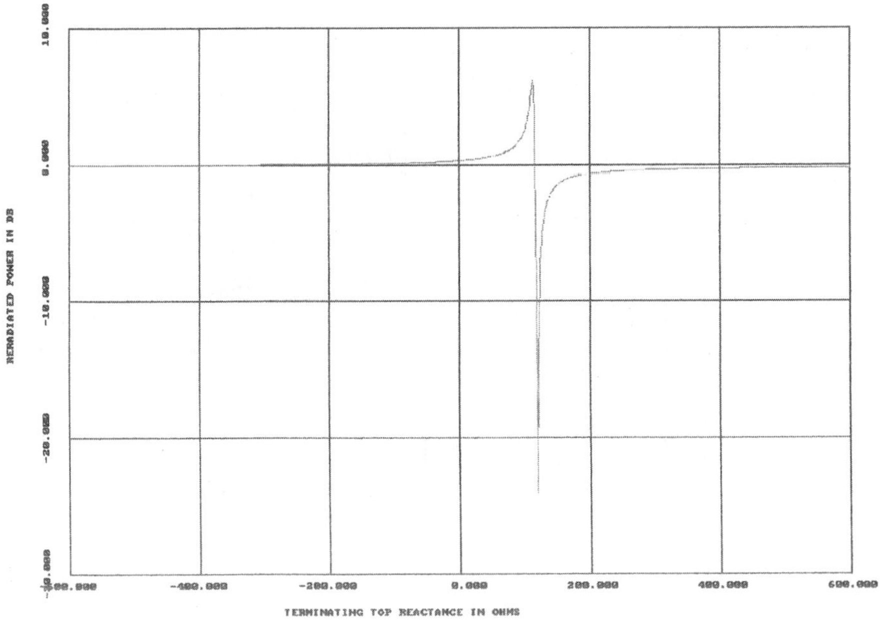

Figure 8-23 Reradiation from the USD configuration as a function of top terminating reactance. The folded section is 225 electrical degrees. The base is terminated with an open circuit.

The equations for voltages and currents that we derived for previous configurations can be used for the configuration of interest here. The currents induced by the incident electromagnetic wave **E** are described by Eqs. 8-23 and 8-24, except that the subscript n ranges over $n = 1,2,$ and 3 instead of just $n = 1,2$. Similarly, common mode currents and voltages exist on all three sections, and they are described by Eqs. 8-25 and 8-26, with the subscript n again ranging from 1 to 3.

A differential mode current and voltage exist on the middle section, and they are described exactly by Eqs. 8-42 and 8-43.

We see that there are eight unknown current amplitudes to determine. They are I_{Fcn} and I_{Rcn}, for $n = 1, 2,$ and 3, and I_{Fd2} and I_{Rd2}. To find them, we must derive eight independent equations. These will follow from constraints, or boundary conditions, at the boundaries of the structure. Most of the boundary conditions will be reminiscent of those formulated for upper and lower section detuning.

At the top of the structure, the common mode current must vanish:

$$I_{c3}(h_3) = 0 \tag{8-56}$$

Equation 8-56 is the same as Eq. 8-29 for lower section detuning, except for a simple change in subscript.

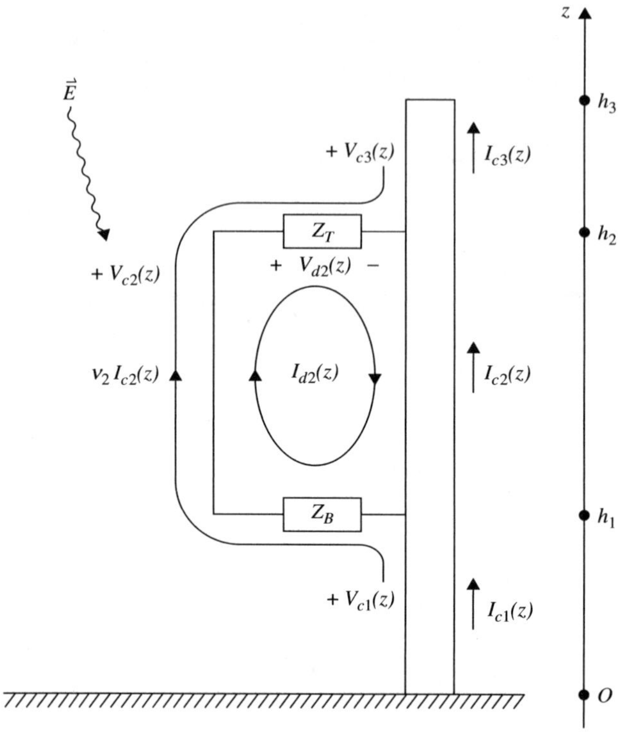

Figure 8-24 Essential features of the midsection detuning (MSD) configuration. It resembles a Tee Match Antenna, and so we might expect useful results using equal terminating impedances.

At the base of the structure, the common mode voltage must vanish:

$$V_{c1}(0) = 0 \tag{8-57}$$

Equation 8-57 is identical to Eq. 8-46 for upper section detuning.

We require six more boundary conditions to uniquely determine the eight unknown current amplitudes. Three of these can be formulated at the boundary located at $z = h_2$. There the terminating impedance Z_T forces a relationship, called a *constituitive relationship*, between the voltage and current:

$$[\nu_2 I_{c2}(h_2) + I_{d2}(h_2)] Z_T = V_{d2}(h_2) \tag{8-58}$$

The common mode current must be continuous across the boundary:

$$(1 + \nu_2) I_{c2}(h_2) = I_{c3}(h_2) \tag{8-59}$$

KVL requires that:

$$V_{d2}(h_2)=V_{c2}(h_2)-V_{c3}(h_2) \tag{8-60}$$

Three more similar boundary conditions can be formulated at the boundary located at $z = h_1$. The constituitive relationship is:

$$[\nu_2 I_{c2}(h_1)+I_{d2}(h_1)]\,Z_B=-V_{d2}(h_1) \tag{8-61}$$

The common mode current must be continuous across the boundary:

$$I_{c1}(h_1)=(1+\nu_2)I_{c2}(h_1) \tag{8-62}$$

KVL requires that:

$$I_{c1}(h_1)=V_{c2}(h_1)-V_{c1}(h_1) \tag{8-63}$$

Equations 8-56 through 8-63 are eight independent equations. We need to determine the unknown current amplitudes. They can be consolidated into a single matrix equation:

$$\mathbf{M}\cdot\mathbf{I}=\begin{pmatrix} -(1+\nu_2)I_{i02}\Psi_R(h_2)+I_{i03}\Psi_R(h_2) \\ -\nu_2 I_{i02}\Psi_R(h_2)Z_T \\ -I_{i01}\Psi_R(h_1)+(1+\nu_2)I_{i02}\Psi_R(h_1) \\ -\nu_2 I_{i02}\Psi_R(h_1)\,Z_B \\ -I_{i03}\Psi_R(h_3) \\ 0 \\ 0 \\ 0 \end{pmatrix} \tag{8-64}$$

In Eq. 8-64, the vector of unknown current amplitudes is:

$$\mathbf{I}=\begin{pmatrix} I_{\mathrm{Fc1}} \\ I_{\mathrm{Rc1}} \\ I_{\mathrm{Fc2}} \\ I_{\mathrm{Rc2}} \\ I_{\mathrm{Fd2}} \\ I_{\mathrm{Rd2}} \\ I_{\mathrm{Fc3}} \\ I_{\mathrm{Rc3}} \end{pmatrix} \tag{8-65}$$

In Eq. 8-64, the matrix **M** has dimensions 8 × 8. So that the individual elements will be legible on the printed page, **M** must be partitioned in the same way prescribed by Eq. 8-37. The submatrices are:

$$\mathbf{A}_{11} = \begin{pmatrix} 0 & 0 & (1+\nu_2)\psi_F(h_2) & (1+\nu_2)\psi_R(h_2) \\ 0 & 0 & \nu_2 Z_T \psi_F(h_2) & \nu_2 Z_T \psi_R(h_2) \\ \psi_F(h_1) & \psi_R(h_1) & -(1+\nu_2)\psi_F(h_1) & -(1+\nu_2)\psi_R(h_1) \\ 0 & 0 & \nu_2 Z_B \psi_F(h_1) & \nu_2 Z_B \psi_R(h_1) \end{pmatrix} \tag{8-66}$$

$$\mathbf{A}_{12} = \begin{pmatrix} 0 & 0 & -\psi_F(h_2) & -\psi_R(h_2) \\ (Z_T - Z_{0d2})\psi_F(h_2) & (Z_T + Z_{0d2})\psi_R(h_2) & 0 & 0 \\ 0 & 0 & 0 & 0 \\ (Z_B + Z_{0d2})\psi_F(h_1) & (Z_B + Z_{0d2})\psi_R(h_1) & 0 & 0 \end{pmatrix} \tag{8-67}$$

$$\mathbf{A}_{21} = \begin{pmatrix} 0 & 0 & 0 & 0 \\ Z_{0c1}\psi_F(h_1) & -Z_{0c1}\psi_R(h_1) & -Z_{0c2}\psi_F(h_1) & Z_{0c2}\psi_R(h_1) \\ 0 & 0 & Z_{0c2}\psi_F(h_2) & -Z_{0c2}\psi_R(h_2) \\ Z_{0c1} & -Z_{0c1} & 0 & 0 \end{pmatrix} \tag{8-68}$$

$$\mathbf{A}_{22} = \begin{pmatrix} 0 & 0 & \psi_F(h_3) & \psi_R(h_3) \\ Z_{0d2}\psi_F(h_1) & -Z_{0d2}\psi_R(h_1) & 0 & 0 \\ -Z_{0d2}\psi_F(h_2) & Z_{0d2}\psi_R(h_2) & -Z_{0c3}\psi_F(h_2) & Z_{0c3}\psi_R(h_2) \\ 0 & 0 & 0 & 0 \end{pmatrix} \tag{8-69}$$

8.10 Examples of MidSection Detuning

Compared with lower and upper section detuning, we expect that isolating the midsection for purposes of reradiation reduction will be more difficult. The reason is that there are two boundaries (i.e., at $z = h_1$ and $z = h_2$) instead of one. If a common mode current maximum coincides with either of those boundaries, then we expect that reradiation reduction will be especially challenging. Nonetheless, we will find that reradiation can be reduced by tens of decibels in many cases.

The design parameters that were fixed at "ballpark" values in Secs. 8.4, 8.6, and 8.8 will retain those values in this section. In addition, the bottom section of the configuration is fixed at 144 electrical degrees, or 400 feet at 984 kHz. The choice of frequency is entirely for the benefit of numerical simplicity. A wavelength at 984 kHz is 1,000 feet, and that is an easy value to discuss without resorting to pencil, paper, or calculator. In fact, our examples would result in the same conclusions if we had selected any frequency from the ELF through EHF bands.

Most of our examples in this section will be presented differently from those in Secs. 8.4, 8.6, and 8.8. Instead of detuning a section with a fixed

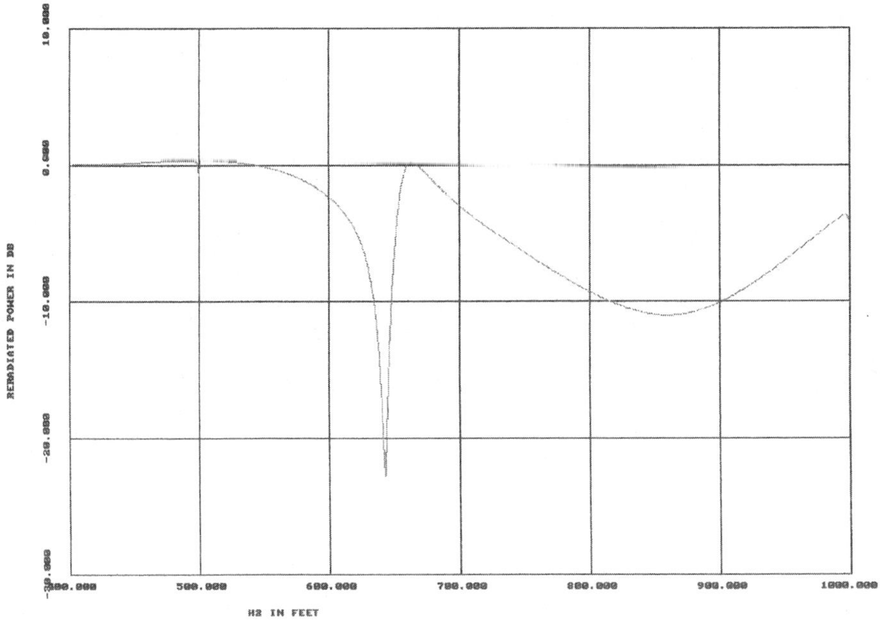

Figure 8-25 Reradiation from the MSD configuration as a function of the upper z-coordinate h_2 of the folded section. The length of the folded section is h_2-400 feet, and the wavelength is 1,000 feet. The folded section is terminated with a short circuit at the base and an open circuit at the top.

length, we will fix the terminating impedances and vary the length. For e.g., Fig. 8-25 shows results using the conventional terminations, that is, an open circuit at one end and a short circuit at the other. Not surprisingly, it is seen that maximum radiation reduction, by about 22 dB is achieved when the midsection is nearly one-quarter wavelength.

Interesting and useful results are obtained when the two terminations are identical. This is reminiscent of the Tee Match antenna. Figure 8-26 shows the results when the both terminating reactances are –36 ohms. It is seen that reradiation is reduced by almost 20 dB when the midsection is much shorter than one-quarter wavelength, about 32 electrical degrees.

Figure 8-27 shows the results when the terminating reactances are –75 ohms. It is seen that reradiation is reduced by about 28 dB when the midsection is about 65 electrical degrees. This is still shorter than one-quarter wavelength.

When the terminating reactances are both –150 ohms, then Fig. 8-28 shows that reradiation is reduced by about 24 dB when the midsection is 103 electrical degrees. Fig. 8-29 shows the results when the terminating reactances are –300 ohms. It is seen that reradiation is reduced by 17 dB when the midsection is 135 electrical degrees. These examples demonstrate

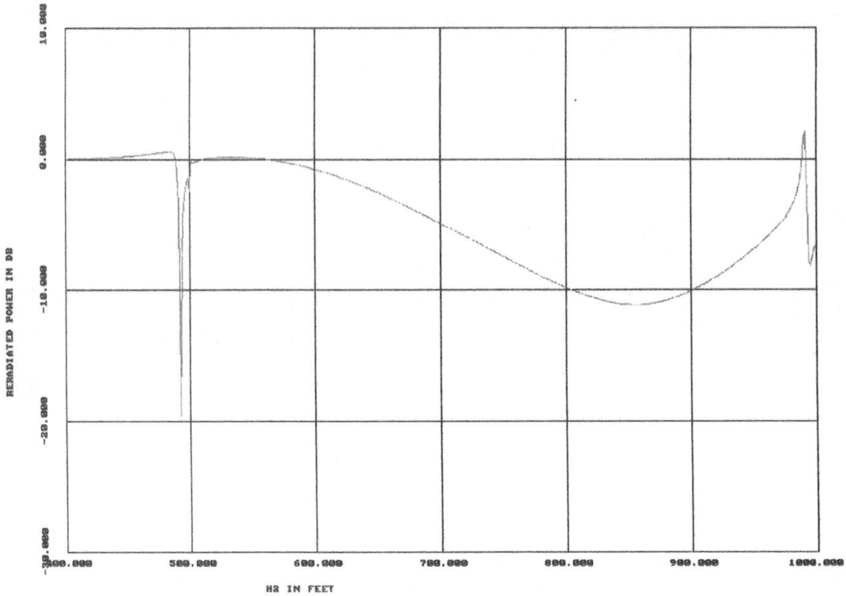

Figure 8-26 Reradiation from the MSD configuration as a function of the upper z-coordinate h_2 of the folded section. The length of the folded section is h_2-400 feet, and the wavelength is 1,000 feet. The folded section is terminated at both ends with a capacitive reactance of –36 ohms.

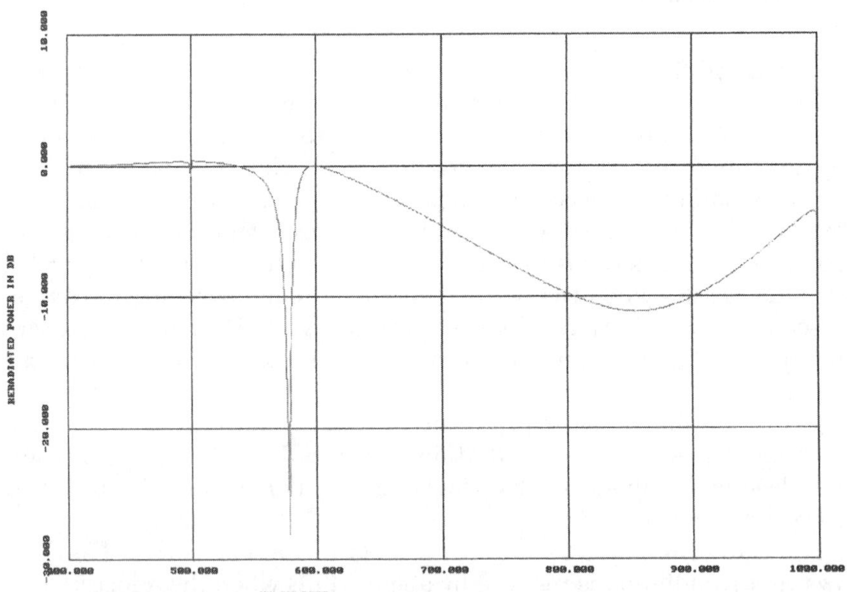

Figure 8-27 Reradiation from the MSD configuration as a function of the upper z-coordinate h_2 of the folded section. The length of the folded section is h_2-400 feet, and the wavelength is 1,000 feet. The folded section is terminated at both ends with a capacitive reactance of –75 ohms.

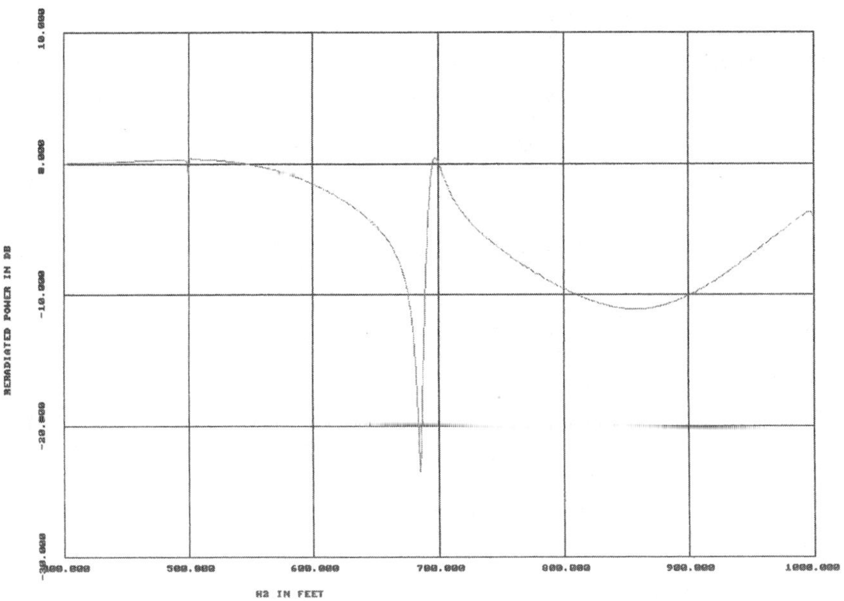

Figure 8-28 Reradiation from the MSD configuration as a function of the upper z-coordinate h_2 of the folded section. The length of the folded section is h_2-400 feet, and the wavelength is 1,000 feet. The folded section is terminated at both ends with a capacitive reactance of –150 ohms.

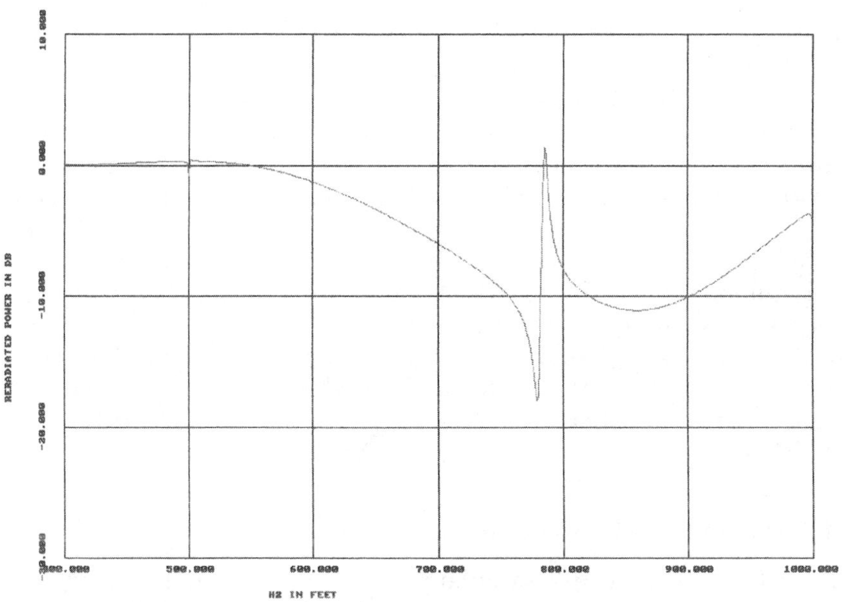

Figure 8-29 Reradiation from the MSD configuration as a function of the upper z-coordinate h_2 of the folded section. The length of the folded section is h_2-400 feet, and the wavelength is 1,000 feet. The folded section is terminated at both ends with a capacitive reactance of –300 ohms.

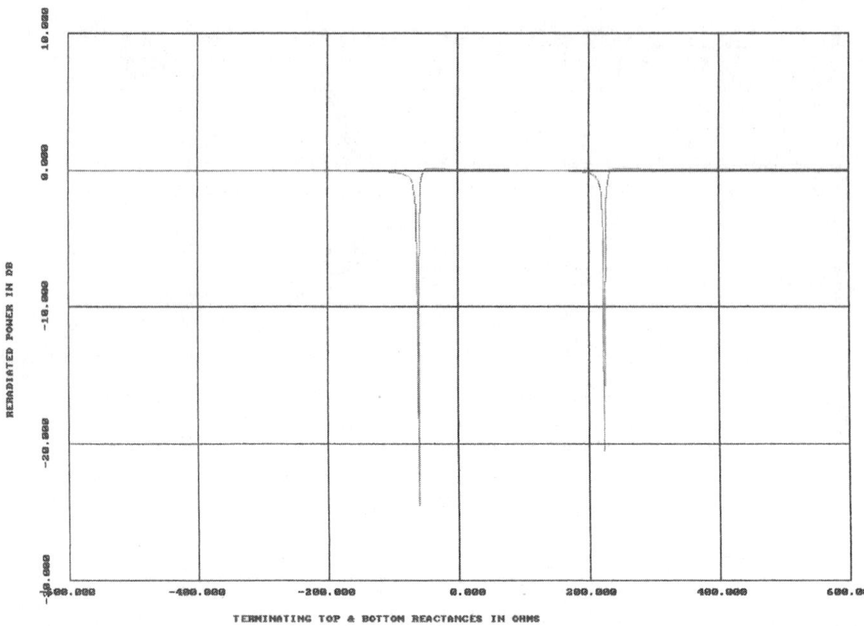

Figure 8-30 Reradiation from a 54 degree MSD configuration as a function of terminating reactances. Both top and base terminating reactances are identical. The total length of the structure is one wavelength. The length of the lower section is 144 electrical degrees.

that reradiation reduction using equal terminating reactances is possible for midsections longer than one-quarter wavelength.

Figures 8-30 and 8-31 are more conventional presentations, similar to those of previous sections. Fig. 8-30 shows the results for a midsection fixed in length at 54 electrical degrees. It is seen that for equal terminating reactances of –63 ohms, reradiation is reduced by about 24 dB. Figure 8-31 shows the results for a midsection fixed at one-quarter wavelength. For equal terminating reactances of –122 ohms, reradiation is reduced by 23 dB. This is slightly better than using the conventional terminations of Fig. 8-25.

In the above examples, we have seen that detuning is not limited to quarter wavelength sections. Nor is it limited to the conventional terminations of open and short circuits. This is consistent with our findings for full length, lower section, and upper section detuning.

8.11 Additional Detuning Configurations

Sections 8.3 through 8.10 only scratched the surface of detuning theory. There are practically endless possibilities for different detuning configurations. We could model multiple detuning sections, analogous to the n-segment folded unipole (nSFU) discussed in Chap. 5. We could

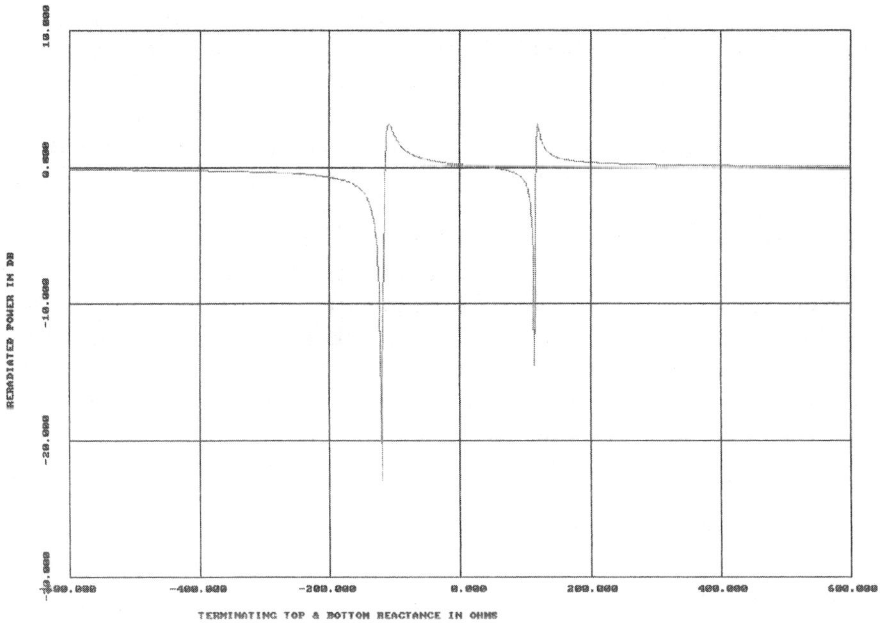

Figure 8-31 Reradiation from a 90 degree, or quarter wavelength, MSD configuration as a function of terminating reactances. Both top and base terminating reactances are identical. The total length of the structure is one wavelength. The length of the lower section is 144 electrical degrees.

terminate the detuning sections with two-port networks instead of two-terminal impedances, analogous to the most general configuration (MGC) of folded unipole discussed in Chap. 6. The mathematical models would be constructed in much the same way that has been demonstrated in this chapter.

We have not yet discussed detuning arrays of structures, *e.g.*, a line of utility poles. The appropriate mathematical model would combine the techniques of this chapter with those of the next chapter concerning arrays of folded unipoles.

Finally, we can envision creating an electronically invisible wall by detuning an entire surface. There are practical applications for such a wall, both commercial and military. An electronically invisible wall is desirable for architectural applications. Its complement, an electronically opaque wall, also has architectural uses. For the military, the stealth applications are enticing.

8.12 Conclusions

An entire book could easily be written exclusively about detuning theory and practice. Practically any structure can be detuned by reconfiguring it as some variation of the folded unipole antenna. Today, detuning is one of the most widely deployed applications of the folded unipole.

In this chapter, we established the foundations for mathematical models of detuning configurations. A simple formula was derived for the current induced by an incident electromagnetic wave. We also derived matrix equations which show how the induced current in turn excites both common and differential mode currents on the structure. Then we can readily calculate the reradiated, or scattered, electromagnetic field. The challenge of detuning is to minimize the reradiation.

We investigated four different configurations for which we can successfully reduce reradiation by at least 20 dB, and sometimes by almost 65 dB! Those detuning configurations are: full length, lower section, upper section, and midsection. They are but a small sample of the full range of possibilities. Other configurations include multiple sections, analogous to the n-segment folded unipole discussed in Chap. 5. They also include sections terminated in two-port networks, analogous to the most general configuration discussed in Chap. 6.

Our investigations ventured beyond the conventional wisdom concerning detuning. That wisdom tells us that the best way to detune a structure is to attach a fold one-quarter wavelength long. The fold is terminated at one end by an open circuit and at the other end by a short circuit. In contrast, our examples demonstrated that folds with lengths both shorter and longer than a quarter wavelength could reduce reradiation just as well if not better. In many cases, the folds were terminated with finite, nonzero reactances.

In some cases, our examples contradicted the conventional wisdom. That is, a structure configured with a quarter wavelength fold was very difficult to detune. The reason is that the ends of the fold coincided with a current maxima along the overall structure. This made it very difficult to isolate the folded section from the rest of the structure and to reduce the reradiation from it. In those cases, it is better to reconfigure the structure with a fold of a different length.

Besides individual structures, other possibilities include arrays of detuned structures. These include commercial applications such as lines of utility poles. Detuning the poles would reduce interference to licensed broadcasting. They also include military applications such as the leading wings of aircraft. Detuning parts of military vehicles would obviously reduce their radar cross sections.

The above applications are essentially linear or one-dimensional, that is, we would detune a *line* of utility poles, or the *edge* of a military vehicle. Detuning is not inherently limited to one dimension, however. An intriguing application is the creation of an electronically invisible wall, that is, detuning an entire surface. This would be a two-dimensional application of detuning.

In view of all the possible applications, we have clearly only scratched the surface of this fascinating subject. We have, however, established a solid foundation for its future exploration.

Reference

1. Schelkunoff, S. A., Theory of Antennas of Arbitrary Size and Shape, *Proc. of the IRE*, **29**, 1941, pp. 493–521.

Symbols Used in This Chapter

$\mathbf{A}_{11}$ Submatrix of **M**
$\mathbf{A}_{12}$ Submatrix of **M**
$\mathbf{A}_{21}$ Submatrix of **M**
$\mathbf{A}_{22}$ Submatrix of **M**
$\mathbf{E}$ Vector electric field component of an incident electromagnetic wave
E_0 Amplitude of **E**
E_z z-directed component of **E**
h z-coordinate of the top of full length detuned structure
h_1 z-coordinate of the top of the first segment of the detuned structure
h_2 z-coordinate of the top of the second segment of the detuned structure
h_3 z-coordinate of the top of the third segment of the detuned structure
$\mathbf{I}$ Vector of the complex current amplitudes
$I_c(z)$ Common mode current
$I_{c1}(z)$ Common mode current on the first segment of the detuned structure
$I_{c2}(z)$ Common mode current on the second segment of the detuned structure
$I_{c3}(z)$ Common mode current on the third segment of the detuned structure
$I_{\text{cn}}(z)$ Common mode current on the nth segment of the folded detuned structure
$I_d(z)$ Differential mode current
$I_{d1}(z)$ Differential mode current on the first segment of the detuned structure
$I_{d2}(z)$ Differential mode current on the second segment of the detuned structure
$I_{d3}(z)$ Differential mode current on the third segment of the detuned structure
$I_{\text{dn}}(z)$ Differential mode current on the nth segment of the detuned structure
I_{Fc} Amplitude of the common mode forward traveling wave current
I_{Fcn} Amplitude of the common mode forward traveling wave current on the nth segment
I_{Fd} Amplitude of the differential mode forward traveling wave current
I_{Fdn} Amplitude of the differential mode forward traveling wave current on the nth segment
$I_i(z)$ Induced current caused by **E**
I_{i0} Amplitude of $I_i(z)$
I_{i01} Amplitude of the induced current on the first segment of the detuned structure
I_{i02} Amplitude of the induced current on the second segment of the detuned structure
I_{i03} Amplitude of the induced current on the third segment of the detuned structure
$I_{\text{in}}(z)$ Induced current on the nth segment of the detuned structure
I_{i0n} Amplitude of the induced current on the nth segment of the detuned structure
I_{Rc} Amplitude of the common mode reverse traveling wave current

I_{Rcn}	Amplitude of the common mode reverse traveling wave current on the *n*th segment
I_{Rd}	Amplitude of the differential mode reverse traveling wave current
I_{Rdn}	Amplitude of the differential mode reverse traveling wave current on the *n*th segment
j	Square root of −1
k	Wave number
$\mathbf{M}$	Matrix in the equation for the complex current amplitudes
$V_c(z)$	Common mode voltage
$V_{c1}(z)$	Common mode voltage on the first segment of the detuned structure
$V_{c2}(z)$	Common mode voltage on the second segment of the detuned structure
$V_{c3}(z)$	Common mode voltage on the third segment of the detuned structure
$V_{\text{cn}}(z)$	Common mode voltage on the *n*th segment of the detuned structure
$V_{d1}(z)$	Differential mode voltage on the first segment of the detuned structure
$V_{d2}(z)$	Differential mode voltage on the second segment of the detuned structure
$V_{d3}(z)$	Differential mode voltage on the third segment of the detuned structure
$V_{\text{dn}}(z)$	Differential mode voltage on the *n*th segment of the detuned structure
X_B	Terminating reactance at the base of the fold on the detuned structure
X_T	Terminating reactance at the top of the fold on the detuned structure
z	Vertical coordinate
Z_{0c}	Characteristic impedance for the common mode
Z_{0c1}	Characteristic impedance for the common mode on the first segment
Z_{0c2}	Characteristic impedance for the common mode on the second segment
Z_{0c3}	Characteristic impedance for the common mode on the third segment
$Z_{0\text{cn}}$	Characteristic impedance for the common mode on the *n*th segment
Z_{0d}	Characteristic impedance for the differential mode
Z_{0d1}	Characteristic impedance for the differential mode on the first segment
Z_{0d2}	Characteristic impedance for the differential mode on the second segment
Z_{0d3}	Characteristic impedance for the differential mode on the third segment
$Z_{0\text{dn}}$	Characteristic impedance for the differential mode on the *n*th segment
Z_B	Terminating impedance at the base of the fold on the detuned structure
Z_T	Terminating impedance at the top of the fold on the detuned structure
ν	Current transformation ratio
ν_1	Current transformation ratio for first segment of the detuned structure
ν_2	Current transformation ratio for second segment of the detuned structure
ν_3	Current transformation ratio for the third segment on the detuned structure
ν_n	Current transformation ratio for the *n*th segment of the detuned structure
$\psi_F(z)$	Forward traveling wave function
$\psi_R(z)$	Reverse traveling wave function

Chapter

9

Arrays of Folded Unipoles

9.1 Introduction

The folded unipole antenna can function as an extremely versatile element of an antenna array. In previous chapters, we explored how the performance of individual folded unipoles was more flexible and controllable compared with that of series-fed monopoles. The reason was the substantially greater number of independent design variables for the folded unipole. In this chapter, we will see that the comparison holds true for arrays as well.

We will explore both active and passive arrays. In active arrays, such as those used in the broadcast industry, each element is connected to a power source. By "connected" we mean a hardwire connection from the antenna input terminals to the output terminals of a transmitter. In passive arrays, one or more of the elements do not have a hardwire connection. In a very real sense, however, this is an artificial distinction because in all arrays the elements are connected to each other via electromagnetic coupling. That is, the electromagnetic fields of one element induce significant currents on all the other elements. One of the chief results of this chapter will be an efficient and useable mathematical model of that coupling.

We will find that active arrays of folded unipoles have special advantages. They can provide shaped radiation patterns just like arrays of any type of antenna. In addition, shaped radiation patterns can be formed by close spaced folded unipoles, that is, with elements much closer to one another than a quarter wavelength. In Chap. 2, we found that individual folded unipoles offered extreme compactness in the vertical dimension. In the present chapter, we will find that arrays of folded unipoles offer compactness in the horizontal dimensions. This saves expensive real estate for long wavelength applications. At shorter wavelengths, compactness enables applications in cramped quarters such as might be encountered in instrumentation and medicine.

Special advantages are also provided by passive arrays of folded unipoles. Passive arrays may be regarded as either intentional or unintentional. Obvious examples of the former include Yagi antennas. We often encounter the latter in practical situations. For example, lines of electric power poles close to a broadcast array can substantially distort its desired radiation pattern. In those cases, the interfering towers can be reconfigured as folded unipole antennas. Then, the independent design parameters can be adjusted in such a way that the interference largely or entirely disappears. That is, the interfering towers can be made electronically invisible. This is the more general problem of detuning which we explored in the Chap. 8. In the present chapter, we will extend our mathematical modeling from detuning a single tower in a plane wave electromagnetic field to detuning multiple towers in a near electromagnetic field.

As explained earlier, the key to understanding arrays of folded unipoles is a useful mathematical model of the electromagnetic coupling between elements. To that end, we will depart from the traditional approach used for arrays of series-fed monopoles. That approach was firmly established in a landmark 1932 paper, by P. S. Carter [1], and it works splendidly for series-fed dipoles and monopoles. It essentially reduces antennas and antenna arrays to circuit models composed of self and mutual impedances. After that step, ordinary circuit analysis can be applied to the array. At the time of Carter's paper, impedance was a relatively new and very popular concept in electronic engineering. Many useful advances were elegantly formulated and developed in terms of impedance models. They included electronic filters and impedance matching networks. In view of those successes, the trend at the time was to reduce antennas mathematically to equivalent impedances. In the case of series-fed monopoles, the results met or exceeded expectations. The elegance and relative simplicity of the impedance concept made arrays of series-fed monopoles the standard for shaping radiation patterns in the radio broadcast industry. With respect to arrays of folded unipole antennas, however, this approach proved more awkward than useful. This awkwardness, no doubt, has discouraged the wider deployment of folded unipole arrays. An alternate approach is clearly indicated for the design and analysis of arrays.

In fact, reducing elements of antenna arrays to equivalent circuits of impedances and mutual impedances is an intermediate step that is not necessary. The fundamental physical mechanism by which array elements couple is induced current. That is, each element induces currents on all of the others. The radiation pattern of the array is the superposition of contributions to the electromagnetic field from all currents, both primary and induced. So, if we can compute the induced currents directly, then there is no need to bother with equivalent impedances. In Sec. 8.2, we found this was readily accomplished, at least for the simple case of a single antenna in a plane wave field. Will we be as fortunate, in the case of multiple antennas in the near fields of one another?

After some persistence with electromagnetic theory, we will be rewarded with some elegant formulas for the induced currents of interest. An added result will be formulas for the near fields of the folded unipole antenna. This is of interest for many reasons, including the evaluation of hazards relating to high power electromagnetic radiation patterns.

As in Chap. 8, we will only scratch the surface of an immense topic. The possibilities for arrays of folded unipole antennas are practically endless. To explore them all in any comprehensive way would require an entire book dedicated to that subject alone. We will, however, develop a mathematical model suitable for exploring any of those possibilities as the need arises.

9.2 The Complete Electromagnetic Field of the Folded Unipole Antenna

With respect to mathematical modeling, the chief difference between an isolated folded unipole antenna and an array of them is the electromagnetic coupling between the elements. That is, each element of the array induces significant currents on all the other elements. We must derive a useable formula for that coupling.

To that end, we will generalize the approach used in Sec. 8.2. In that special case, we derived a simple formula for the current induced on a single folded unipole antenna by a plane electromagnetic wave. Implicitly, there were two antennas in that problem. The source of the electromagnetic wave was, in fact, a very distant antenna. It was sufficiently far away that its electromagnetic field could be very closely approximated by a simple plane wave. In the present chapter, we will remove the restriction of great distance. The source antenna can be arbitrarily close. So, we will require a more accurate representation for its electromagnetic field. Not surprisingly, it will also be more complicated. The representation is often called the *near field*. In fact, it will be accurate at any distance.

We begin our derivation with the classic formulas for the electromagnetic field in terms of the magnetic vector potential. The electric field is:

$$\mathbf{E} = \frac{1}{j\omega\varepsilon_0} \nabla \times \mathbf{H} \tag{9-1}$$

The magnetic field is:

$$\mathbf{H} = \frac{1}{\mu_0} \nabla \times \mathbf{A} \tag{9-2}$$

The magnetic vector potential is:

$$\mathbf{A}(\mathbf{r}) = \mu_0 \iint G(\mathbf{r}, \mathbf{r}')\mathbf{J}(\mathbf{r}')d^2r' \tag{9-3}$$

In Eqs. 9-1 through 9-3, ε_0 and μ_0 are the permittivity and permeability of free space, respectively. In Eq. 9-3, $\mathbf{J}(\mathbf{r}')$ is the current density, in amperes per square meter, at the vector position $\mathbf{r}'$. The double integration is performed over all surfaces for which the current density is nonzero. The vector position $\mathbf{r}$ is the location of the observer, or the location at which the electric and magnetic fields are being evaluated.

The weighting function on the right side of Eq. 9-3 is called the *Green's Function*, and in free space it is:

$$G(\mathbf{r}, \mathbf{r}') = \frac{e^{-jk|\mathbf{r}-\mathbf{r}'|}}{4\pi\,|\mathbf{r} - \mathbf{r}'|} \tag{9-4}$$

In Eq. 9-4, k is the wave number.

Equations 9-1 through 9-4 are general and elegant. They show in a very concise way how to find the electromagnetic field at any distance from any current distribution. As a practical matter, however, they are not convenient to evaluate until we specialize them to a specific coordinate system. For our purposes, cylindrical coordinates (ρ, ϕ, z) will do just fine. For that choice of coordinate system, and for the vertical conductors which make up the antennas of interest here, the magnetic vector potential is entirely z-directed. Also, the current densities are one-dimensional because they flow along good conductors. So, Eq. 9-3 becomes:

$$A_z = \mu_0 \int_{\text{all wires}} G(\mathbf{r}, \mathbf{r}')I(z')dz' \tag{9-5}$$

The Green's Function in Eq. 9-4 becomes:

$$G(\mathbf{r}, \mathbf{r}') = \frac{e^{-jkR}}{4\pi R} \tag{9-6}$$

In Eq. 9-6, the distance between the current and the observer position is:

$$R = \sqrt{\rho^2 + (z - z')^2} \tag{9-7}$$

There is only a φ-directed component of the magnetic field, and Eq. 9-2 becomes:

$$H_\varphi = -\frac{1}{\mu_0}\frac{\partial A_z}{\partial \rho} \tag{9-8}$$

There are two nonzero components of the electric field, and they both follow from Eq. 9-1:

$$E_\rho = -\frac{1}{j\omega\varepsilon_0}\frac{\partial H_\varphi}{\partial z} \tag{9-9}$$

$$E_z = \frac{1}{j\omega\varepsilon_0\rho}\frac{\partial}{\partial \rho}(\rho H_\varphi) \tag{9-10}$$

Substituting from Eq. 9-8 into Eqs. 9-9 and 9-10, we obtain formulas for the electric field explicitly in terms of the magnetic vector potential:

$$E_\rho = \frac{1}{j\omega\varepsilon_0\mu_0}\frac{\partial^2 A_z}{\partial z \partial \rho} \tag{9-11}$$

$$E_z = \frac{j}{\omega\varepsilon_0\mu_0\rho}\left(\frac{\partial A_z}{\partial \rho} + \rho\frac{\partial^2 A_z}{\partial \rho^2}\right) \tag{9-12}$$

Equations 9-8, 9-11, and 9-12 are explicit instructions for evaluating the near field of any system of vertical conductors, including the folded unipole antenna. They do require mathematically tractable expressions for the magnetic vector potential A_z, described by Eqs. 9-5 through 9-7. In the following two sections, we will derive those expressions, and the resultant electromagnetic field, for traveling waves of current along the folded unipole antenna.

9.3 Complete Fields of Forward Traveling Current Waves

In the previous section, we derived straightforward formulas for the electromagnetic field of any vertical conductor, including folded unipole antennas. Those formulas are in terms of the *magnetic vector potential* A_z. It is seen from Eq. 9-5 that A_z is very much a function of the current $I(z')$ along the conductor. So, if we specify the current on a folded unipole, then we should obtain the magnetic vector potential, and the resultant electromagnetic field, for that antenna. That will be our strategy in this section.

To that end, Fig. 9-1 shows a segment of vertical conductor plus its mirror image in the ground. The relative directions of propagation are correct as shown, for the reasons discussed in Sec. 2.9.

Figure 9-1 also illustrates an important subtlety. Though the propagation vectors are mirror images as shown, the polarization or direction of the currents are not. That is, on both the actual segment and its image, the current is z-directed. There is an intuitively satisfying way to see why this must be true. Imagine tilting the segment and its image ninety degrees until they become horizontal and parallel. Then, the mirror image propagation vectors maintain the currents on both segments in the proper phase relationship. Because the currents have the same polarization, however, they are perfectly balanced, just as they must be in a horizontal transmission line.

Ultimately, we will obtain the electromagnetic field for the complete folded unipole antenna; however, to keep the mathematics as simple as possible, we will attack the problem in smaller pieces. For starters,

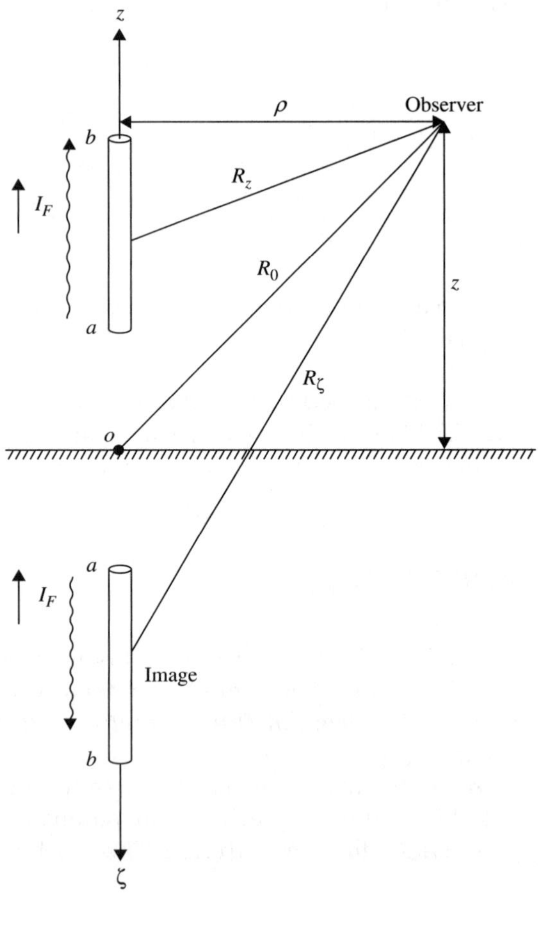

Figure 9-1 Essential features of a conducting segment and its image in a conducting ground. Forward traveling current waves flow along both of them. The directions of propagation are opposite, but the polarizations of the currents are the same.

we will focus on the forward traveling wave current with amplitude I_F. Also, we will focus on the conducting segment above ground. For that current on that conductor, the current is:

$$I(z') = I_F e^{-jkz'} \tag{9-13}$$

Substituting Eq. 9-13 into Eq. 9-5,

$$A_F = \frac{\mu_0}{4\pi} I_F \int_a^b \frac{e^{-jk(R+z')}}{R} dz' \tag{9-14}$$

On the left side of Eq. 9-14, we have substituted the subscript F for z to remind us that this is the magnetic vector potential for a forward traveling wave current. Also,

$$R = \sqrt{\rho^2 + (z' - z)^2} \tag{9-15}$$

The integration on the right side of Eq. 9-14 appears formidable; however, it is our good fortune that it can be performed exactly. It requires some cleverness. If we make the change of variable,

$$w = R + z' - z \tag{9-16}$$

it follows that,

$$dz' = \frac{R}{w} dw \tag{9-17}$$

Using Eqs. 9-16 and 9-17, Eq. 9-14 becomes:

$$A_F = \frac{\mu_0 I_F e^{-jkz}}{4\pi} \int_{R_a+a-z}^{R_b+b-z} \frac{e^{-jkw}}{w} dw \tag{9-18}$$

One more change of variable will transform the integral on the right side of Eq. 9-18 into a classic function from mathematical physics. Let,

$$w' = -jkw \tag{9-19}$$

Using Eq. 9-19, Eq. 9-18 becomes:

$$A_F = \frac{\mu_0 I_F e^{-jkz}}{4\pi} \int_{-jk(R_a+a-z)}^{-jk(R_b+b-z)} \frac{e^{w'}}{w'} dw' \tag{9-20}$$

In Eqs. 9-18 and 9-20 the distances R_a and R_b are:

$$R_a = \sqrt{\rho^2 + (z-a)^2} \tag{9-21}$$

$$R_b = \sqrt{\rho^2 + (z-b)^2} \tag{9-22}$$

The integral on the right side of Eq. 9-20 can be evaluated in terms of the *exponential integral function* $Ei(x)$

$$Ei(x) = \int_{-\infty}^{x} \frac{e^{w'}}{w'} dw' \tag{9-23}$$

$Ei(x)$ is a classic mathematical function from applied physics. It was used extensively by antenna engineers during the Golden Age of Antenna Analysis. It will be found in numerous papers authored by P. S. Carter [1] and S. A. Schelkunoff [2]. In terms of this function, the magnetic vector potential is:

$$A_F = \frac{\mu_0 I_F e^{-jkz}}{4\pi} \{Ei[-jk(R_b + b - z)] - Ei[-jk(R_a + a - z)]\} \tag{9-24}$$

Equation 9-24 is the explicit formula in terms of the forward wave traveling current that we desired. Now we can apply it to Eqs. 9-8, 9-11, and 9-12 to obtain the electromagnetic field. We see from those equations, that we will need some derivatives of A_F. The formulas for those derivatives can be obtained directly from Eq. 9-24. They are:

$$\frac{\partial A_F}{\partial \rho} = \frac{-jk\mu_0 \rho I_F e^{-jkz}}{4\pi} \left\{ \frac{Ei'[-jk(R_b + b - z)]}{R_b} - \frac{Ei'[-jk(R_a + a - z)]}{R_a} \right\} \tag{9-25}$$

$$\frac{\partial^2 A_F}{\partial \rho^2} = \frac{-jk\mu_0 I_F e^{-jkz}}{4\pi} \left\{ \begin{array}{l} \frac{Ei'[-jk(R_b + b - z)]}{R_b} \left[1 - \left(\frac{\rho}{R_b} \right)^2 \right] - \\ \frac{Ei'[-jk(R_a + a - z)]}{R_a} \left[1 - \left(\frac{\rho}{R_a} \right)^2 \right] - \\ jk \left(\frac{\rho}{R_b} \right)^2 Ei''[-jk(R_b + b - z)] + \\ jk \left(\frac{\rho}{R_a} \right)^2 Ei''[-jk(R_a + a - z)] \end{array} \right\} \tag{9-26}$$

$$\frac{\partial^2 A_F}{\partial z \partial \rho} = \frac{-jk\mu_0 \rho I_F e^{-jkz}}{4\pi} \left\{ \begin{array}{l} \dfrac{Ei'[-jk(R_b + b - z)]}{R_b} \left[-jk - \left(\dfrac{z-b}{R_b^2} \right)^2 \right] - \\ \dfrac{Ei'[-jk(R_a + a - z)]}{R_a} \left[-jk - \left(\dfrac{z-a}{R_a^2} \right) \right] + \\ \dfrac{jkEi''[-jk(R_b + b - z)]}{R_b} \left(\dfrac{R_b + b - z}{R_b} \right) - \\ \dfrac{jkEi''[-jk(R_a + a - z)]}{R_a} \left(\dfrac{R_a + a - z}{R_a} \right) \end{array} \right\} \tag{9-27}$$

In Eqs. 9-26 and 9-27, the derivatives of the exponential integral function are:

$$Ei'(x) = \frac{e^x}{x} \tag{9-28}$$

$$Ei''(x) = \frac{e^x}{x} \left(1 - \frac{1}{x} \right) \tag{9-29}$$

It is seen that the derivatives of $Ei(x)$ are easier to evaluate than the function itself. Fortunately, we need only the derivatives to compute the electromagnetic field.

To complete the description of the electromagnetic field, we must compute the contribution from the mirror image of the conducting segment. Fortunately, that will not require a lengthy repetition of the previous steps. We need only rewrite Eqs. 9-13 through 9-15 in terms of the mirror coordinate ζ. The current on the image is:

$$I(\zeta') = I_F e^{-jk\zeta'} \tag{9-30}$$

The magnetic vector potential of the currents flowing on the mirror image is:

$$A_{F,\text{image}} = \frac{\mu_0 I_F}{4\pi} \int_a^b \frac{e^{-jk(R+\zeta')}}{R} d\zeta' \tag{9-31}$$

In Eq. 9-31, the distance between the current and the observer is:

$$R = \sqrt{\rho^2 + (\zeta' + z)^2} \tag{9-32}$$

Comparing Eqs. 9-13 through 9-15 with Eqs. 9-30 through 9-32, it is seen that $A_{F,\text{image}}$ will be identical with A_F if we replace the observer coordinate z with $-z$, that is:

$$A_{F,\text{image}}(\rho,z) = A_F(\rho,-z) \tag{9-33}$$

In Eq. 9-33, we have explicitly written the dependence of the magnetic vector potentials on the observer coordinates ρ and z. A similar relationship can be shown between the derivatives of $A_{F,\text{image}}$ and A_F.

We are almost done. The total magnetic vector potential is:

$$A_{F\Sigma}(\rho,z) = A_F(\rho,z) + A_F(\rho,-z) \tag{9-34}$$

Similar formulas apply to the derivatives of the total magnetic vector potential:

$$\frac{\partial A_{F\Sigma}(\rho,z)}{\partial\rho} = \frac{\partial A_F(\rho,z)}{\partial\rho} + \frac{\partial A_F(\rho,-z)}{\partial\rho} \tag{9-35}$$

$$\frac{\partial^2 A_{F\Sigma}(\rho,z)}{\partial\rho^2} = \frac{\partial^2 A_F(\rho,z)}{\partial\rho^2} + \frac{\partial^2 A_F(\rho,-z)}{\partial\rho^2} \tag{9-36}$$

$$\frac{\partial^2 A_{F\Sigma}(\rho,z)}{\partial\rho\partial z} = \frac{\partial^2 A_F(\rho,z)}{\partial\rho\partial z} + \frac{\partial^2 A_F(\rho,-z)}{\partial\rho\partial z} \tag{9-37}$$

If we apply Eqs. 9-35 through 9-37 to Eqs. 9-8 , 9-11, and 9-12, then we will obtain the complete electromagnetic field of a forward traveling current wave and its image.

In the next section, we will obtain similar results for a reverse traveling current wave.

9.4 Complete Fields of Reverse Traveling Current Waves

The currents on folded unipole antennas are the superposition of forward and reverse traveling waves. In the previous section, we derived exact formulas for the magnetic vector potential $A_{F\Sigma}$ of the former. Those formulas, in turn, can be used to calculate the exact electromagnetic field of forward traveling waves. In this section, we will derive similar formulas for the magnetic vector potential $A_{R\Sigma}$ of reverse traveling waves of current. Then, we will have the complete mathematical description of the exact electromagnetic field of folded unipole, valid at any distance near or far.

To that end, Fig. 9-2 shows the geometry of interest in a fashion very similar to Fig. 9-1. The figure again illustrates a subtle point. That is, though the directions of propagation for the traveling wave currents are mirror images, the polarizations of those currents are not.

The mathematical steps will be nearly identical to those taken in the previous section, and so we can proceed with relative brevity. As before, we will focus on the conducting segment above ground, and then extend our results to its image in the ground.

The reverse traveling wave current on the segment above ground is:

$$I(z') = I_R e^{+jkz'} \tag{9-38}$$

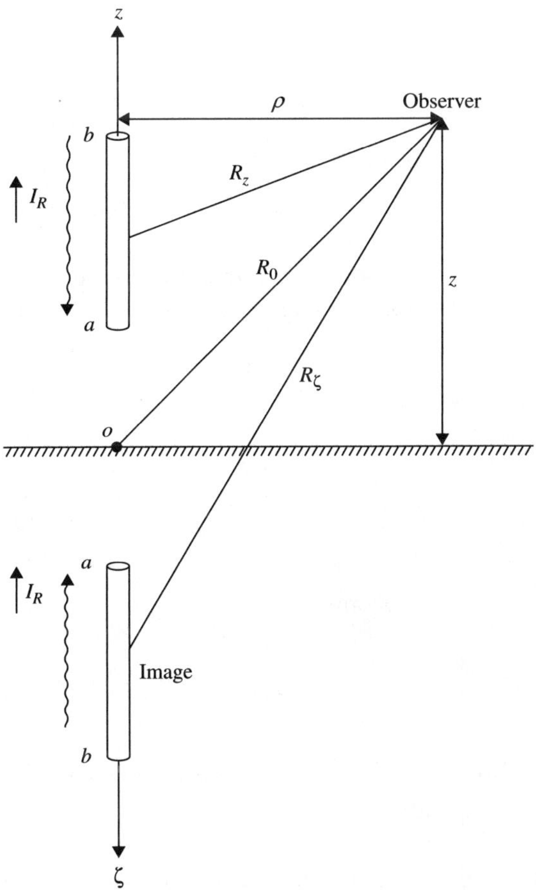

Figure 9-2 Essential features of a conducting segment and its image in a conducting ground, with reverse traveling current waves flowing on them. The directions of propagation are opposite, but the polarizations of the currents are the same.

Substituting Eq. 9-38 into Eq. 9-5, the magnetic vector potential of the reverse traveling wave current is:

$$A_R = \frac{\mu_0 I_R}{4\pi} \int_a^b \frac{e^{-jk(R-z')}}{R} dz' \tag{9-39}$$

In Eq. 9-39, the distance R is described by Eq. 9-15. As before, with some cleverness we can evaluate the integral on the right side of Eq. 9-39 exactly. We begin by making the change of variable,

$$w = R - z' + z \tag{9-40}$$

Equation 9-40 is similar to but not quite the same as Eq. 9-16. It follows that,

$$dz' = -\frac{R}{w} dw \tag{9-41}$$

Using Eqs. 9-40 and 9-41, Eq. 9-39 becomes:

$$A_R = -\frac{\mu_0 I_R e^{+jkz}}{4\pi} \int_{R_a - a + z}^{R_b - b + z} \frac{e^{-jkw}}{w} dw \tag{9-42}$$

In Eq. 9-42, the distances R_a and R_b are described by Eqs. 9-21 and 9-22.

One more change of variable will transform the integral in Eq. 9-42 into one that can be evaluated in terms of the classic exponential integral function $Ei(x)$.

$$w' = -jkw \tag{9-43}$$

It follows from Eq. 9-43 that,

$$\frac{dw}{w} = \frac{dw'}{w'} \tag{9-44}$$

Using Eqs. 9-43 and 9-44, Eq. 9-42 becomes:

$$A_R = -\frac{\mu_0 I_R e^{+jkz}}{4\pi} \int_{-jk(R_a - a + z)}^{-jk(R_b - b + z)} \frac{e^{w'}}{w'} dw' \tag{9-45}$$

As in the previous section, Eq. 9-45 can be rewritten in terms of the exponential integral $Ei(x)$:

$$A_R = -\frac{\mu_0 I_R e^{+jkz}}{4\pi} \{Ei[-jk(R_b - b + z)] - Ei[-jk(R_a - a + z)]\} \tag{9-46}$$

In. Eq. 9-46, $Ei(x)$ is described by Eq. 9-23. Equation 9-46 is analogous to Eq. 9-24 for forward traveling current waves. Comparing the two equations shows striking similarities, but also expected differences due to the different directions of propagation between forward and reverse traveling waves.

According to Eqs. 9-8, 9-11 and 9-12, we need the derivatives of A_R to calculate the electromagnetic field, not A_R itself. Those derivatives are:

$$\frac{\partial A_R}{\partial \rho} = \frac{jk\mu_0 \rho I_R e^{+jkz}}{4\pi}\left\{\frac{Ei'[-jk(R_b - b + z)]}{R_b} - \frac{Ei'[-jk(R_a - a + z)]}{R_a}\right\} \tag{9-47}$$

$$\frac{\partial^2 A_R}{\partial \rho^2} = \frac{jk\mu_0 I_R e^{+jkz}}{4\pi}\left\{\begin{array}{l} \dfrac{Ei'[-jk(R_b - b + z)]}{R_b}\left[1-\left(\dfrac{\rho}{R_b}\right)^2\right] \\ \dfrac{Ei'[-jk(R_a - a + z)]}{R_a}\left[1-\left(\dfrac{\rho}{R_a}\right)^2\right] - \\ jk\left(\dfrac{\rho}{R_b}\right)^2 Ei''[-jk(R_b - b + z)] + \\ jk\left(\dfrac{\rho}{R_a}\right)^2 Ei''[-jk(R_a - a + z)] \end{array}\right\} \tag{9-48}$$

$$\frac{\partial^2 A_R}{\partial z \partial \rho} = \frac{jk\mu_0 \rho I_R e^{+jkz}}{4\pi}\left\{\begin{array}{l} \dfrac{Ei'[-jk(R_b - b + z)]}{R_b}\left[jk-\left(\dfrac{z-b}{R_b^2}\right)\right] - \\ \dfrac{Ei'[-jk(R_a - a + z)]}{R_a}\left[jk-\left(\dfrac{z-a}{R_a^2}\right)\right] - \\ jk\dfrac{Ei''[-jk(R_b - b + z)]}{R_b}\left(\dfrac{R_b - b + z}{R_b}\right) + \\ jk\dfrac{Ei''[-jk(R_a - a + z)]}{R_a}\left(\dfrac{R_a - a + z}{R_a}\right) \end{array}\right\} \tag{9-49}$$

In Eqs. 9-47 through 9-49, the derivatives of the exponential integral function, $Ei'(x)$ and $Ei''(x)$, are described by Eqs. 9-28 and 9-29.

Equations 9-47 through 9-49 are analogous to Eqs. 9-25 through 9-27 for forward traveling current waves. Continuing the analogy, and using the same logic as in the previous section, the total magnetic vector potential of the conductor and its ground image is:

$$A_{R\Sigma}(\rho,z) = A_R(\rho,z) + A_R(\rho,-z) \tag{9-50}$$

The derivatives of the total magnetic vector potential are:

$$\frac{\partial A_{R\Sigma}(\rho,z)}{\partial\rho} = \frac{\partial A_R(\rho,z)}{\partial\rho} + \frac{\partial A_R(\rho,-z)}{\partial\rho} \tag{9-51}$$

$$\frac{\partial^2 A_{R\Sigma}(\rho,z)}{\partial\rho^2} = \frac{\partial^2 A_R(\rho,z)}{\partial\rho^2} + \frac{\partial^2 A_R(\rho,-z)}{\partial\rho^2} \tag{9-52}$$

$$\frac{\partial^2 A_{R\Sigma}(\rho,z)}{\partial z\partial\rho} = \frac{\partial^2 A_R(\rho,z)}{\partial z\partial\rho} + \frac{\partial^2 A_R(\rho,-z)}{\partial z\partial\rho} \tag{9-53}$$

Equations 9-50 through 9-53 are analogous to Eqs. 9-34 through 9-37 for forward traveling current waves.

Combined with the results of the previous section, we now have formulas for the magnetic vector potential of all common mode currents flowing on the folded unipole antenna:

$$A_{\text{total}} = A_{F\Sigma} + A_{R\Sigma} \tag{9-54}$$

Similar formulas describe the derivatives of A_{total}. Applying them to Eqs. 9-8, 9-11, and 9-12, we can calculate the exact electromagnetic field of the folded unipole antenna at any distance, near or far. These include potentially hazardous fields for antennas connected to high power transmitters.

9.5 Current Induced by the Exact Electromagnetic Field

In the previous three sections, much mathematical effort was dedicated to obtaining exact formulas for the electromagnetic field of folded unipole antennas. The main motivation was to evaluate the currents induced by each element of an array on all of the others. In this section, we carry out the next step by deriving a formula for the currents induced by the exact fields. It is our very good fortune that the formula will turn out to be simple and elegant.

For new results in electromagnetic theory, intuition can precede mathematical rigor or *vice versa*. When intuition takes the first step, measured data and/or numerical experiments focus our thoughts on a formula that seems plausible, if not compelling, and then careful mathematical steps verify the formula. Without intuition, we never would have stumbled across that formula. In contrast, when mathematical rigor takes the lead, we may randomly happen upon unexpected results while exploring the Maxwell Equations with an open mind. In the case of induced currents, intuition takes the first step.

In this section, we will derive a formula for induced currents using analogies and intuitive arguments first. After that we will rederive the formula using rigorous electromagnetic theory.

To that end, Fig. 9-3 shows how free space is a three-dimensional transmission line for electromagnetic fields. For simplicity in the drawing, only a two-dimensional slice is shown; however, it is easy to visualize the capacitance per unit length and inductance per unit length extending above and below the plane of the paper. The value of the capacitance per unit length is simply ε_0, the permittivity of free space. The value of the inductance per unit length is μ_0, the permeability of free space. The observables in this three-dimensional transmission line are the electric and magnetic fields. They are analogous to voltage and current, respectively, on the one-dimensional transmission line.

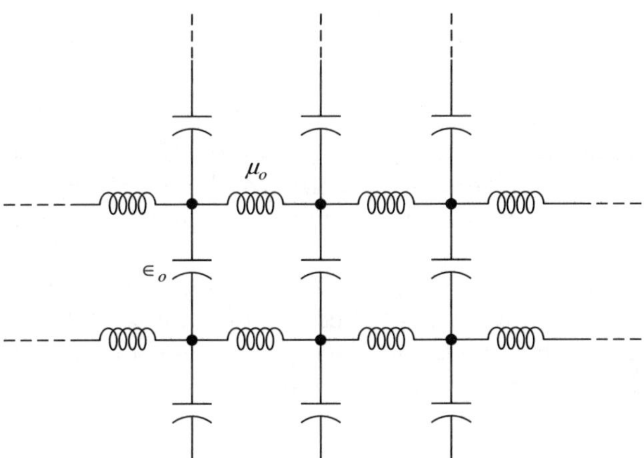

Figure 9-3 Free space functions as a three-dimensional transmission line for electromagnetic waves. The capacitance per unit length is the permittivity of free space, and the inductance per unit length is the permeability of free space.

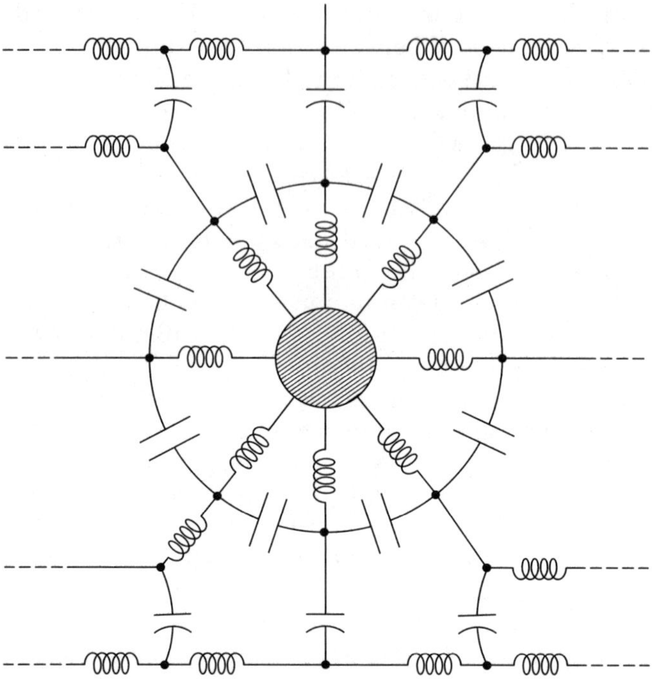

Figure 9-4 In the vicinity of a good conductor and other materials capable of carrying electric currents, the free space transmission line is distorted. Solutions of the Maxwell Equations for standard shapes and cross sections support this intuitive point of view.

Figure 9-4 shows how the three-dimensional transmission line becomes distorted in the vicinity of a good conductor. If we were to solve the Maxwell Equations rigorously for this conductor, we would find that the electromagnetic fields change direction qualitatively in the same way.

Figure 9-5 is a more detailed view of Fig. 9-4. We confine our attention to a one-radian arc of the conductor. That slice of conductor terminates a segment of the three-dimensional transmission line in a short circuit. The electric field in that segment is analogous to a voltage source. So, according to ordinary circuit theory:

$$E_z = -j\omega\mu_0 I \qquad (9\text{-}55)$$

On the right side of Eq. 9-55, a minus sign is necessary because the polarization of E_z is opposite to that of the voltage across the terminals of the transmission line segment.

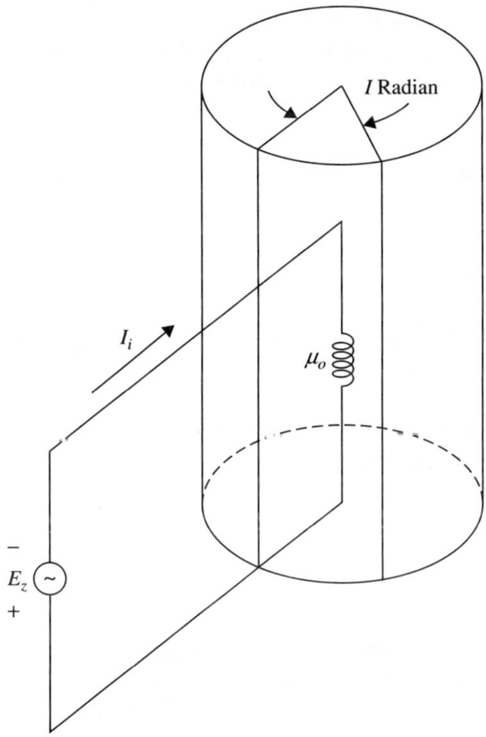

Figure 9-5 Using the free space transmission line analogy in the vicinity of a good conductor, current induced by the electric field flows through the inductance per unit length. The electric field can be regarded as a voltage source, and the permeability of free space is its output impedance.

If the conductor has a significant internal impedance per unit length z_{int}, then Eq. 9-55 must be modified to account for it:

$$E_z = -(j\omega\mu_0 + z_{\text{int}})I \tag{9-56}$$

As the conductor becomes infinitely thin, z_{int} becomes an open circuit, and the field does not distort in its vicinity. In that way, the formula satisfies our intuition because it will not permit current to be induced on an infinitely thin conductor. We will neglect z_{int} in our derivation, however, because we are not concerned with infinitely thin conductors.

Solving Eq. 9-55 for the induced current,

$$I = \frac{jE_z}{\omega\mu_0} \tag{9-57}$$

We explicitly performed this minor bit of algebra for an important reason. A dimensional analysis of the right side of Eq. 9-57 shows that the current I is not in units of amperes, but in units of amps/radian.

This is consistent with Fig. 9-5 in which we focused on a one-radian slice of the conductor. Evidently, we must replace I with a quantity measured in amps/radian. Formally, we can define the total induced current I_i as,

$$\frac{\partial I_i}{\partial \varphi} = I \tag{9-58}$$

Using Eq. 9-58, Eq. 9-57 becomes:

$$\frac{\partial I_i}{\partial \varphi} = \frac{jE_z}{\omega \mu_0} \tag{9-59}$$

In Eqn. 9-59, both sides have dimensions of amps/radian. To obtain the total induced current, we need only integrate with respect to φ:

$$I_i = \frac{j}{\omega \mu_0} \int_0^{2\pi} E_z \, d\varphi \tag{9-60}$$

Equation. 9-60 is a very general result. If the conductor has a cross section not small compared with a wavelength, then E_z will vary around its perimeter. That is, E_z is a function of φ, with respect to both phase and amplitude. For the antennas of interest here, however, that is not the case. So, E_z is a constant with respect to φ, and Eq. 9-60 becomes:

$$I_i = j \frac{2\pi}{\omega \mu_0} E_z \tag{9-61}$$

Equation 9-61 is the desired result, a formula relating induced current to an incident electromagnetic field. In the derivation, there was no restriction on the form of E_z. It might be a component of a plane wave originating at distant source, or it might be a component of the near field from a nearby antenna.

As explained earlier in this section, when a formula is derived intuitively, ultimately it must be validated with rigorous electromagnetic theory. To that end, we begin with the Maxwell Equation,

$$\nabla \times \mathbf{E} = -j\omega \mu_0 \mathbf{H} \tag{9-62}$$

In cylindrical coordinates (ρ, φ, z), and retaining only the nonzero components of **E** and **H**, Eq. 9-62 becomes:

$$\frac{\partial E_z}{\partial \rho} = j\omega \mu_0 H_\varphi \tag{9-63}$$

Let's integrate both sides of Eq. 9-63, from the center of the conductor at $\rho=0$, to its surface at $\rho=a$:

$$\int_0^a \frac{\partial E_z}{\partial \rho} d\rho = j\omega\mu_0 \int_0^a H_\varphi d\rho \tag{9-64}$$

The left side of Eq. 9-64 is a perfect differential, and so the equation becomes:

$$E_z(a) = -j\omega\mu_0 \int_0^a H_\varphi d\rho \tag{9-65}$$

since $E_z(0)=0$ inside the conductor. Inside the conductor, the magnetic field H_φ attenuates exponentially:

$$H_\varphi = H_\varphi(a) e^{\frac{(\rho-a)}{\delta}} \tag{9-66}$$

where δ is the skin depth. Using Eq. 9-66, Eq. 9-65 becomes:

$$E_z(a) = -j\omega\mu_0 H_\varphi(a) e^{-\frac{a}{\delta}} \int_0^a e^{\frac{\rho}{\delta}} d\rho \tag{9-67}$$

The integral on the right side of Eq. 9-67 is an elementary form that is readily revaluated so that the equation becomes:

$$E_z(a) = -j\omega\mu_0 \delta\, H_\varphi(a) \left(1 - e^{-\frac{a}{\delta}}\right) \tag{9-68}$$

At the surface of the conductor the magnetic field is related by the boundary conditions of Maxwell's Equations to a surface current density:

$$H_\varphi(a) = K_z \tag{9-69}$$

The surface current density is simply related to the total surface current by:

$$K_z = \frac{I_i}{2\pi a} \tag{9-70}$$

Using Eqs. 9-69 and 9-70, Eq. 9-68 becomes:

$$E_z(a) = -\frac{j\omega\mu_0}{2\pi} I_i \frac{\delta}{a} \left(1 - e^{-\frac{a}{\delta}}\right) \tag{9-71}$$

On the right side of Eq. 9-71, the expression in parentheses is subject to some useful approximations. For solid rods made from good conductors like copper or brass, the exponential term vanishes because the skin depth δ is only a millimeter or less. Aside from those cases, there are many practical antennas for which the approximation,

$$e^{-\frac{a}{\delta}} \approx 1 - \frac{a}{\delta} \tag{9-72}$$

is more appropriate. These certainly include AM broadcast antennas and reradiating structures such as communication towers and electric power poles. These structures are hollow lattices. Inside them there are measurable electromagnetic fields. Also, because the only conducting part is a latticework around the perimeter, these structures carry surface currents. Thus, we have a curious combination, from the point of view of the Maxwell Equations. We have a conductor that supports a surface current, and at the same time it supports a gently attenuating interior electromagnetic field. These are just the sort of configurations to which Eq. 9-72 applies, and Eq. 9-71 becomes:

$$E_z = -\frac{j\omega\mu_0}{2\pi} I_i \tag{9-73}$$

In Eq. 9-73 we have dropped explicit reference to the argument a in E_z.

From Eq. 9-73, it is a quick step to solve for the induced current I_i:

$$I_i = j\frac{2\pi}{\omega\mu_0} E_z \tag{9-74}$$

Equation 9-74 is identical to Eq. 9-61. Evidently, the Maxwell Equations and our intuition lead to the same result, at least for some very broad classes of radiating and reradiating structures.

For those structures to which Eq. 9-72 does not apply precisely, Eq. 9-74 can be modified by a coupling factor κ:

$$I_i = j\frac{2\pi}{\omega\kappa\mu_0} E_z \tag{9-75}$$

For the examples in this chapter, we will use Eq. 9-74, that is, Eq. 9-75 with $\kappa = 1$; however, we will revisit the coupling factor for other values in Sec. 9.13.

The intuitive process leading up to Eq. 9-61 is a very powerful one. It provided a useful visualization of how incident electromagnetic fields interact with conductors. That result was sufficient for our purposes in this chapter; however, it was only a starting point. For example, we could pursue the intuitive process further and extend the formulas to conductors that are clad with dielectric and magnetic materials.

In the next section, we will digress briefly to present a companion formula to Eq. 9-61.

9.6 Voltage Induced by the Exact Electromagnetic Field

In addition to current, an electromagnetic potential is also induced along each element in the array by all of the other elements. This potential is the same as common mode voltage. Calculating it is straightforward, beginning with the rigorous Lorentz Gage from electromagnetic theory:

$$\Phi(z) = \frac{j}{\omega\mu_0\varepsilon_0}\nabla\cdot\mathbf{A}(z) \tag{9-76}$$

For vertical conductors, the only nonzero component of $\mathbf{A}$ is A_z, and Eq. 9-76 becomes:

$$\Phi(z) = -\frac{1}{j\omega\varepsilon_0}\frac{1}{\mu_0}\frac{\partial A_z}{\partial z} \tag{9-77}$$

In Eq. 9-77, we have separated the expression $j\omega\varepsilon_0$ from the other terms to emphasize that it is the admittance per meter of free space. This suggests the physical interpretation shown in Fig. 9-6. It is seen that the weighted derivative of the magnetic vector potential $\mathbf{A}$ plays the role of a current source. In that sense, it is the *dual* of the electric field $\mathbf{E}$ in Fig. 9-5. The permeability μ_0 in Fig. 9-5 and the permittivity ε_0 in Fig. 9-6 are also duals of each other.

From Eq. 9-77, we see that it is necessary to evaluate the derivative of A_z for both forward and reverse traveling waves. From Eq. 9-24, the derivative of the forward traveling wave is:

$$\frac{\partial A_F}{\partial z} = -jkA_F - \frac{jk\mu_0 I_F e^{-jkz}}{4\pi}\left\{(z-b-R_b)\frac{Ei'[-jk(R_b+b-z)]}{R_b} - (z-a-R_a)\frac{Ei'[-jk(R_a+a-z)]}{R_a}\right\} \tag{9-78}$$

From Eq. 9-46, the derivative of the reverse traveling wave is:

$$\frac{\partial A_R}{\partial z} = -jkA_R + \frac{jk\mu_0 I_R e^{jkz}}{4\pi}\left\{(z-b+R_b)\frac{Ei'[-jk(R_b-b+z)]}{R_b} - (z-a+R_a)\frac{Ei'[-jk(R_a-a+z)]}{R_a}\right\} \tag{9-79}$$

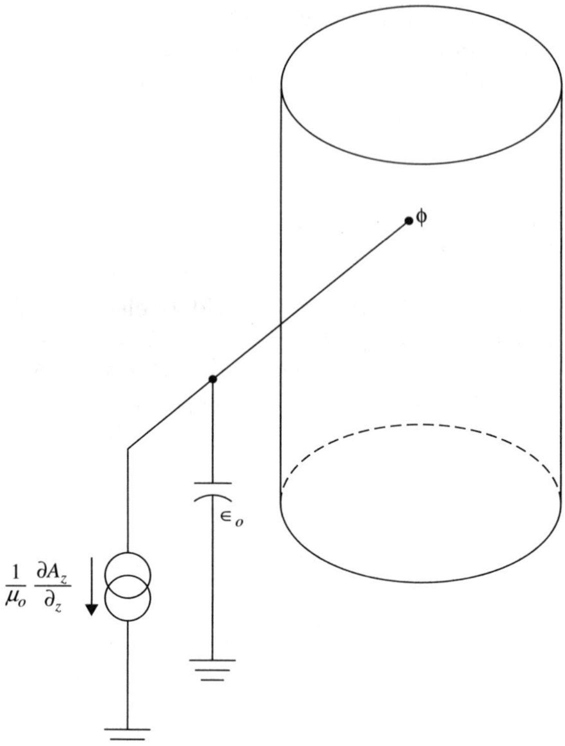

Figure 9-6 Continuing the free space transmission line analogy of Fig. 9-5, a common mode voltage is induced along the conductor by the magnetic vector potential A. The weighted derivative of A_z acts as a current source with an output admittance determined by the permittivity of free space.

Equations 9-78 and 9-79 describe the contribution from conductors above ground. We must add the contribution from their images below the ground as we did in Secs. 9.3 and 9.4, in similar fashion to Eqs. 9-34 through 9-37.

Finally, it is helpful to rewrite Eq. 9-77 so that the contribution to the potential $\mathbf{\Phi}$ from the mth element is separated from the contributions from all other elements in the array. That is, the potential $\mathbf{\Phi}_m$ along the mth array element is:

$$\Phi_m(z) = V_{\text{cm}}(z) + \frac{j}{\omega\mu_0\varepsilon_0}\sum_{\substack{n=1 \\ n\neq m}}^{N}\frac{\partial A_z}{\partial z} \tag{9-80}$$

On the right side of Eq. 9-80, the first term is the common mode voltage that would be observed if the mth element were the only one in the array. The second term is the voltage induced by the other elements.

Equation 9-80 will prove especially useful for mathematically modeling what happens at the base of each element in the array. It is readily seen that a voltage can occur across the base even in the absence of a hardwired source.

9.7 Electromagnetic Coupling Between Array Elements

In this section we bring together the results of Secs. 9.2 through 9.5. In Sec. 9.5, Eq. 9-75 tells us how the component E_z of the incident electromagnetic field induces a current I_i. In Sec. 9.2, Eq. 9-12 tells us how to compute E_z in terms of the magnetic vector potential A_z. In Sec. 9.3, Eqs. 9-35 and 9-36 tell us how to compute A_z for forward traveling current waves. In Sec. 9.4, Eqs. 9-51 and 9-52 tell us how to compute A_z for reverse traveling current waves. So, we have all the ingredients necessary to construct a single formula for the current induced on one antenna by the traveling waves of current on another antenna. We will derive that formula here.

In Sec. 9.6, we supplied the ingredients necessary for a companion formula describing the common mode voltage induced along an antenna by the traveling waves of current on another antenna. That formula will also be derived.

We begin by combining Eq. 9-75 and 9-12 to obtain:

$$I_i = -\frac{2\pi}{\omega^2 \mu_0^2 \kappa}\left(\frac{1}{\rho}\frac{\partial A_z}{\partial \rho} + \frac{\partial^2 A_z}{\partial \rho^2}\right) \tag{9-81}$$

Equation 9-81 can be rewritten in terms of the wave number k:

$$k = \omega\sqrt{\mu_0 \varepsilon_0} \tag{9-82}$$

Using Eq. 9-82, Eq. 9-81 becomes:

$$I_i = -\frac{2\pi}{k^2 \mu_0 \kappa}\left(\frac{1}{\rho}\frac{\partial A_z}{\partial \rho} + \frac{\partial^2 A_z}{\partial \rho^2}\right) \tag{9-83}$$

We can use Eq. 9-83 to find the current induced by a forward traveling current wave on another antenna. To that end, Eqs. 9-35 and 9-36 for the derivatives of the magnetic vector potential can be rewritten as:

$$\frac{\partial A_{F\Sigma}}{\partial \rho} = -\frac{jk\mu_0 I_F}{4\pi}\rho[\psi_F(z)P_F(\rho,z) + \psi_R(z)P_F(\rho,-z)] \tag{9-84}$$

$$\frac{\partial^2 A_{F\Sigma}}{\partial \rho^2} = -\frac{jk\mu_0 I_F}{4\pi}[\psi_F(z)Q_F(\rho,z) + \psi_R(z)Q_F(\rho,-z)] \tag{9-85}$$

In Eqs. 9-84 and 9-85, the functions P_F and Q_F are:

$$P_F(\rho,z)=\frac{Ei'[-jk(R_b+b-z)]}{R_b}-\frac{Ei'[-jk(R_a+a-z)]}{R_a} \tag{9-86}$$

$$\begin{aligned}Q_F(\rho,z)=&\frac{Ei'[-jk(R_b+b-z)]}{R_b}\left[1-\left(\frac{\rho}{R_b}\right)^2\right]-\frac{Ei'[-jk(R_a+a-z)]}{R_a}\left[1-\left(\frac{\rho}{R_a}\right)^2\right]\\&-jk\left(\frac{\rho}{R_b}\right)^2 Ei''[-jk(R_b+b-z)]+jk\left(\frac{\rho}{R_a}\right)^2 Ei''[-jk(R_a+a-z)]\end{aligned} \tag{9-87}$$

All of the terms and functions on the right sides of Eqs. 9-86 and 9-87 have been defined in Sec. 9.3. Equations 9-84 through 9-87 contain nothing that we have not already seen in Sec. 9.3. We have simply regrouped terms for the purpose of separating out the forward and reverse traveling wave functions, $\psi_F(z)$ and $\psi_R(z)$, respectively.

If we combine Eqs. 9-83 through 9-85, we finally obtain a single formula for the current induced by a forward traveling current wave with complex amplitude I_F:

$$I_i(z)=I_F[F(\rho,z)\psi_F(z)+G(\rho,z)\psi_R(z)] \tag{9-88}$$

In Eq. 9-88, the functions F and G are:

$$F(\rho,z)=\frac{j}{2k\kappa}[P_F(\rho,z)+Q_F(\rho,z)] \tag{9-89}$$

$$G(\rho,z)=\frac{j}{2k\kappa}[P_F(\rho,-z)+Q_F(\rho,-z)] \tag{9-90}$$

From Eq. 9-88, we see that a forward traveling current wave on one antenna induces both forward and reverse current waves on another antenna. This is not unexpected because the electromagnetic field of a forward traveling wave propagates away from its source in both directions.

We also see that the complex amplitude of the induced current is a function of the vertical coordinate z along the antenna. Again, this is not unexpected because different positions along the receiving antenna are at different distances from the source antenna.

A similar result to Eq. 9-88 can be derived for the case of a reverse traveling current wave with complex amplitude I_R on the source antenna. In Sec. 9.4, Eqs. 9-51 and 9-52 for the derivatives of $A_{R\Sigma}$ can be used together with Eq. 9-83 to obtain:

$$I_i(z)=I_R[R(\rho,z)\psi_R(z)+S(\rho,z)\psi_F(z)] \tag{9-91}$$

In Eq. 9-91 the functions R and S are:

$$R(\rho,z) = -\frac{j}{2k\kappa}[P_R(\rho,z)+Q_R(\rho,z)] \tag{9-92}$$

$$S(\rho,z) = -\frac{j}{2k\kappa}[P_R(\rho,-z)+Q_R(\rho,-z)] \tag{9-93}$$

Equations 9-92 and 9-93 are analogous to Eqs. 9-89 and 9-90 for forward traveling current waves. The functions P_R and Q_R are:

$$P_R(\rho,z) = \frac{Ei'[-jk(R_b-b+z)]}{R_b} - \frac{Ei'[-jk(R_a-a+z)]}{R_a} \tag{9-94}$$

$$\begin{aligned} Q_R(\rho,z) = {} & \frac{Ei'[-jk(R_b-b+z)]}{R_b}\left[1-\left(\frac{\rho}{R_b}\right)^2\right] - \frac{Ei'[-jk(R_a-a+z)]}{R_a}\left[1-\left(\frac{\rho}{R_a}\right)^2\right] \\ & - jk\left(\frac{\rho}{R_b}\right)^2 Ei''[-jk(R_b-b+z)] + jk\left(\frac{\rho}{R_a}\right)^2 Ei''[-jk(R_a-a+z)] \end{aligned} \tag{9-95}$$

Equations 9-94 and 9-95 are analogous to Eqs. 9-86 and 9-87. We have seen all of the terms on the right side of those equations previously in Sec. 9.4.

Observations similar to those made for Eq. 9-88 also apply to Eq. 9-91. A reverse traveling current wave on the source antenna induces both reverse and forward traveling waves of current on the receiving antenna. Along the z-coordinate of the latter, the complex amplitude (i.e., both magnitude and phase) of the induced current is not constant because the distance of each location from the source antenna is not constant.

We are almost done! To find the total induced current we simply combine Eqs. 9-88 and 9-91:

$$I_i(z) = [I_F F(\rho,z) + I_R S(\rho,z)]\psi_F(z) + [I_R R(\rho,z) + I_F G(\rho,z)]\psi_R(z) \tag{9-96}$$

Equation 9-96 is a relatively concise formula that tells us how forward and reverse traveling current waves, with complex amplitudes I_F and I_R, respectively, induce traveling waves of current on another conductor. The induced currents do not necessarily have constant amplitudes along the conductor. The conductor need not be in the far field zone of the traveling wave currents. It can be arbitrarily close by.

Equation 9-96 is a general result that applies to any wire conductor, not just to folded unipole antennas. In that sense, it was perhaps worth all the extra mathematical exertion!

In the later sections of this chapter, it will prove helpful to change the notation of Eq. 9-96. In order to keep track on the current induced on the mth antenna by the traveling waves on the nth antenna, let us write,

$$I_{\mathrm{im}}(z)=[I_{\mathrm{Fcn}}F_{\mathrm{nm}}(z)+I_{\mathrm{Rcn}}S_{\mathrm{nm}}(z)]\,\psi_F(z)+[I_{\mathrm{Rcn}}R_{\mathrm{nm}}(z)+I_{\mathrm{Fcn}}G_{\mathrm{nm}}(z)]\,\psi_R(z) \tag{9-97}$$

In Eq. 9-97,

$$F_{\mathrm{nm}}(z)=F(\rho_{\mathrm{nm}},z) \tag{9-98}$$

and similarly for S_{nm}, R_{nm}, and G_{nm}. The argument ρ_{mn} is the horizontal distance between the nth and mth antennas.

Before leaving this section, it will prove helpful to derive a companion formula to Eq. 9-97 for the common mode voltage induced along the antenna. As mentioned in Sec. 9.6, such a formula is especially useful for mathematically articulating boundary conditions at the base of the antenna. To that end, Eq. 9-78 can be readily rewritten as:

$$\frac{\partial A_F}{\partial z}=-\frac{jk\mu_0 I_F}{4\pi}\psi_F(z)W_F(\rho,z) \tag{9-99}$$

In Eq. 9-99, we have introduced a function analogous to the functions P_F and Q_F:

$$\begin{aligned}W_F(\rho,z)&=Ei[-jk(R_b+b-z)]-Ei[-jk(R_a+a-z)]-\\&\quad(z-b-R_b)\frac{Ei'[-jk(R_b+b-z)]}{R_b}+(z-a-R_a)\frac{Ei'[-jk(R_a+a-z)]}{R_a}\end{aligned} \tag{9-100}$$

In Eq. 9-100, the dependence on the horizontal distance ρ is implicit. While that variable does not appear explicitly in the equation, it does appear in the formulas for the distances R_a and R_b.

Using the arguments developed in Sec. 9.3, Eq. 9-99 is readily extended to include the potential from the image of the conductor in the ground:

$$\frac{\partial A_{F\Sigma}}{\partial z}=-\frac{jk\mu_0 I_F}{4\pi}[\psi_F(z)W_F(\rho,z)+\psi_R(z)W_F(\rho,-z)] \tag{9-101}$$

The analogous equation for reverse traveling current waves is derived using the same steps:

$$\frac{\partial A_{R\Sigma}}{\partial z}=-\frac{jk\mu_0 I_R}{4\pi}[\psi_R(z)W_R(\rho,z)+\psi_F(z)W_R(\rho,-z)] \tag{9-102}$$

In Eq. 9-102, we have introduced the function:

$$W_R(\rho,z) = Ei[-jk(R_b - b + z)] - Ei[-jk(R_a - a + z)] - (z - b - R_b)\frac{Ei'[-jk(R_b - b + z)]}{R_b} + (z - a - R_a)\frac{Ei'[-jk(R_a - a + z)]}{R_a} \tag{9-103}$$

As in Eq. 9-100, the dependence of W_R upon the horizontal distance ρ is implicitly contained within the distances R_a and R_b, according to Eqs. 9-21 and 9-22.

Using Eqs. 9-101 and 9-102, the second term on the right side of Eq. 9-80 can be written:

$$\Phi_i(z) = \frac{\eta}{4\pi}\{[I_F W_F(\rho,z) + I_R W_R(\rho,-z)]\psi_F(z) + [I_R W_R(\rho,z) + I_F W_F(\rho,-z)]\psi_R(z)\} \tag{9-104}$$

Equation 9-104 describes the common mode voltage induced on one antenna by another. Just as we changed the notation between Eqs. 9-96 and 9-97, we introduce the same notation to Eq. 9-104:

$$\Phi_{\mathrm{im}}(z) = \frac{\eta}{4\pi}\{[I_{\mathrm{Fcn}} W_{\mathrm{Fnm}}(z) + I_{\mathrm{Rcn}} W_{\mathrm{Rnm}}(-z)]\psi_F(z) + [I_{\mathrm{Rcn}} W_{\mathrm{Rnm}}(z) + I_{\mathrm{Fcn}} W_{\mathrm{Fnm}}(-z)]\psi_R(z)\} \tag{9-105}$$

In Eq. 9-105,

$$W_{\mathrm{Fnm}}(z) = W_F(\rho_{\mathrm{nm}}, z) \tag{9-106}$$

$$W_{\mathrm{Rnm}}(z) = W_R(\rho_{\mathrm{nm}}, z) \tag{9-107}$$

In Eqs. 9-106 and 9-107, and as in Eqn. 9-98, ρ_{nm} is the horizontal distance between the nth and mth antenna.

Equation 9-105 is a compact description of the common mode voltage induced along the mth antenna by the forward and reverse traveling current waves on the nth antenna.

With Equations 9-97 and 9-105 for the induced current and common mode voltage, we now have the essential mathematical tools required to analyze and design arrays of folded unipole antennas. In the following sections, we will apply those tools.

9.8 Transformation for the Radiated Electromagnetic Field

Before investigating some actual antenna arrays, we will briefly digress to derive some convenient formulas for their radiated electromagnetic fields.

In Secs. 9.2 through 9.4, we derived formulas for the exact electromagnetic fields of traveling waves of current; however, they were expressed in cylindrical components, that is, as E_ρ, E_z, and H_φ. At distances far from the array, we would prefer to express the fields in spherical components, E_R, E_θ, and H_φ. Of these, only the magnetic field is in the desired form. So, we must derive a transformation for the electric field.

It turns out that the transformation is simple, elegant, and worth documenting here. To that end, Fig. 9-7 shows the geometry of interest. It is seen that,

$$E_R = E_\rho \sin\theta + E_z \cos\theta \tag{9-108}$$

$$E_\theta = E_z \sin\theta + E_\rho \cos\theta \tag{9-109}$$

We could stop with Eqs. 9-108 and 9-109; however, if we take the derivation one more step, then results become more convenient. The trigonometric functions are:

$$\sin\theta = \frac{\rho}{\sqrt{\rho^2 + z^2}} \tag{9-110}$$

$$\cos\theta = \frac{z}{\sqrt{\rho^2 + z^2}} \tag{9-111}$$

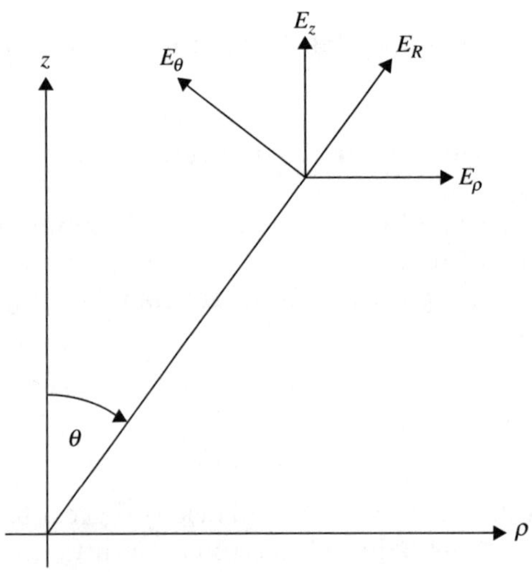

Figure 9-7 Geometry for transforming electromagnetic field components from a cylindrical coordinate system to a spherical coordinate system. The former system is useful for expressing near fields, that is, those close to the antenna or array. The latter system is most useful for expressing radiation patterns.

Using Eqs. 9-110 and 9-111, Eqs. 9-108 and 9-109 become:

$$E_R = \frac{\rho E_\rho + zE_z}{\sqrt{\rho^2 + z^2}} \tag{9-112}$$

$$E_\theta = \frac{\rho E_z - zE_\rho}{\sqrt{\rho^2 + z^2}} \tag{9-113}$$

9.9 An Array of Series-Fed Monopole Antennas

We begin our examples with a pair of series-fed monopoles, for at least three reasons: (1) they are the simplest introduction to the use of our new mathematical tools, (2) the results will be a good sanity check for our cumulative mathematical endeavors thus far in this chapter, and (3) the results will demonstrate the limitation of monopole arrays compared with arrays of folded unipoles.

Figure 9-8 shows the geometry of interest. It is seen that one monopole is active, that is, hardwired to a voltage source with amplitude V_1. The other is passive and terminated in an impedance Z_{B2}. They are separated by an arbitrary distance ρ_{12}, and they may have different heights h_1 and h_2.

There are only common mode currents on monopoles. On the active one, the current is the superposition of primary forward and reverse

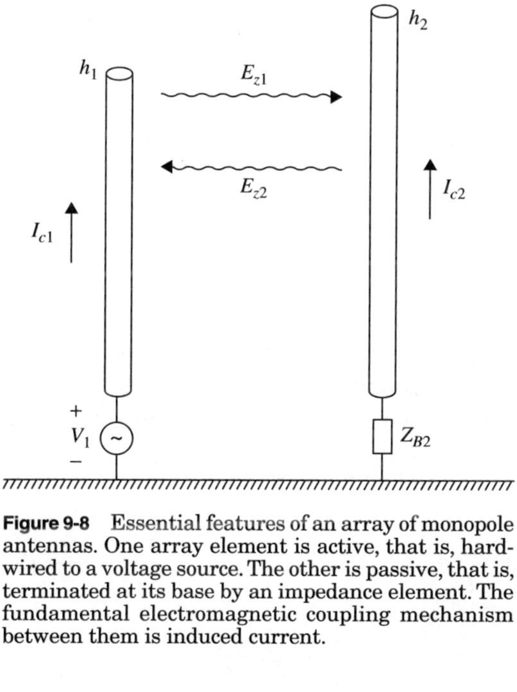

Figure 9-8 Essential features of an array of monopole antennas. One array element is active, that is, hardwired to a voltage source. The other is passive, that is, terminated at its base by an impedance element. The fundamental electromagnetic coupling mechanism between them is induced current.

traveling current waves, I_{Fc1} and I_{Rc1}, and induced currents from the passive monopole, according to Eq. 9-97:

$$I_{c1}(z)=[I_{\mathrm{Fc1}}+I_{\mathrm{Fc2}}F_{21}(z)+I_{\mathrm{Rc2}}S_{21}(z)]\,\psi_F(z)+[I_{\mathrm{Rc1}}+I_{\mathrm{Rc2}}R_{21}(z)+I_{\mathrm{Fc2}}G_{21}(z)]\,\psi_R(z) \tag{9-114}$$

On the passive monopole, the current is a similar superposition:

$$I_{c2}(z)=[I_{\mathrm{Fc2}}+I_{\mathrm{Fc1}}F_{12}(z)+I_{\mathrm{Rc1}}S_{12}(z)]\,\psi_F(z)+[I_{\mathrm{Rc2}}+I_{\mathrm{Rc1}}R_{12}(z)+I_{\mathrm{Fc1}}G_{12}(z)]\,\psi_R(z) \tag{9-115}$$

In Eqs. 9-114 and 9-115, the functions F_{21} and F_{12} are described in Eq. 9-98, and similarly for the functions S_{12}, R_{12}, G_{12}, and so forth.

The common mode voltages on the two monopoles are:

$$V_{c1}(z)=Z_{0c1}[I_{\mathrm{Fc1}}\psi_F(z)-I_{\mathrm{Rc1}}\psi_R(z)]+\Phi_{i1}(z) \tag{9-116}$$

$$V_{c2}(z)=Z_{0c2}[I_{\mathrm{Fc2}}\psi_F(z)-I_{\mathrm{Rc2}}\psi_R(z)]+\Phi_{i2}(z) \tag{9-117}$$

In Eqs. 9-114 through 9-117, there are four unknown complex current amplitudes, I_{Fc1}, I_{Rc1}, I_{Fc2}, and I_{Rc2}. To determine them, we must formulate four independent equations. Two of them are boundary conditions at the tops of the monopoles. There, if we neglect antenna end effect, the currents must vanish because they have no place to flow:

$$I_{c1}(h_1)=0 \tag{9-118}$$

$$I_{c2}(h_2)=0 \tag{9-119}$$

The remaining two independent equations are boundary conditions at the bases of the monpoles. At the base of the active monopole:

$$V_{c1}(0)=V_1 \tag{9-120}$$

$$V_{c2}(0)=-Z_{B2}I_{c2}(0) \tag{9-121}$$

Equations 9-118 through 9-121 are four independent equations. We need to determine the four unknown current amplitudes. As we have done many times earlier in previous chapters, the equations may be consolidated into a single matrix equation:

$$\mathbf{M}\cdot\begin{pmatrix} I_{\mathrm{Fc1}} \\ I_{\mathrm{Rc1}} \\ I_{\mathrm{Fc2}} \\ I_{\mathrm{Rc2}} \end{pmatrix}=\begin{pmatrix} V_1 \\ 0 \\ 0 \\ 0 \end{pmatrix} \tag{9-122}$$

In Eq. 9-122, the matrix **M** is:

$$\mathbf{M} = \left(\begin{array}{c|c|c|c} Z_{0c1} & -Z_{0c1} & \frac{\eta}{4\pi} W_{F21}(0) & \frac{\eta}{4\pi} W_{R21}(0) \\ \hline Z_{B2}[F_{12}(0)+G_{12}(0)] & Z_{B2}[S_{12}(0)+R_{12}(0)] & & \\ +\frac{\eta}{2\pi} W_{F12}(0) & +\frac{\eta}{2\pi} W_{R12}(0) & Z_{0c2}+Z_{B2} & -Z_{0c2}+Z_{B2} \\ \hline \psi_F(h_1) & \psi_R(h_1) & F_{21}(h_1)\psi_F(h_1) & S_{21}(h_1)\psi_F(h_1) \\ & & +G_{21}(h_1)\psi_R(h_1) & +R_{21}(h_1)\psi_R(h_1) \\ \hline F_{12}(h_2)\psi_F(h_2) & S_{12}(h_2)\psi_F(h_2) & \psi_F(h_2) & \psi_R(h_2) \\ +G_{12}(h_2)\psi_R(h_2) & +R_{12}(h_2)\psi_R(h_2) & & \end{array} \right) \tag{9-123}$$

While Eq. 9-123 may appear intimidating at first glance, in fact, it is a compact and efficient formulation of a two-element array. The matrix has dimensions of only 4×4. In contrast, a general purpose wire antenna numerical analysis computer program would have dimensions on the order of 100×100. That is, in terms of numbers of elements, our matrix is at least *three orders of magnitude smaller* than that of the general-purpose program. The advantages are the same as we have articulated in previous chapters, namely, faster solutions and less numerical noise. While these advantages might seem marginal for a two-element array, for larger arrays they will be substantial.

Furthermore, Eq. 9-123 is a complete formulation in the sense that its solutions provide us with all the information we need for this array. The solutions are the complete currents on both antennas. So, we have obtained the input currents $I_{c1}(0)$ and $I_{c2}(0)$ directly. We can also obtain the exact electromagnetic fields everywhere, near and far, according to Secs. 9.2 through 9.4. We can obtain the traditional far field radiation pattern, according to Sec. 9.8. We can obtain the input impedance, according to Sec. 2.10.

Figure 9-9 shows the radiation pattern for two quarter-wave monopoles, with quarter-wave spacing between them. The active monopole is at the origin of the coordinate system. The passive monopole is terminated in a short circuit. The common mode characteristic impedance for both is a "ballpark" value, $Z_{0c1} = Z_{0c2} = 1{,}600$ ohms. The pattern is what we might expect from an active antenna plus a reflector a quarter-wave behind it. The reflected field has a 180-degree phase shift during the round trip from the active antenna to the reflector. There is an additional 180-degree phase shift at the reflector itself. So, the reflected field is in phase in front of the array, and out of phase behind it. That is exactly what we see in the figure. Behind the array, the field is attenuated by almost 8 dB.

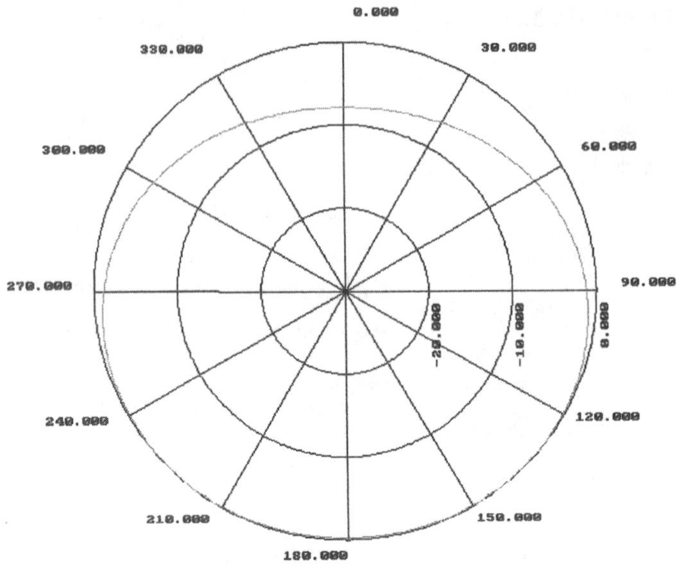

Figure 9-9 Electric field radiation pattern of a two-element monopole array. Separation between them is 90 degrees, or one-quarter wavelength. The passive element is terminated with a short circuit. The pattern shows that the passive element functions as a reflector as expected.

Figure 9-10 shows the radiation pattern for the same array, except that the passive element has been terminated in an open circuit.

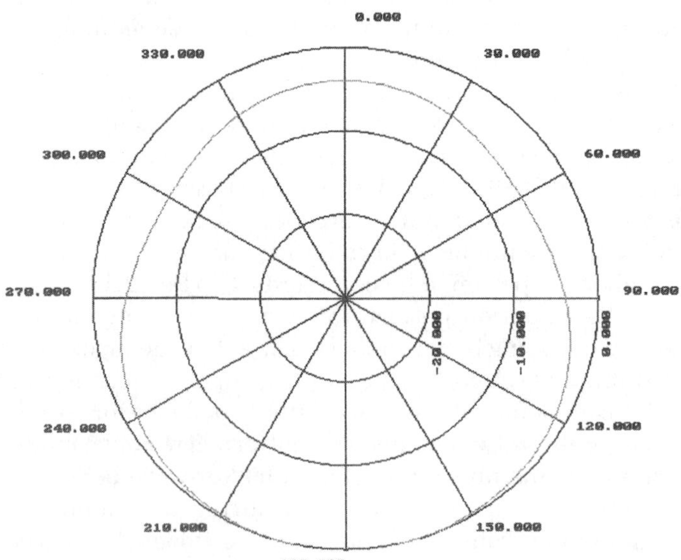

Figure 9-10 Pattern of the same two-element monopole array, except that the passive element is terminated with an open circuit. Because the separation between array elements is small compared with a wavelength, the passive element still functions noticeably as a reflector.

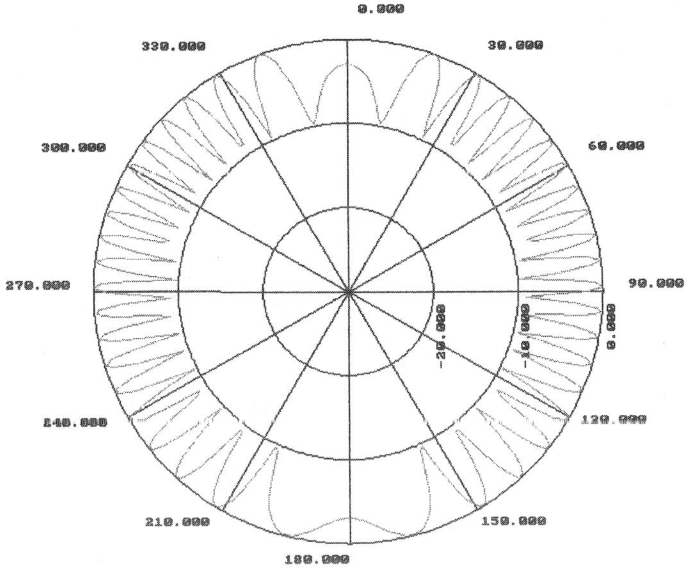

Figure 9-11 If the separation in the two-element monopole array is increased to 10 wavelengths, then an interferometric pattern is observed, as expected. The passive antenna is terminated in a short circuit, and so it is still a good scatterer of electromagnetic waves.

The conventional wisdom suggests that this should completely detune a quarter-wave monopole; however, we see that is not the case. A quarter-wave conductor in the near field intercepts and scatters a significant amount of electromagnetic energy, whether it is open or short-circuited at its base. Behind the array, the field is attenuated by about 4 dB. There is also a dimple at bearings of 60 and 300 degrees of almost 6 dB.

The difference between an open and short circuit termination is more distinctive when the passive element is in the far field of the active one. Figure 9-11 shows the resultant pattern when the separation is ten wavelengths, and the termination is a short circuit. There are multiple minima of almost –10 dB, or scalloping. The number of minima is consistent with the wide spacing between the array elements.

Figure 9-12 shows the resultant pattern when the termination is an open circuit. The separation is still ten wavelengths. Compared with Fig. 9-11, the minima are barely noticeable ripples, less than –1 dB. This agrees with the conventional wisdom that a quarter-wave monopole can be detuned by terminating it in an open circuit. We have learned from Fig. 9-10, however, that this conventional wisdom is true only if the passive element is sufficiently far away!

In Fig. 9-13, we vary the terminating reactance of the passive element and examine its effect on the field at a fixed bearing of 60 degrees, and, by symmetry, also at 300 degrees. It is seen that some sort of resonance

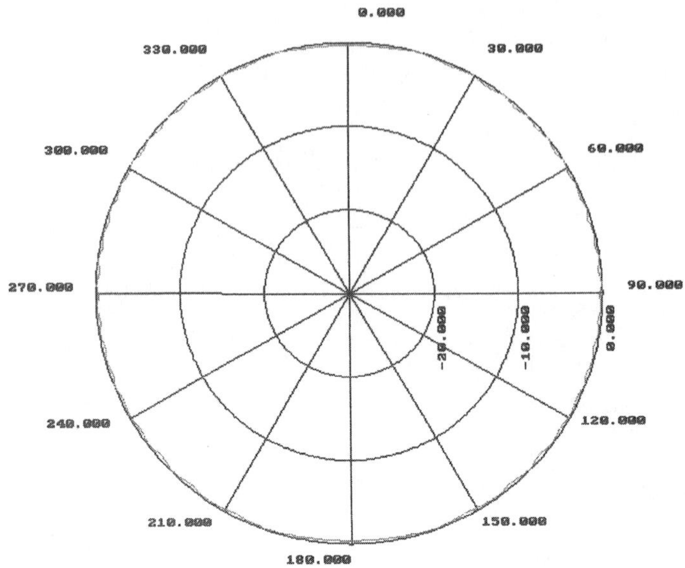

Figure 9-12 If the passive antenna is terminated in an open circuit, then it is a poor scatterer. For a separation of ten wavelengths, the pattern is determined almost entirely by the active antenna alone. The contribution from the passive antenna is less than 1 dB. The results agrees with the conventional wisdom that to detune a quarter-wave or shorter monopole, terminate it in an open circuit.

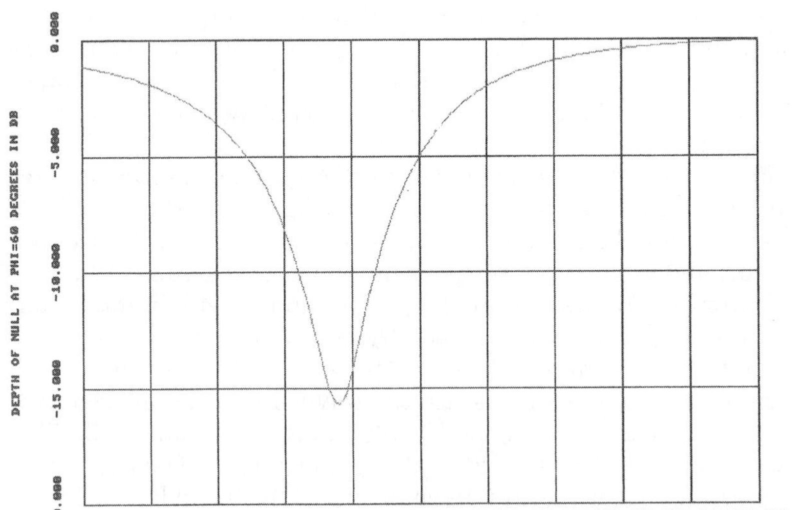

Figure 9-13 The passive element can be resonated to produce the most significant effect on the radiation pattern. The observer is at a fixed location, one mile distant, and at a bearing of 60 degrees, while the base terminating reactance is varied.

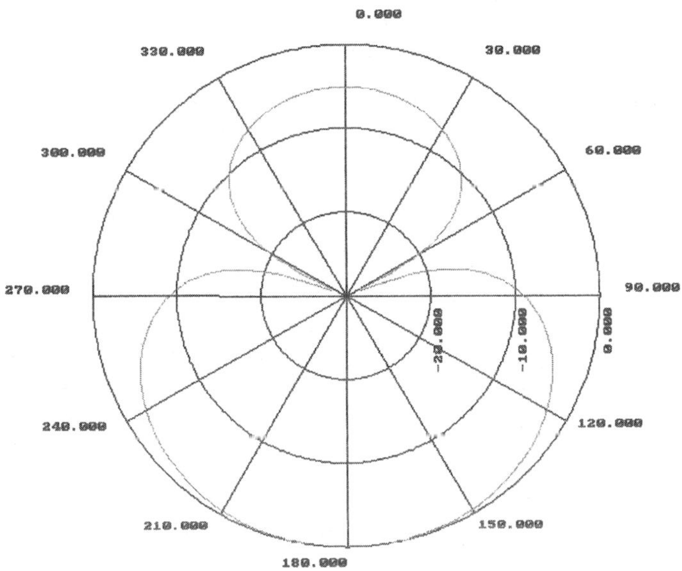

Figure 9-14 When the passive element is resonated, a highly directional radiation pattern results.

occurs for a terminating reactance of about –250 ohms. Figure 9-14 shows the radiation pattern when the passive antenna is terminated in that reactance. The nulls at 60 and 300 degrees are almost 30 dB deep.

Depending upon its spacing and terminating reactance, we have seen that a passive monopole as an element in an array can have a variety of effects. But unless it is located far from the active elements, it evidently cannot be detuned and still remain configured as a monopole.

9.10 An Array of Folded Unipole Antennas

The previous section served as a gentle introduction to our mathematical model of antenna arrays. In this section, we advance to the topic of main interest, namely, arrays of folded unipole antennas. We will begin with some simple examples. To that end, Fig. 9-15 shows a two-element array. Both elements are the most simple form of folded unipole. One element is active, and the other is passive.

In contrast to the monopole array of the previous section, the folded unipole array has twice as many currents because there are differential modes in addition to common modes. The common mode currents and voltages are still described by Eqs. 9-114 through 9-117. The differential mode currents and voltages are:

$$I_{\mathrm{dn}}(z) = I_{\mathrm{Fdn}}\psi_F(z) + I_{\mathrm{Rdn}}\psi_R(z) \qquad n = 1,2 \qquad (9\text{-}124)$$

$$V_{\mathrm{dn}}(z) = Z_{0\mathrm{dn}}[I_{\mathrm{Fdn}}\psi_F(z) - I_{\mathrm{Rdn}}\psi_R(z)] \qquad n = 1,2 \qquad (9\text{-}125)$$

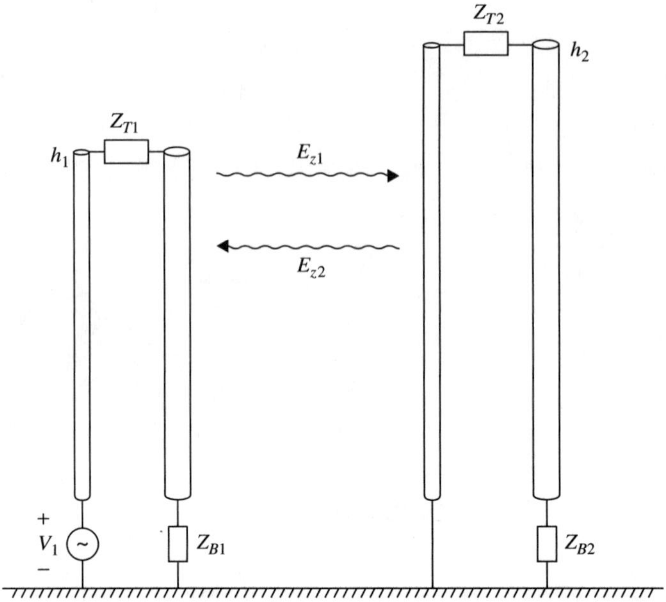

Figure 9-15 Essential features of an array of folded unipole antennas. One array element is active, that is, hardwired to a voltage source. The other is passive, that is, the fold is grounded. The fundamental electromagnetic coupling mechanism between them is induced current. Compared with the monopole array, we have four independently variable terminating impedances instead of only one. They will give us more flexibility for pattern design.

In Eqs. 9-114 through 9-117, and Eqs. 9-124 and 9-125, we have eight unknown complex current amplitudes. They are I_{Fcn}, I_{Rcn}, I_{Fdn}, and I_{Rdn}, for $n=1$ and $n=2$. To determine them, we must formulate eight independent equations. Four of them will be boundary conditions at the tops of the antennas, and four of them will boundary conditions at the bases.

At the tops of the antennas, neglecting antenna end effect, the common mode currents must vanish:

$$(1+\nu_1)I_{c1}(h_1)=0 \tag{9-126}$$

$$(1+\nu_2)I_{c2}(h_2)=0 \tag{9-127}$$

The differential mode voltages and currents are related by the terminating top impedances:

$$V_{d1}(h_1)=Z_{T1}I_{d1}(h_1) \tag{9-128}$$

$$V_{d2}(h_2)=Z_{T2}I_{d2}(h_2) \tag{9-129}$$

At the antenna bases, Kirchhoff's voltage law requires that:

$$[-I_{c1}(0)+I_{d1}(0)]Z_{B1}+V_{d1}(0)=V_{c1}(0) \tag{9-130}$$

$$[-I_{c2}(0)+I_{d2}(0)]Z_{B2}+V_{d2}(0)=0 \tag{9-131}$$

$$[-I_{c2}(0)+I_{d2}(0)]Z_{B2}+V_{d2}(0)=0 \tag{9-132}$$

$$V_{c2}(0)[-I_{c2}(0)+I_{d2}(0)]Z_{B2} \tag{9-133}$$

Equations 9-126 through 9-133 are eight independent equations. We need to find the eight unknown complex current amplitudes. The equations can be consolidated into a single matrix equation:

$$\mathbf{M}\cdot\begin{pmatrix}\mathbf{I}_1\\ \mathbf{I}_2\end{pmatrix}=\begin{pmatrix}V_1\\0\\0\\0\\0\\0\\0\\0\end{pmatrix} \tag{9-134}$$

In Eq. 9-134, the unknown current vectors are:

$$\mathbf{I_n}=\begin{pmatrix}I_{\text{Fcn}}\\ I_{\text{Rcn}}\\ I_{\text{Fdn}}\\ I_{\text{Rdn}}\end{pmatrix}\qquad n=1,2 \tag{9-135}$$

The matrix **M** has dimensions 8×8. For its elements to be legible on the printed page, it must be partitioned into four submatrices:

$$\mathbf{M}=\begin{pmatrix}\mathbf{A}_{11} & \mathbf{A}_{12}\\ \mathbf{A}_{21} & \mathbf{A}_{22}\end{pmatrix} \tag{9-136}$$

The submatrices are:

$$\mathbf{A}_{11} = \begin{pmatrix} Z_{0c1} & -Z_{0c1} & 0 & 0 \\ [F_{12}(0)+G_{12}(0)]Z_{B2} + \dfrac{\eta}{2\pi}W_{F12}(0) & [R_{12}(0)+S_{12}(0)]Z_{B2} + \dfrac{\eta}{2\pi}W_{R12}(0) & 0 & 0 \\ \psi_F(h_1) & \psi_R(h_1) & 0 & 0 \\ F_{12}(h_2)\psi_F(h_2) + G_{12}(h_2)\psi_R(h_2) & R_{12}(h_2)\psi_R(h_2) + S_{12}(h_2)\psi_F(h_2) & 0 & 0 \end{pmatrix} \tag{9-137}$$

$$\mathbf{A}_{12} = \begin{pmatrix} \dfrac{\eta}{2\pi}W_{F21}(0) & \dfrac{\eta}{2\pi}W_{R21}(0) & 0 & 0 \\ Z_{0c2}+Z_{B2} & -Z_{0c2}+Z_{B2} & -Z_{B2} & -Z_{B2} \\ F_{21}(h_1)\psi_F(h_1) + G_{21}(h_1)\psi_R(h_1) & R_{21}(h_1)\psi_R(h_1) + S_{21}(h_1)\psi_F(h_1) & 0 & 0 \\ \psi_F(h_2) & \psi_R(h_2) & 0 & 0 \end{pmatrix} \tag{9-138}$$

$$\mathbf{A}_{21} = \begin{pmatrix} -Z_{B1}-Z_{0c1} & -Z_{B1}+Z_{0c1} & Z_{B1}+Z_{0d1} & Z_{B1}-Z_{0d1} \\ -Z_{B2}F_{12}(0) - Z_{B2}G_{12}(0) & -Z_{B2}R_{12}(0) - Z_{B2}S_{12}(0) & 0 & 0 \\ 0 & 0 & (Z_{0d1}-Z_{T1})\psi_F(h_1) & (-Z_{0d1}-Z_{T1})\psi_R(h_1) \\ 0 & 0 & 0 & 0 \end{pmatrix} \tag{9-139}$$

$$\mathbf{A}_{22} = \begin{pmatrix} -Z_{B1}F_{21}(0) - Z_{B1}G_{21}(0) & -Z_{B1}R_{21}(0) - Z_{B1}S_{21}(0) & 0 & 0 \\ -Z_{B2} & -Z_{B2} & Z_{B2}+Z_{0d2} & Z_{B2}-Z_{0d2} \\ 0 & 0 & 0 & 0 \\ 0 & 0 & (Z_{0d2}-Z_{T2})\psi_F(h_2) & (-Z_{0d2}-Z_{T2})\psi_R(h_2) \end{pmatrix} \tag{9-140}$$

Formulating the elements in Eqs. 9-137 through 9-140 may seem like a lot of tedious work for just a two-element array; however, a repetitive procedure is being established. In the next section, we will generalize our matrix equation Eq. 9-134 to an arbitrary number of N elements.

For now, however, let's examine some examples that result from solving the matrix equation. Figures 9-16 through 9-19 show the radiation patterns for the four possible combinations of extreme impedance terminations (i.e., open circuit or short circuit) for the top and base of the passive antenna. Both active and passive antennas are one quarter-wave tall, and they are spaced a quarter-wave apart.

Figures 9-16 and 9-17 are practically indistinguishable from each other, and from Fig. 9-10. Evidently, when the base termination of the passive element is an open circuit, it does not matter whether it is a monopole or a folded unipole.

Figures 9-18 and 9-19 show the radiation pattern when the base termination of the passive element is a short circuit. In Fig. 9-18 the top termination is a short circuit. In Fig. 9-19, it is an open circuit. Comparing Fig. 9-18 with Fig. 9-9 for the monopole array, we see no difference in the radiation pattern. Evidently, when the top termination is a short circuit, it does not matter whether the element is a monopole or folded unipole.

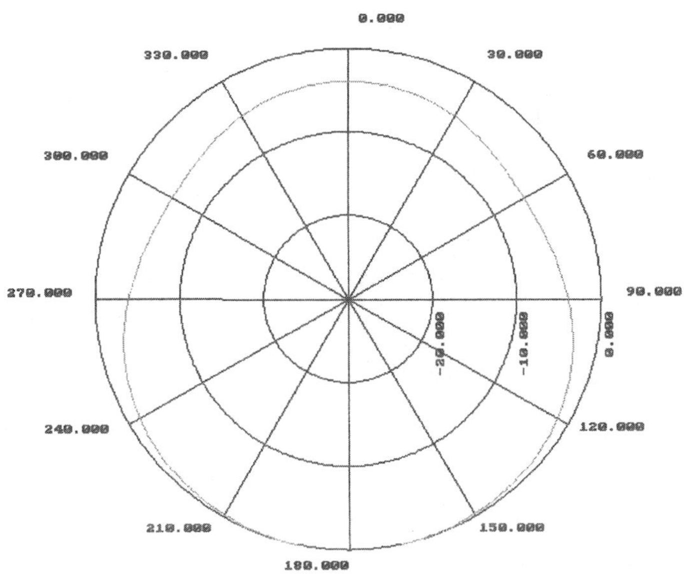

Figure 9-16 Resulting pattern when the passive unipole is terminated with a short circuit at the top and an open circuit at the base. The active unipole is terminated with short circuits. The pattern is practically identical to that of the monopole array when the passive element is also terminated with an open circuit at its base.

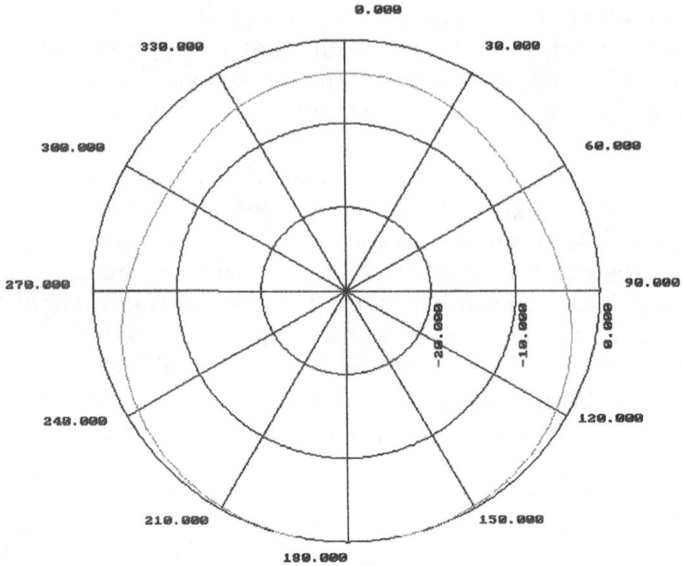

Figure 9-17 Resulting pattern when the passive unipole is terminated with open circuits at both top and base.

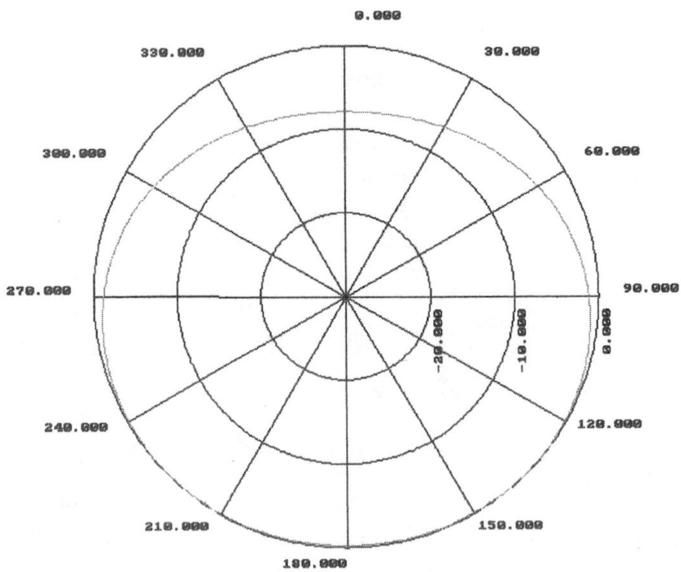

Figure 9-18 Resulting pattern when the passive unipole is terminated with short circuits at both top and base. The result is practically identical to the pattern of a monopole array when the passive element is terminated with a short circuit at its base.

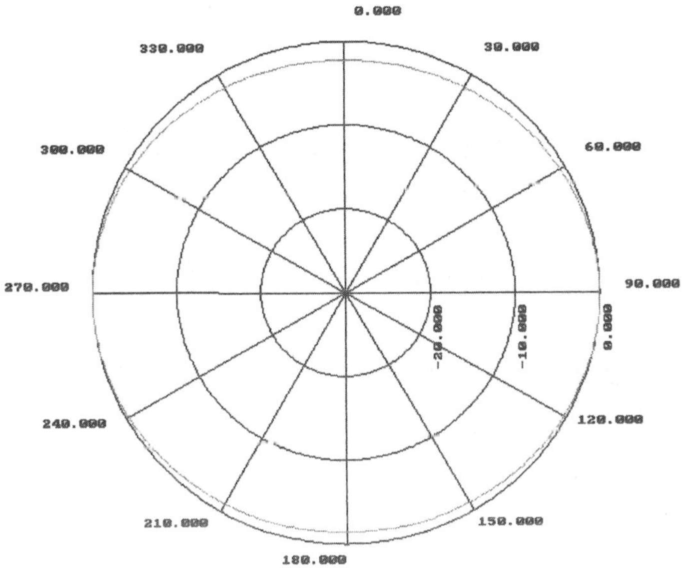

Figure 9-19 Resulting pattern when the passive unipole is terminated with an open circuit at the top and a short circuit at the base. It is almost detuned, a result we could not obtain with the monopole array when the elements were separated by only 90 electrical degrees.

The pattern of Fig. 9-19, however, is much more interesting. The base termination is a short circuit, and the top termination is open. As a result, the passive element is more detuned than any example we have seen thus far. So, a folded unipole array element does make a difference after all!

Can we do any better than Fig. 9-19 with respect to detuning? Fig. 9-20 shows the field at fixed bearing of 60 degrees, and by symmetry also at 300 degrees, while the top terminating reactance is varied. The base terminating impedance is an open circuit. We see a distinct minimum for $X_{T2}=0$. As X_{T2} becomes increasingly negative, the field reaches a maximum and stays there. Fig. 9-21 shows the radiation pattern for $X_{T2}=-100$ ohms. Consistent with Fig. 9-20, the field is not attenuated much at those bearings. Moreover, the passive element appears to be even more detuned than in Fig. 9-19.

Can we obtain the opposite of detuning, that is, a deep null at those bearings? Figure 9-22 is a contour plot that addresses that question. The two independent variables are the top and base terminating reactances. The dark blue contour is the deepest null. We see that there are many combinations of X_{B2} and X_{T2} that provide a deep null. Figure 9-23 is the radiation pattern the results from just one of those combinations. It is practically indistinguishable from Fig. 9-14 for the monopole array.

A simple two-element array of folded unipoles can provide any radiation pattern obtainable from a two-element monopole array, and then some!

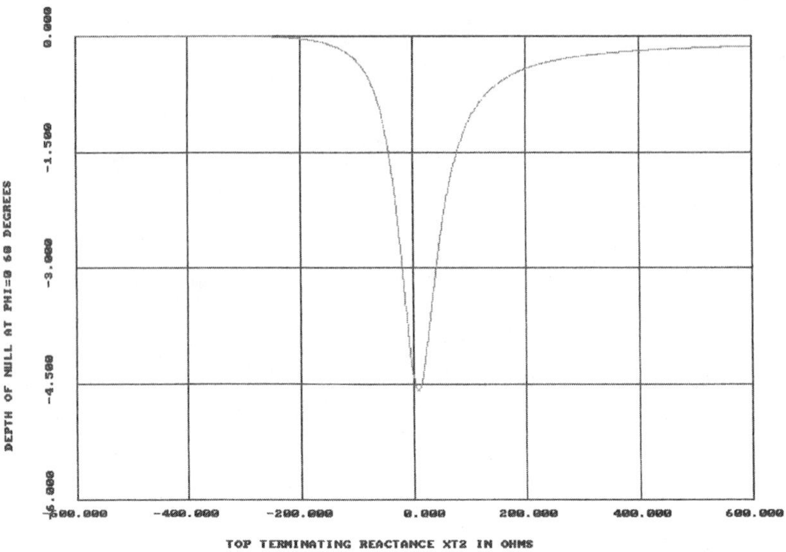

Figure 9-20 We can further detune the passive unipole with a more optimal choice of top terminating reactance. In this graph, the observer is at a fixed location, one mile away and at a bearing of 60 degrees, while the top terminating reactance is varied. It is seen that the destructive wave interference vanishes as the reactance becomes more negative.

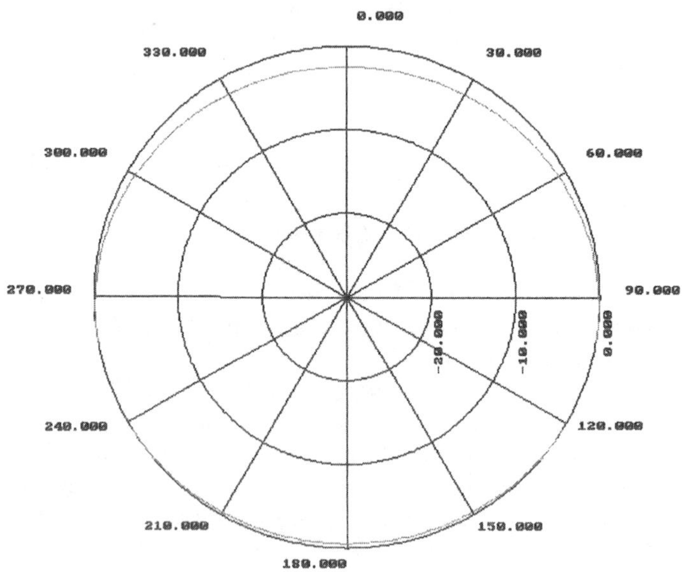

Figure 9-21 Resulting pattern when the passive unipole is terminated in $-j100$ ohms at the top and a short circuit at the base. It is even more detuned than when the top was terminated in an open circuit.

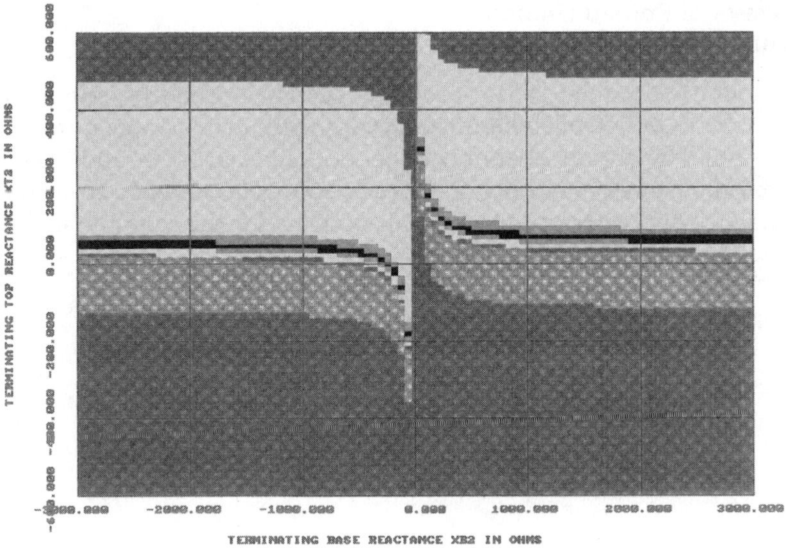

Figure 9-22 For this plot, the observer is at a fixed location, one mile away and at a bearing of 60 degrees, while both terminations of the passive antenna are continuously varied. The darkest contour shows the deepest null in the observed radiation pattern. There are many combinations that will produce that null. Note: See Preface, Notes Concerning the Examples for explanation of the gray scale which has been subsituted for color.

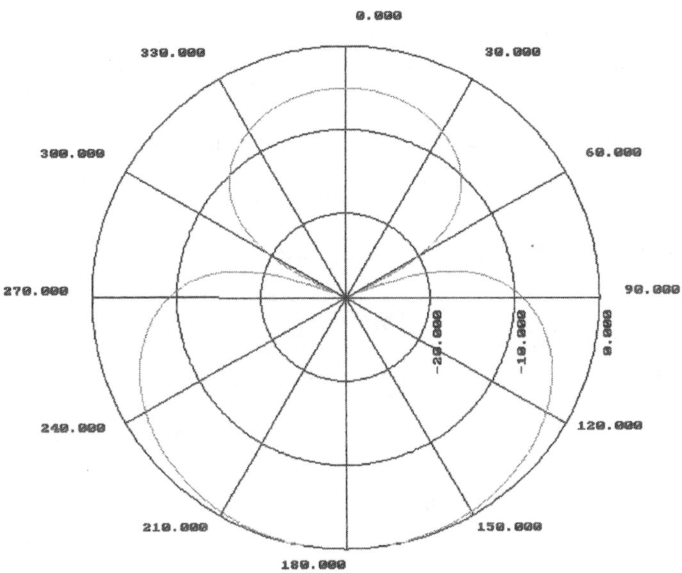

Figure 9-23 The resulting pattern when the passive unipole is terminated in a short circuit at the top, and with $-j250$ ohms at the base.

9.11 Arrays of Folded Unipoles with *N* Elements

The mathematical model of the previous section is readily extended from arrays with only two elements to arrays with an arbitrary number of *N* elements. The essential step is to generalize Eqs. 9-114 and 9-115 for the common mode currents to include induced currents from all other elements in the array. The general equation for the common mode current on the *m*th element is:

$$I_{\text{cm}}(z) = \left\{ I_{\text{Fcm}} + \sum_{n=1, n\neq m}^{N} [I_{\text{Fcn}} F_{\text{nm}}(z) + I_{\text{Rcn}} S_{\text{nm}}(z)] \right\} \psi_F(z) + \left\{ I_{\text{Rcm}} + \sum_{n=1, n\neq m}^{N} [I_{\text{Rcn}} R_{\text{nm}}(z) + I_{\text{Fcn}} G_{\text{nm}}(z)] \right\} \psi_R(z) \tag{9-141}$$

In Eq. 9-141, the terms within brackets are the induced currents from all of the other elements in the array. The functions $F_{\text{nm}}(z)$ are described in Eq. 9-98, and similarly for $S_{\text{nm}}(z)$, $R_{\text{nm}}(z)$, and $G_{\text{nm}}(z)$.

Equations 9-116 and 9-117 for the common mode voltage can also be generalized. The common mode voltage along the *m*th element is:

$$V_{\text{cm}}(z) = \left\{ Z_{0\text{cm}} I_{\text{Fcm}} + \frac{\eta}{4\pi} \sum_{n=1, n\neq m}^{N} [I_{\text{Fcn}} W_{\text{Fnm}}(z) + I_{\text{Rcn}} W_{\text{Rnm}}(-z)] \right\} \psi_F(z) + \left\{ -Z_{0\text{cm}} I_{\text{Rcm}} + \frac{\eta}{4\pi} \sum_{n=1, n\neq m}^{N} [I_{\text{Rcn}} W_{\text{Rnm}}(z) + I_{\text{Fcn}} W_{\text{Fnm}}(-z)] \right\} \psi_R(z) \tag{9-142}$$

The matrix Eq. 9-134 can be generalized to be:

$$\mathbf{M} \cdot \begin{pmatrix} \mathbf{I}_1 \\ \vdots \\ \mathbf{I}_\mathbf{n} \\ \vdots \\ \mathbf{I}_\mathbf{N} \end{pmatrix} = \begin{pmatrix} \mathbf{V} \\ \vdots \\ 0 \\ \vdots \\ 0 \end{pmatrix} \tag{9-143}$$

In Eq. 9-143, the unknown current vectors are:

$$\mathbf{I}_\mathbf{n} = \begin{pmatrix} I_{\text{Fcn}} \\ I_{\text{Rcn}} \\ I_{\text{Fdn}} \\ I_{\text{Rdn}} \end{pmatrix} \qquad n = 1, 2, \ldots n, \ldots N \tag{9-144}$$

On the right side of Eq. 9-143, the subvector **V** contains the complex amplitudes (i.e., magnitudes and phases) of the voltage sources hardwired to each of the N elements of the array:

$$\mathbf{V} = \begin{pmatrix} V_1 \\ \vdots \\ V_n \\ \vdots \\ V_N \end{pmatrix} \qquad n = 1,2,\ldots n,\ldots N \tag{9-145}$$

On the right side of Eq. 9-145, V_n=0 for passive antennas, that is, for antennas in the array that are not hardwired to a voltage source.

For generalizing the matrix **M**, it is conceptually convenient to partition it into four submatrices:

$$\mathbf{M} = \begin{pmatrix} \leftarrow & \mathbf{A} & \rightarrow \\ \hline \leftarrow & \mathbf{B} & \rightarrow \\ \hline \leftarrow & \mathbf{C} & \rightarrow \\ \hline \leftarrow & \mathbf{D} & \rightarrow \end{pmatrix} \tag{9-146}$$

Each of the submatrices has N rows and $4N$ columns.

The submatrix **A** results from generalizing Eq. 9-132, a boundary condition at the base of each antenna in the array. Each row m of **A** may be written as a row vector:

$$\mathbf{A_m} = (\cdots\ \mathbf{a_{mn}}\ \cdots\ \mathbf{a_{mm}}\ \cdots\ \mathbf{a_{mn}}\ \cdots) \qquad n = 1,5,9,\ldots,4N-3 \tag{9-147}$$

On the right side of Eq. 9-147, the $\mathbf{a_{mn}}$ and $\mathbf{a_{mm}}$ are four-element row vectors. The diagonal row vector is:

$$\mathbf{a_{mm}} = \left(Z_{0\mathrm{cm}} \;\vdots\; -Z_{0\mathrm{cm}} \;\vdots\; 0 \;\vdots\; 0 \right) \tag{9-148}$$

The off-diagonal row vectors are:

$$\mathbf{a_{mn}} = \left(\frac{\eta}{2\pi} W_{\mathrm{Fnm}}(0) \;\vdots\; \frac{\eta}{2\pi} W_{\mathrm{Rnm}}(0) \;\vdots\; 0 \;\vdots\; 0 \right) \tag{9-149}$$

The submatrix **B** results from generalizing Eq. 9-126, a boundary condition on the common mode current at the top of each antenna in the array. Each row m of **B** may be written as a row vector:

$$\mathbf{B_m} = (\cdots\ \mathbf{b_{mn}}\ \cdots\ \mathbf{b_{mm}}\ \cdots\ \mathbf{b_{mn}}\ \cdots) \qquad n = 1,5,9,\ldots,4N-3 \tag{9-150}$$

On the right side of Eq. 9-150, the diagonal elements are:

$$\mathbf{b}_{\mathbf{mm}} = \left(\psi_F(h_m) \;\vdots\; \psi_R(h_m) \;\vdots\; 0 \;\vdots\; 0 \right) \tag{9-151}$$

The off-diagonal elements are:

$$\mathbf{b}_{\mathbf{mn}} = \begin{pmatrix} F_{\text{nm}}(h_m)\psi_F(h_m) & R_{\text{nm}}(h_m)\psi_R(h_m) & & \\ +G_{\text{nm}}(h_m)\psi_R(h_m) & +S_{\text{nm}}(h_m)\psi_F(h_m) & 0 & 0 \end{pmatrix} \tag{9-152}$$

The submatrix **C** results from generalizing Eq. 9-130, a boundary condition required by KVL at the base of each antenna in the array. Each row m of **C** may be written as a row vector:

$$\mathbf{C}_{\mathbf{m}} = (\cdots\ \mathbf{c}_{\mathbf{mn}} \cdots\ \mathbf{c}_{\mathbf{mm}} \cdots\ \mathbf{c}_{\mathbf{mn}} \cdots) \qquad n = 1,5,9,\ldots,4N-3 \tag{9-153}$$

On the right side of Eq. 9-153, the diagonal subvectors are:

$$\mathbf{c}_{\mathbf{mm}} = \left(-Z_{\text{Bm}} - Z_{\text{0cm}} \;\vdots\; -Z_{\text{Bm}} + Z_{\text{0cm}} \;\vdots\; Z_{\text{Bm}} + Z_{\text{0dm}} \;\vdots\; Z_{\text{Bm}} - Z_{\text{0dm}} \right) \tag{9-154}$$

The off-diagonal subvectors are:

$$\mathbf{c}_{\mathbf{mn}} = \begin{pmatrix} -Z_{\text{Bm}}[F_{\text{nm}}(0) + G_{\text{nm}}(0)] & -Z_{\text{Bm}}[R_{\text{nm}}(0) + S_{\text{nm}}(0)] & & \\ -\dfrac{\eta}{2\pi} W_{\text{Fnm}}(0) & -\dfrac{\eta}{2\pi} W_{\text{Rnm}}(0) & 0 & 0 \end{pmatrix} \tag{9-155}$$

Finally, the submatrix **D** results from generalizing Eq. 9-128, a boundary condition on the differential mode current required by the constituitive relation at the top of each antenna in the array:

$$\mathbf{D}_{\mathbf{m}} = (\cdots\ \mathbf{d}_{\mathbf{mn}} \cdots\ \mathbf{d}_{\mathbf{mm}} \cdots\ \mathbf{d}_{\mathbf{mn}} \cdots) \qquad n = 1, 5, 9,\ldots,4N-3 \tag{9-156}$$

On the right side of Eq. 9-156, the off-diagonal subvectors are:

$$\mathbf{d}_{\mathbf{mn}} = (\mathbf{0} \;\vdots\; \mathbf{0} \;\vdots\; \mathbf{0} \;\vdots\; \mathbf{0}) \tag{9-157}$$

The diagonal subvectors are:

$$\mathbf{d}_{\mathbf{mm}} = \left(0 \;\vdots\; 0 \;\vdots\; (Z_{\text{0dm}} - Z_{\text{Tm}})\psi_F(h_m) \;\vdots\; (-Z_{\text{0dm}} - Z_{\text{Tm}})\psi_R(h_m) \right) \tag{9-158}$$

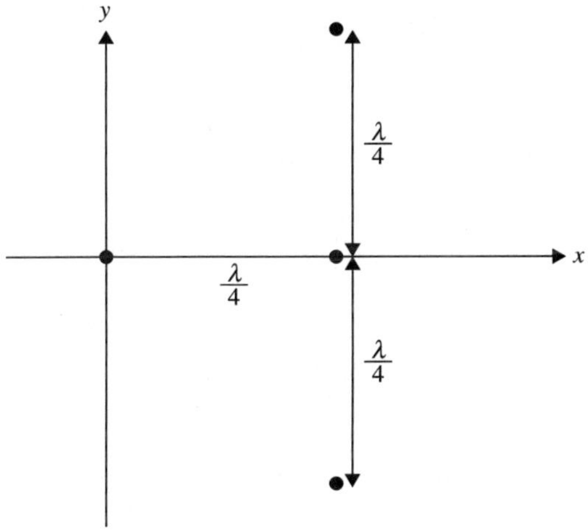

Figure 9-24 Essential features of an array of four folded unipole antennas. The array element at the origin of the coordinate system is active, that is, hard-wired to a voltage source. The other three are passive, that is, the folds are grounded. The fundamental electromagnetic coupling mechanism between all elements is induced current. There are eight independently variable terminating impedances, providing considerable pattern design flexibility.

Equations 9-146 through 9-158 show us how to construct a matrix equation for an arbitrarily large array of folded unipole antennas. In the remainder of this section, we will investigate some examples.

Figure 9-24 shows a four-element array. One active element is at the origin of the coordinate system. The remaining elements form a line of equally spaced passive elements. All elements are 90 electrical degrees (i.e., one-quarter wavelength) tall. The common mode characteristic impedances are 1,200 ohms. The differential mode characteristic impedances are 120 ohms. The current transformation ratios are 0.5. The active element is terminated with short circuits at both top and base.

Figures 9-25 through 9-28 show the radiation patterns for the four combinations of extreme terminations (i.e., short circuit or open circuit) when all three passive elements are identically terminated. It is seen that the patterns are the same except for Fig. 9-28. In that case, all bases and tops are terminated in short circuits.

Let's see if we can detune the line of passive elements. For example, they might be electric power poles interfering with a nondirectional broadcast antenna. In the previous section, we found that if the passive element was terminated in a short circuit at the base, and a reactance of

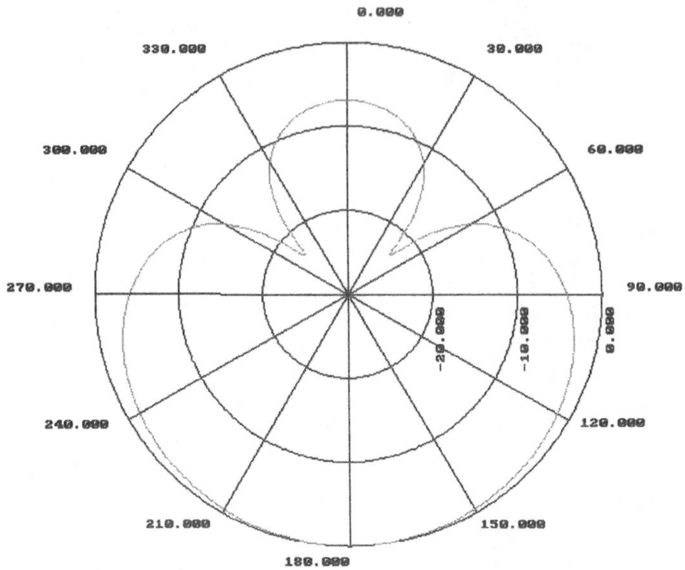

Figure 9-25 Resulting pattern when all three passive unipoles are terminated in the same way. The tops are terminated in short circuits. The bases are terminated with open circuits.

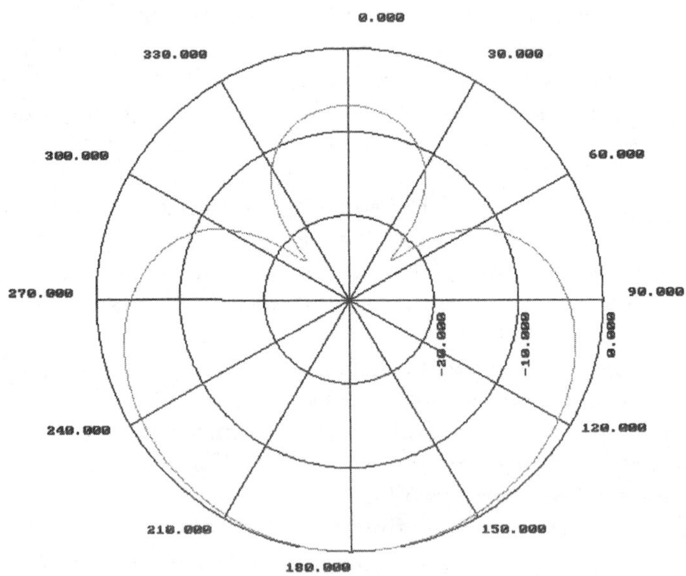

Figure 9-26 Resulting pattern when all passive elements are terminated with open circuits at both top and base.

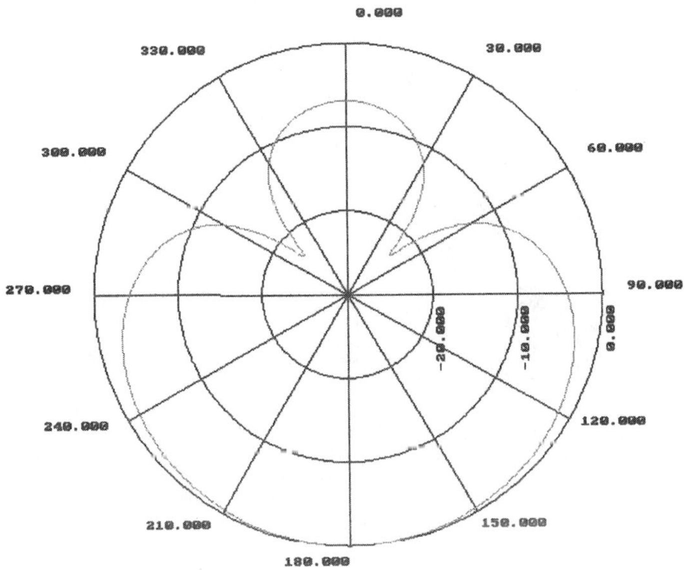

Figure 9-27 Resulting pattern when all passive elements are terminated with open circuits at the top, and short circuits at the base.

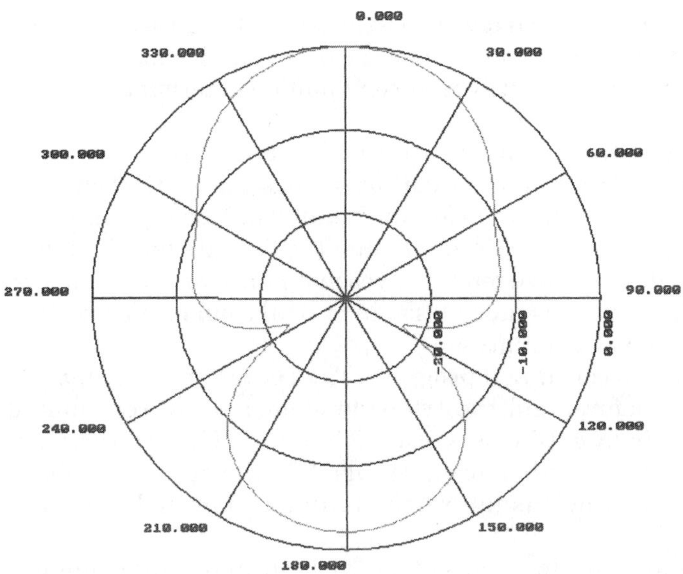

Figure 9-28 Resulting pattern when all passive elements are terminated with short circuits at both top and base.

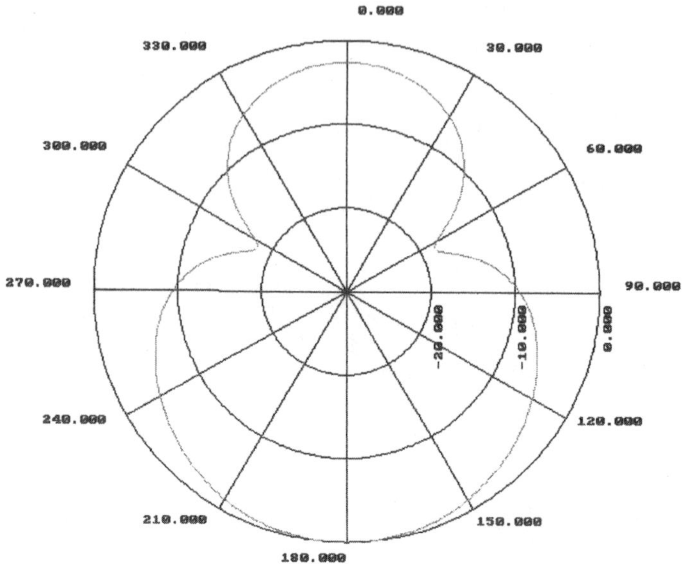

Figure 9-29 Resulting pattern when we attempt to detune the passive elements using the same terminations that worked for the two-element array. The tops are terminated in $-j100$ ohms, and the bases at terminated with short circuits. Obviously, different terminations will be required for successful detuning.

$X_T=-100$ ohms at the top, then it was mostly detuned. Figure 9-29 shows the radiation pattern that results when all three passive elements are terminated in that way. Evidently, that combination of terminations will not work here.

Figure 9-30 shows the radiation pattern when the center passive element is terminated with a short circuit at the base, and $-j100$ ohms at the top, but the two-side elements are terminated differently. For them, the bases are terminated in $-j250$ ohms, and the tops are terminated in short circuits. That is, we have tended to resonate the two-side elements. It is seen that the pattern is much less directional than in the previous figure. So, we are making progress.

Figure 9-31 shows continued progress. The center passive element is terminated as before, and the two-side elements are terminated at their bases as before, that is, with $-j250$ ohms. The top terminations of the side elements has been changed, however, to $-j34$ ohms. Compared with the previous figure, the improvement in detuning is significant.

We might be able to do even better with so many independent design variables at our disposal; however, it is time to move along to another topic.

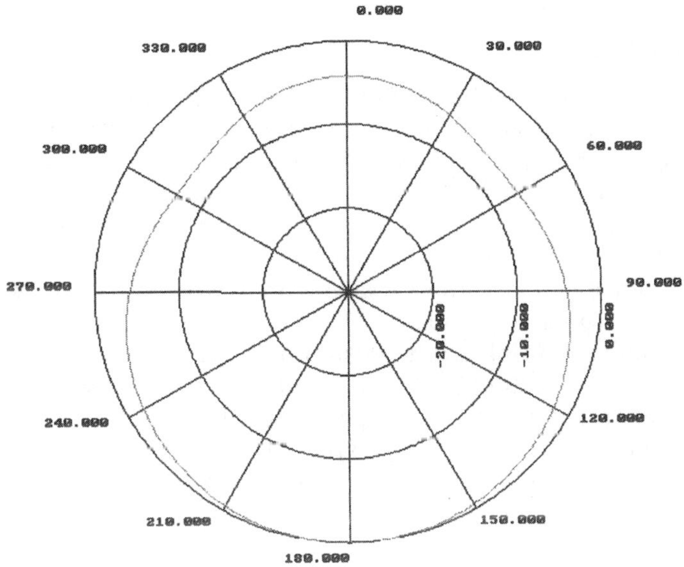

Figure 9-30 Resulting pattern when the closest passive element is terminated in the way that worked for the two-element array, that is, with $-j100$ ohms at the top, and a short circuit at the base. The remaining two passive elements are terminated with short circuits at the top, and with $-j250$ ohms at the base. Evidently, resonating them helps with respect to detuning, but we can obviously do better.

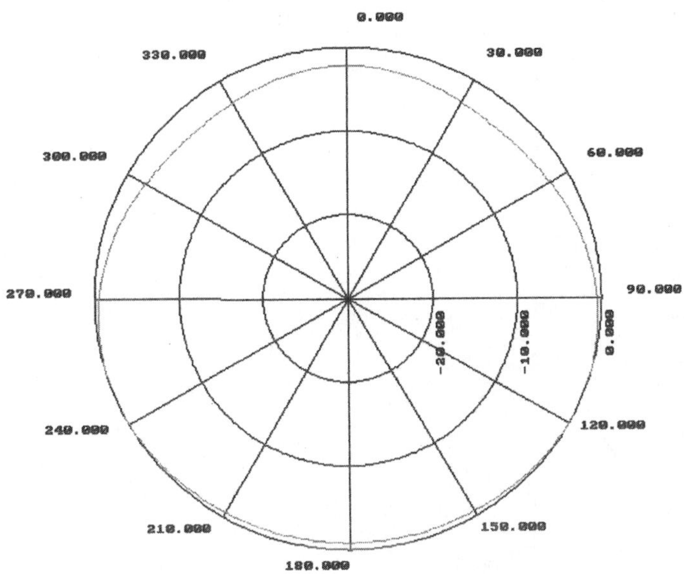

Figure 9-31 Resulting pattern when the outer two passive elements are terminated with $-j34$ ohms at the top, and $-j250$ ohms at the base. Detuning has much improved.

9.12 An Array of Active Folded Unipole Antennas

In the previous two sections, we focused on the use of passive elements in arrays of folded unipoles. In the present section, we shift our attention to an array of only active elements. We will find that folded unipoles can be elements in close-spaced arrays, that is, in arrays where the spacing between elements is much less than a quarter wave. In Chap. 2, we found that individual folded unipoles can be very compact in the vertical dimension. Here we will find that arrays of folded unipoles can be compact in the horizontal dimension as well.

We will confine our attention to the design of a cardioid radiation pattern using two active elements. To that end, Fig. 9-32 shows our first attempt, using straightforward design values. That is, the array elements are spaced 90 electrical degrees apart, and the input voltage to one element is phased 90 degrees with respect to the other. It is seen that the results are not even close to what is desired. Evidently, electromagnetic coupling between the elements has changed the radiating currents in such a way that they bear little resemblance to the input voltage, with respect to either amplitude or phase. So basing our design upon input voltage alone is not a useful approach.

Figure 9-33 is a contour plot showing the intensity of the radiated electromagnetic field at a location directly behind the line of elements, where we desire the null in the cardioid shaped pattern. The two independent

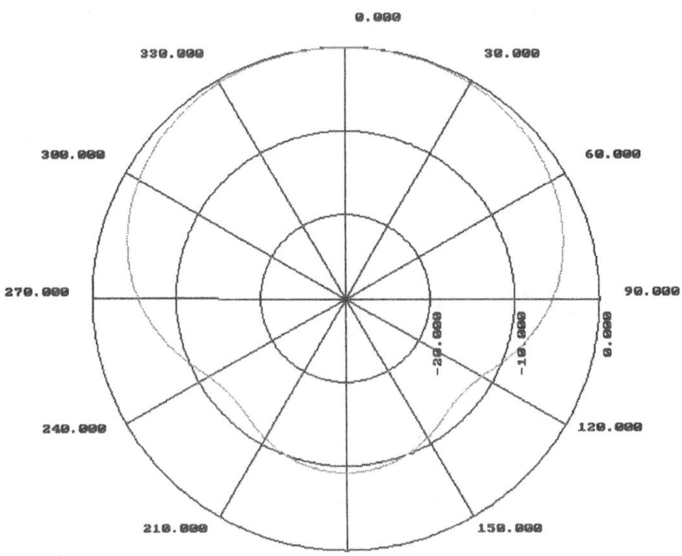

Figure 9-32 First attempt to form a cardioid radiation pattern using two active folded unipoles separated by 90 electrical degrees, or one-quarter wavelength. The input voltage amplitudes and phases were specified without considering the effects of electromagnetic coupling between the array elements. Evidently, those effects are significant.

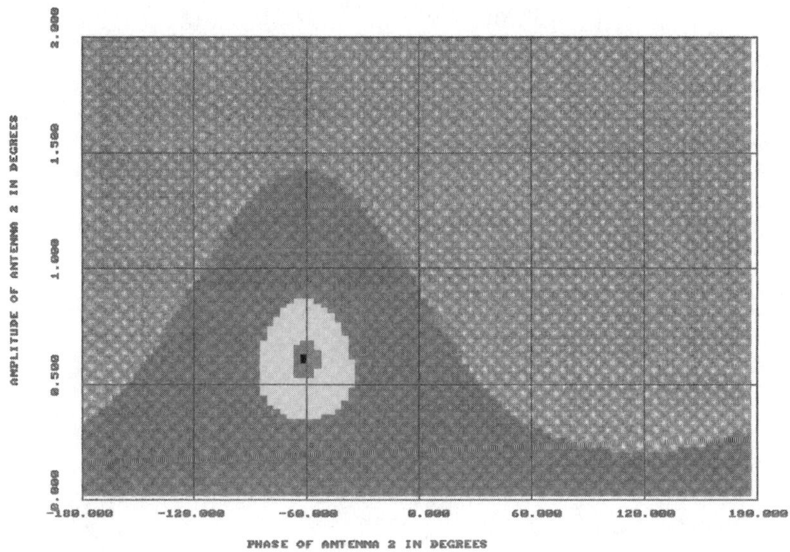

Figure 9-33 Contour plot showing the radiation pattern behind the array, at a fixed location one mile distant and at a bearing of 0 degrees. The two independent variables are the amplitude and phase of one of the input voltages. The darkest region on the plot shows the deepest null.

design variables are the amplitude and phase of the input voltage of the front element. The dark blue region shows the deepest null. It is seen that it occurs for an input voltage amplitude of 0.61, and phase of –61 degrees. This is quite different from our initially assumed amplitude of 1.0 and phase of +90 degrees.

Figure 9-34 shows the resulting pattern using the values suggested by the contour plot. We have obtained a robust cardioid pattern as desired.

Let us attempt to decrease the separation between the elements, that is, to construct a close-spaced array. Figure 9-35 shows the radiation pattern when we use straightforward design values for the input voltages, that is, an amplitude of 1.0 and phase of +135 degrees for elements separated by only 45 degrees. As earlier, the pattern is not acceptable. Electromagnetic coupling between the elements has again played a significant role that cannot be neglected.

Figure 9-36 is a contour plot similar to Fig. 9-33, except that the spacing between elements is 45 degrees. It is seen that the deepest null behind the line of elements occurs for an input voltage amplitude of 0.42 and phase of 86 degrees. Figure 9-37 shows the resulting radiation pattern. We have obtained a robust cardioid shape for a close-spaced array.

Can we move the elements even closer together? Figure 9-38 shows the pattern elements separated by only 22.5 degrees, when neglecting electromagnetic coupling between them. We have used straightforward values for the input voltage amplitude of 1.0, and 157.5 degrees for the phase.

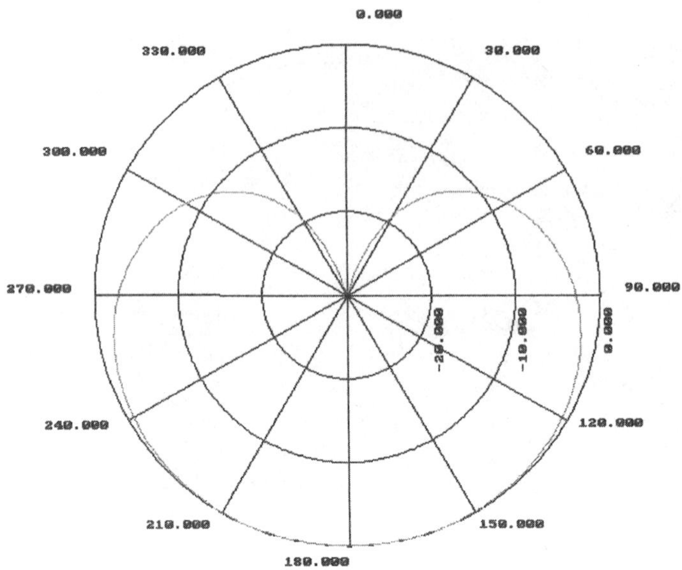

Figure 9-34 Resulting radiation pattern obtained using the contour plot. A robust cardioid was successfully obtained by changing the voltage amplitude from 1.0 to 0.61, and the phase from 90 to −61 degrees.

Curiously, this pattern is closer to the desired cardioid than in the previous examples. It is not yet acceptable, however.

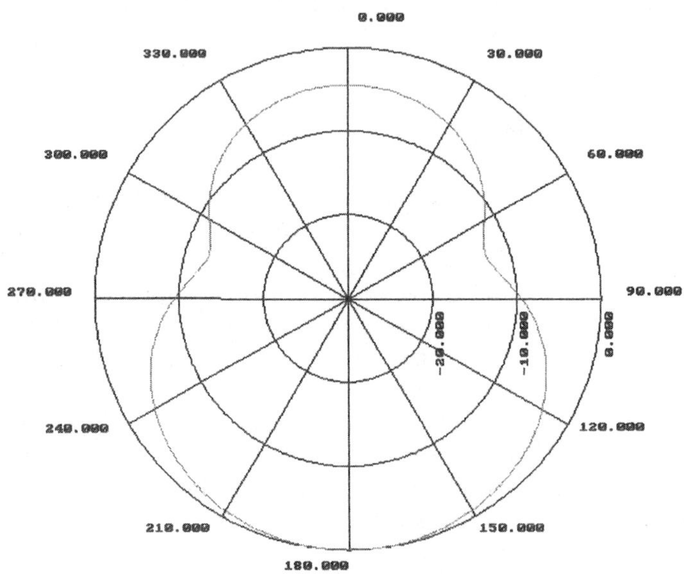

Figure 9-35 First attempt to form a cardioid radiation pattern using two active folded unipoles separated by 45 electrical degrees, or 1/8 wavelength. The input voltage amplitudes and phases were specified without considering the effects of electromagnetic coupling between the array elements. As in the previous example, those effects are evidently significant.

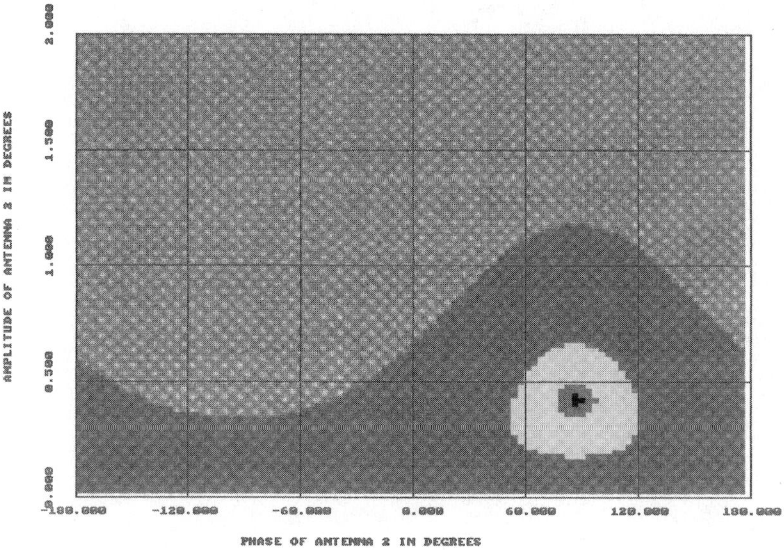

Figure 9-36 Contour plot showing the radiation pattern behind the array, at a fixed location one mile distant and at a bearing of 0 degrees. As in the previous example, the two independent variables are the amplitude and phase of one of the input voltages. The darkest region on the plot shows the deepest null.

Figure 9-39 is a contour plot similar to Figs. 9-33 and 9-36, except that the separation between elements is only 22.5 degrees. It is seen

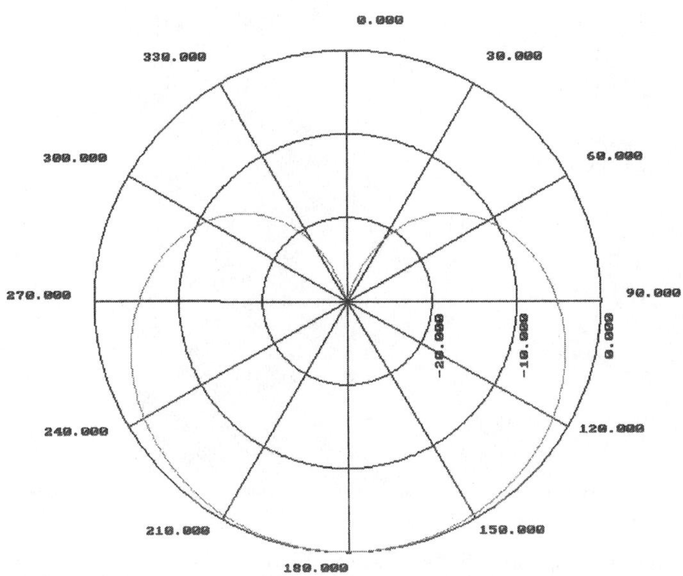

Figure 9-37 Resulting radiation pattern obtained using the contour plot. A robust cardioid was successfully obtained by changing the voltage amplitude from 1.0 to 0.42, and the phase from 135 to 86 degrees.

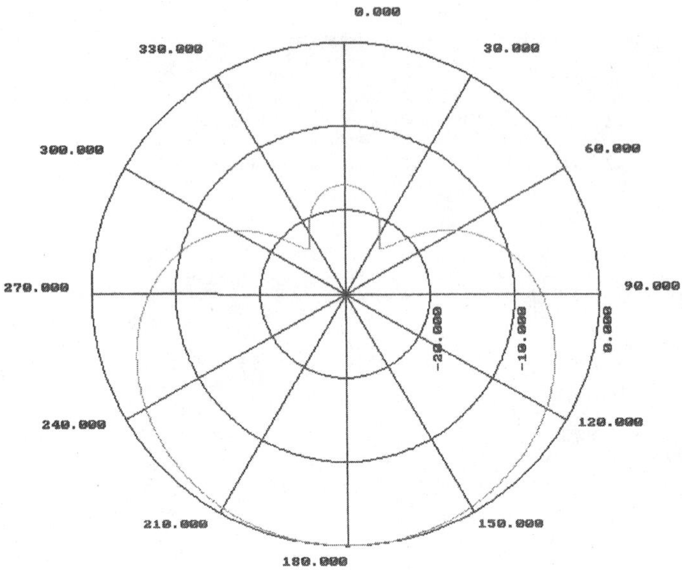

Figure 9-38 First attempt to form a cardioid radiation pattern using a close-spaced array of two active folded unipoles separated by 22.5 electrical degrees, or 1/16 wavelength. The input voltage amplitudes and phases were specified without considering the effects of electromagnetic coupling between the array elements. As in the previous examples, those effects are evidently significant. Curiously, though, they seem less significant than for elements more widely spaced.

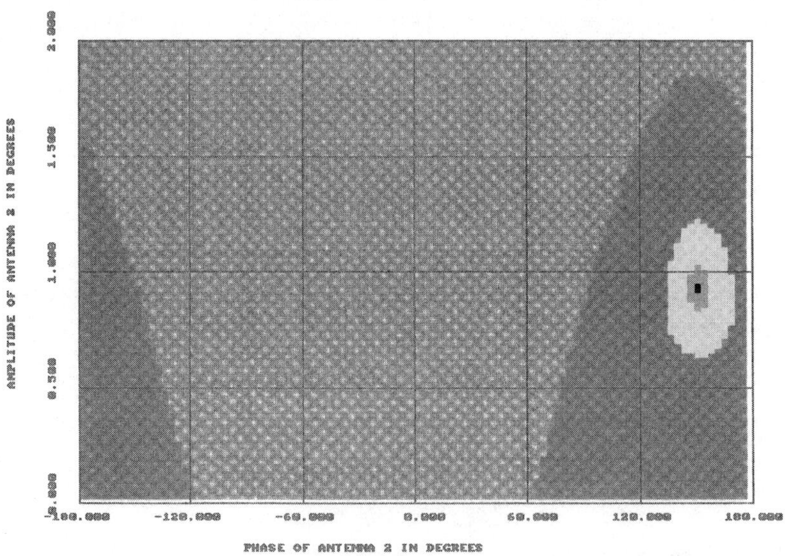

Figure 9-39 Contour plot showing the radiation pattern behind the array, at a fixed location one mile distant and at a bearing of 0 degrees. As in the previous examples, the two independent variables are the amplitude and phase of one of the input voltages. The darkest region on the plot shows the deepest null.

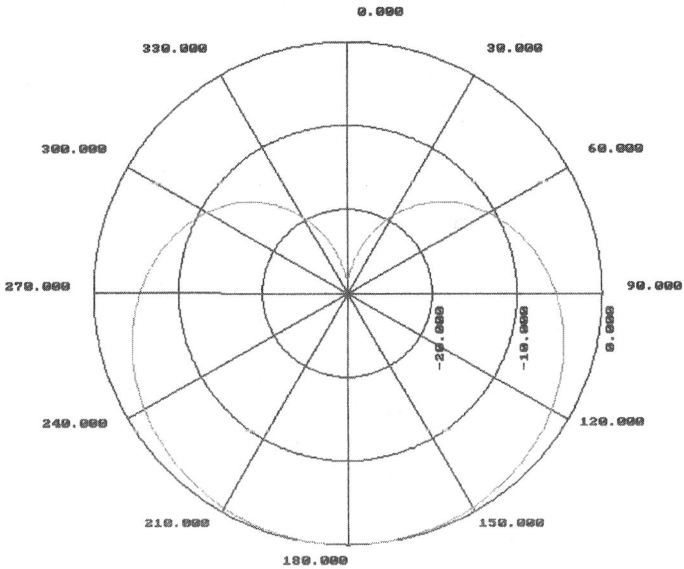

Figure 9-40 Resulting radiation pattern obtained using the contour plot. A robust cardioid was successfully obtained even for a close-spaced array by changing the voltage amplitude from 1.0 to 0.92, and the phase from 157.5 to 150 degrees.

that the deepest null behind the line of elements is obtained for a voltage amplitude of 0.92 and phase of 150 degrees. Figure 9-40 shows the resulting robust cardioid radiation pattern.

9.13 Electromagnetic Coupling Factor for Arbitrary Cross Sections

In Sec. 9.5, we intuitively derived Eq. 9-61 for the induced current on array elements constructed from hollow lattices, including most broadcast and communications towers. We might be interested in elements with different cross sections, however, to which Eq. 9-61 does not apply. With that in mind, we rewrote the formula as Eq. 9-75, introducing a coupling factor κ. In the present section, we return to examine that factor more quantitatively.

We begin with a classic result from electromagnetic theory for the current induced on an isolated, solid, conducting cylinder with circular cross section of radius a [3]. That result is:

$$I_i = j\frac{2\pi}{\omega\mu_0 \ln(1/ka)}E_z \tag{9-159}$$

Comparing Eq. 9-159 with Eq. 9-75, the coupling factor for a single cylinder is:

$$\kappa = \ln(1/ka) \tag{9-160}$$

It is readily seen from Eq. 9-160 that, in general, κ is different from unity. In fact, $\kappa = 1$ only if the radius of the conductor is:

$$a = 1/ke = \lambda/2\pi e \approx \frac{\lambda}{17} \tag{9-161}$$

In general, the radius of a cylinder will not satisfy Eq. 9-161. This confirms that for solid cylinders, and for a variety of other antenna array elements, we can not use Eq. 9-61 to find the induced current. It will be necessary to use Eq. 9-75 with the correct coupling factor κ.

To that end, if we compare Eq. 9-160 with Eq. 7-54 for the characteristic impedance of a single wire:

$$\kappa = \frac{2\pi}{\eta} Z_{0c} \tag{9-162}$$

Shortly, we will show that Eq. 9-162 applies to not only single cylinders but to antennas with any cross section.

If we substitute Eq. 9-162 into Eq. 9-75, and use the relations,

$$\eta = \sqrt{\frac{\mu_0}{\varepsilon_0}} \tag{9-163}$$

and

$$k = \omega\sqrt{\mu_0 \varepsilon_0} \tag{9-164}$$

then an especially simple formula is obtained for the induced current:

$$I_i = \frac{jE_z}{kZ_{0c}} \tag{9-165}$$

Equation 9-165 agrees exactly with Eq. 8-10.

We could have derived Eq. 9-165 directly by considering an incremental segment of transmission line, as shown in Fig. 9-41. In response to the incident electric field, the induced current flows through the inductance per unit length according to the constituitive relation,

$$I_i = \frac{-E_z}{j\omega L'} \tag{9-166}$$

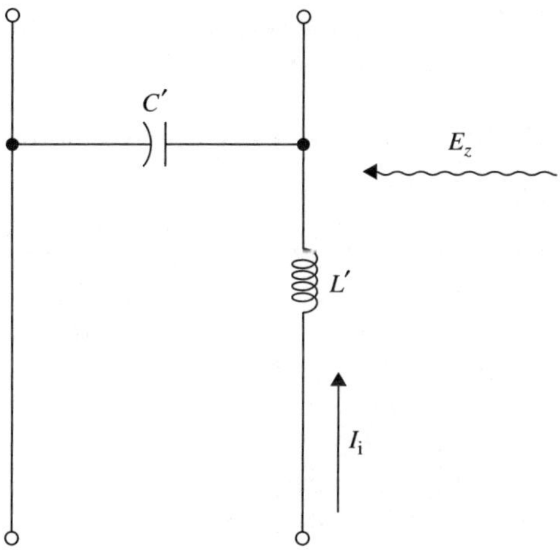

Figure 9-41 The transmission line model of an antenna can be used to readily obtain the current induced by an incident electromagnetic field. In response to that field, the induced current flows through the inductance per unit length. This model applies to antennas with any cross section that can be represented by a common mode characteristic impedance Z_{0c}.

As we showed in Eq. 7-33, L' is related to the characteristic impedance:

$$L' = \sqrt{\mu_0 \varepsilon_0} Z_{0c} \tag{9-167}$$

Using Eqs. 9-164 and 9-167, Eq. 9-166 becomes:

$$I_i = \frac{jE_z}{kZ_{0c}} \tag{9-168}$$

Equation 9-168 is identical to Eq. 9-165.

In retrospect, we could have omitted the lengthy arguments of Sec. 9-5 entirely in favor of the simple derivation of Eq. 9-168; however, there were benefits of starting with the fundamental electromagnetic fields. First, we confirmed Eq. 9-168 using an entirely different approach, and this provided extra confidence both in the result and in the transmission line model of antennas. Second, we developed an intuitive view of electromagnetic fields that may serve us well in developing models of other phenomena.

9.14 Conclusions

This chapter was a lengthy journey along two intertwined paths. The first path explored the advantages of the folded unipole antenna as an array element. The second was an odyssey through electromagnetic theory for the purpose of deriving the mathematical tools necessary to analyze and design arrays of folded unipoles. Fortunately, both paths ultimately led to useful destinations.

In Chaps. 2 through 6, we found that the folded unipole offers advantages as an individual antenna. In this chapter, we found that it offers additional advantages as an element in both active and passive arrays. In passive arrays, elements are often unintentional. For example, they might be a line of electrical power poles. In those cases, they can be reconfigured as folded unipoles and be detuned, or rendered almost electronically invisible. In active arrays, folded unipoles can be short spaced, that is, separated by distances much less than 90 electrical degrees, or one-quarter wavelength. In that compact configuration, they can still produce distinctly shaped radiation patterns. So, the folded unipole can provide compactness in both vertical and horizontal dimensions. In this chapter, we presented examples that demonstrate all of these advantages.

In the past, many arrays of folded unipoles have not been deployed simply because the design procedure was too awkward. The methods that worked quickly and elegantly for arrays of monopole antennas do not readily apply to folded unipoles. So, it was necessary for us to embark on a fresh approach.

To that end, we developed new formulas for the electromagnetic coupling between array elements. The currents induced on one element by all of the others are calculated directly. We bypassed the traditional but intermediate steps of computing mutual and self impedances. One can argue that our approach is more logical. Each element in the array is a receiving antenna for the fields radiated from all of the other elements. As S. A. Schelkunoff stated:

> To obtain the current in a [receiving] antenna, we note first that the current in an infinitely long wire in a uniform field parallel to the wire is constant. In a finite wire, the current must vanish at the ends, which must, therefore, act as virtual sources of waves on the wire. [4].

In our approach, we have translated Schelkunoff's explanation into convenient mathematical formulas.

The new formulas also apply to monopoles, and we used a simple array of those for a test. The results readily passed all sanity checks.

As we have emphasized in previous chapters, our approach has two chief computational advantages over general purpose wire antenna analysis programs. Both approaches are based upon matrix equations. The dimensions of our matrix, however, are many orders of magnitude less than those for the general-purpose programs. One resulting advantage is faster computation speed. We can examine 10,000 different combinations of design variables in just a few seconds. The second advantage is lower numerical noise which tends to compromise the accuracy of results. This is a special problem associated with the cross section of folded unipoles, in which many wire segments are located inside a perimeter small compared with a wavelength.

Extra benefits of our approach are a useful collection of formulas that have application beyond antenna arrays. We derived a fairly simple and elegant formulation for the induced current on a complicated conductor with any cross section that can be represented by a common mode characteristic impedance.

We also derived a set of formulas for the complete electromagnetic fields of any forward or reverse traveling current wave. These waves include, of course, the currents on the folded unipole antenna. The field formulas are accurate both close to an antenna or array, that is, in the *near field*, and far away, in the *radiation zone*. So they are useful for evaluating potential hazards from antennas connected to high power transmitters.

As in Chap. 8, we have barely scratched the surface of an immense topic with practically unlimited applications. We have, however, developed the mathematical tools to explore far beyond the limited space available here.

References

1. Carter, P. S., Circuit Relationships in Radiating Systems and Applications to Antenna Problems, New York: *Proc. of the IRE*, **20**, 1932, pp. 1004–1041.
2. Schelkunoff, S. A., Theory of Antennas of Arbitrary Size and Shape, New York: *Proc. of the IRE,* **29**, 1941, pp. 493–521.
3. Harrington, R. F., Time Harmonic Electromagnetic Fields, McGraw-Hill, 1961, pp. 232–234.
4. Schelkunoff, S. A., and H. T. Friis., Antennas Theory and Practice, John Wiley & Sons, 1952, p. 242.

Symbols Used in This Chapter

$\mathbf{A}$	1. Magnetic vector potential. 2. Submatrix of $\mathbf{M}$
$\mathbf{A}_{11}$	Submatrix of $\mathbf{M}$
$\mathbf{A}_{12}$	Submatrix of $\mathbf{M}$
$\mathbf{A}_{21}$	Submatrix of $\mathbf{M}$
$\mathbf{A}_{22}$	Submatrix of $\mathbf{M}$
A_F	Magnetic vector potential of z-directed, forward traveling current wave
$A_{F,\,\text{image}}$	Magnetic vector potential of the ground image of a z-directed, forward traveling current wave
$A_{F\Sigma}$	Sum of A_F and $A_{F,\,\text{image}}$
$\mathbf{A}_\mathbf{m}$	Row vector of submatrix $\mathbf{A}$
A_R	Magnetic vector potential of a z-directed, reverse traveling current wave
$A_{R\Sigma}$	Sum of A_R and its ground image
A_{total}	Sum of $A_{F\Sigma}$ and $A_{R\Sigma}$

A_z	z-component of magnetic vector potential $\mathbf{A}$
a	1. Radius of a cylinder. 2. z-coordinate of the lower end of a conducting wire or antenna
$\mathbf{a_{mn}}$	Subvector of row vector $\mathbf{A_m}$
a_n	Radius of nth cylinder
$\mathbf{B}$	Submatrix of $\mathbf{M}$
$\mathbf{B_m}$	Row vector of submatrix $\mathbf{B}$
b	z-coordinate of the upper end of a conducting wire or antenna
$\mathbf{b_{mn}}$	Subvector of row vector $\mathbf{B_m}$
$\mathbf{C}$	Submatrix of $\mathbf{M}$
$\mathbf{C_m}$	Row vector of submatrix $\mathbf{C}$
$\mathbf{c_{mn}}$	Subvector of row vector $\mathbf{C_m}$
$\mathbf{D}$	Submatrix of $\mathbf{M}$
$\mathbf{D_m}$	Row vector of submatrix $\mathbf{D}$
$\mathbf{d_{mn}}$	Subvector of row vector $\mathbf{D_m}$
$\mathbf{E}$	Electric field
E_R	R-directed component of $\mathbf{E}$
E_z	z-directed component of $\mathbf{E}$
$E_z^{\ s}$	Scattered electric field from a cylinder
E_θ	θ-directed component of $\mathbf{E}$
E_ρ	ρ-directed component of $\mathbf{E}$
$Ei(x)$	Exponential integral function
$Ei'(x)$	First derivative of $Ei(x)$
$Ei''(x)$	Second derivative of $Ei(x)$
e	Naperian base, 2.718281…
$F(\rho, z)$	Function used to calculate current induced on an antenna by E_z incident from a forward traveling current wave on another antenna
$F_{\mathrm{nm}}(z)$	$F(\rho_{\mathrm{nm}}, z)$
$G(\boldsymbol{r}, \boldsymbol{r}')$	Green's function used to calculate magnetic vector potential $\mathbf{A}$
$G(\rho, z)$	Function used to calculate current induced on an antenna by E_z incident from a forward traveling current wave on another antenna
$G_{\mathrm{nm}}(z)$	$G(\rho_{\mathrm{nm}}, z)$
$\mathbf{H}$	Magnetic field
H_φ	φ-directed component of $\mathbf{H}$
h_n	Height of nth array element
$I(z)$	Current along antenna or conductor
$I_c(z)$	Common mode current along antenna
$I_{\mathrm{cn}}(z)$	Common mode current along nth antenna or array element
$I_{\mathrm{dn}}(z)$	Differential mode current along nth antenna or array element
I_F	Complex amplitude of forward traveling current wave
I_{Fcn}	Complex amplitude of common mode, forward traveling current wave along nth antenna or array element
I_{Fdn}	Complex amplitude of differential mode, forward traveling current wave along nth antenna or array element
I_i	Induced current
$I_{\mathrm{im}}(z)$	Induced current on mth antenna
$\mathbf{I_n}$	Vector of unknown complex current amplitudes on the nth antenna in the matrix equation for currents in an antenna array

I_R	Complex amplitude of reverse traveling current wave
I_{Rcn}	Complex amplitude of common mode, reverse traveling current wave along nth antenna or array element
I_{Rdn}	Complex amplitude of differential mode, reverse traveling current wave along nth antenna or array element
$\mathbf{J}$	Current density in amperes per square meter
j	Square root of -1
K_z	Surface current density in amperes per meter
k	Wave number, $2\pi/\lambda$
L'	Inductance per unit length of an antenna or transmission line
$\mathbf{M}$	Matrix in the equation for currents in an antenna array, or in the equation for induced currents on an ensemble of cylinders
N	Number of elements in an antenna array
$P_F(\rho, z)$	Function used to compute $F(\rho, z)$ and $G(\rho, z)$
$P_R(\rho, z)$	Function used to compute $R(\rho, z)$ and $S(\rho, z)$
$Q_F(\rho, z)$	Function used to compute $F(\rho, z)$ and $G(\rho, z)$
$Q_R(\rho, z)$	Function used to compute $R(\rho, z)$ and $S(\rho, z)$
R	Distance between a location on an antenna and an observer
$R(\rho, z)$	Function used to calculate induced current on an antenna by E_z incident from a reverse traveling current wave on another antenna
R_a	Distance between the base of an antenna and an observer
R_b	Distance between the top of an antenna and an observer
$R_{\mathrm{nm}}(z)$	$R(\rho_{\mathrm{nm}}, z)$
$S(\rho, z)$	Function used to calculate induced current on an antenna by E_z incident from a reverse traveling current wave on another antenna
$S_{\mathrm{nm}}(z)$	$S(\rho_{\mathrm{nm}}, z)$
$\mathbf{V}$	Vector of voltages hardwired to the bases of elements in an antenna array.
$V_c(z)$	Common mode voltage along an antenna
$V_{\mathrm{cn}}(z)$	Common mode voltage along the nth antenna or nth array element
$V_{\mathrm{dn}}(z)$	Differential mode voltage along the nth antenna or nth array element
V_{Fc}	Complex amplitude of common mode, forward traveling voltage wave
V_i	Induced voltage
V_n	Voltage hardwired to the base of nth element in an antenna array.
V_{Rc}	Complex amplitude of common mode, reverse traveling voltage wave
w	Transformed variable of integration
w'	Transformed variable of integration
$W_F(\rho, z)$	Function used to calculate common mode voltage along an antenna by a forward traveling current wave on another antenna
$W_{\mathrm{Fnm}}(z)$	$W_F(\rho_{\mathrm{nm}}, z)$
$W_R(\rho, z)$	Function used to calculate common mode voltage along an antenna by a reverse traveling current wave on another antenna
$W_{\mathrm{Rnm}}(z)$	$W_R(\rho_{\mathrm{nm}}, z)$
X_{Bn}	Terminating base reactance on nth antenna, imaginary part of Z_{Bn}
X_{Tn}	Terminating top reactance on nth antenna, imaginary part of Z_{Tn}
Y_{0c}	Common mode characteristic admittance, $1/Z_{0c}$
$Z_{0\mathrm{cn}}$	Common mode characteristic impedance for nth antenna
$Z_{0\mathrm{dn}}$	Differential mode characteristic impedance for nth antenna
Z_{Bn}	Terminating base impedance on nth antenna

Z_{Tn}	Terminating top impedance on nth antenna
z	Vertical coordinate
z'	Variable of integration
z_{int}	Internal or intrinsic impedance of conductor
δ	Skin depth
ε_0	Permittivity of free space
ζ	Vertical coordinate in image space, $-z$
ζ'	Variable of integration
η	Impedance of free space
θ	Angular coordinate in spherical coordiante system
κ	Electromagnetic coupling factor used to calculate current induced by an incident electric field
λ	Wavelength
μ_0	Permeability of free space
ν_n	Current transformation ratio for nth folded unipole antenna
ρ	Horizontal distance in the cylindrical coordinate system
ρ_{mn}	1. Horizontal distance between positions m and n. 2. Horizontal distance between cylinders m and n
$\mathbf{\Phi}$	Electromagnetic potential
$\mathbf{\Phi_{im}}$	Common mode voltage induced along the mth element in an antenna array
$\mathbf{\Phi_m}$	Electromagnetic potential along the mth element in an antenna array
φ	Angular coordinate in cylindrical coordinate system
$\Psi_F(z)$	Forward traveling wave function
$\Psi_R(z)$	Reverse traveling wave function
ω	Frequency in radians per second, $2\pi f$

Chapter

10

Synergy

10.1 Introduction

Many alternative titles for this final chapter came to mind. The choices ranged from the staid and scholarly *Selected Topics,* to the melodramatic *Epilogue*, to the whimsical and humorous *Unlimited Potential (Pun Intended).* In the final analysis, however, *Synergy* was clearly the best choice because that sums up what is unique and useful about the folded unipole antenna.

The recurring theme throughout this book has been that the folded unipole antenna is the integration of linear radiators and transmission lines into a single electromagnetic device. It exhibits phenomena and performance not found in linear antennas alone, nor in transmission lines alone. We have seen that two modes of current propagate simultaneously, and that one of those modes is distributed across the antenna cross section according to a current transformation ratio. These phenomena can be controlled in the design process to provide compact and broadband antennas and antenna arrays. They can also be controlled to detune structures or make them electronically invisible.

Lumped impedances are also integrated into the folded unipole antenna, up to three per segment in the most general configuration. Combined with two characteristic impedances and the current transformation ratio, there are six independent design variables for each folded segment. In contrast, an entire series-fed monopole has only two independent design variables. With so many more variables, the folded unipole antenna obviously offers more versatility and design options.

Though we have filled an entire book with the details concerning the above phenomena and applications, they are only a beginning. In Chap. 8, concerning Detuning, and in Chap. 9, concerning Arrays of

Folded Unipoles, we suggested that entire books could be written about those topics. There are still other books that could be written about additional properties and applications of the folded unipole antenna. We will take an introductory look at some of them in this chapter.

With the above as prelude, we will now embark upon a technical smorgasbord of applications both existing and pending. Some of them involve variations in geometry, such as deforming the component wires and cylinders of the folded unipole from straight lines into helices. Other applications involve immersing the folded unipole into complex environments with arbitrary conductivity and permittivity. We have established the physical principles and mathematical foundations for all of them in previous chapters. The mathematical models will be outlined in sufficient detail so that the motivated reader should be able to supply the remaining steps.

10.2 Extreme Top Loading

In previous chapters, we have seen that the folded unipole antenna lends itself to compact size, both for individual radiators and for arrays. In some applications, that compactness can be further improved with top loading. In this section, we will investigate how top loading can be incorporated into our mathematical model of the folded unipole, and how it might be expected to improve its performance.

Top loading is a technique applied to many different antennas. The goal is to make antennas exhibit the performance of taller or longer ones. In that way, antennas that are small compared with a wavelength can compete with larger, less practical, and less economical structures. The design logic behind top loading is to provide the common mode current with a path at the top of the antenna so that it may continue to flow from there. Otherwise, boundary conditions at the top force it to vanish there, decreasing the overall current moment of the antenna.

In fact, top loading is a logical extension or generalization of the *antenna end effect*, which was discussed in Sec. 5.7. Any antenna with a nonzero radius exhibits some capacitance at its end, and so the common mode current does not quite vanish there. A practical consequence is that the current distribution along the antenna is not a pure standing wave with perfect nulls. In particular, a half-wave monopole will not have exactly zero input current. That is why its input impedance is large but not infinite.

Figure 10-1 shows how top loading can be readily added to our mathematical model of folded unipole antennas. It is similar to Fig. 5-7, which concerned antenna end effect. It is seen that a deliberate top loading admittance Y_{TL} is connected from the top of the antenna to the ground. As a result, the antenna end effect is modified because Y_{TL} is in parallel with the intrinsic capacitance C_{cap} of the top of the antenna. With an appropriate choice of Y_{TL}, the common mode current at the top of the antenna,

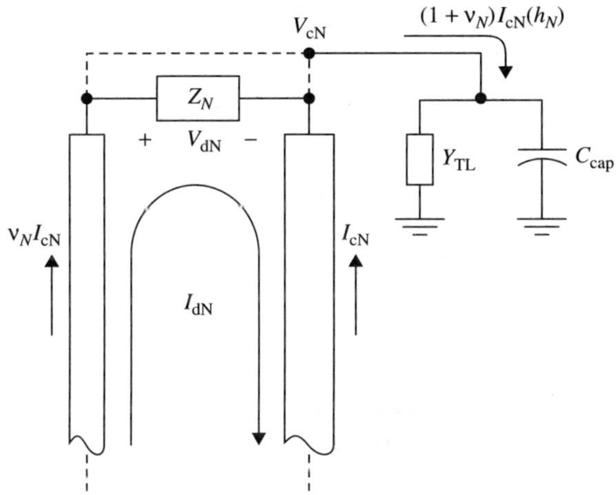

Figure 10-1 In addition to the intrinsic antenna top capacitance due to a finite wire radius, top loading may be added by creating additional paths from the top of the antenna to ground. That path usually includes lumped admittances. In the case of the NORD LF antenna, this was accomplished with umbrella wires sloping from the top of the antenna toward ground. At the far end of the umbrella wires, lumped admittances completed the path to ground.

and consequently the current distribution everywhere along the antenna, can be varied over a wide range.

We have already derived the mathematical formulas for describing the effects of top loading. Equations 5-33 and 5-34 describe the boundary condition at the top of the antenna, taking into account only the intrinsic capacitance C_{cap}; however, Eq. 5-33 is readily extended to account for an arbitrary admittance Y_{TL}:

$$(1+\nu_N)\, I_{cN}(h_N) = (Y_{cap} + Y_{TL})\, V_{cN}(h_N) \tag{10-1}$$

With Eq. 10-1, we can use our existing models of folded unipole antennas to predict input impedance and radiation pattern.

The top loaded folded unipole antenna was reduced to practice in the LF band by Multronics, Inc., Rockville, Maryland, circa 1965. Figure 10-2 shows one of the actual configurations. It is seen that the intrinsic capacitance of the antenna was enhanced using three wires, called *umbrella wires*, extending from the top. It is also seen that near the end of the umbrella wires, vertical wires extend to the ground through an enclosure containing an electronic network. This network was called the GTU, or *guy termination unit*. Actually, the GTU was none other than the admittance Y_{TL} in Eq. 10-1.

By all accounts, the design resulted in a successful working antenna which was deployed throughout the Pacific basin and in the Mediterranean.

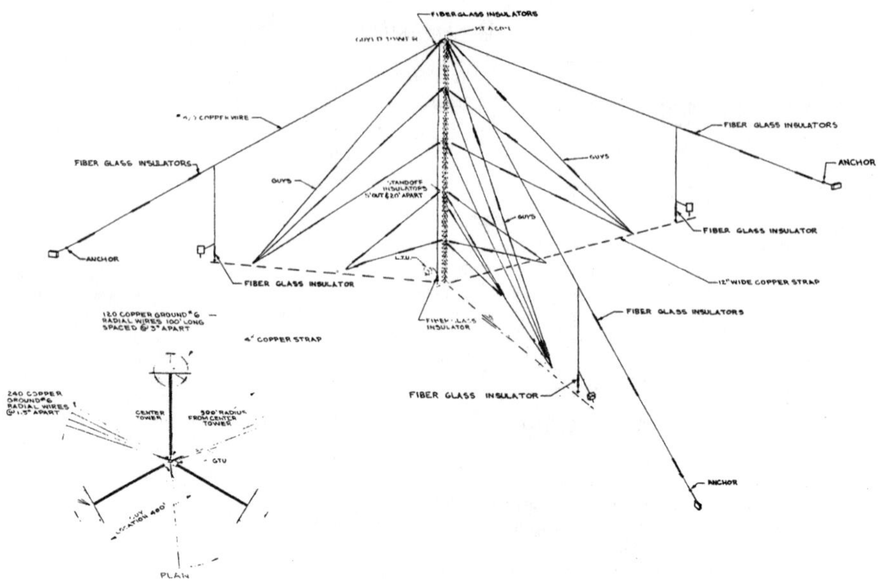

Figure 10-2 One actual configuration of the NORD LF antenna. The top loading consisted of three umbrella wires radiating from the top of the antenna. About two-thirds of the distance along the umbrella wires, a vertical wire connected to ground through a guy terminating network, consisting of lumped admittances.

It operated at frequencies as low as 50 kHz. According to test results, it exhibited a bandwidth of over 13 percent. Keep in mind that the overall dimensions of the antenna were tiny compared with a wavelength. At 50 kHz, a wavelength is 6,000 meters, or 3.73 miles. In contrast, the height of the Multronics designed, top loaded, folded unipole antenna was only about 400 feet, or 0.02 wavelength, or 7.3 electrical degrees.

Multronics took time to document its experience with this unique antenna in a formal course manual [1]. As a result, we are fortunate to have a description of the working antenna, called the NORD, by the engineers who actually designed it:

> The NORD antenna, based on the folded unipole principle, is basically a vertical tower radiator which is electrically grounded at its base and fed as a folded unipole by one or more feed wires (or folds) which are connected to the very top of the tower. In addition to feeding the antenna at its base as a folded unipole type, three wires attached to the top of the tower and spaced at 120° intervals are used for top loading. The antenna is completely over-top-loaded...From the end of each of the three top loading wires a connection is run down to a Guy Termination Unit. The G.T.U. is used for controlling the feed point impedance of the antenna. If desired the three G.T.Us can also be used to control the phase angles and magnitudes of the termination currents to provide a limited directional radiation effect.

This section has been a gentle introduction, in terms of numbers of equations, to applications of our model of the folded unipole. In addition to the extremely compact and extremely top loaded folded unipole, other applications have also been reduced to practice. We will continue with one of those in the next section.

10.3 Multiplexing

The previous section emphasized the extreme compactness compared with a wavelength that can be obtained using the folded unipole configuration. There is yet another way to save space with the folded unipole, and that is the subject here.

The folded unipole is especially well suited to multiplexing, that is, hardwiring the same antenna to multiple transmitters. Multiplexing using the folded unipole was reduced to practice by George P. Howard (1931–1993.) His innovative efforts in the AM broadcast band made efficient use of valuable real estate in such locations as Hawaii.

Figure 10-3 shows a general multiplexing configuration for N separate transmitters. Each transmitter is hardwired to its own fold

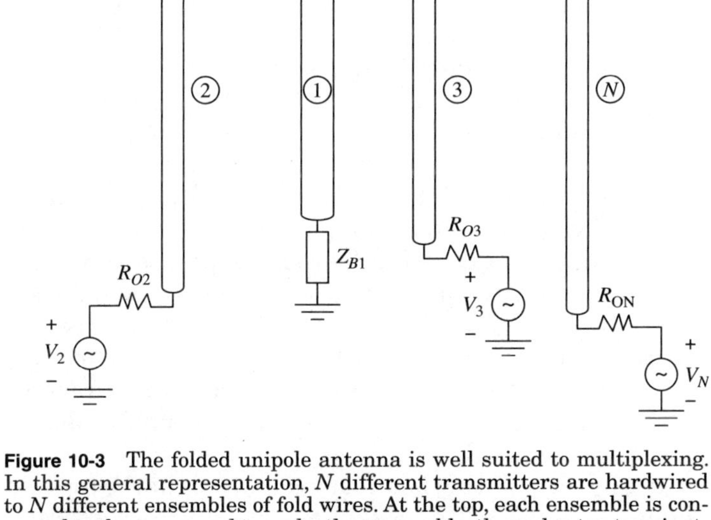

Figure 10-3 The folded unipole antenna is well suited to multiplexing. In this general representation, N different transmitters are hardwired to N different ensembles of fold wires. At the top, each ensemble is connected to the tower, and to each other ensemble, through a top terminating impedance Z_{Tmn}. As a practical matter, many of the top terminating impedances are simply open circuits.

constructed from one or more parallel wires. Each fold is connected separately to the top of the tower or main conductor through an arbitrary impedance Z_{T1m} with $m = 2,3,4,\ldots N$. Further, the top of each fold m is connected to the top of each other fold n through an impedance Z_{Tmn}. As a practical matter, most of these will be open circuits; however, we will maintain them as arbitrary design variables in our model. The total number of top terminating impedances is given by the familiar combinatorial formula for the total number of pairs formed within a group of N items:

$$\binom{N}{2} = \frac{N!}{2!(N-2)!} = \frac{N(N-1)}{2} \tag{10-2}$$

We will return to Eq. 10-2 shortly in this section.

The separate folds are not connected to each other at the base, and this is why they do not form a single fold. Instead, each fold is connected to a transmitter with output voltage V_m and output impedance R_{0m}. Though we are labeling the output impedance as a resistance, in fact, it may be complex because it can include bandpass, band reject, and other filters constructed from reactive components. Our model will allow for this possibility.

The basic operating principles and mathematical model are extensions of our efforts in Chap. 2, for the simplest configuration of the folded unipole. In the case of multiplexing, the chief difference is the number of differential modes of current. As in Chap. 2, there is only one common mode of current for each transmitter. The number of differential modes excited by each transmitter, however, is determined in much the same way as above for the number of top terminating impedances. That is, for each pair of folds and for each pair formed by a fold and the tower, there is a separate differential mode of current. Therefore, the number of separate differential modes is given by Eq. 10-2.

To demonstrate the above principles and mathematical procedures, let's work out the details for the simple configuration shown in Fig. 10-4. There are two separate transmitters hardwired to separate folds. It is readily seen that there are three different top terminating impedances, as predicted by Eq. 10-2. For each transmitter, there will also be one common mode current and three distinct differential mode currents. Since each current mode has a forward traveling wave and a reverse traveling wave, there will be eight unknown current amplitudes to determine. As in Chap. 2, we will derive eight independent equations for those amplitudes by formulating boundary conditions at the top and base of the antenna.

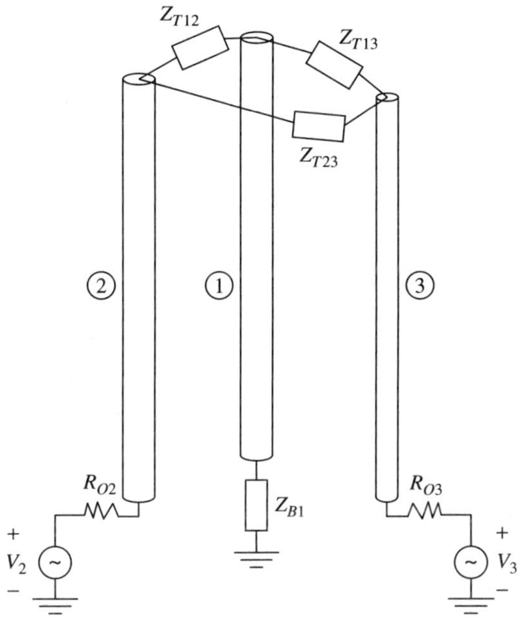

Figure 10-4 In one of the simplest examples of a multiplexed folded unipole, two transmitters are hardwired to separate ensembles of fold wires. There are three top terminating impedances. For each transmitter, there will be one common mode current and three differential mode currents.

To that end, Fig. 10-5 shows the top of the antenna. We can formulate four boundary conditions there. As usual, the common mode current must vanish there, if we neglect the antenna end effect:

$$(\nu_2 + \nu_3 + 1)I_c(h) = 0 \tag{10-3}$$

In Eq. 10-3, note that there are two different current transformation ratios, ν_2 and ν_3. These result because the two folds, wires 2 and 3, are distinct and not connected to each other. The separate current transformation ratios follow from computing the common mode characteristic impedance according to the procedures described in Sec. 7.7.

The three differential mode currents and voltages are constrained by the top terminating impedances:

$$I_{d12}(h)Z_{T12} = V_{d12}(h) \tag{10-4}$$

$$I_{d13}(h)Z_{T13} = V_{d13}(h) \tag{10-5}$$

$$I_{d23}(h)Z_{T23} = V_{d23}(h) \tag{10-6}$$

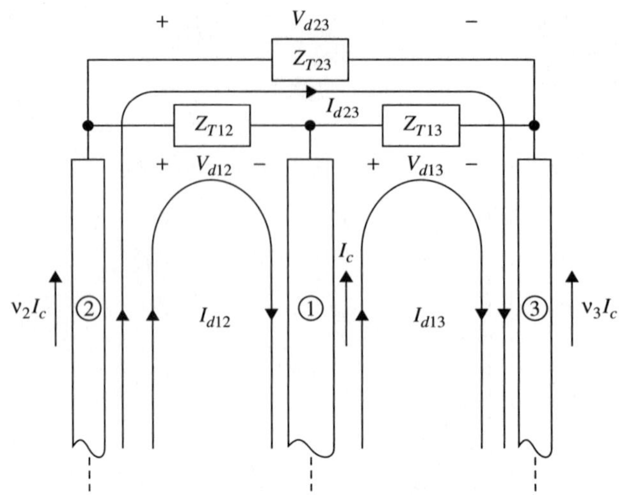

Figure 10-5 For each of two transmitters, one common mode current and three differential mode currents flow at the top of the multiplexed folded unipole. Four boundary conditions on those currents and related voltages can be formulated there.

We can formulate the remaining four boundary conditions at the base of the antenna. Figure 10-6 shows the essential geometry, currents, and voltages there. Kirchhoff's voltage law requires that:

$$R_{03}[-\nu_3 I_c(0)+I_{d13}(0)+I_{d23}(0)]+V_{d13}(0)+V_{d12}(0)=V_c(0) \tag{10-7}$$

$$Z_{B1}[-I_c(0)-I_{d13}(0)+I_{d12}(0)]+V_{d12}(0)=V_c(0) \tag{10-8}$$

$$V_2+R_{02}[-\nu_2 I_c(0)-I_{d12}(0)-I_{d23}(0)]=V_c(0) \tag{10-9}$$

$$V_{d13}(0)+V_{d12}(0)=V_{d23}(0) \tag{10-10}$$

In Eqs. 10-3 through 10-10, the currents are the superposition of forward and reverse traveling waves:

$$I_c(z)=I_{\mathrm{Fc}}\psi_F(z)+I_{\mathrm{Rc}}\psi_R(z) \tag{10-11}$$

$$I_{\mathrm{dmn}}(z)=I_{\mathrm{Fdmn}}\psi_F(z)+I_{\mathrm{Rdmn}}\psi_R(z) \quad \mathrm{mn}=12,13,23 \tag{10-12}$$

The voltages are related to the currents by the usual transmission line equations:

$$V_c(z)=Z_{0c}[I_{\mathrm{Fc}}\psi_F(z)-I_{\mathrm{Rc}}\psi_R(z)] \tag{10-13}$$

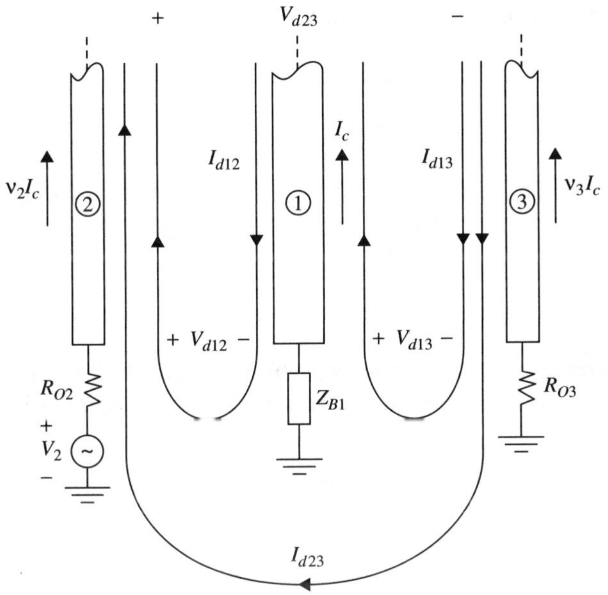

Figure 10-6 At the base of the multiplexed folded unipole, four additional boundary conditions can be formulated, bringing the total to eight. That is sufficient to determine the amplitudes of the common and differential mode currents for each transmitter.

$$V_{\mathrm{dmn}}(z) = Z_{0\mathrm{dmn}}\left[I_{\mathrm{Fdmn}}\psi_F(z) - I_{\mathrm{Rdmn}}\psi_R(z)\right] \qquad \mathrm{mn} = 12, 13, 23 \tag{10-14}$$

In Eqs. 10-11 through 10-14, the forward and reverse traveling wave functions are described by Eqs. 1-2 and 1-3.

In Eq. 10-14, it is seen that there are three distinct differential mode characteristic impedances $Z_{0\mathrm{dmn}}$. These can be calculated using the method described in Sec. 7.7.

As we have done many times in previous chapters, Eqs. 10-11 through 10-14 can be substituted into Eqs. 10-3 through 10-10 to obtain eight explicit equations for the eight unknown current amplitudes I_{Fc}, I_{Rc}, I_{Fdmn}, and I_{Rdmn}. Those eight equations may then be consolidated into a single matrix equation:

$$\mathbf{M}\cdot\mathbf{I} = \begin{pmatrix} V_2 \\ 0 \\ \vdots \\ 0 \end{pmatrix} \tag{10-15}$$

In Eq. 10-15, the vector of unknown current amplitudes is:

$$\mathbf{I} = \begin{pmatrix} I_{\mathrm{Fc}} \\ I_{\mathrm{Rc}} \\ I_{\mathrm{Fd12}} \\ I_{\mathrm{Rd12}} \\ I_{\mathrm{Fd13}} \\ I_{\mathrm{Rd13}} \\ I_{\mathrm{Fd23}} \\ I_{\mathrm{Rd23}} \end{pmatrix} \tag{10-16}$$

The matrix **M** has dimensions 8 × 8. So that the individual elements are legible on the printed page, **M** must be partitioned into four submatrices:

$$\mathbf{M} = \begin{pmatrix} \mathbf{A}_{11} & \mathbf{A}_{12} \\ \mathbf{A}_{21} & \mathbf{A}_{22} \end{pmatrix} \tag{10-17}$$

In Eq. 10-17, the submatrices are:

$$\mathbf{A}_{11} = \begin{pmatrix} Z_{0c} + v_2 R_{02} & Z_{0c} - R_{02} & R_{02} & R_{02} \\ -Z_{0c} - v_3 R_{03} & Z_{0c} - v_3 R_{03} & Z_{0d12} & -Z_{0d12} \\ -Z_{0c} - Z_{B1} & Z_{0c} - Z_{B1} & Z_{0d12} + Z_{B1} & -Z_{0d12} + Z_{B1} \\ 0 & 0 & Z_{0d12} & -Z_{0d12} \end{pmatrix} \tag{10-18}$$

$$\mathbf{A}_{12} = \begin{pmatrix} 0 & 0 & R_{02} & R_{02} \\ Z_{0d13} + R_{03} & -Z_{0d13} + R_{03} & R_{03} & R_{03} \\ -Z_{B1} & -Z_{B1} & 0 & 0 \\ Z_{0d13} & -Z_{0d13} & -Z_{0d23} & Z_{0d23} \end{pmatrix} \tag{10-19}$$

$$\mathbf{A}_{21} = \begin{pmatrix} \psi_F(h) & \psi_R(h) & 0 & 0 \\ 0 & 0 & (Z_{0d12} - Z_{T12})\psi_F(h) & -(Z_{0d12} + Z_{T12})\psi_R(h) \\ 0 & 0 & 0 & 0 \\ 0 & 0 & 0 & 0 \end{pmatrix} \tag{10-20}$$

$$\mathbf{A}_{22} = \begin{pmatrix} 0 & 0 & 0 & 0 \\ 0 & 0 & 0 & 0 \\ (Z_{0d13} - Z_{T13})\psi_F(h) & -(Z_{0d13} + Z_{T13})\psi_R(h) & 0 & 0 \\ 0 & 0 & (Z_{0d23} - Z_{T23})\psi_F(h) & -(Z_{0d23} + Z_{T23})\psi_R(h) \end{pmatrix} \tag{10-21}$$

As already mentioned, in Eqs. 10-18 through 10-21, the resistances R_{02} and R_{03} may, in fact, be complex to account for band pass, band reject, and other filters constructed from reactive components.

The solutions to matrix Eq. 10-15 describe the common and differential mode currents everywhere on the antenna in response to the transmitter excitation V_2 on wire 2. The solutions to a similar equation will describe the response to the transmitter excitation V_3 on wire 3. The total current is the superposition of those solutions.

In this section, we have considered only the most simple configuration of folded unipole. More complicated configurations can also be multiplexed. The mathematical model becomes commensurately more complicated, not so much conceptually, but in terms of straightforward algebra and larger matrices. Note, however, that the matrices are still very small compared with those needed by the general purpose, wire antenna, numerical analysis computer programs. In any case, the benefits, in terms of saving costly real estate and materials, may well be worth the extra mathematical effort!

10.4 Helical Folded Unipole Antennas

Until this section, the constituent wires and cylinders of the folded unipole have been linear or straight. There has long existed conventional wisdom, especially among microwave design engineers, that if the geometry of a device is deformed gradually enough from a straight line, then most, if not all, of the design principles are retained. This wisdom was successfully used to design curved and twisted waveguides. It ought to apply to the folded unipole antenna as well. If it is deformed from an ensemble of straight wires to a helix, then its remarkable properties ought to be preserved, while at the same time acquiring new ones.

From another point of view, reconfiguring the helix as a folded unipole should result in a helical antenna with extra desirable properties.

The helix geometry for antennas has been well documented since circa 1947 [2]. One of the most prolific writers and contributors was John D. Kraus (1910–2004), and he summarized the versatile geometry eloquently [3]:

> The helical antenna is the general form of antenna of which the linear and loop antennas are special cases. Thus, a helix of fixed diameter collapses to a loop as the spacing approaches zero. On the other hand, a helix of fixed spacing between turns straightens out into a linear conductor as the diameter approaches zero… A helix may radiate in many modes… In the axial mode of radiation the field is a maximum in the direction of the helix axis and is circularly polarized or nearly so… In the normal mode of radiation, the field is a maximum in a direction normal to the helix axis, and for a certain relation between the spacing and diameter the field is, in theory, circularly polarized.

From Kraus' description, it is seen that the helix geometry offers at least two new design possiblities. First, the polarization of the electromagnetic radiation can be circular, in contrast to linear for the straight geometry.

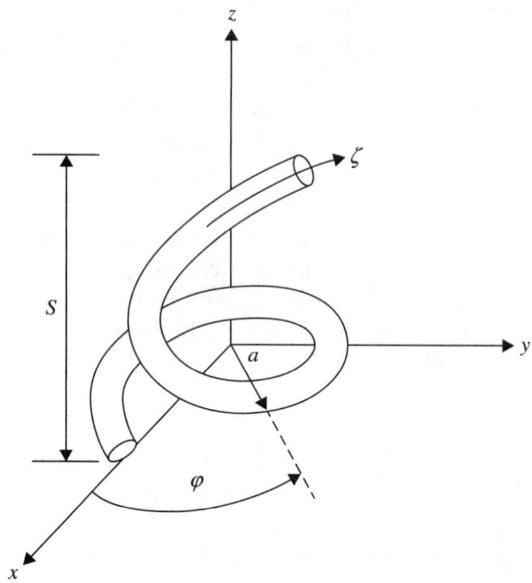

Figure 10-7 The geometry of one turn of a helical antenna is determined by the radius of the turn a and the spacing between turns S.

Second, to the extent that the helix is a loop antenna, it is a magnetic dipole. This is in contrast to the purely electric dipole provided by the straight geometry. Electric and magnetic dipoles exhibit different radiation patterns and other properties, especially in the vicinity of complex environments.

Based upon the above discussion, the most striking difference between the linear and helical folded unipole will be in the shape and polarization of the electromagnetic radiation pattern. Accordingly, in this section, we will focus on its calculation. To that end, Fig. 10-7 shows a single turn of a helix. For purposes of radiation pattern calculation, the turn is characterized by radius a and spacing S. A coordinate ζ runs through the center of the wire.

Figure 10-8 results if we unroll the helix and flatten it onto a plane, conveniently showing the relationship among the geometric parameters. In addition to radius and spacing discussed earlier, it also shows the total length of the turn L, and the pitch angle α. It is seen that:

$$\zeta = \sqrt{(a\phi)^2 + z^2} = \frac{a\phi}{\cos\alpha} = \frac{z}{\sin\alpha} \tag{10-22}$$

In Eq. 10-22, when $\zeta = L$, then $z = S$.

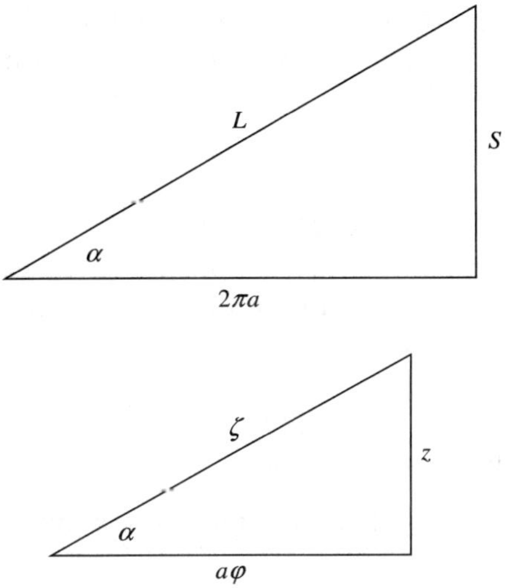

Figure 10-8 If a single turn of the helical antenna is unrolled and flattened onto a plane, then the relationship between the essential geometric parameters is clarified. The total length of the turn is L, and the pitch angle is α. The coordinate ζ follows the wire and ranges in value from 0 to L.

Figure 10-9 shows the geometry convenient for calculating the electromagnetic field in the radiation zone, that is, the far field. The observer is at vector postion $\mathbf{r}_0$ with respect to the origin of the coordinate system:

$$\mathbf{r_0} = \mathbf{u_x} r_0 \sin\theta\cos\phi + \mathbf{u_y} r_0 \sin\theta\sin\phi + \mathbf{u_z} r_0 \cos\theta \tag{10-23}$$

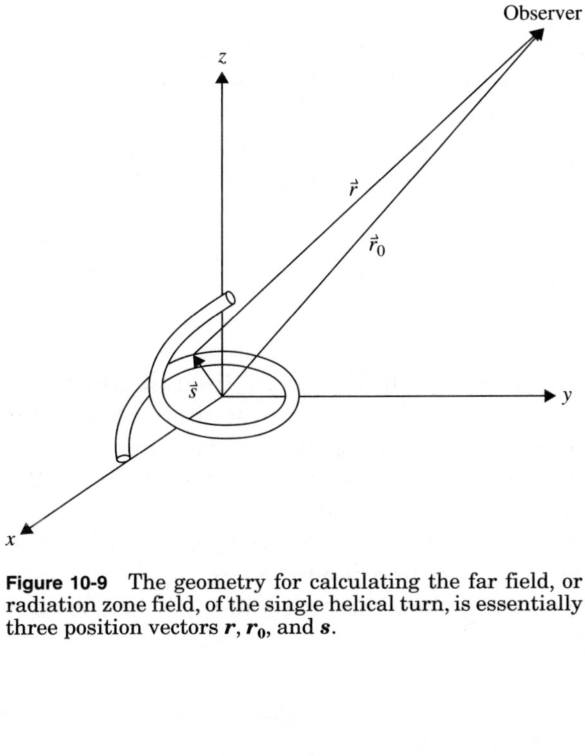

Figure 10-9 The geometry for calculating the far field, or radiation zone field, of the single helical turn, is essentially three position vectors $\boldsymbol{r}$, $\boldsymbol{r}_0$, and $\boldsymbol{s}$.

In Eq. 10-23, **u** is a unit vector in the direction indicated by the subscript. A small segment of the helical wire is at vector position **s** with respect to the origin:

$$\mathbf{s} = \mathbf{u_x} a\cos\phi' + \mathbf{u_y} a\sin\phi' + \mathbf{u_z}\frac{S}{2\pi}\phi' \tag{10-24}$$

In Eq. 10-24, the prime will distinguish the bearing of the wire segment ϕ' from the bearing of the observer ϕ. From Eqs. 10-23 and 10-24, the position of the observer with respect to the each wire segment of the helix is:

$$\mathbf{r} = \mathbf{r_0} - \mathbf{s} \tag{10-25}$$

With Eqs. 10-22 through 10-25, we can begin to compute the magnetic vector potential of the helical turn:

$$\mathbf{A(r)} = \mu_0 \int_0^{2\pi} \frac{e^{-\mathrm{jkr}}}{4\pi r}\mathbf{I}(\phi')a\,d\phi' \tag{10-26}$$

From Eq. 10-26, we see that **A** will in general have nonzero components in all three directions **x**, **y**, and **z** because the current on the helix also is nonzero in those directions. In contrast, the straight or linear antenna has only a nonzero A_z. So, we can expect some extra bookkeeping in the case of the helix.

In the integral on the right side of Eq. 10-26, the distance r is given by the magnitude of Eq. 10-25. We can evaluate the integral analytically if we make a far field approximation to r. The details of this approximation are useful in many antenna calculations, and so we articulate them carefully here. Using Eqs. 10-23 and 10-24, the exact value for r is:

$$r = |\mathbf{r}| = \sqrt{(r_0\sin\theta\cos\phi - a\cos\phi')^2 + (r_0\sin\theta\cos\phi - a\sin\phi')^2 + \left(r_0\cos\theta - \frac{S}{2\pi}\phi'\right)^2} \tag{10-27}$$

If we were to use Eq. 10-27 directly in Eq. 10-26, then the resulting integral would require numerical evaluation at best, and in the worst case it might prove completely intractable due to precision requirements. Fortunately, we can approximate the right side of Eq. 10-27, and the approximation will result in an integral that can be evaluated analytically. If we expand each of the squared terms within the square root operator, then an approximation becomes apparent. Since a is much less than r_0, we can neglect the terms in a^2. Also, the sum of the terms in r_0^2 is unity. Applying these observations, Eq. 10-27 becomes:

$$r = r_0\sqrt{1 - 2\frac{a\sin\theta\cos\phi\cos\phi'}{r_0} - 2\frac{a\sin\theta\sin\phi\sin\phi'}{r_0} - 2\frac{S\phi'\cos\theta}{2\pi r_0}} \tag{10-28}$$

Finally, we can use the binomial expansion for the square root,

$$\sqrt{1+x} \approx 1 + \frac{x}{2} \tag{10-29}$$

so that Eq. 10-28 becomes:

$$r = r_0 - a\sin\theta\cos\phi\cos\phi' - a\sin\theta\sin\phi\sin\phi' - \frac{S\cos\theta}{2\pi}\phi' \tag{10-30}$$

Using Eq. 10-30, Eq. 10-26 becomes:

$$\mathbf{A(r)} = \frac{a\mu_0 e^{-\mathrm{j}kr_0}}{4\pi r_0}\int_0^{2\pi} \mathbf{I}(\phi')e^{\mathrm{j}ka(\sin\theta\cos\phi\cos\phi' + \sin\theta\sin\phi\sin\phi' + \phi'\cos\theta\tan\alpha)}d\phi' \tag{10-31}$$

In Eq. 10-31, we have used the geometric relation,

$$\frac{S}{2\pi a} = \tan\alpha \tag{10-32}$$

Equation 10-31, is a concise description of the magnetic vector potential, provided that we can describe the vector current distribution I(ϕ') on the helix. As with the straight folded unipole antenna, there will be forward and reverse traveling current waves. The forward wave is:

$$\mathbf{I_F}(\phi') = (\mathbf{u_\varphi}\cos\alpha + \mathbf{u_z}\sin\alpha)I_{\mathrm{Fc}}e^{-\mathrm{j}k\zeta} \tag{10-33}$$

Before we can apply Eq. 10-33 to 10-32, we must rewrite ζ in terms of ϕ'. Also, it is important to realize that the unit vector $\mathbf{u_\phi}$ is not a constant vector; however, it can be rewritten in terms of constant unit vectors from the Cartesian or rectangular coordinate system:

$$\mathbf{u_\varphi} = -\mathbf{u_x}\sin\phi' + \mathbf{u_y}\cos\phi' \tag{10-34}$$

Using Eqs. 10-22 and 10-34, Eqn. 10-33 becomes:

$$\mathbf{I_F}(\phi') = (-\mathbf{u_x}\cos\,\alpha\sin\phi' + \mathbf{u_y}\cos\,\alpha\cos\phi' + \mathbf{u_z}\sin\alpha)I_{\mathrm{Fc}}e^{-j\frac{ka\phi'}{\cos\alpha}} \tag{10-35}$$

Similarly, the reverse traveling wave current is:

$$\mathbf{I_R}(\phi') = (-\mathbf{u_x}\cos\alpha\sin\phi' + \mathbf{u_y}\cos\alpha\cos\phi' + \mathbf{u_z}\sin\alpha)I_{Rc}e^{j\frac{ka\phi'}{\cos\alpha}} \tag{10-36}$$

In Eqs. 10-35 and 10-36 the subscript c on the current amplitudes is a reminder that the only radiating currents are common mode currents.

Equations 10-31, 10-35, and 10-36 are a complete description of the magnetic vector potential for one turn of the helical antenna. The magnetic vector potential for the other turns are obtained simply by translating the results along the z-axis. The total magnetic vector potential is the superposition of those from all of the turns.

From Eqs. 10-35 and 10-36, it is seen that the magnetic vector potential has nonzero components in all three directions **x**, **y**, and **z**, as expected.

Once the total magnetic vector potential **A** is obtained, the magnetic field follows readily from:

$$\mathbf{H} = \frac{1}{\mu_0}\nabla\times\mathbf{A} \tag{10-37}$$

In the radiation zone, or far field, the electric field is simply related to the magnetic field:

$$\mathbf{E} = -\frac{\mathbf{k}\times\mathbf{H}}{\omega\varepsilon_0} \tag{10-38}$$

In the far field, the vector wave number is:

$$\mathbf{k} = \mathbf{u_r}\omega\sqrt{\mu_0\varepsilon_0} \tag{10-39}$$

Using Eq. 10-39, Eq. 10-38 becomes:

$$\mathbf{E} = -\eta\mathbf{u_r}\times\mathbf{H} \tag{10-40}$$

We can perform a sanity check on Eq. 10-31 by showing that it reduces to the classic result for a small loop antenna. In that case, the pitch angle $\alpha = 0$, and the current $\mathbf{I}(\phi')$ is a constant:

$$\mathbf{I}(\phi') = (-\mathbf{u_x}\sin\phi' + \mathbf{u_y}\cos\phi')I_0 \qquad \textit{small loop} \tag{10-41}$$

Using Eq. 10-41, let's evaluate Eq. 10-31 in the xz plane, that is, in the plane $\phi = 0$. Actually, we could calculate the magnetic vector potential in any plane and by symmetry we would obtain the same result; however, the xz plane is less mathematically complicated. The result is:

$$\mathbf{A}=\frac{a\mu_0 e^{-jkr_0} I_0}{4\pi r_0}\left[\frac{(\mathbf{u_x}-j\mathbf{u_y})}{2}\int_0^{2\pi} d\phi' e^{j(ka\sin\theta\cos\phi'+\phi')}+\frac{(\mathbf{u_x}+j\mathbf{u_y})}{2}\int_0^{2\pi} d\phi' e^{j(ka\sin\theta\cos\phi'-\phi')}\right] \tag{10-42}$$

We can evaluate the integrals on the right side of Eq. 10-42 exactly, using a formula for Bessel functions [4]:

$$J_n(x)=\frac{1}{\pi}\int_0^{\pi}\cos(x\sin\phi'-n\phi')d\phi'=\frac{1}{2\pi}\int_0^{2\pi}\cos(x\sin\phi'-n\phi')d\phi' \tag{10-43}$$

Using Eq. 10-43, Eq. 10-42 becomes:

$$\mathbf{A}=\frac{a\mu_0 e^{-jkr_0} I_0}{2r_0}\left[\frac{(\mathbf{u_x}-j\mathbf{u_y})}{2}J_{-1}(ka\sin\theta)+\frac{(\mathbf{u_x}+j\mathbf{u_y})}{2}J_1(ka\sin\theta)\right] \tag{10-44}$$

The Bessel functions are asymmetric with respect to order:

$$J_{-n}(x)=(-1)^n J_n(x) \tag{10-45}$$

Using Eq. 10-45, Eq. 10-44 becomes:

$$\mathbf{A}=\mathbf{u_y}\frac{ja\mu_0 e^{-jkr_0} I_0}{2r_0}J_1(ka\sin\theta) \tag{10-46}$$

In the xz plane, the unit vector $\mathbf{u_y}$ is the same as the unit vector $\mathbf{u}_\phi$, and so Eq 10-46 finally becomes:

$$A_\phi=j\frac{a\mu_0 e^{-jkr_0} I_0}{2r_0}J_1(ka\sin\theta) \qquad \textit{small loop} \tag{10-47}$$

Equation 10-47 is the classic result from the literature for a small loop antenna [5]. This gives us confidence about using Eqs. 10-31, 10-35, and 10-36 for the more general helical antenna.

In this section, we have focused on predicting the radiation pattern of the helical folded unipole antenna. The current distribution is determined in the same manner as for straight folded unipoles. Boundary conditions are formulated at the top, base, and segment boundaries along the antenna. Instead of the linear coordinate z, we use the coordinate ζ that follows the helix.

The helical folded unipole has been reduced to practice in the UHF band as a Wi-Fi antenna, and, no doubt, there will be many more applications of this configuration in the future. If the straight folded unipole is the superset of linear antennas and transmission lines, then the helical folded unipole is the superset of these plus loop antennas. It is a very general and versatile electromagnetic device, indeed.

10.5 The Folded Unipole Antenna in Medicine, Geophysics, and Complex Environments

Its extreme compactness makes the folded unipole antenna especially well suited to instrumentation. As part of a tiny probe, it could be used to sense diverse environments ranging from the interior of blood vessels to oil drilling shafts. In this section, we will explore how such a probe might interact with those surroundings.

Figure 10-10 shows the cross section for such an application. Space is divided into two regions, each characterized by its own permeability, permittivity, and conductivity. The values may be arbitrary. We do not require that either region be lossless. Region 1 is depicted as circular; however, we can generalize beyond that if necessary. The circular shape

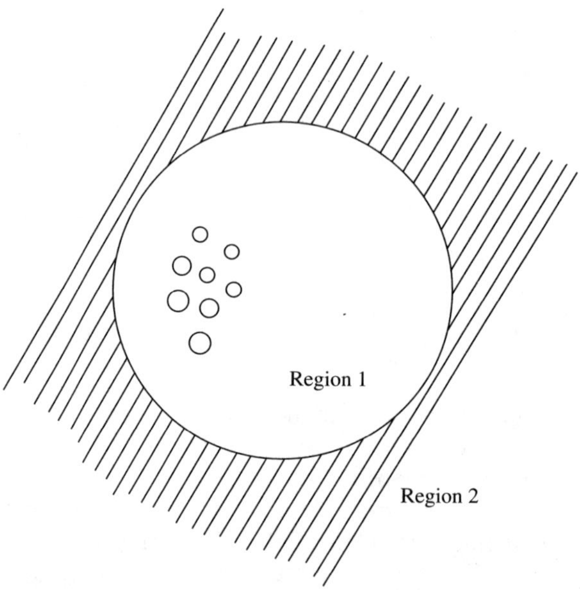

Figure 10-10 The cross section of a folded unipole antenna inside a complex environment consists of an arbitrary number of wires with arbitrary positions inside a cylindrical region. Each of the two regions has arbitrary permittivity, permeability, and conductivity. They could represent environments as diverse as blood vessels or geophysical exploration shafts.

is purely for mathematical convenience. In the analysis that follows, if we are willing to deal with Mathieu functions instead of Bessel functions, then we can readily exchange the circular region for an elliptical one. If we are willing to resort to numerical methods, then the shape of Region 1 may be completely irregular.

The folded unipole is depicted as an ensemble of parallel wires. The number, radius, and position of each wire is arbitrary. In particular, the antenna is not necessarily centered within the region.

With the geometry defined as above, we are ready to develop a mathematical model of the antenna. Much of the work has already been accomplished in previous chapters, especially in Chap. 7 concerning Characteristic Impedance. The three independent design parameters Z_{0d} (differential mode characteristic impedance), Z_{0c} (common mode characteristic impedance), and ν(current transformation ratio) will be sensitive to the environment. By extending the Quasi Static Log Potential (QSLP) theory described in Sec. 7.5, from free space to complex environments, we can anticipate how the design parameters will change.

The building block of QSLP theory is the potential of a single line charge in free space, given by Eq. 7-51. How will that equation change for a single line charge in a complex environment? We can calculate the answer exactly in terms of special functions often used in antenna theory. To that end, Fig. 10-11 shows the most simple case, a single wire

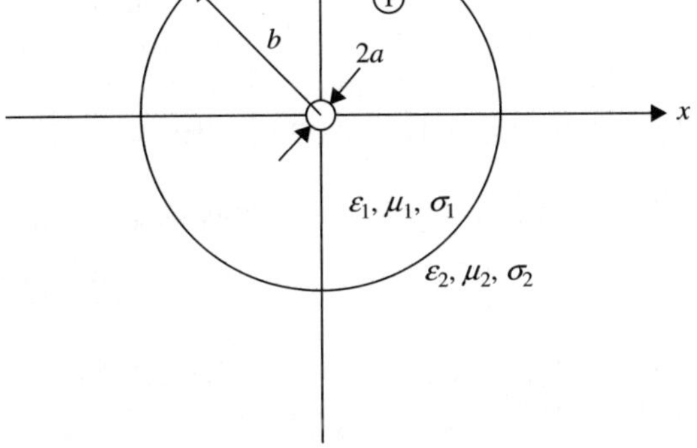

Figure 10-11 The mathematical model of the folded unipole in a complex environment begins with the quasi static potential of a single wire perfectly centered within Region 1.

centered within the two regions. To find the QSLP, it is sufficient to solve the scalar wave equation in Regions 1 and 2, and to apply appropriate boundary conditions at the inteface between them.

We will digress for a moment to justify our approach in further detail, for at least two reasons. First, one might challenge using a *scalar* wave equation for potential to model a system in which there are, in fact, *vector*, time varying electric and magnetic fields. Second, the approach comes from the brilliant mind of William Webster Hansen (1909–1949) who contributed to electromagnetic theory in the 1930s [6,7,8]. His results were summarized in the classic 1941 reference, by Stratton [9]. They are still valuable today, but, unfortunately, they seem to have been largely forgotten in modern text and reference books. Nonetheless, we will make good use of them here.

Hansen showed that the solution to any vector wave equation,

$$\nabla^2\mathbf{C} + k^2\mathbf{C} = \mathbf{0} \tag{10-48}$$

is the superposition of three independent solutions, **L**, **M**, and **N**, given by:

$$\mathbf{N} = \frac{1}{k}\nabla \times \mathbf{M} \tag{10-49}$$

$$\mathbf{L} = \nabla V \tag{10-50}$$

$$\mathbf{M} = \nabla \times \mathbf{u}V \tag{10-51}$$

In Eq. 10-48, the vector **C** can be any vector field, such as the electric field **E**, the magnetic field **H**, or the magnetic vector potential **A**. They all must satisfy the vector wave equation. In our QSLP theory, the electric field **E** is the vector of interest.

In Eq. 10-51, the unit vector **u** is called the *pilot vector*. Typically, it is one of the three ordinate directions from the coordinate system of interest, for example, **x**, **y**, or **z** in a Cartesian system. The choice is arbitrary, provided that it leads to satisfying the appropriate boundary conditions.

In Eqs. 10-49 and 10-50, V is the solution to the scalar wave equation:

$$\nabla^2 V + k^2 V = 0 \tag{10-52}$$

Equations 10-49 and 10-52 are the foundations for our QSLP theory. They explain how we can segue from the solution of a scalar wave equation to the potential for a vector electric field.

With confidence based upon the above discussion, we can write down the solutions to the wave equation for the geometry shown in Fig. 10-11. By symmetry, the solutions have no angular dependence, only a radial one. They are:

$$V_1(\rho) = \frac{H_0^{(2)}(k_1\rho)}{j4\boldsymbol{\varepsilon}_1} Q + AJ_0(k_1\rho) \tag{10-53}$$

$$V_2(\rho) = BH_0^{(2)}(k_2\rho) \tag{10-54}$$

In Eqs. 10-53 and 10-54, $H_0^{(2)}$ is the Hankel function of the second kind, and J_0 is the Bessel function of the first kind. The first term in Eq. 10-53 is the particular solution to Eq. 10-52, corresponding to the line charge Q. The remaining terms are homogeneous solutions. A and B are unknown constants to be determined by boundary conditions at the interface between Regions 1 and 2. The complex wave numbers are:

$$k_n = \omega\sqrt{\mu_n\left(\varepsilon_n - j\frac{\sigma_n}{\omega}\right)} \qquad n = 1, 2 \tag{10-55}$$

In Eq. 10-55, the permeability μ_n, permittivity ε_n, and conductivity σ_n are arbitrary complex numbers. Thus, we can model practically any environment.

To find the unknown constants A and B, we need two independent boundary conditions. These follow from the Maxwell Equations for the **E** field at the interface between the two regions. First, the tangential field, i.e., the component parallel to the interface, must be continuous across the boundary. In terms of the potential V, that means:

$$V_1(b) = V_2(b) \tag{10-56}$$

The second boundary condition requires that the displacement current $j\omega\boldsymbol{\varepsilon}\mathbf{E}$ be continuous across the interface. In terms of the potential V,

$$\boldsymbol{\varepsilon}_1 \frac{\partial V_1(b)}{\partial \rho} = \boldsymbol{\varepsilon}_2 \frac{\partial V_2(b)}{\partial \rho} \tag{10-57}$$

In Eqs. 10-53 and 10-57, the conductivity is included in the symbol for complex permittivity:

$$\boldsymbol{\varepsilon}_\mathbf{n} = \varepsilon_n - j\frac{\sigma_n}{\omega} \qquad n = 1, 2 \tag{10-58}$$

If Eqs. 10-53 and 10-54 are substituted into Eqs. 10-56 and 10-57, then, after some algebra, we end up with formulas for the unknown constants A and B:

$$A = B\frac{H_0^{(2)}(k_2 b)}{J_0(k_1 b)} - \frac{Q}{j4\varepsilon_1}\frac{H_0^{(2)}(k_1 b)}{J_0(k_1 b)} \tag{10-59}$$

$$B = \frac{k_1 Q}{j4}\frac{J_0(k_1 b)H_0^{(2)\prime}(k_1 b) - J_0{}'(k_1 b)H_0^{(2)}(k_1 b)}{\varepsilon_2 k_2 J_0(k_1 b)H_0^{(2)\prime}(k_2 b) - \varepsilon_1 k_1 J_0{}'(k_1 b)H_0^{(2)}(k_2 b)} \tag{10-60}$$

We can greatly simplify Eq. 10-60. Since their official debut in 1824, more has been studied and written about Bessel functions than any other special function of applied physics. In particular, the numerator of the second factor on the right side of Eq. 10-60 is known as a *Wronskian*. The Wronskian is reducible to a simple expression [4]:

$$J_n(x)H_n^{(2)\prime}(x) - J_n{}'(x)H_n^{(2)}(x) = -j\frac{2}{\pi x} \qquad n = 0, 1, 2, \ldots \tag{10-61}$$

Using Eq. 10-61, Eq. 10-61 simplifies to:

$$B = -\frac{Q}{2\pi b}\frac{1}{\varepsilon_2 k_2 J_0(k_1 b)H_0^{(2)\prime}(k_2 b) - \varepsilon_1 k_1 J_0{}'(k_1 b)H_0^{(2)}(k_2 b)} \tag{10-62}$$

As a sanity check on our considerable algebra, it is readily verified that, as Regions 1 and 2 become identical, then, $k_1 \to k_2$, $\varepsilon_1 \to \varepsilon_2$, and:

$$\lim_{Region\,1 \to 2} B = \frac{Q}{j4\varepsilon_1} \tag{10-63}$$

$$\lim_{Region\,1 \to 2} A = 0 \tag{10-64}$$

When Eqs. 10-63 and 10-64 are true, then the solution to the scalar wave equation, described by Eq. 10-53, reduces to the potential for a line charge in free space.

Small argument approximations may be applied to Eqs. 10-53, 10-54, 10-59 and 10-62, eliminating the need to evaluate Bessel functions with complex arguments.

Before we can apply QSLP theory to an arbitrary ensemble of wires, we need to extend our results from the perfectly centered wire in Fig. 10-11 to an offset wire at radial position ρ_0 and angular position

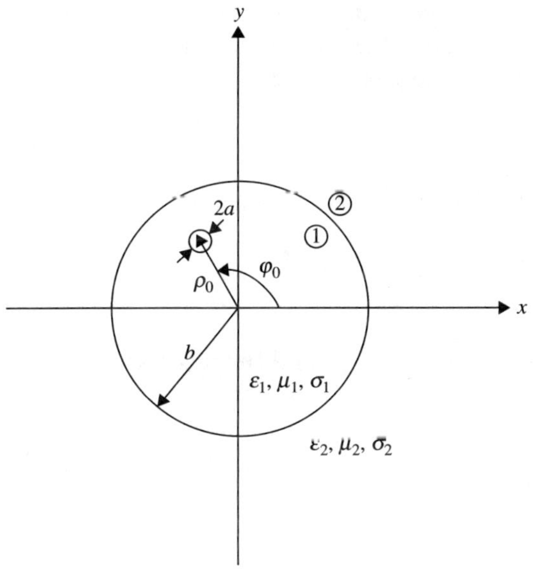

Figure 10-12 The mathematical model of the folded unipole in a complex environment can be generalized by moving the wire away from the center. Its potential can be that of any of the arbitrary wires in the ensemble that compose the complete unipole.

ϕ_0 as shown in Fig. 10-12. The mathematical steps will be similar. The solutions to the scalar wave equation in Regions 1 and 2 are:

$$V_1(\rho) = \frac{Q}{j4\varepsilon_1} H_0^{(2)}(k_1|\boldsymbol{\rho} - \boldsymbol{\rho}_0|) + \sum_{-\infty}^{\infty} A_n J_n(k_1\rho)\, e^{jn\phi} \tag{10-65}$$

$$V_2(\rho) = \sum_{-\infty}^{\infty} B_n H_n^{(2)}(k_2\rho)\, e^{jn\phi} \tag{10-66}$$

Comparing Eqs. 10-65 and 10-66 to Eqs. 10-53 and 10-54, the differences are easy to explain. In both Eqs. 10-65 and 10-53, the first terms are the particular solutions to the scalar wave equation due to the line charge Q. In Eq. 10-53 we only needed one homogeneous solution, the fundamental mode, because of angular symmetry. In Eq. 10-65, however, the angular symmetry was destroyed when the wire was offset from the center. As a result, we now need all of the higher order modes, with unknown coefficients A_n, to complete the solution. Similarly, in Eq. 10-66, we also need the higher order modes, with unknown coefficients B_n, because the angular symmetry was destroyed. Fortunately, all of these higher order modes will not add substantially to our mathematical efforts.

As mentioned earlier, more has been studied and written about Bessel functions than any other special function in mathematical physics. One of the results we will use now is the *generating function* for Hankel functions [10, 11]:

$$H_0^{(2)}(k_1|\boldsymbol{\rho}-\boldsymbol{\rho}_0|)=\begin{cases}\sum_{-\infty}^{\infty} H_n^{(2)}(k_1\rho_0)J_n(k_1\rho)\,e^{\mathrm{j}n(\phi-\phi_0)} & \rho<\rho_0\\ \sum_{-\infty}^{\infty} J_n(k_1\rho_0)H_n^{(2)}(k_1\rho)\,e^{\mathrm{j}n(\phi-\phi_0)} & \rho>\rho_0\end{cases} \tag{10-67}$$

Using Eq. 10-67, in the space between $\rho=\rho_0$ and the interface at $\rho=b$, Eq. 10-65 becomes:

$$V_1(\rho)=\sum_{-\infty}^{\infty}\left[\frac{Q}{j4\varepsilon_1}J_n(k_1\rho_0)H_n^{(2)}(k_1\rho)\,e^{\mathrm{j}n(\phi-\phi_0)}+A_nJ_n(k_1\rho)\,e^{\mathrm{j}n\phi}\right] \tag{10-68}$$

The pair of equations, Eq. 10-66 and 10-68, are almost as easy to solve as the pair for the perfectly centered wire, Eqs. 10-53 and 10-54. The reason is the summation operators. Because they operate identically on all terms in both equations, that means we can solve for the unknown constants A_n and B_n separately for each n, just as we solved for A and B in Eqs. 10-53 and 10-54. As a matter of formal procedure, we simply ignore the summation operator, and then solve for A_n and B_n.

The boundary conditions on Eqs. 10-66 and 10-68 are the very same ones we applied to Eqs. 10-53 and 10-54. That is, the boundary conditions are Eqs. 10-56 and 10-57. When those boundary conditions are applied separately to each set of terms with like subscript n in Eqs. 10-66 and 10-68, the solutions are:

$$A_n=B_n\frac{H_n^{(2)}(k_2b)}{J_n(k_1b)}-\frac{Q}{j4\varepsilon_1}\frac{J_n(k_1\rho_0)}{J_n(k_1b)}H_n^{(2)}(k_1b)\,e^{-\mathrm{j}n\phi_0} \tag{10-69}$$

$$B_n=-\frac{J_n(k_1\rho_0)\,e^{-\mathrm{j}n\phi_0}Q}{2\pi b}\,\frac{1}{\varepsilon_2k_2J_n(k_1b)H_n^{(2)\prime}(k_2b)-\varepsilon_1k_1J_n{}'(k_1b)H_n^{(2)}(k_2b)} \tag{10-70}$$

Just as we used the Wronskian, Eq. 10-61, to obtain Eq. 10-62 for the perfectly centered wire, we also used it in the process of solving for B_n in Eq. 10-70.

Equations 10-65, 10-69, and 10-70 allow us to calculate the potential everywhere inside Region 1 for a wire arbitrarily positioned within that region. As a practical matter, when the region is small compared with a wavelength, we may use small argument approximations to evaluate the Bessel functions, and we may also truncate the summations after just a few terms. So the computational burden is not that severe.

We may use that potential to find the characteristic impedances and current transformation ratio using the procedures described in Sec. 7.7.

10.6 The Human Body as a Folded Unipole Antenna

Anyone who has grabbed hold of a rabbit ears type antenna connected to a television set has observed how the human body can function as a receiving antenna. Our investigations in previous chapters, and especially in the previous section, have established the principles and mathematical tools to pursue this topic in a quantitative way.

Figure 10-13 shows a human body, including various cross sections. The latter are clearly recognizable as cross sections of a folded unipole antenna. Thus, the body may be modeled as a multisegment folded unipole. No wonder it functions as a receiving antenna!

The chief difference between the human body and a deliberately constructed folded unipole is the choice of materials. The human body is made from organic matter that exhibits a wide range of dielectric constants, and poor to moderate conductivity. A commercial antenna, of course, is fabricated from highly conducting metal. Both carry electrical currents, however. Currents in the metal are conduction currents, that is, electrons moving through a thin lattice near the surface. In contrast, currents in the human body are polarization currents, that is, bound charges in microscopic dipoles that oscillate in response to an incident electromagnetic field.

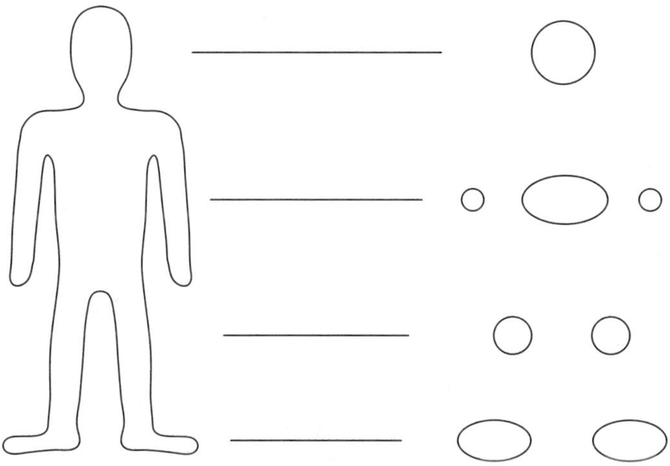

Figure 10-13 If the various cross sections are considered, the human body can be readily regarded as a multisegment folded unipole antenna constructed from organic wires. The wires are concentric layers of skin, muscle, bone, and other tissue. The wires are obviously not metal but they carry electrical currents just the same. The currents are polarization currents rather than conduction currents.

Our prior mathematical efforts, especially in the previous section, have set the stage for calculating the polarization currents in the human body induced by an incident electromagnetic wave. For example, Fig. 10-14 shows the cross section of a limb. We have chosen a three-layer cylinder of skin, muscle, and bone as a model. Alternatively, we could choose more layers or fewer layers, depending upon the wavelength and level of mathematical detail desired. Each layer is distinguished by a different value for permittivity, permeability, and conductivity. Also, we are not constrained to a circular cylinder. If we are willing to deal with the added complexity of Mathieu functions instead of Bessel functions, then the cross section could be elliptical. If we are willing to resort to purely numerical methods, then the cross section could be irregular. For purposes of discussion here, however, we prefer a simple example.

We follow much the same mathematical procedure as in the previous section. Due to the work of W. W. Hansen, it is sufficient to solve the scalar wave equation for the potentials V_i, with $i = 1, 2, 3, 4$. The scalar wave equation is Eq. 10-52. The solutions will involve unknown coefficients, and these will be determined by satisfying boundary conditions at the interfaces between layers.

In Region 1, the solutions to Eq. 10-52 are:

$$V_1(\rho) = \sum_{-\infty}^{\infty} A_n J_n(k_1 \rho)\, e^{jn\phi} \tag{10-71}$$

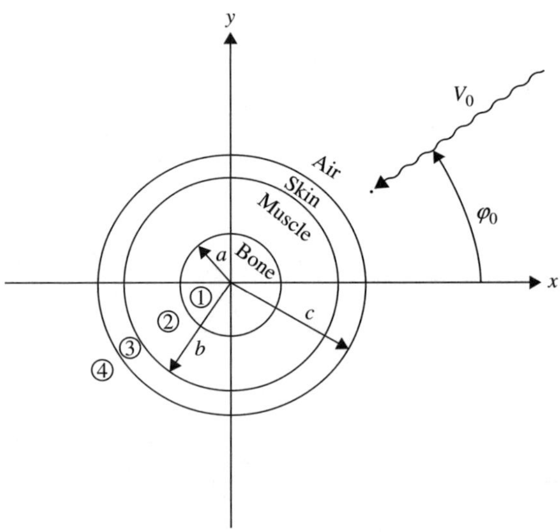

Figure 10-14 The three-region model of an organic wire, such as a limb in the human body, can be used to calculate the polarization current induced by an incident electromagnetic wave.

In Eq. 10-71, only Bessel functions of the first kind J_n are allowed as solutions for V_1 because others, which will "blow up" at the origin. A_n are the unknown coefficients.

In Regions 2 and 3, the solutions to the scalar wave equation are:

$$V_2(\rho) = \sum_{-\infty}^{\infty} [B_n J_n(k_2\rho) + C_n Y_n(k_2\rho)]\, e^{jn\phi} \tag{10-72}$$

$$V_3(\rho) = \sum_{-\infty}^{\infty} [D_n J_n(k_3\rho) + E_n Y_n(k_3\rho)]\, e^{jn\phi} \tag{10-73}$$

In Eqs. 10-72 and 10-73, Bessel functions of the second kind Y_n are permitted because Regions 2 and 3 do not include the origin.

The scalar potential in Region 4, which is free space, is:

$$V_4(\rho) = V_0\, e^{-jk_4\rho\cos\phi_0} + \sum_{-\infty}^{\infty} F_n H_n^{(2)}(k_4\rho)\, e^{jn\phi} \tag{10-74}$$

On the right side of Eq. 10-74, the first term is the incident plane wave. The remaining terms, with unknown coefficients F_n, are outward traveling waves scattered from the cylindrical Regions 1 through 3.

In Eqs. 10-71 through 10-74, the wave numbers k_n are:

$$k_n = \sqrt{\omega^2 \mu_n \boldsymbol{\varepsilon}_\mathbf{n} - k_z^2} \tag{10-75}$$

Equation 10-75 is called a *dispersion relation*. On the right side, the term k_z is the propagation constant in the longitudinal, or z, direction. It is determined by the incident plane wave, which may arrive at a nonzero angle with respect to the xy plane, that is, the two-dimensional plane containing the cross section shown in Fig. 10-14. As a result, all of the potentials V_n implicitly have a longitudinal dependence of the form $exp(-jk_z z)$.

Also on the right side of Eq. 10-75, the complex permittivity $\boldsymbol{\varepsilon}_\mathbf{n}$ includes the conductivity of the region, and is described by Eq. 10-58.

As discussed in the previous section, more has been studied and written about Bessel functions than any other special function of applied physics. We now make use of one of many *generating functions* for Bessel functions [12],

$$e^{-jk\rho\cos\phi} = \sum_{-\infty}^{\infty} j^{-n} J_n(k\rho)\, e^{jn\phi} \tag{10-76}$$

Using Eq. 10-76, Eq. 10-74 becomes:

$$V_4(\rho) = \sum_{-\infty}^{\infty} \left[V_0 j^{-n} J_n(k_4\rho)\, e^{jn\phi_0} + F_n H_n^{(2)}(k_4\rho)\, e^{jn\phi} \right] \tag{10-77}$$

Comparing Eqs. 10-71, 10-72, 10-73, and 10-77, it is seen that they all have the same summation operator in common. That means we can solve for the unknown coefficients A_n, B_n, C_n, D_n, E_n and F_n, separately for each n. This simplifies our mathematical labors substantially. As a matter of formal procedure, we ignore the summation operator, and solve the equations as though they have only six unknowns instead of $6n$ unknowns.

To determine the six unknown coefficients for each n, we will need six independent boundary conditions. These can be formulated at the interfaces between regions. Between Regions 1 and 2, both the tangential electric field and the normal displacement current must be continuous. These requirements translate into the two boundary conditions,

$$V_1(a) = V_2(a) \tag{10-78}$$

$$\boldsymbol{\varepsilon}_1 \frac{\partial V_1(a)}{\partial \rho} = \boldsymbol{\varepsilon}_2 \frac{\partial V_2(a)}{\partial \rho} \tag{10-79}$$

Similarly, at the interface between Regions 2 and 3:

$$V_2(b) = V_3(b) \tag{10-80}$$

$$\boldsymbol{\varepsilon}_2 \frac{\partial V_2(b)}{\partial \rho} = \boldsymbol{\varepsilon}_3 \frac{\partial V_3(b)}{\partial \rho} \tag{10-81}$$

At the interface between Region 3 and Region 4:

$$V_3(c) = V_4(c) \tag{10-82}$$

$$\boldsymbol{\varepsilon}_3 \frac{\partial V_3(c)}{\partial \rho} = \varepsilon_0 \frac{\partial V_4(c)}{\partial \rho} \tag{10-83}$$

Equations 10-78 through 10-83 can be applied to Eqs. 10-71, 10-72, 10-73, and 10-77, separately for each n. In that way, the unknown coefficients A_n, B_n, C_n, D_n, E_n, and F_n will be determined. Then the potentials V_n in all regions of the cross section will be known.

It is a short logical step from determining the potentials to determining the polarization currents flowing through the cross section:

$$I_{\text{polarization}} = -j\omega \iint_{\text{cross section}} (\boldsymbol{\varepsilon}_{\mathbf{n}} - \varepsilon_0) \frac{\partial V_n(\rho)}{\partial z} dA \tag{10-84}$$

Since the z-dependence for all the potentials is $exp(-jk_z z)$, Eq. 10-84 becomes:

$$I_{\text{polarization}} = -k_z \omega \iint_{\text{cross section}} (\boldsymbol{\varepsilon}_\mathbf{n} - \varepsilon_0) V_n(\rho) dA \qquad (10\text{-}85)$$

In Eq. 10-84 and 10-85 the subscript n is 1, 2, or 3, as the integration progresses through regions 1, 2, and 3, respectively.

The steps outlined in this section have limitations with respect to frequency or wavelength; however, these can be overcome if we are willing to accept extra mathematical complexity. The boundary conditions, Eqs. 10-78 through 10-83, imply that we have selected the quasi static solution to the vector wave equation. That is, we have selected the function **L** in Eq. 10-50 as the solution to Eq. 10-48, with the vector wave function **C** representing the electric field **E**. This is a perfectly acceptable choice provided dimensions of the cross section of the human body are small compared with the wavelength. This requirement is certainly satisfied for frequencies ranging from ELF to the lower part of the UHF band.

If we wish to extend our mathematical model to shorter wavelengths, then we must choose the functions **M** and **N** in Eqs. 10-49 and 10-51 as the solutions to Eq. 10-48. In that case, the vector wave function **C** will be the complete electromagnetic field with all components of both **E** and **H**. Accordingly, we must also formulate boundary conditions different from the ones we have used in this section. We must require that the tangential components of both **E** and **H** be continuous across each interface between regions. We can formulate those boundary conditions in terms of a scalar potential V; however, they will be more complicated than Eqs. 10-78 through 10-83. Using those boundary conditions to determine the complete electromagnetic field will be substantially more complicated than determining the quasi static electric field as outlined in this section.

We have described a procedure for calculating the current induced in the human body by an incident plane wave. This places us squarely in the fascinating, and often controversial, discipline of bioelectromagnetics. It is just one more unique point of view provided by the folded unipole antenna.

10.7 Conclusions

"Compact" and "versatile" are adjectives used frequently in this book, and they apply especially to the applications discussed in this chapter. Most of the ones documented here have been reduced to actual practice. Countless others are waiting to be explored.

Of those that have been reduced to practice, we briefly described an LF antenna that was only 0.02 wavelengths tall. Multifrequency operation was described in a section about multiplexing, and this has been accomplished in the AM broadcast band. We discussed bending the folded unipole into a helix shape to provide circular polarization while preserving its desirable input impedance and broadband properties, and commercial examples are available for the UHF band.

Still awaiting development are folded unipoles immersed in such complex environments as an oil well or a blood vessel. Due to their compactness, they most likely will make excellent probes for instrumentation and diagnostic purposes. We set down the mathematical framework that might guide their design.

The folded unipole antenna even occurs in nature. We need look no further than the human body which, as a matter of common experience, functions as a receiving antenna in the broadcast radio and television bands. We described how the currents induced by an incident electromagnetic wave might be calculated in organic wires composed of layers of skin, muscle, and bone. This is a fascinating, and sometimes controversial, topic within the discipline of bioelectromagnetics.

We outlined the mathematical procedures for investigating the above examples in varying degrees of detail. A complete exposition would require separate and lengthy chapters for each topic, and some could be expanded into entire books. In three of the five applications, we benefited from the great wealth of accumulated knowledge concerning Bessel functions. More has been studied and written about them than any other special function in applied physics. Sadly, they are disappearing from the undergraduate engineering curriculum. They have been displaced by courses in numerical methods. The applications presented here, however, should demonstrate that Bessel functions still have an important place in the design and understanding of antennas. They can model environments and phenomena that remain well beyond the reach of general purpose, numerical, wire antenna computer programs.

"Universal" is the final adjective that comes to mind, although that might be stretching things a bit. To be sure, our investigations and models of the folded unipole integrated the concepts of impedance, transmission lines, and electromagnetic radiation. We ended up with antennas that can be broadband, linearly polarized, or circularly polarized. Actual and potential applications span the spectrum, certainly from VLF through UHF, and very likely beyond that.

"Universal" might also describe our odyssey through electromagnetic theory while exploring the folded unipole antenna. The journey would make a good *second* course in electromagnetics. The usual first course at the undergraduate level is well structured and orderly, and this is as it should be. On the other hand, the course is usually over before the student arrives at the study of antennas and electromagnetic radiation. In contrast, that is where our investigation began. As necessary, we returned to the fundamental Maxwell Equations, for example to compute the exact fields of traveling current waves, or to formulate a quasi static approximation for characteristic impedance. We resurrected the work of W. W. Hansen to unify the scalar wave equation and vector electromagnetic fields. We utilized many of the special functions of mathematical physics as needed, including the exponential integral and Bessel functions. These are a broad assortment of tools and techniques that a practitioner of electromagnetics ought to know.

As befitting a near-universal device, much remains to be discovered and written about the folded unipole antenna. In that respect, we hope this book has stimulated the imagination in addition to presenting useful design information.

References

1. Mullaney, J. H., *et al.*, NORD LF Antenna Course (Second Edition), Rockville, Maryland: Multronics, Inc., 1965.
2. Kraus, J. D., Helical Beam Antenna, *Electronics*, **20**, April 1947, pp. 109–111.
3. Kraus, J. D., Antennas, New York: McGraw-Hill, 1950, p.173.
4. Abramowitz, A., and L. A. Stegun, Handbook of Mathematical Functions, Washington, D.C.: U.S. Government Printing Offices, 1972, p. 360.
5. Kraus, J. D., Antennas, New York: McGraw-Hill, 1950, p. 163.
6. Hansen, W. W., A New Type of Expansion in Radiation Problems, *Physical Review*, **47**, January, 1935, pp. 139–143.
7. Hansen, W. W., Directional Characteristics of Any Antenna Over a Plane Earth, *Physics*, **7**, December, 1936, pp. 460–465.
8. Hansen, W. W., Transformations Useful in Certain Antenna Calculations, *Journal of Applied Physics*, **8**, April, 1937, pp. 284–286.
9. Stratton, J. A., Electromagnetic Theory, New York: McGraw-Hill, 1941, p. 393.
10. Morse, P. M., and H. Feshbach, Methods of Theoretical Physics, New York: McGraw-Hill, 1953, p. 621.
11. Harrington, R. F., Time-Harmonic Electromagnetic Fields, New York: McGraw-Hill, 1961, p. 232.
12. *Ibid.*, p. 231.

Symbols Used in This Chapter

$\mathbf{A}$	Magnetic vector potential
$\mathbf{A}_{11}$	Submatrix of $\mathbf{M}$
$\mathbf{A}_{12}$	Submatrix of $\mathbf{M}$
$\mathbf{A}_{21}$	Submatrix of $\mathbf{M}$
$\mathbf{A}_{22}$	Submatrix of $\mathbf{M}$
A	Coefficient in wave expansion for scalar potential in Region 1
A_n	Coefficient in wave expansion for scalar potential in Region 1
A_ϕ	ϕ-directed component of magnetic vector potential $\mathbf{A}$
a	1. Radius of helix antenna. 2. Radius of Region 1
B	Coefficient in wave expansion for scalar potential in Region 2
B_n	Coefficient in wave expansion for scalar potential in Region 2
b	Radius of Region 2
$\mathbf{C}$	General solution to the vector wave equation
C_n	Coefficient in wave expansion for scalar potential in Region 2

c	Radius of Region 3
dA	Infinitesimal area, used in integration formulas
D_n	Coefficient in wave expansion for scalar potential in Region 3
$\mathbf{E}$	Electric field vector
E_n	Coefficient in wave expansion for scalar potential in Region 3
F_n	Coefficient in wave expansion for scalar potential in Region 4
$\mathbf{H}$	Magnetic field vector
$H_0^{(2)}$	Hankel function of the second kind and zero order
$H_0^{(2)\prime}$	First derivative of $H_0^{(2)}$
$H_n^{(2)}$	Hankel function of the second kind and order n
$H_n^{(2)\prime}$	First derivative of $H_n^{(2)}$
h	Height of folded unipole antenna
h_N	Height of top or Nth segment of folded unipole antenna
$\mathbf{I}$	1. Vector of unknown currents in matrix equation for multiplexed folded unipole. 2. Vector current on a helical folded unipole
$\mathbf{I_R}$	Vector reflected traveling wave of current on helical folded unipole
$\mathbf{I_F}$	Vector forward traveling wave of current on helical folded unipole
I_0	Amplitude of vector current on small loop
I_c	Common mode current
I_{cN}	Common mode current on Nth segment of folded unipole
I_{d12}	Differential mode current flowing on wire ensembles 1 and 2 in multiplexed folded unipole
I_{d13}	Differential mode current flowing on wire ensembles 1 and 3 in multiplexed folded unipole
I_{d23}	Differential mode current flowing on wire ensembles 2 and 3 in multiplexed folded unipole
I_{dmn}	Differential mode current flowing on wire ensembles m and n in multiplexed folded unipole
I_{Fc}	Amplitude of common mode forward traveling wave of current
I_{Fdmn}	Amplitude of differential mode forward traveling current flowing on wire ensembles m and n in multiplexed folded unipole
$I_{\mathrm{polarization}}$	Polarization current induced in the human body
I_{Rc}	Amplitude of common mode reverse traveling wave of current
I_{Rdmn}	Amplitude of differential mode reverse traveling current flowing on wire ensembles m and n of multiplexed folded unipole
J_0	Bessel function of the first kind and order 0
J_0'	First derivative of J_0
J_1	Bessel function of the first kind and order 1
J_{-1}	Bessel function of the first kind and order −1
J_n	Bessel function of the first kind and order n
J_n'	First derivative of J_n
j	Square root of −1
$\mathbf{k}$	Vector wave number
k	Wave number, magnitude of $\mathbf{k}$
k_1	Wave number in Region 1
k_2	Wave number in Region 2
k_3	Wave number in Region 3
k_4	Wave number in Region 4
k_n	Wave number in Region n
$\mathbf{L}$	One of three independent solutions to the vector wave equation, derivable from solutions to the scalar wave equation
L	Total length of one turn of helical antenna

$\mathbf{M}$	1. Matrix in the equation for currents on the multiplexed folded unipole. 2. One of three independent solutions to the vector wave equation, derivable from solutions to the scalar wave equation
$\mathbf{N}$	One of three independent solutions to the vector wave equation, derivable from solutions to the scalar wave equation
N	Total number of items, in the formula for number of possible combinations
Q	Charge per unit length
$\mathbf{r}$	Vector position of observer in three-dimensional space, with respect to incremental length of antenna wire
$\mathbf{r_0}$	Vector position of observer in three-dimensional space, with respect to the origin of the coordinate system
R_{02}	Output impedance of the transmitter hard wired to wire ensemble 2 of a multiplexed folded unipole antenna
R_{03}	Output impedance of the transmitter hard wired to wire ensemble 3 of a multiplexed folded unipole antenna
r_0	Magnitude of $\mathbf{r_0}$
$\mathbf{s}$	Vector position of an incremental length of wire on a helical antenna, with respect to the origin of the coordinate system
S	Spacing between turns of a helical antenna
$\mathbf{u}$	Vector of unit length
$\mathbf{u_r}$	r-directed vector of unit length
$\mathbf{u_x}$	x-directed vector of unit length
$\mathbf{u_y}$	y-directed vector of unit length
$\mathbf{u_z}$	z-directed vector of unit length
$\mathbf{u_\phi}$	ϕ-directed vector of unit length
V	Scalar potential, solution to scalar wave equation
V_0	Amplitude of incident scalar potential wave
V_1	Scalar potential in Region 1
V_2	1. Scalar potential in Region 2. 2. Output voltage of transmitter hard wired to wire ensemble 2 in the multiplexed folded unipole antenna
V_3	Scalar potential in Region 3.
V_4	Scalar potential in Region 4
V_{cN}	Common mode voltage on Nth segment of folded unipole
V_{d12}	Differential mode voltage between wire ensembles 1 and 2 on a multiplexed folded unipole antenna
V_{d13}	Differential mode voltage between wire ensembles 1 and 3 on a multiplexed folded unipole antenna
V_{d23}	Differential mode voltage between wire ensembles 2 and 3 on a multiplexed folded unipole antenna
V_{dmn}	Differential mode voltage between wire ensembles m and n on a multiplexed folded unipole antenna
V_n	Scalar potential in Region n
x	Arbitrary or dummy variable
Y_{cap}	Admittance of top capacitance in antenna end effect
Y_n	Bessel function of the second kind and order n
Y_{TL}	Admittance of top loading
$Z_{0\mathrm{c}}$	Common mode characteristic impedance
$Z_{0\mathrm{dmn}}$	Differential mode characteristic impedance for wire ensembles m and n on a multiplexed folded unipole antenna
Z_{B1}	Base terminating impedance for the tower or main conductor
Z_{T12}	Top terminating impedance connected between wire ensembles 1 and 2 on a multiplexed folded unipole antenna

Z_{T13}	Top terminating impedance connected between wire ensembles 1 and 3 on a multiplexed folded unipole antenna
Z_{T23}	Top terminating impedance connected between wire ensembles 2 and 3 on a multiplexed folded unipole antenna
z	Vertical, longitudinal, or z-coordinate
α	Pitch angle on a helical antenna
ε_1	Complex permittivity, including conductivity, for Region 1
ε_2	Complex permittivity, including conductivity, for Region 2
ε_3	Complex permittivity, including conductivity, for Region 3
ε_0	Permittivity of free space
ε_n	Permittivity of Region n
ζ	Coordinate that follows the wire of a helical antenna
η	Impedance of free space
θ	Angular coordinate in the spherical coordinate system
μ_0	Permeability of free space
μ_n	Permeability of Region n
ν_2	Current transformation ratio for wire ensemble 2 in multiplexed folded unipole antenna
ν_3	Current transformation ratio for wire ensemble 3 in multiplexed folded unipole antenna
ν_N	Current transformation ratio for Nth segment of folded unipole antenna
$\boldsymbol{\rho}$	Vector position of observer in xy-plane, with respect to the origin of a polar coordinate system
$\boldsymbol{\rho}_0$	Vector position of line charge in xy-plane, with respect to the origin of a polar coordinate system
ρ	Radial coordinate in a polar or cylindrical coordinate system
σ_n	Conductivity of Region n
ϕ	1. Angular coordinate in a polar or cylindrical coordinate system. 2. Angular position of observer in polar or cylindrical coordinate system
ϕ'	Angular position of incremental segment of helical antenna
ϕ_0	Angular position of line charge
ψ_F	Forward traveling wave function
ψ_R	Reverse traveling wave function
ω	Frequency in radians per second

Index